MW01285018

Air Conditioning and Refrigeration Troubleshooting Handbook

Second Edition

Billy C. Langley

Upper Saddle River, New Jersey
Columbus, Ohio

Library of Congress Cataloging-in-Publication Data

Langley, Billy C.
Air conditioning and refrigeration troubleshooting handbook/Billy C. Langley—2nd ed.
p. cm.
Includes index.
ISBN 0-13-578741-6
1. Air conditioning—Equipment and supplies—Maintenance and repair. 2. Refrigeration and refrigerating machinery—Maintenance and repair. I. Title.

TH7687.5 .L36 2003
697.9'3'0288—dc21 2002016915

Editor in Chief: Stephen Helba
Editor: Ed Francis
Production Editor: Christine M. Buckendahl
Production Coordination: Carlisle Publishers Services
Design Coordinator: Diane Ernsberger
Cover Designer: Thomas Mack
Production Manager: Brian Fox
Marketing Manager: Mark Marsden

This book was set in Times by Carlisle Communications, Ltd.
The cover was printed by The Lehigh Press, Inc.

Pearson Education Ltd.
Pearson Education Australia Pty. Limited
Pearson Education Singapore Pte. Ltd.
Pearson Education North Asia Ltd.
Pearson Education Canada, Ltd.
Pearson Educación de Mexico, S.A. de C.V.
Pearson Education-Japan
Pearson Education Malaysia Pte. Ltd.
Pearson Education, *Upper Saddle River, New Jersey*

10 9 8
ISBN: 0-13-578741-6

Preface

How to Use This Handbook

When trouble is experienced with a particular type of equipment, turn to the troubleshooting chart for that type of unit in Chapter 7, Troubleshooting Charts. The action of the unit can then be found in the column labeled *condition.* The next column lists the *possible* cause for the problem encountered. The third column lists the *corrective action* that may be taken to correct the problem. The *reference* column directs the reader to the specific sections in Chapter 2 and/or Chapter 3 where the component is introduced and where checkout procedures for the component are given. Specific procedures described in Chapter 4, "Standard Service Procedures," will help the beginner in accomplishing any unfamiliar tasks.

Example: You are having trouble with a refrigeration unit. After checking the unit you find that the compressor fails to start (no hum). All the fans are running and the lights are on. When you refer to the troubleshooting chart on Refrigeration, the first three possible causes are readily checked and there is electrical power to the unit. The fourth possible cause is a burned-out compressor motor—somewhat more difficult to diagnose. The fourth corrective action indicates a replacement is in order. The reference column indicates the specific sections in the text where the proper procedure for this job is discussed. The specific sections are: Chapter 2, Sections 4-1; 4-2A, B, C, and D. Chapter 4 will provide additional information on changing burned-out compressor motors.

Also included in this handbook are start-up procedures for different types of units (Chapter 3), representative wiring diagrams for some types of equipment (Chapter 5), safety procedures (Chapter 6), and engineering data including charts and tables of useful information (Chapters 7 & 8).

Table of Contents

1

Component Troubleshooting

The primary intention of this handbook is to assist the service engineer in diagnosing problems and repairing air conditioning and refrigeration equipment. The use of this handbook should allow more competent and economical service of the equipment.

The problems encountered in servicing air conditioning and refrigeration equipment can be divided into the following categories:

1. When the refrigeration unit does not operate at all, the problem is in the *electrical circuit.*
2. When the refrigeration system will run but not refrigerate, the problem is in the *mechanical components.*
3. The problem can also be the result of a *combination* of electrical and mechanical malfunctions.

Therefore, if the service technician can determine whether the trouble is electrical or mechanical in nature, he has eliminated half of the possible causes of the problem. An example of a combination of electrical and mechanical problems would be if the bearings in a compressor became stuck and caused the compressor motor to burn out.

1–1 POWER FAILURE:

When a power failure occurs, the electric company must be contacted to make the necessary repairs. The repairs may consist of replacing a fuse on the electric pole, a transformer replacement, or any one of many reasons. However, it is the electric company that must make the repairs. After the repairs have been made, the air conditioning and refrigeration service technician should check all the equipment to be certain that it is operating properly. At times, an electrical failure will cause damage to the electric motors, and they will need to be repaired or replaced.

2–1 DISCONNECT SWITCHES:

Disconnect switches are electrical switches used to interrupt electrical power to condensing units, large electric motors, and other equipment that requires a heavy current flow. These switches are usually installed within an arm's reach of the equipment that they control. Some of these switches contain only a set of electrical contacts, whereas others contain fuses to the equipment in addition to the contacts. Located on the outside of the switch box is a lever used to open and close the contacts. See Figure 1–1. This lever may be accidentally bumped, causing the contacts to open, or it may be accidentally left in the "off" position. In either case the lever must be returned to the "on" position before operation can be resumed.

3–1 FUSE:

An electrical fuse is a protective device placed in the electrical line to an electric circuit. There are basically two types of fuses: the plug type and the cartridge type. See Figure 1–2. The purpose of a fuse is to protect the electric circuit in case of an electrical overload. This overload may be due to tight bearings, shorted motor winding, breakdown of electrical insulation, fan belt too tight, burned contactor contacts, etc. The problem that caused the blown fuse must be found and be eliminated before the job is complete.

Defective fuses may be found with either a voltmeter or an ohmmeter. To check fuses with a voltmeter, select a scale that is high enough to prevent damage to the meter. Check the voltage across each fuse. See Figure 1–3. If a voltage is found between the ends of the fuse, it is defective and must be replaced with one of the proper size. Obviously, this procedure requires the electrical power to be on. Use caution to avoid an electrical shock.

To check with an ohmmeter, remove the fuse from the holders and check for continuity between the ends of the fuse. See Figure 1–4. If no continuity is found, replace the fuse with one of the proper size.

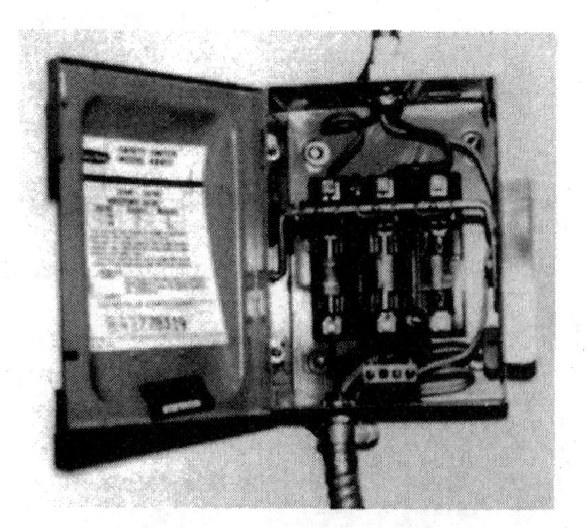

Figure 1–1. Mounted disconnect box.

Buss one-time fuses
Non-renewable

Buss Fustat fuses and adapters
Dual element - Time delay - Type S base

For motor protection, take amperage of motor and add 25%. Adapter makes Fustat fuses fit any plug fuse holder.

Buss clear window plug fuses

Figure 1–2. Plug and cartridge-type fuses. (Courtesy of Buss)

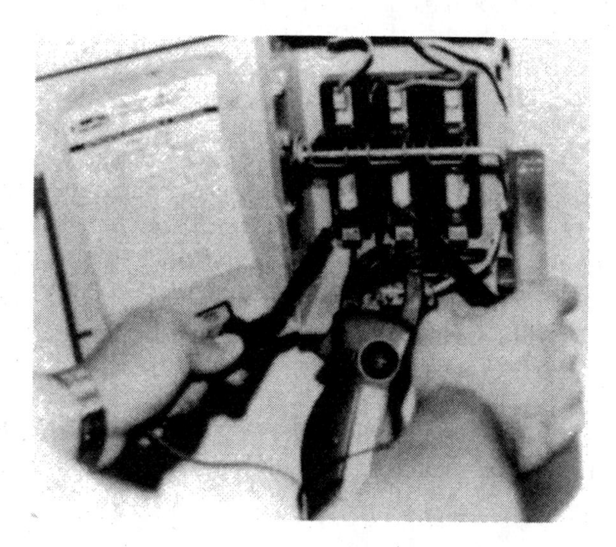

Figure 1–3. Checking fuses in disconnect.

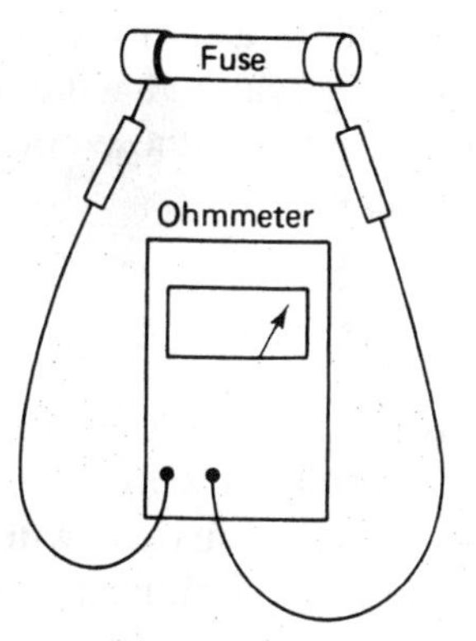

Good fuse will indicate "0" resistance

Blown fuse will indicate infinite resistance

Figure 1–4. Checking fuse out of disconnect box with ohmmeter.

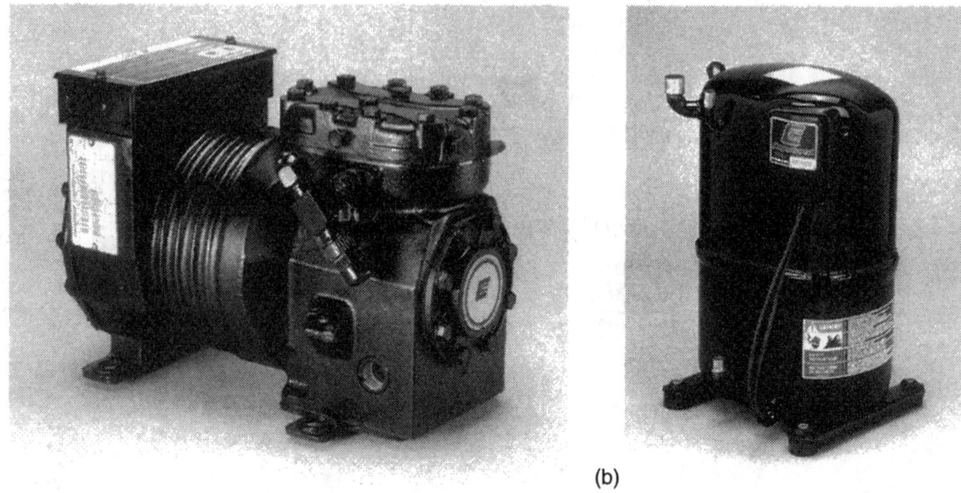

(a) (b)

Figure 1–5. (a) semihermetic compressor; (b) hermetic compressor. (Courtesy of Copeland Corp.)

4–1 COMPRESSOR:

The compressor is a device used to circulate the refrigerant through the system. See Figure 1–5. It has two functions: (1) It draws the refrigerant vapor from the evaporator and lowers the pressure of the refrigerant in the evaporator to the desired evaporating temperature; (2) the compressor raises the pressure of the refrigerant vapor in the condenser high enough so that the saturation temperature is higher than the temperature of the cooling medium used to cool the condenser and condense the refrigerant.

The compressors in use today are usually of the hermetic or semihermetic type. Therefore, the problems encountered could be electrical, mechanical, or a combination of the two.

4–2 COMPRESSOR ELECTRICAL PROBLEMS:

The problems encountered in the electrical circuit of a compressor may be divided into the following classifications: open winding, shorted winding, or grounded winding. An accurate ohmmeter is needed to check for these conditions. The following checks are good for any type of electric motor.

4–2A

Open compressor motor windings occur when the path for electrical current is interrupted. This interruption occurs when the insulation wire becomes bad and allows the wire to overheat and burn apart. To check for an open winding, remove all external wiring from the motor terminals. Using the ohmmeter, check the continu-

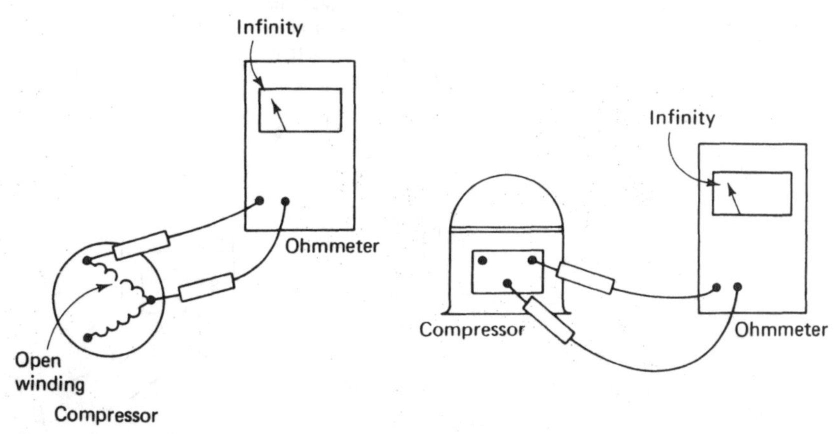

Figure 1–6. Checking for open compressor motor winding.

ity from one terminal to another terminal. See Figure 1–6. Be sure to zero the ohmmeter. The open winding will be indicated by an "infinity" resistance reading on the ohmmeter. There should be no continuity from any terminal to the motor case.

4–2B

Shorted compressor motor windings occur when the insulation on the winding becomes bad and allows a shorted condition (two wires to touch), which allows the electrical current to bypass part of the winding. See Figure 1–7. In some instances, depending on how much of the winding is bypassed, the motor may continue to operate but will draw excessive amperage. To check for a shorted winding, remove all external wiring from the motor terminals. Using the ohmmeter, check the continuity from one terminal to another terminal. Be sure to zero the ohmmeter. See Figure 1–8. The shorted winding will be indicated by a less than normal resistance. In some cases it will be necessary to consult the motor manufacturer's data for that particular motor to determine the correct resistance requirements. There should be no continuity from any terminal to the motor case.

4–2C

Grounded compressor motor windings occur when the insulation on the winding is broken down and the winding becomes shorted to the housing. See Figure 1–9. In such cases the motor will rarely run and will immediately blow fuses or trip the circuit breaker. To check for a grounded winding, remove all external wiring from the motor terminals. Using the ohmmeter, check the continuity from each terminal to the motor case. Be sure to zero the ohmmeter. See Figure 1–10. A grounded winding will be indicated by a low resistance reading. It may be necessary to remove some paint or scale from the motor case so that an accurate reading can be obtained.

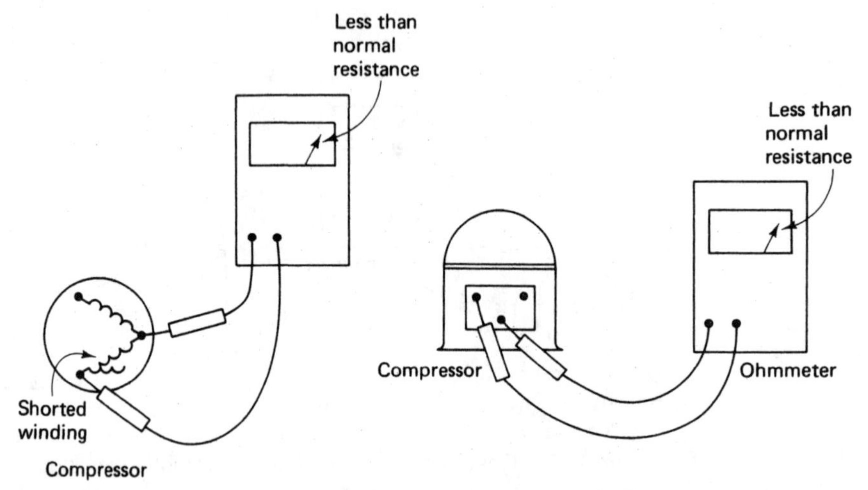

Figure 1–7. Checking for shorted compressor motor winding.

Figure 1–8. Checking compressor motor winding resistance.

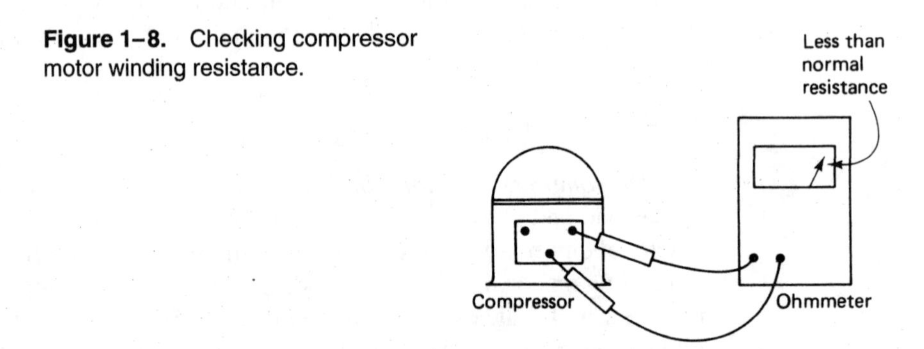

Figure 1–9. Grounded compressor motor winding.

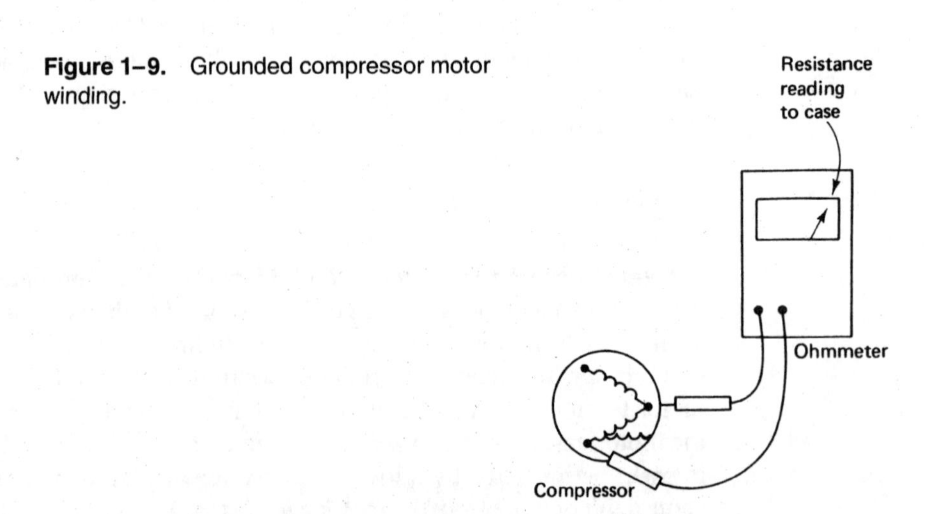

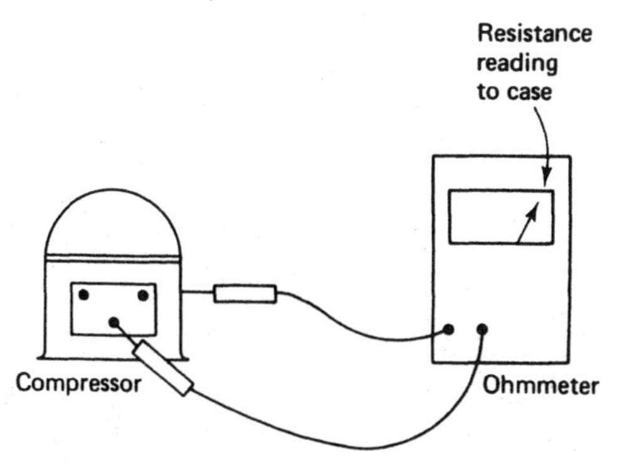

Figure 1–10. Checking for grounded compressor motor winding.

4–2D

Determining the common, run, and start terminals of a compressor is a relatively simple process. Be sure that all external wiring is removed from the compressor so that no false readings are indicated. Draw the terminal configuration on a piece of paper. Measure the resistance between each terminal with an ohmmeter. Be sure to zero the ohmmeter. Record the resistance found on the diagram. See Figure 1–11. Apply the following formula: The least resistance indicated is between the run and common terminals; the medium resistance indicated is between the common and start terminals; the most resistance indicated is between the start and run terminals. The compressor motor should now be properly wired.

4–2E

Compressor motor overloads are used to protect the compressor motor from damage that might occur from overcurrent, overtemperature, or both. Motor overloads may be mounted externally or internally depending on the design of the compressor. They are generally mounted near the hottest part of the motor winding.

4–2F

External mounted overloads are manufactured in three configurations: (1) the two-terminal, (2) the three-terminal, and (3) the four-terminal. See Figure 1–12. To check the two-terminal overload, place an ammeter on the common electric line to the compressor. Start the compressor while observing the ammeter. The ammeter should indicate a momentary current flow of approximately six times the running amperage of the compressor motor, then drop back to the rated amperage draw of the motor or below. If the overload then cycles the motor, the problem is in the overload. If the amperage remains above the rated amperage of the motor, the trouble is not in the overload. To check that the overload is cycling the compressor motor, check across the overload terminals with a voltmeter while the compressor is off. See Figure 1–13. If the overload is open, a voltage reading will be indicated.

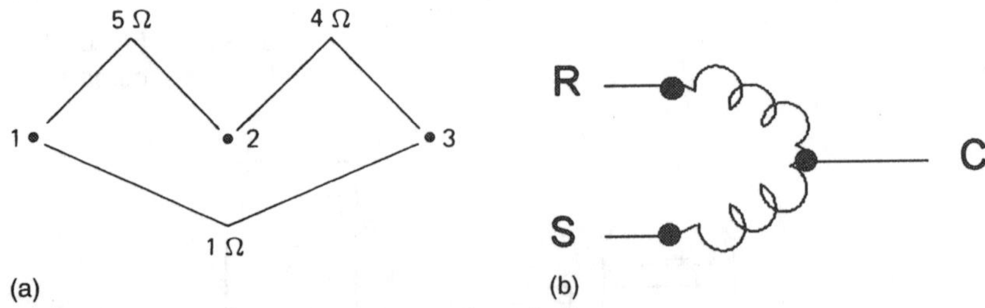

Figure 1–11. Locating compressor terminals.

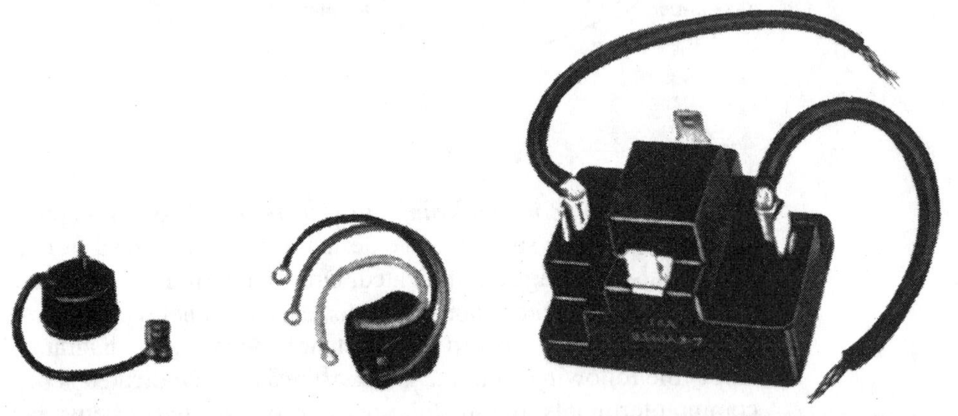

Figure 1–12. Two-, three-, and four-terminal compressor overloads. (Courtesy of Klixon Controls Division, Texas Instruments, Inc.)

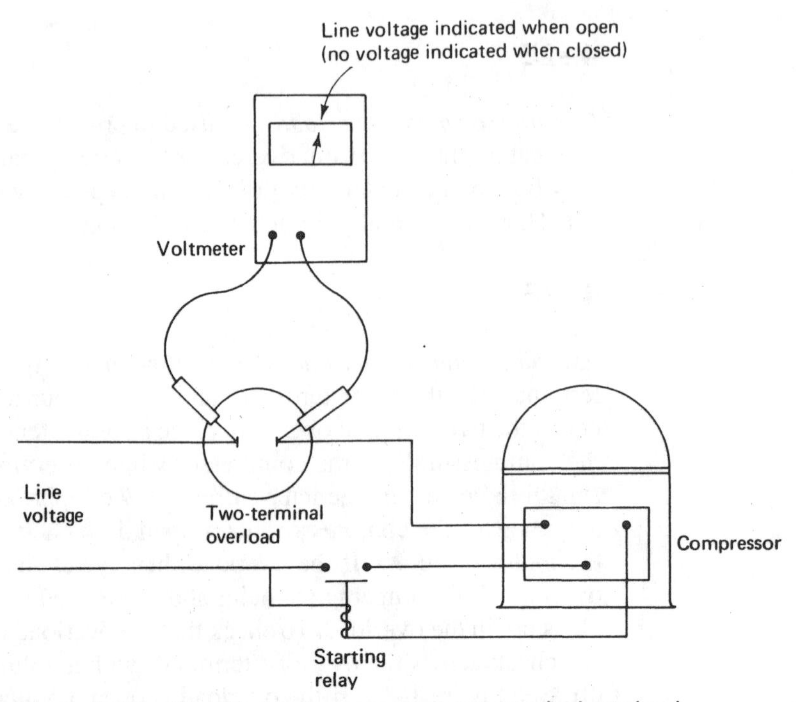

Figure 1–13. Checking voltage across a two-terminal overload.

No voltage reading indicates that the overload has not opened the circuit and the trouble is elsewhere. Be sure to replace these overloads with the type recommended by the manufacturer.

4–2G

The three-terminal type external overload is used on compressor motors where it is desirable to protect the starting winding in addition to the running winding. The terminals are numbered 1, 2, and 3. Terminal 1 is connected to the electrical line that goes to the compressor. Terminal 2 is connected to the run terminal of the compressor. Terminal 3 is connected to the start capacitor. This will allow closer protection of the compressor motor in the event of a bad starting component. To check the three-terminal overload, place an ammeter on the line connected to terminal 1 and start the compressor. See Figure 1–14. Start the compressor while observing the ammeter. The ammeter should indicate a momentary flow of approximately six times the running amperage of the motor, then drop back to the amperage rating of the motor or below. If the overload then cycles the motor, the problem is in the overload. If the amperage remains above the rated amperage of the motor, the trouble is not in the overload. To check to see if the trouble is in the starting or running circuit, measure the amperage draw through the wire connected to terminal 2, then the wire to terminal 3, while the compressor is running. The circuit with the high amperage draw is where the fault is. The external components must be checked. To be certain that the overload is cycling the compressor motor, check the voltage between terminals 1 and 2, then between terminals 1 and 3, while the compressor is off. If voltage is indicated, replace overload. See Figure 1–15. If the overload must be replaced be sure to use an exact replacement to provide proper protection.

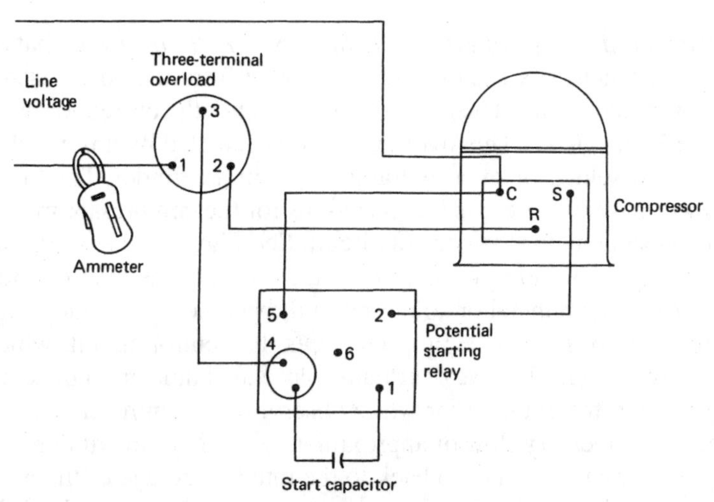

Figure 1–14. Checking amperage draw through a three-terminal overload.

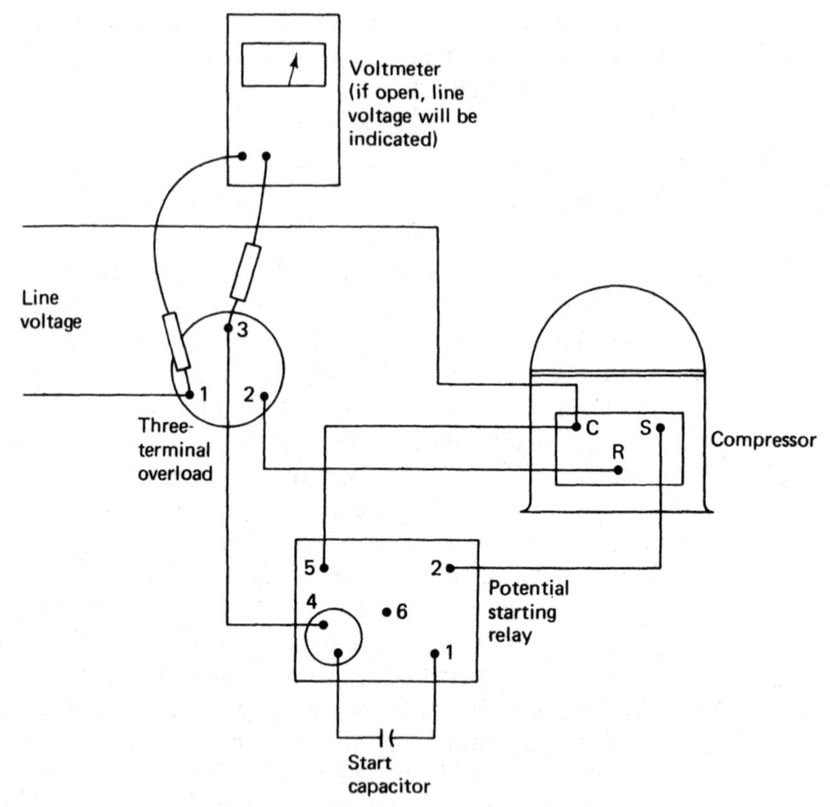

Figure 1–15. Checking voltage across terminals 1 and 3 on a three-terminal overload.

4–2H

The four-terminal type of external compressor motor overload that is used on larger compressor motors is usually mounted away from the compressor in the control panel or on the starter. This overload senses only the current draw to the compressor. See Figure 1–16. This overload will be actuated by a bimetal, by the melting of a special solder, or by a hydraulic fluid in a cylinder. The bimetal and solder types will have two electrical connections for the compressor motor and two electrical connections for the control circuit. See Figure 1–17. When the current flow to the compressor motor remains over the rating of the overload for a definite period of time, the bimetal or solder pot will become heated and will allow the control circuit to open. This action interrupts the control circuit, which will stop the compressor. To check these overloads, place an ammeter on the compressor motor wire and start the compressor while observing the ammeter. The ammeter should indicate a momentary flow of approximately six times the running amperage of the compressor motor, then drop back to the rated amperage of the motor or below. If the overload then cycles the motor, the problem is in the overload. If the amperage

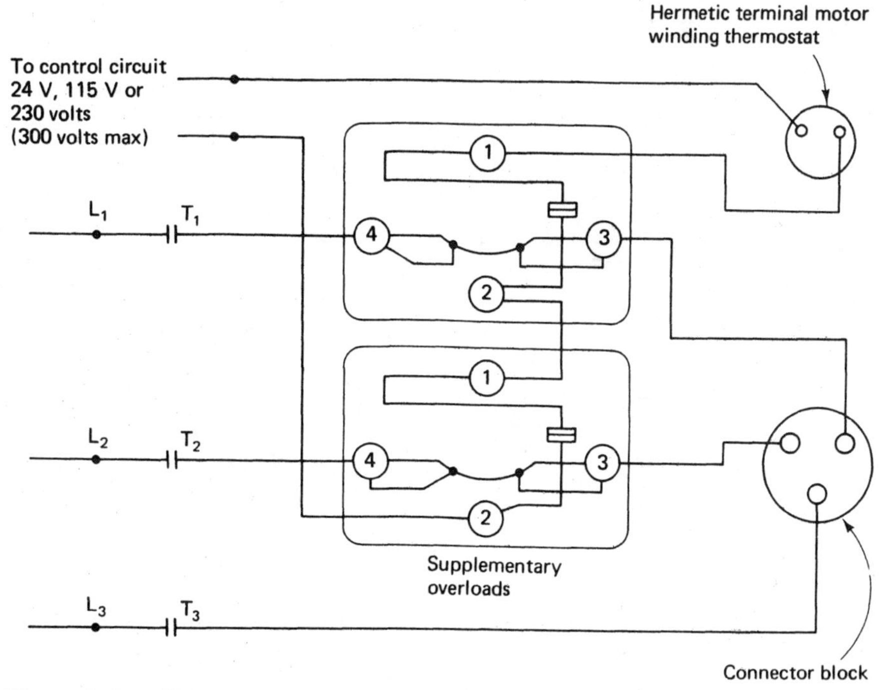

Figure 1–16. External thermostat overload connection.

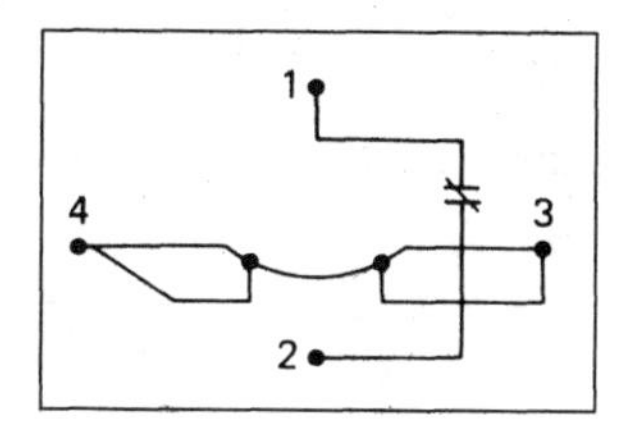

Figure 1–17. Internal view of a four-terminal overload.

remains above the rated amperage of the motor, the trouble is not in the overload. To check to be sure that the overload is cycling the compressor motor, check across the control circuit terminals of the overload with a voltmeter while the compressor is off. See Figure 1–18. If the overload is open, a voltage reading will be indicated. No voltage reading indicates that the overload has not opened the circuit and the trouble is elsewhere. These overloads may be either manual or automatic reset types. Be sure to replace them with the proper size for the current flow.

4–2I

The hydraulic fluid-type compressor motor overload operates on current flow to the compressor only and is used on larger compressor motors. It is located in the control panel away from the compressor. The current passing through the overload is

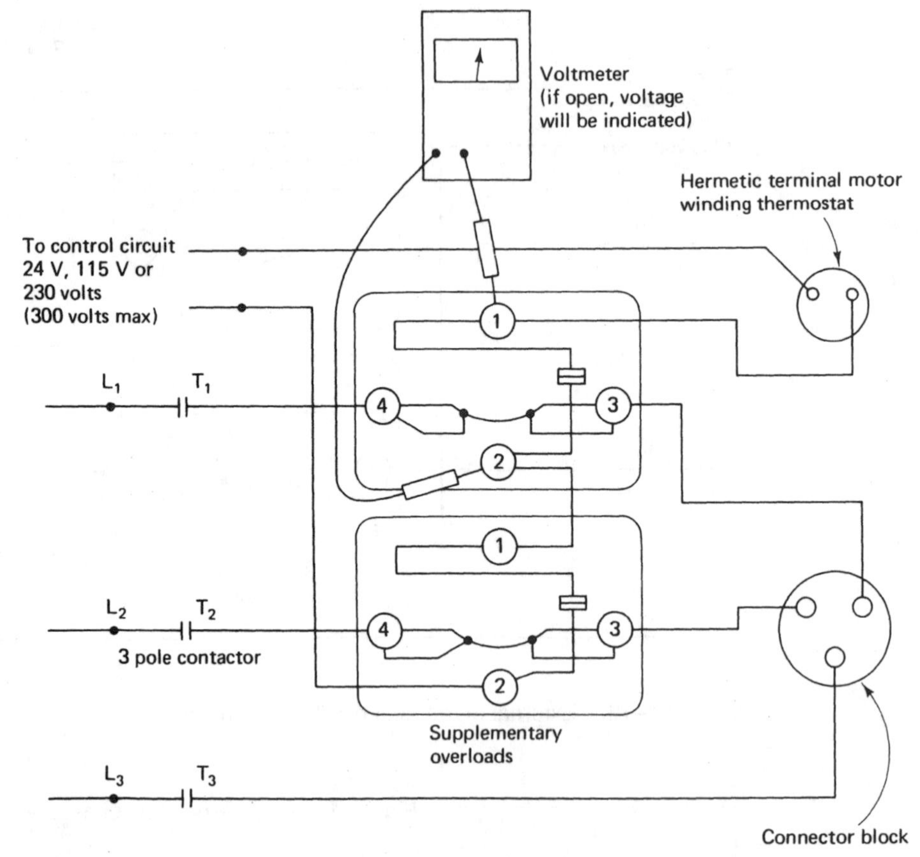

Figure 1–18. Checking voltage across control circuit terminals of a four-terminal overload.

routed through a heater, which causes the hydraulic fluid to warm. This warming action increases the pressure of the fluid, which causes the electrical circuit to the compressor to be interrupted. When the hydraulic fluid has cooled down, the overload may be reset and the compressor started again. To check these overloads, place an ammeter on the compressor motor wire and start the compressor while observing the ammeter. See Figure 1–19. The ammeter should indicate a momentary current flow of approximately six times the running amperage of the compressor motor, then drop back to the rated amperage of the motor or below. If the overload then cycles the motor, the problem is in the overload. If the amperage remains above the rated amperage of the motor, the trouble is not in the overload. To check to be sure that the overload is cycling the compressor motor, check across the line terminals of the overload with a voltmeter while the compressor is off. See Figure 1–20. If the overload is open, a voltage reading will be indicated. No voltage reading indicates that the overload has not opened the circuit and the trouble is elsewhere. These overloads are of the manual reset type. Be sure to replace them with the proper size for the current flow.

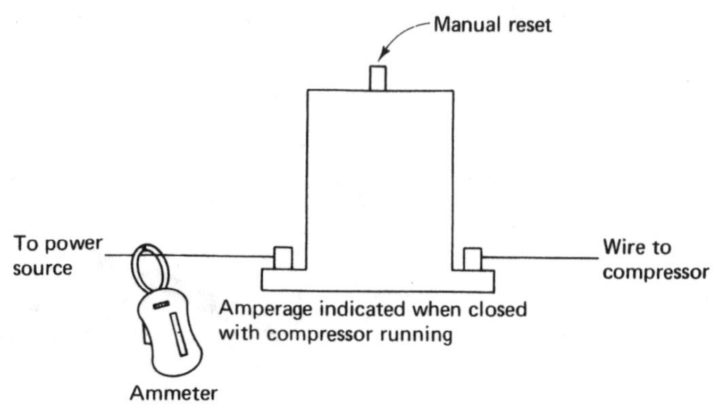

Figure 1–19. Checking amperage through a hydraulic-type compressor overload.

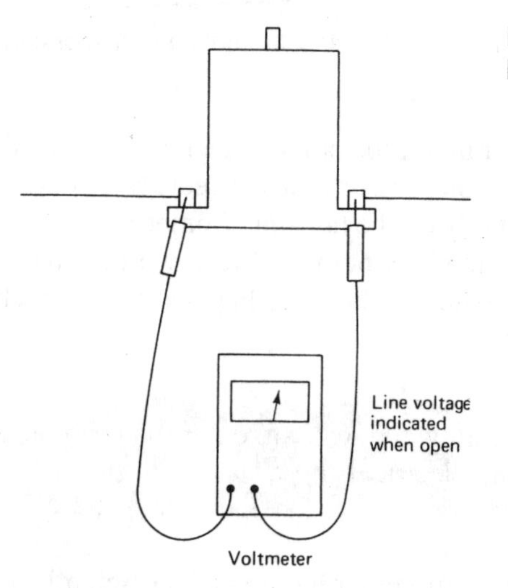

Figure 1–20. Checking voltage across a hydraulic-type compressor overload.

4–2J

Internal motor thermostatic overloads operate like the external two-terminal overload. They are usually wired in the control circuit in electrical series with the contactor or starter holding coil. However, sometimes they are wired in the line voltage circuit to the compressor motor common terminal. They are placed inside the compressor housing and are imbedded in the motor winding and have separate terminals to the outside of the housing. See Figure 1–21.

To check these types of overloads, turn off all electrical power to the compressor and remove all wiring from both the compressor motor terminals and the internal thermostat terminals. Next, check across the internal thermostatic overload terminals with an ohmmeter set on the R × 100 scale. Be sure to properly zero the ohmmeter before making these tests. When no continuity is indicated, the internal thermostatic overload contacts are open. Remember that no continuity is not always an indication of a defective internal thermostat. Sometimes it takes at least

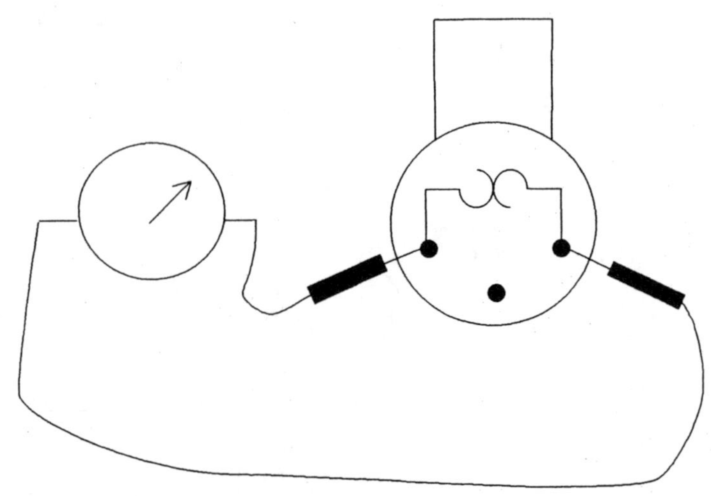

Figure 1–21. Checking internal compressor motor thermostatic overload.

45 minutes for the compressor motor to cool down enough for the thermostat to automatically reset itself. If the compressor housing feels hot to the touch, wait until the hand can be held comfortably on the housing. Sometimes a small trickle of water over the compressor housing will help cool it down quicker. This indicates a temperature of about 120°F. After the housing has cooled, recheck the continuity of the thermostatic overload.

CAUTION: Do not allow the water to enter the terminal box and cause an electrical short.

As a last resort, when the compressor has cooled sufficiently, the internal thermostatic overload terminals can be jumpered. Leave the jumper on the terminals only for a short period of time to prevent possibly burning out the compressor motor. If the compressor should start, it can be reasonably assumed that the internal thermostatic overload is defective. If so, the compressor will need to be replaced or the unit will be operating without this protection. The compressor should not be operated for long periods of time with the thermostatic overload jumpered; however, it can be done until a replacement compressor is obtained.

Another type of internal thermostatic compressor overload is one that is placed inside the compressor motor winding and is wired in electrical series with the compressor motor common. This internal overload has no external terminals. It simply breaks the line voltage electrical circuit to the compressor when it becomes overheated. See Figure 1–22.

Normally, compressors that are equipped with internal overload protection are marked near the terminal box to indicate their use.

To check this type of internal overload, turn off all electricity to the compressor and remove all wiring from the terminals. Then check the continuity with

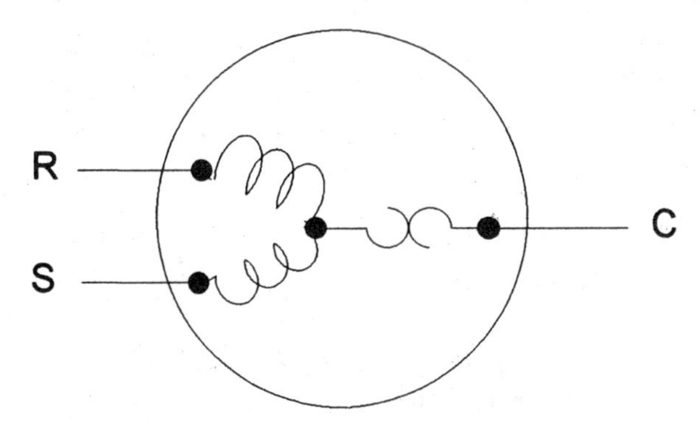

Figure 1–22. Internal thermal overload wiring connections.

an ohmmeter set on the R × 100 scale between the compressor common and run terminals. Be sure to properly zero the ohmmeter. If no continuity is indicated, check for continuity between the start and run terminals. If no continuity is found with either of these checks, it is probable that the internal overload has opened the electrical circuit to the motor common connection. The compressor must be cooled to the reset temperature of the internal thermostat before it will automatically reset. It may help to cool the compressor by using a small trickle of water over the housing.

CAUTION: Do not allow the water to enter the compressor terminal box and cause an electrical short when the power is turned back on. If the compressor has cooled so that the hand can be comfortably held on the housing and there is still no continuity indicated by repeating the above steps, the internal overload is defective. The compressor must be replaced before the unit will operate again. This type of internal overload cannot be jumpered to permit operation. However, before replacing the compressor allow plenty of time for it to cool.

4–3 COMPRESSOR LUBRICATION:

Compressors, like any moving mechanical device, require lubrication. The proper oil level should be maintained in the crankcase to provide the necessary lubrication. The oil level in a sight glass should be at or slightly above the center of the sight glass. See Figure 1–23. On compressors not equipped with an oil sight glass, the manufacturers' recommendations should be followed. These recommendations will generally refer to the amount of oil in ounces for a given model. When replenishing the oil in hermetic compressors, it may be necessary to remove all the oil from the housing and measure in the correct amount. The proper type of oil for the operating temperature should be used to insure proper oil return and good

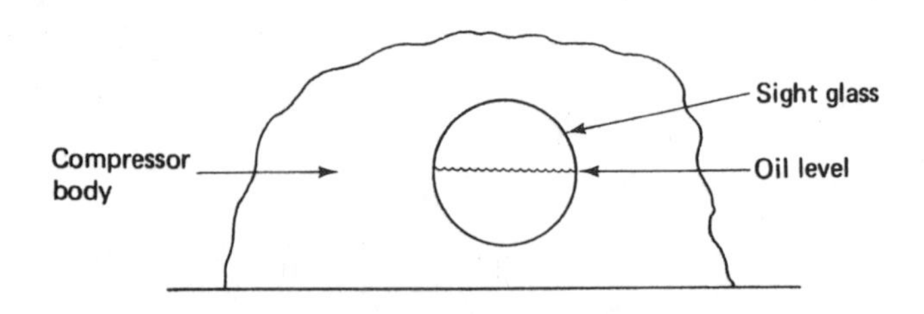

Figure 1–23. Oil level in compressor sight glass.

compressor lubrication. Refrigeration oil containers must be kept sealed to prevent moisture and contaminants getting into the oil.

There are three types of lubrication systems used on compressors: the splash, pressure, and a combination of both. The splash system is used on 3 hp compressors and smaller. Compressors of 3 hp and larger are pressure-lubricated. The oil pressure may be checked to determine if proper lubrication is being provided on forced lubrication compressors. See Figure 1–24. A net oil pressure of 30 to 40 psi is normal; however, adequate lubrication will be provided at pressures down to 10 psi. To obtain the net oil pressure, subtract the suction pressure from the oil pump pressure. *Example:* 90 psig pump pressure –50 psig suction pressure = 40 psi net oil pump pressure (620.528 Kp (Kilo-pascal) –344.738 Kp = 275.79 Kp).

4–3A

Compressor bearings that have not received proper lubrication will become tight and sometimes will "freeze" the shaft and prevent operation. If this condition occurs, the compressor motor will become overloaded and will trip out on the overload or the circuit breaker, or will blow the electrical fuses. In any case, the amperage draw of the compressor motor will be higher than normal. When the compressor motor is not allowed to turn, it will draw locked rotor amperage. This amperage rating is indicated by LR on the motor nameplate. The compressor must be replaced when drawing LR amperage. The compressor oil should be checked and if it is found insufficient, the system must be checked to determine the reason for the lack of oil. The loss of oil may be due to refrigerant leaks, oil logging of the evaporator, low refrigerant charge, etc. The reason must be corrected before the new compressor is placed in operation or it may soon fail. Do not confuse this condition with faulty starting components or a faulty running capacitor.

A malfunctioning oil pump will generally go undetected until the compressor bearings are damaged enough to cause knocking or until the compressor is frozen mechanically. A malfunctioning oil pump may be due to mechanical wear of the pump. It may become vapor-locked with refrigerant, or the inlet screen may become plugged with dirt or sludge. Should a worn pump be the culprit, it must be repaired or replaced at the same time that the compressor is repaired. A vapor-locked oil pump will produce no oil pressure. The vapor lock must be removed by bleeding off the vapor through the gauge connection.

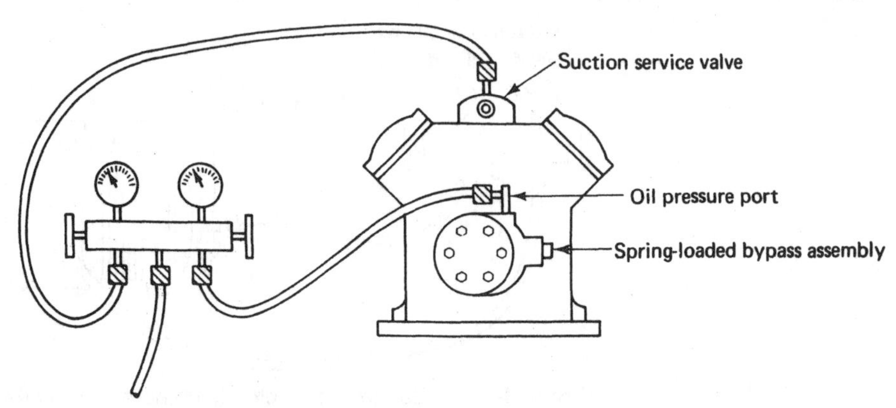

Figure 1–24. Checking oil pressure.

When the oil inlet screen becomes clogged, the oil flow will be restricted and perhaps completely stopped. In this case the screen must either be cleaned or replaced, along with a complete cleaning of the compressor crankcase, replacement of the oil charge, and replacement of the refrigerant filter-drier.

4–3B

Compressor bearings that are worn become loose and noisy. The compressor loses its efficiency, and poor refrigeration results. Loose bearings will be indicated by more noise than usual, and sometimes excessive vibration will occur. When the compressor is equipped with a forced lubrication system, the oil pressure will be low. The amperage draw will be from normal to low. The suction pressure may be high and the discharge pressure low. When this condition occurs, the compressor must be replaced or overhauled depending on the type. This condition is generally a result of age and usage, not faulty lubrication.

4–3C

Compressor valves are the components that control the flow of refrigerant through a compressor. If they are broken or leaking, the compressor will become inefficient. A broken or leaking suction valve will result in a higher than normal suction pressure. To check for a defective compressor suction valve, connect the gauge manifold to the service valves on the compressor and open the service valves. See Figure 1–25. Next, front-seat the suction service valve (screw all the way in), and observe the suction pressure while the compressor is operating. The suction pressure should pull down to at least a 28-in. vacuum in one or two minutes. If it does not pump down to this reading, stop the compressor for two or three minutes; then restart the compressor and

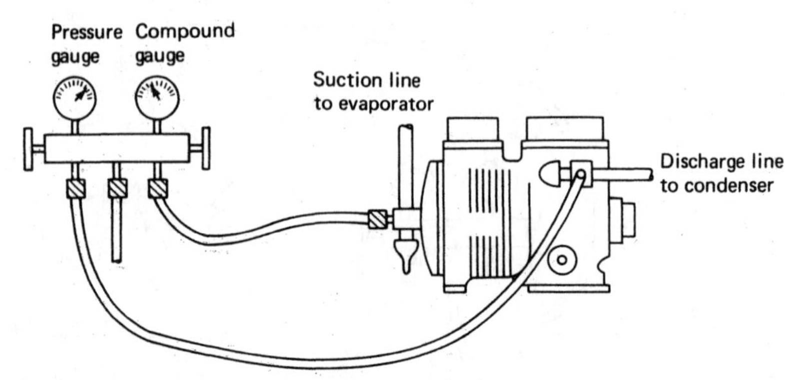

Figure 1–25. Gauges connected to compressor to take refrigerant pressure readings.

allow it to run for one or two minutes. If the desired 28-in. vacuum is not reached, the valves must be replaced. In the case of a hermetic compressor, the compressor must be replaced.

4–3D

A broken or leaking discharge valve will result in a lower than normal discharge pressure. To check for a defective compressor discharge valve, connect the gauge manifold to the service valves on the compressor and open the service valves. See Figure 1–25. Next, front-seat the suction service valve (screw all the way in), and start the compressor and allow it to pump as deep a vacuum as possible. Stop the compressor and observe the compound gauge. If the pressure rises, start the compressor and pump another vacuum. Stop the compressor and again observe the compound gauge. If the pressure increases again, close off the discharge service valve. If the pressure reading on the compound gauge stops rising, the discharge valve is bad and must be replaced. In the case of a hermetic compressor, the compressor must be replaced.

4–4 COMPRESSOR SLUGGING:

Slugging is a noisy condition that occurs when the compressor is pumping oil or liquid refrigerant. When a compressor is slugging, it will sound like the clattering of an automobile engine that is under strain. Continued slugging will probably result in broken valves, scored pistons, and galled bearings. It should be evident that a slugging condition must be corrected.

4–4A

Oil slugging of a compressor occurs when there is too much oil in the compressor crankcase. When this occurs, some of the oil must be removed to provide the oil

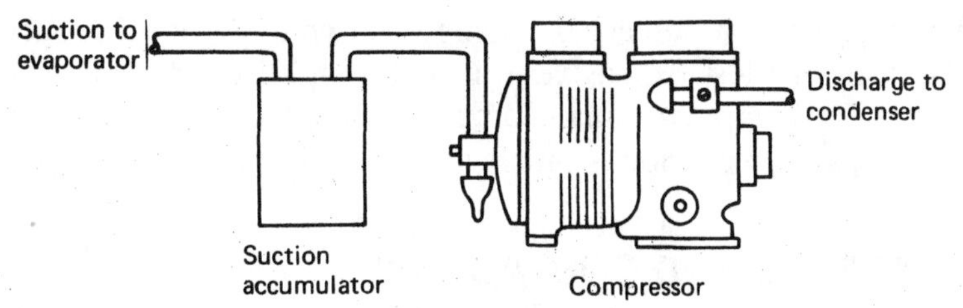

Figure 1–26. Suction line accumulator installation.

level recommended by the manufacturer. The excess oil may be drained through a drain plug, or some installations may require the removal of the compressor so the oil may be poured out. In either case the refrigerant must be pumped from the compressor or purged from the system. Do not attempt to remove oil from the compressor with refrigerant pressure in the crankcase.

To pump refrigerant from the compressor, connect the gauge manifold to the compressor service valves. See Figure 1–25. Front-seat the suction service valve and allow the unit to run until approximately 2 psig pressure is shown on the compound gauge. Do not pump the unit below atmospheric pressure for this procedure. Front-seat the compressor discharge service valve and relieve any remaining pressure through the gauge manifold. The compressor may now be serviced.

4–4B

Refrigerant slugging of a compressor is the result of liquid refrigerant being returned to the compressor. This can be caused by several things and will usually result in moisture condensing on the compressor housing because of the lower temperature, and sometimes ice or frost will form on the compressor housing. Some of the causes are: overcharge of refrigerant, especially on capillary tube systems; superheat setting too low on the thermostatic expansion valves; automatic expansion valves open too much; or a low load condition on the evaporator. The obvious solution is to correct any of these causes. However, if these conditions cannot be remedied, a suction line accumulator may be installed to prevent refrigerant slugging of the compressor. See Figure 1–26. Almost all equipment manufacturers also install or recommend the installation of crankcase healers to reduce slugging due to liquid refrigerant entering the crankcase.

4–5 OVERSIZED COMPRESSOR:

An oversized compressor will generally provide unsatisfactory results. A lower than normal suction pressure will result with a lower than normal evaporator temperature. This lower temperature causes excessive removal of humidity from the space. This is critical and is usually not desired in commercial refrigerators. The best procedure is to replace the compressor with one of the proper capacity. However, it may be less expensive to install a capacity reduction device on the compressor. The compressor manufacturer should be consulted to determine whether or not capacity reduction is satisfactory with the particular unit under question.

5–1 MOTOR STARTERS AND CONTACTORS:

The purpose of a motor starter or contactor is to provide the switching action of the high current and voltage required by a compressor. This is done by a signal given by the control circuit on demand from the thermostat or temperature controller. These are electromagnetic-operated devices.

5–1A

A burned coil will prevent the operation of starters and contactors because there will be no electromagnetic field to operate the device. To check for a burned coil, turn off the electrical power to the unit and remove the electric wiring from the terminals of the coil. Zero an ohmmeter and check the continuity of the coil. See Figure 1–27. There should be no continuity indicated if the coil is open. Many times the coil will be discolored, indicating that it has been overheated.

5–1B

A sticking motor starter or contactor can cause permanent damage to the motor or compressor. A starter or contactor that sticks may prevent the motor from starting or may keep it running when there is no demand for it. When the starter or contactor sticks during the initial start-up, it will usually buzz and either prevent starting of the motor or cause a delayed starting of the motor. When the contactor or starter

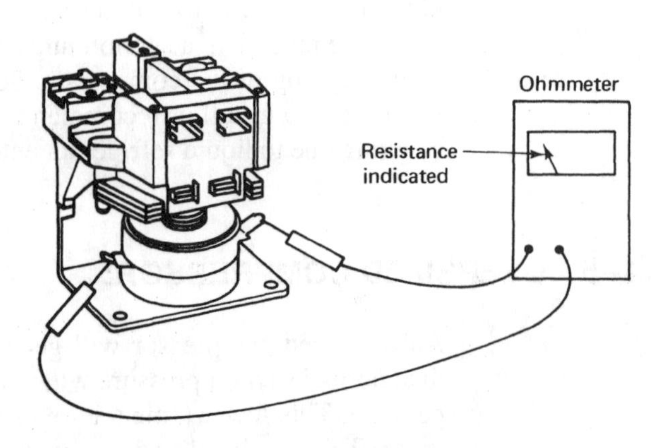

Figure 1–27. Checking continuity of contactor coil.

sticks closed, the compressor or motor will never stop. Although there are several types of sprays on the market that may be used to lubricate these troublesome and dangerous controls, it is generally recommended that faulty starters and contactors be replaced.

5–1C

Burned starter and contactor contacts can cause permanent damage to the motor windings by preventing the proper flow of current through them. These contacts will be severely pitted and will not make good contact, thus causing a higher current draw than normal. In an emergency, these contacts may be lightly filed until the mating surfaces match. See Figure 1–28. However, the damaged contacts should be replaced as soon as possible because they will again burn and become pitted in a very short time.

Sometimes these contacts may become so bad that they will not make contact at all. This can be determined by energizing the starter or contactor and checking across the contact points with a voltmeter. See Figure 1–29. If the contacts are open, the applied voltage will be indicated. However, if the contacts are closed, there will be no indication of voltage.

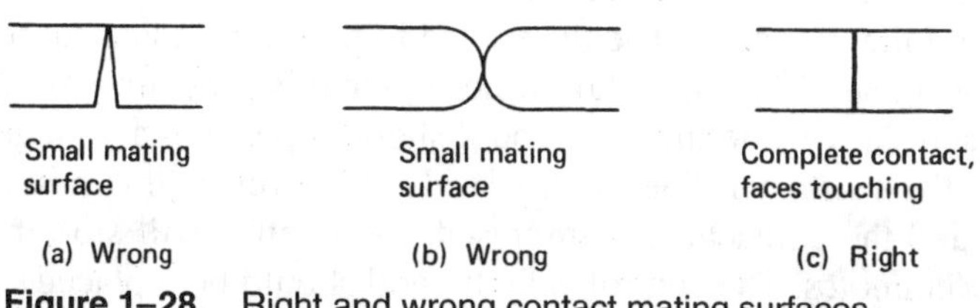

Figure 1–28. Right and wrong contact mating surfaces.

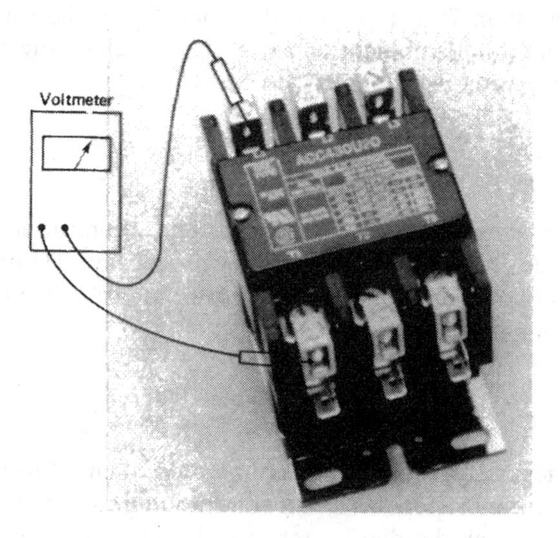

Figure 1–29. Checking voltage across contacts. (Courtesy of Crouse-Hinds Co., Arrow-Hart Division)

6–1 CONTROL CIRCUIT:

The control circuit is an electrical circuit that has a very small current flow through it. It is used to signal the components when to operate. These circuits usually contain various controls, such as the thermostat, contactors, starters, and various safety devices used to protect the compressors, motors, heat exchangers, etc. To check a control circuit, set the thermostat to call for the system to function; then check each component individually for a voltage drop. If the control contacts are closed, no voltage will be indicated on the voltmeter. If the control contacts are open, the applied voltage will be indicated. See Figure 1–30.

7–1 OIL FAILURE CONTROLS:

Oil failure controls are used to protect the compressor from improper lubrication. The control is actuated by the difference in pressure between the oil pump outlet and the crankcase pressure. A time delay switch allows the oil pressure to build up to preset operating pressure on compressor start and also prevents nuisance shutdown of the compressor if the oil pressure drops for a short time.

To check for a faulty control, connect a compound gauge to the oil pump outlet. Be sure to leave the oil failure control connected. Connect a voltmeter across terminals 1 and 2 on the oil failure control. Start the compressor and observe the gauges and the voltmeter. The difference in pressure indicated on the two gauges should be at least 10 psi in a short time. When this pressure differential is reached, the contacts between terminals 1 and 2 should open and a voltage should be indicated on the voltmeter. See Figure 1–31. However, if this minimum pressure is reached and the contacts remain closed, the control will stop the compressor in about two minutes. The control is faulty and should be replaced.

The time delay of these units is based on 120 or 240 volts ac applied in an ambient temperature of 75°F (23.9°C) with the cover in place. If the ambient temperature is much higher than 75°F (23.9°C), the control may be causing nuisance shutdown because of a high ambient temperature rather than low oil pressure. In this case the control must be relocated.

8–1 THERMOSTAT:

Thermostats are temperature-sensitive devices used to control equipment in response to the demands of the space in which they are located. Thermostats may be operated by a bimetal or "feeler" bulb filled with a fluid that expands and contracts in response to temperature changes.

8–1A

A room thermostat generally makes use of a bimetal element for its operation. This type of thermostat is the most popular for air conditioning and heating applications. See Figure 1–32. To check the thermostat, turn it below room temperature and

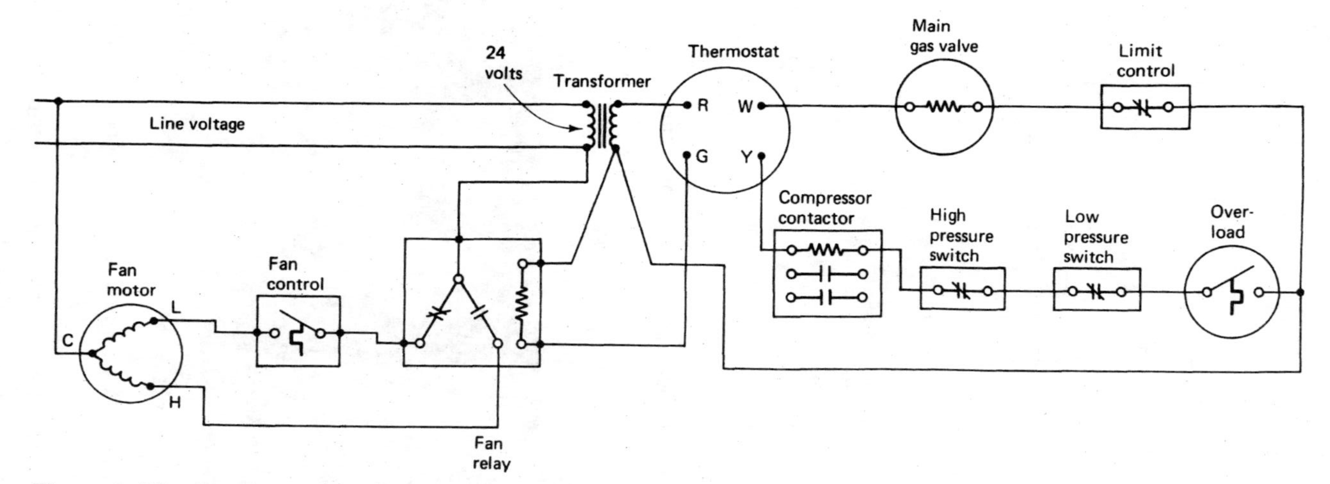

Figure 1–30. Heating and cooling control circuit.

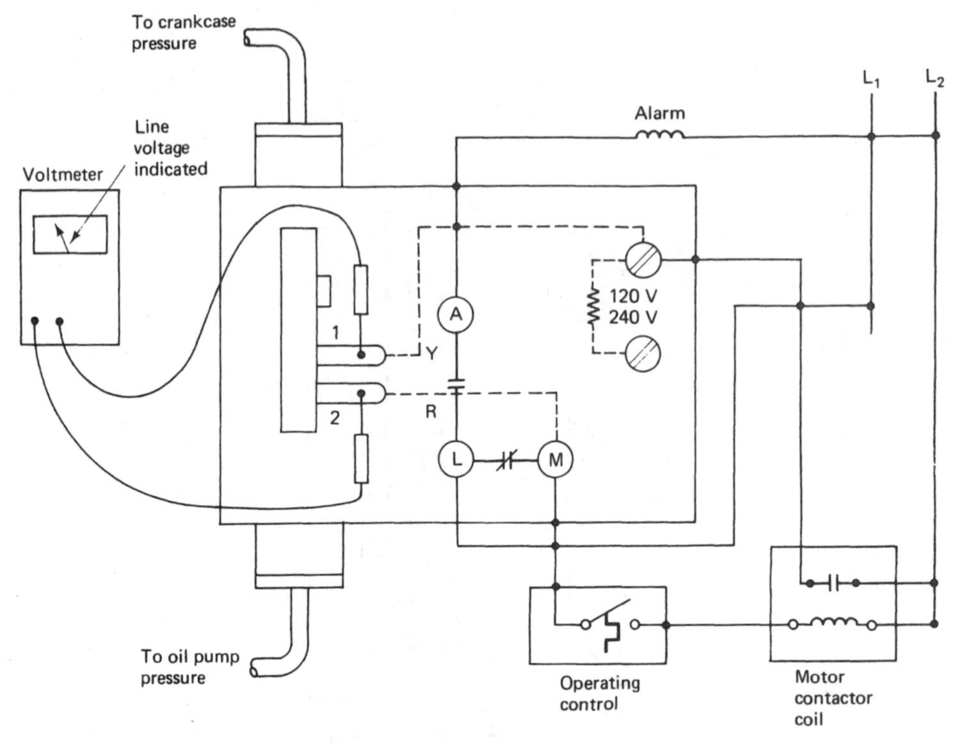

Figure 1–31. Checking operation of oil safety control.

Figure 1–32. Thermostat with cover removed showing bimetal.

place a reliable thermometer as close to the thermostat as possible. Allow the thermometer to remain there for ten minutes. Next, turn the thermostat temperature selector up. The contacts should "make" (close) at no more than 2° above the temperature indicated by the thermometer. If not, the thermostat must be calibrated. There are several means of calibrating a thermostat. Therefore, the manufacturer's specifications should be consulted. If more than 10°F calibration is needed, replace the thermostat. To check for an inoperative room thermostat, check the voltage from the red terminal to the terminal for the section in question (heating or cooling). If the contacts are closed, no voltage will be indicated. If the contacts are open, a voltage will be indicated when the thermostat is demanding.

8–1B

The heat anticipator incorporated in a thermostat is used during the heating cycle only. Most heat anticipators are adjustable. See Figure 1–33. When adjusting a heat anticipator, the total amperage draw of the control circuit must be known. To determine the amperage draw, wrap the tong of the ammeter with one wire to the main gas valve and read the amperage draw while the circuit is in operation. See Figure 1–34. P.26 Divide the amperage indicated by the number of turns taken on the ammeter tong. Set the heat anticipator to match this amperage draw.

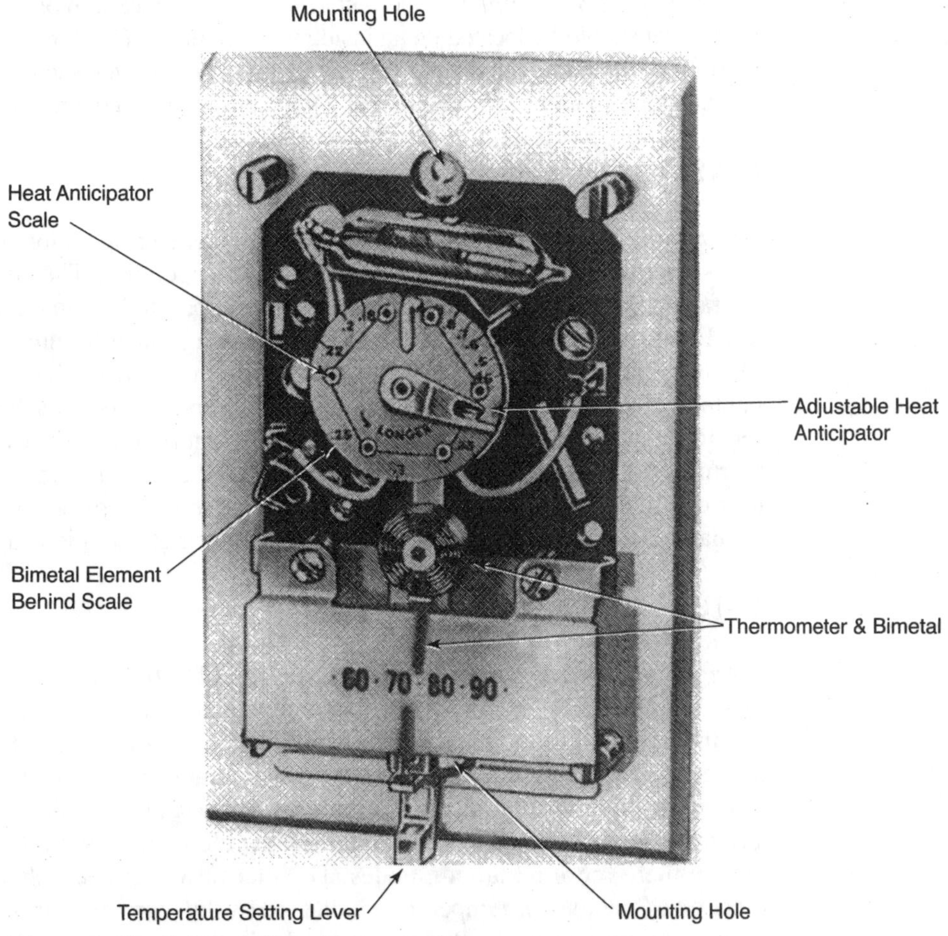

Figure 1–33. Internal view of a thermostat.

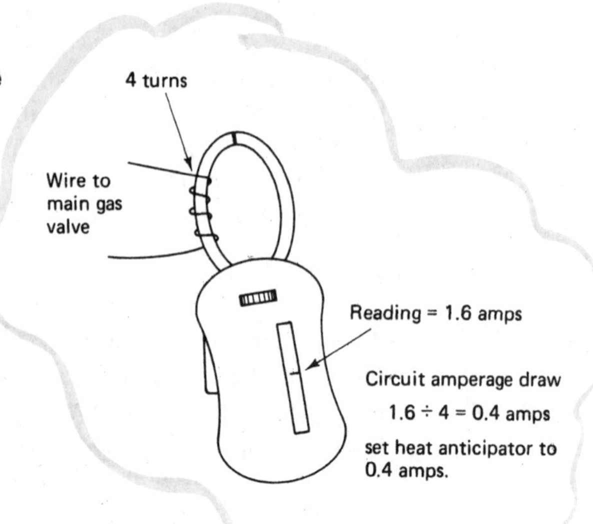

Figure 1–34. Checking amperage in temperature control circuit.

8–1C

Location of the thermostat is important to satisfactory operation of the equipment. The thermostat should be located on an inside wall about 5 ft (1.52 m) from the floor. The thermostat should not be affected by any external heat source such as lights, sun, television, etc. It should be located so that it will sense the average return air temperature.

8–1D

Thermostat switches are placed according to the type of operation desired. The system switch controls heating or cooling equipment operation. The fan switch controls the fan operation. Both switches are usually incorporated in the thermostat subbase. See Figure 1–35. To check out the switches, use a jumper to jump from the "R" or "V" terminal to the "Y" or "C" terminal for cooling, or from the "R" or "V" terminal to the "W" or "H" terminal for heating. The first letters of each of these sets refer to Honeywell thermostats; the second letters refer to General Controls thermostats. When the fan switch is set to "on," the fan will run continuously. The user can select the operation desired. Should these switches become defective, the subbase or thermostat must be replaced before normal operation can be resumed.

8–1E

Refrigeration thermostats generally make use of a fluid-filled bulb-sensing element. See Figure 1–36. The thermostat may be mounted outside the cooled space, and the bulb will be mounted either on the evaporator or near it. There is generally not much calibration required on this type of thermostat. It is usually best to replace refrigeration thermostats that give problems. To check for a faulty thermostat, place a reliable thermometer as close to the sensing element as possible and allow it to remain there for at least ten minutes. The thermostat contacts should open at not more than 2° above the temperature indicated by the thermometer. If not, make only minor adjustments on the thermostat to obtain this temperature. If the thermostat requires excessive calibrating, replace it. To check the contacts, place a voltmeter

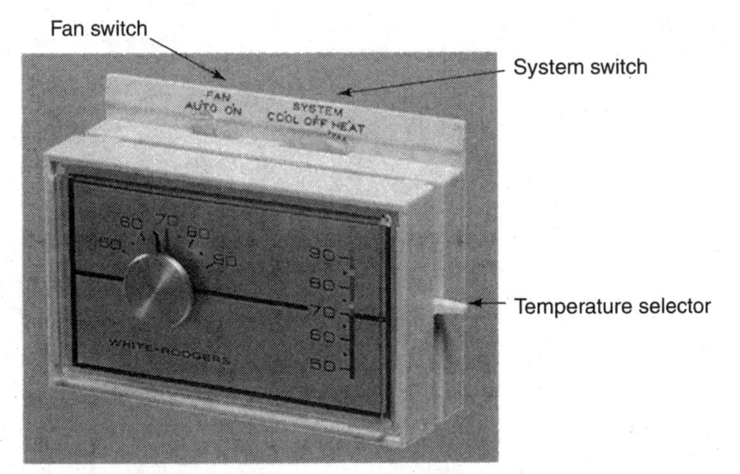

Figure 1–35. Air conditioning thermostat. (Courtesy of White-Rodgers Division, Emerson Electric Co.)

Figure 1–36. Refrigeration-type thermostat.

across the contacts of the thermostat. There should be a voltage reading on the voltmeter if the contacts are open. Turn the temperature selector down below the thermometer reading. When the contacts close, the voltage on the voltmeter should drop back to zero. If these readings are not obtained, replace the thermostat.

8–1F

Outdoor thermostats are remote bulb-type thermostats that are used on heat pump systems. They are sometimes mounted in the terminal box of the condensing unit. Other times they are mounted under the eaves of buildings. When mounted under the eaves, they provide the thermostat protection from the wind, rain, and sun. These thermostats are used to energize the auxiliary heating elements when the outdoor temperature falls below the balance point. The thermostat setting is determined by the designer to provide greater efficiency and economy. There may be

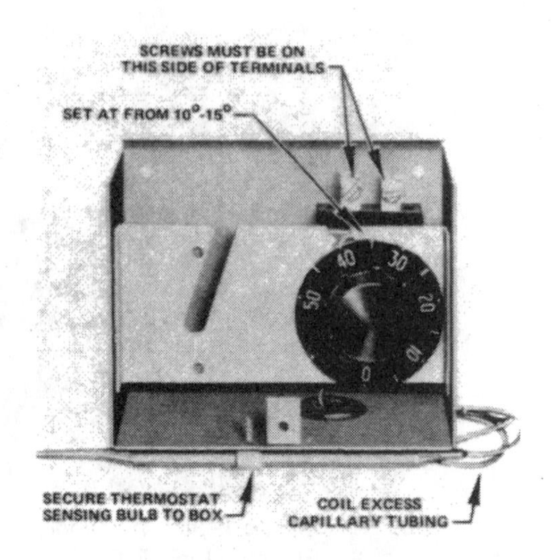

Figure 1–37. Outdoor thermostats. (Courtesy of Lennox Industries, Inc.)

Figure 1–38. Dual pressure control. (Courtesy of Penn Controls, Inc.)

more than one outdoor thermostat. Therefore, each one is set at a temperature equal to the combined heat output balance point of all the other heat strips plus the heat pump. Electric power is provided from the second stage of the indoor thermostat. See Figure 1–37. To check the thermostat, the bulb must be cooled to see if the contacts open and close at the desired temperature. Insert the bulb in an ice and salt solution containing a thermometer. Adjust the thermostat to the desired temperature if possible. If adjustment is not possible, replace the thermostat.

9–1 PRESSURE CONTROLS:

Pressure controls are designed to protect compressors and motors from damage as a result of excessive pressures. See Figure 1–38. Low-pressure controls are used to open the control circuit when the refrigerant pressure in the low side of the system falls below a given pressure. High-pressure controls are used to open the control circuit when the refrigerant pressure in the high side of the system rises to a given

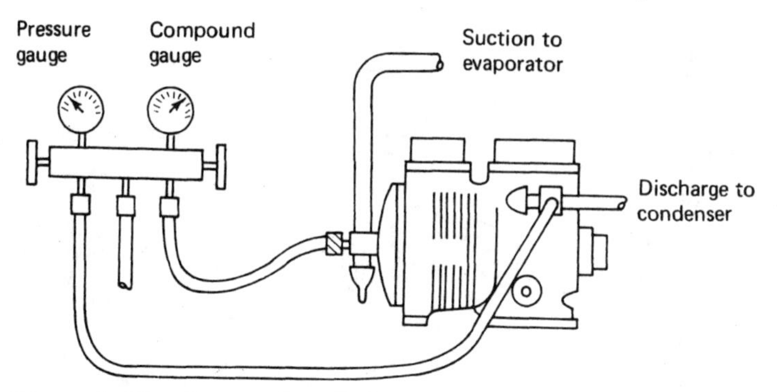

Figure 1–39. Gauges installed on compressor.

pressure. These pressure settings are generally recommended by the equipment manufacturer. Usually when these controls cause the compressor to cycle, the problem is due to some cause other than the control.

9–1A

Low-pressure controls are designed to respond to the refrigerant pressure in the low side of the system. They will stop the compressor motor to protect it from overheating and to prevent the compressor from pumping oil out of the crankcase. On some smaller units the low-pressure control may also be used as a temperature control. To check a low-pressure control, install a compound gauge on the compressor suction service valve. See Figure 1–39. Do not disconnect the low-pressure control or cause it to be inoperative. Crack the service valve off the back seat. With the compressor running, front-seat the suction service valve and observe the pressure on the compound gauge when the pressure control stops the compressor. If the actual pressure does not correspond with the control setting, adjust the control, back-seat the suction service valve, and repeat the above procedure until the desired cut-out and cut-in points are obtained. Rarely do these controls need replacement except in the case of refrigerant leakage; in which case, replacement is preferred to repair.

9–1B

High-pressure controls are designed to respond to the refrigerant pressure in the high side of the system. They will stop the compressor motor to protect it from being overloaded due to excessive discharge pressures. To check a high-pressure control install a pressure gauge on the compressor discharge service valve. Crack the service valve off the back seat. See Figure 1–40. Block the air flow through the condenser, or stop the water pump if a water-cooled condenser. Start the compressor and observe the pressure on the pressure gauge when the pressure control

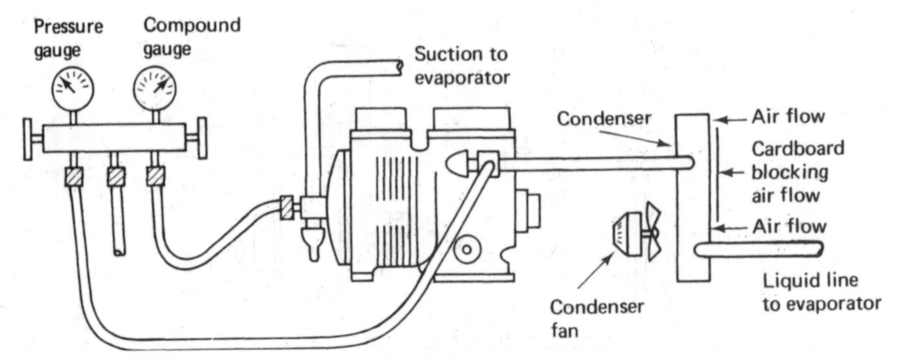

Figure 1–40. Blocking air flow to raise discharge pressure and check high-pressure control.

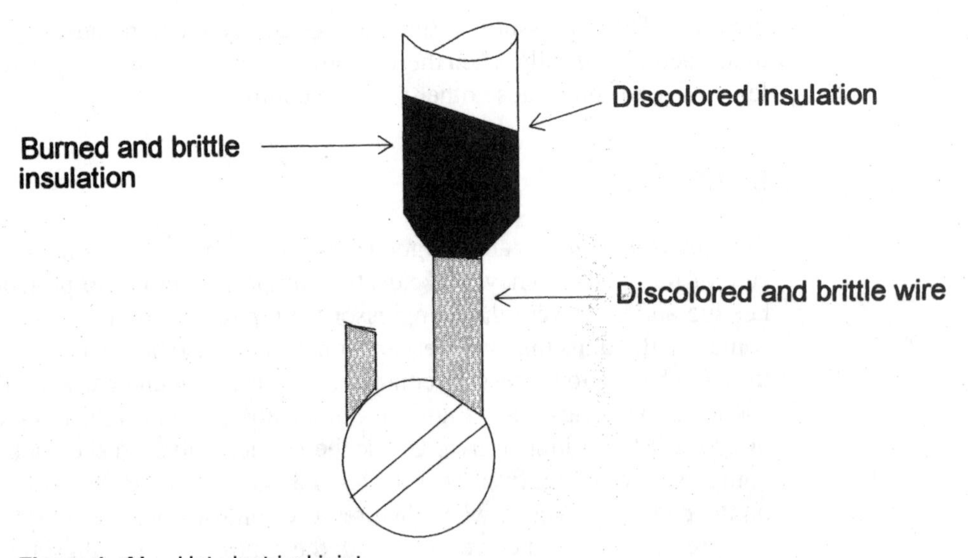

Figure 1–41. Hot electrical joint.

stops the compressor. If the actual pressure does not correspond with the control setting, adjust the control, push the reset, and repeat the above procedure until the desired cut-out point is obtained. Rarely do these controls need to be replaced except in the case of refrigerant leakage; in which case, replacement is preferred to repair.

10–1 LOOSE ELECTRICAL WIRING:

Loose electrical wiring can cause many problems and at times may be extremely difficult to locate. The problems caused by loose wiring do not follow any set pattern. Most loose wiring can be found by visual inspection. See Figure 1–41. However, when a loose wire is suspected, it may be necessary to check each wire and its connections individually. This is usually a time-consuming and a grueling task. However, it must be done before the unit will satisfactorily operate again. Once the bad wire or connection is found, it must be repaired or replaced. Any wiring that is not properly repaired will only cause problems in the future.

11–1 IMPROPERLY WIRED UNITS:

Improperly wired units will operate inefficiently, if at all. They will not produce the results desired of the equipment. Each manufacturer designs their own wiring diagrams for each piece of equipment, designs that will cause the unit to produce the desired results. If a service technician is in doubt about the wiring on a piece of equipment, the recommended wiring diagram should be consulted and the wiring changed to match the diagram.

12–1 LOW VOLTAGE:

Low voltage to a motor can cause it to overheat and will damage the motor windings. This overheating is due to excessive current draw due to low voltage. There are several causes of low voltage, such as too small wires, loose connections, or low voltage provided by the power company. To check for low voltage connect a voltmeter to the common and run terminals of the motor. See Figure 1–42. Start the unit and observe the voltage reading. It should not vary more than 10% from the rated voltage of the unit. If voltage drops more than 10%, check the size of the wire to the unit. Be sure that it is at least as big as the recommendation of the equipment manufacturer. If not, replace it with the proper size. If the wire is of sufficient size and the voltage is still low, check for loose connections. These connections will usually be indicated by the wire insulation being overheated or burned. Repair these connections. If there are no loose connections, check the

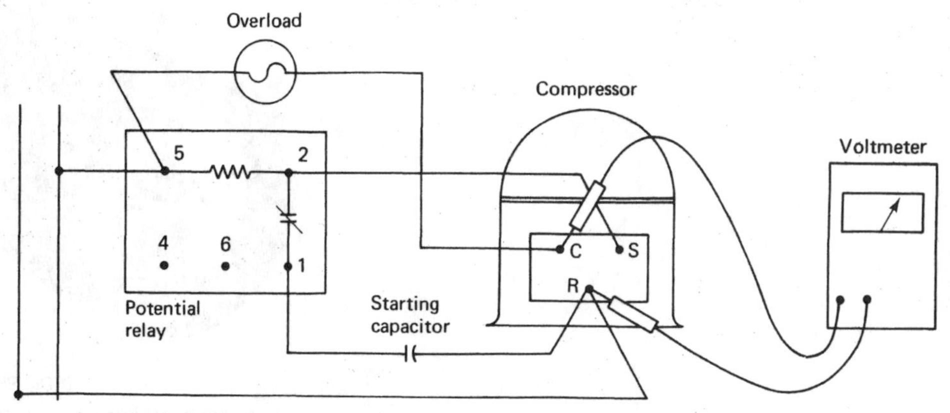

Figure 1–42. Checking starting and running voltage to a compressor.

voltage at the electric meter. If it is found to be low here, contact the power company for assistance.

13–1 STARTING AND RUNNING CAPACITORS:

Capacitors are used by many manufacturers to improve the starting and running characteristics of their motors. Capacitors are manufactured both for use in starting and in running motors. Each manufacturer determines the proper size for their motor. Manufacturer's recommendations should be followed.

13–1A

Starting capacitors are used in the starting circuit of the motor. They are generally round, encased in a plastic casing, and have a relatively high microfarad (mfd) rating. See Figure 1–43. These capacitors are designed for short periods of use only. Prolonged use will usually result in damage to the capacitor. If a starting capacitor is found to be defective, be sure to check the starting relay before the unit is placed back in service or the new capacitor may also be damaged. The best way to check a capacitor is by the use of a capacitor analyzer. This type of meter provides a direct reading of the mfd output of a capacitor without the use of bulky equipment and mathematical formulas. See Figure 1–44. Starting capacitors that are found to be out of the range of 0% to +20% of the mfd rating of the capacitor must be replaced with the proper size.

Replacement capacitors must have the same voltage rating as those being replaced. Also the voltage rating of any capacitor must be equal to or greater than the capacitor being replaced. When replacing starting capacitors, be sure that a 15,000–18,000 ohm, two-watt resistor is soldered across the terminals of the ca-

Figure 1–43. Starting capacitor with bleed-resistor. (Courtesy of Copeland Corp.)

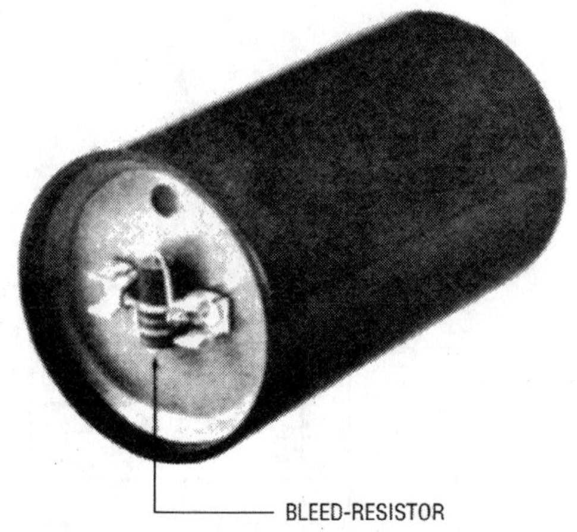

Figure 1–44. Capacitor analyzer. (Courtesy of Nationwide Electronics, Inc.)

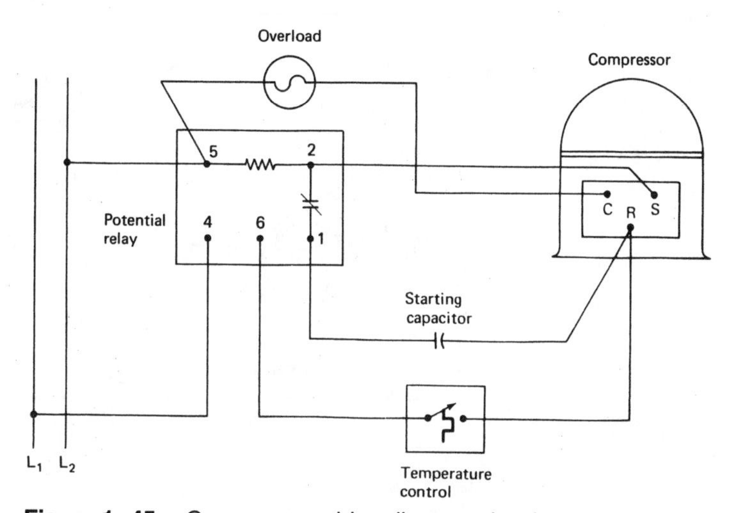

Figure 1–45. Compressor wiring diagram showing starting circuit.

pacitor. See Figure 1–43. This is a precautionary measure to prevent the arcing and burning of the starting relay contacts. To wire a starting capacitor into the circuit, see Figure 1–45.

13–1B

Running capacitors are in the operating circuit continuously. They are normally oil-filled type capacitors. The mfd rating of these capacitors is relatively low, even though they are larger in size. Running capacitors are used to improve the running efficiency of motors. They also provide enough torque to start the PSC (permanent split capacitor) type motors. Running capacitors are provided with a terminal marked with a red dot. See Figure 1–46. Because of the relatively high voltage generated in the starting winding, the unmarked terminal is connected to the starting terminal on the motor. The red dot indicates the terminal that is most likely to short out in case of a capacitor breakdown. If the terminal with the red dot is connected to the motor starting terminal, damage to the winding could result. The best way to check these capacitors is with a capacitor analyzer. This type of meter provides a direct reading of the mfd output of the capacitor without the use of bulky equipment and mathematical formulas. See Figure 1–44. Running capacitors that are found to be out of the range of ± 10% of the mfd rating of the capacitor must be replaced with the proper size. The voltage rating of any capacitor must be equal to or greater than the one being replaced. A higher than normal running amperage usually indicates a weak capacitor. To wire a running capacitor into the circuit, see Figure 1–47.

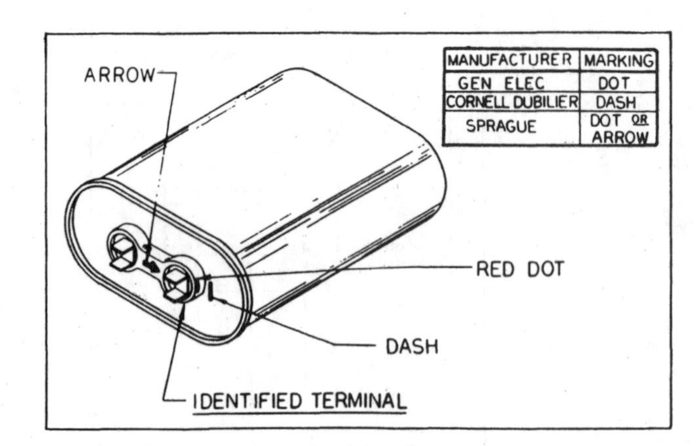

Figure 1–46. Run capacitor. (Courtesy of Copeland Corp.)

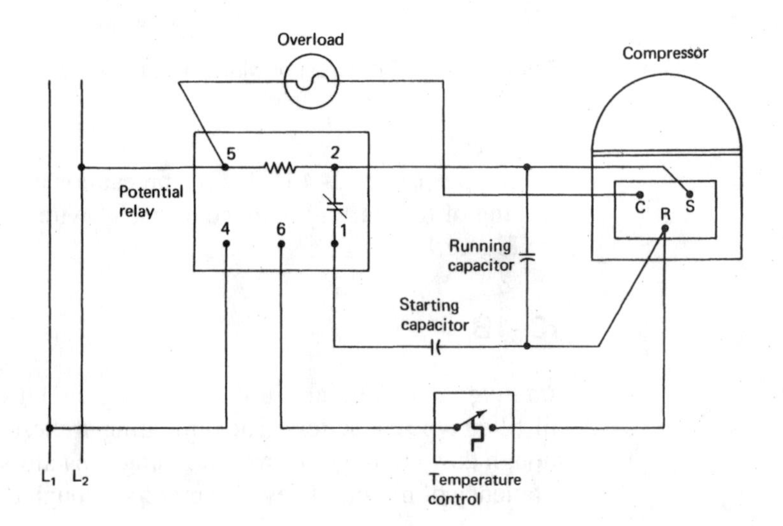

Figure 1–47. Compressor wiring diagram using run and start capacitors.

14–1 STARTING RELAYS:

Starting relays are the devices used to remove the starting circuit from operation when the motor reaches approximately 75% of its normal running speed. Its function is basically the same as the centrifugal switch used in split-phase motors. There are four types of starting relays in use today. They are: (1) amperage (current) relay, (2) hot wire relay, (3) solid-state relay, and (4) potential (voltage) relay. The horsepower size and the design of the equipment regulate which type of starting relay is used.

14–1A

The amperage (current) relay is an electromagnetic-type relay, which is normally used on 1/2 hp units and smaller. See Figure 1–48. These relays are positional types and must be properly mounted for satisfactory operation. They must be sized for each motor horsepower and amperage rating. To check an amperage relay, turn off the electricity to the unit and remove the wire from the "S" terminal and touch it to the "L" terminal on the relay. See Figure 1–49. Place an ammeter on the common wire to the compressor. Start the compressor and immediately remove the "S" wire from the "L" terminal. If the compressor continues to run and the amperage draw is within the rating of the compressor, replace the relay.

CAUTION: Do not allow the loose wire to come in contact with anything or anyone so as to prevent electrical shock.

An amperage relay that is too large for a motor may not allow the relay contacts to close, thus leaving out the starting circuit. The motor will not start under these conditions. A relay that is rated too small for a motor may keep the contacts closed at all times while electrical power is applied, leaving the starting circuit

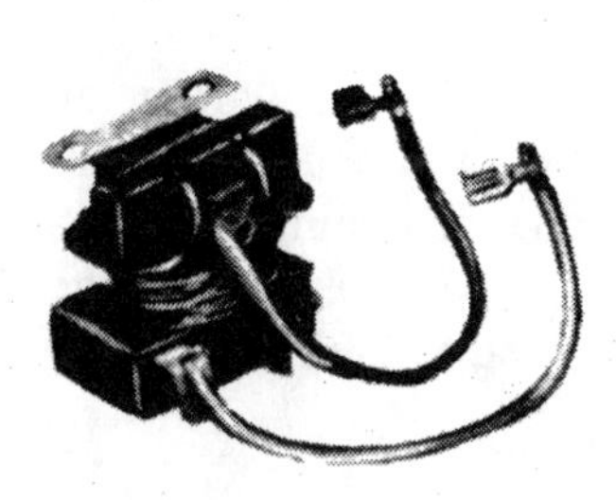

GENERAL ELECTRIC
3ARR2
BRACKET TYPE

Figure 1–48. Amperage relay.

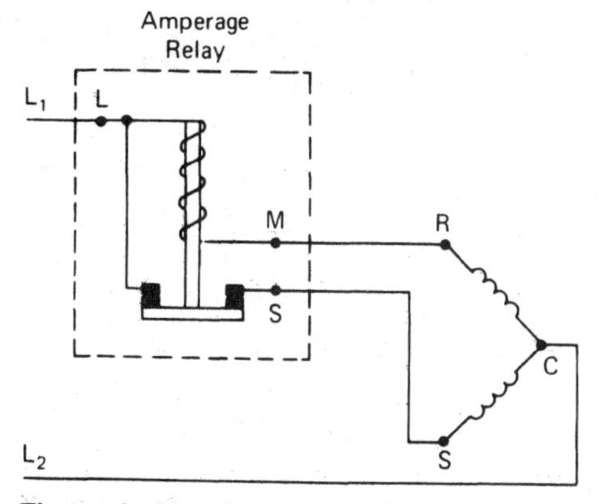

Figure 1–49. Amperage relay connections.

engaged continuously. Damage to the starting circuit may occur under these conditions. A motor protector must be used with this type of relay.

14–1B

Solid-state starting relays use a self-regulating conductive ceramic developed by Texas Instruments, Inc., which increases in electrical resistance as the compressor starts, thus quickly reducing the starting winding current flow to a milliamp level. The relay switches in approximately .35 second. This allows this type of relay to be applied to refrigerator compressors without being tailored to each particular system within the specialized current limitations. These relays will start virtually all split-phase 115-volt hermetic compressors up to 1/3 hp. An overload must be used with these relays. See Figure 1–50. Since these relays are push-on type devices, the easiest method of checking their operation is to simply install a new one. Be sure to check the amperage draw of the compressor motor.

14–1C

Potential relays operate on the electromagnetic principle. They incorporate a coil of very fine wire wound around a core. These starting relays are used on motors of almost any size. They are non-positional. The contacts are normally closed and are caused to open when a plunger is pulled into the relay coil. These relays have three connections to the inside in order for the relay to perform its function. These terminals are numbered 1, 2, and 5. Other terminals numbered 4 and 6 are sometimes used as auxiliary terminals. See Figure 1–51. To check a potential relay, turn off the electrical current and remove the wire from terminal 2 on the relay and touch it to terminal 1 on the relay. Place an ammeter on the common wire to the motor. Start the motor and immediately remove the 2 wire from the 1 terminal on the relay.

CAUTION: Do not allow the loose wire to touch anything or anyone to prevent electrical shock. If the compressor continues to run and the amperage draw is within the rating of the compressor, replace the relay.

The sizing of potential relays is not as critical as with the amperage and hot wire relays. A good way to determine what relay is required is to manually start the motor and check the voltage between the start and common terminals while the motor is operating at full speed. See Figure 1–52. Multiply the voltage obtained by .75 and this will be the pick-up voltage of the required relay.

14–1D

A starting relay that is mounted so that it vibrates will cause the contacts to arc excessively and become burned. When a relay is mounted on such a surface, it must be remounted on a more solid surface. Generally, the relay will need to be replaced dur-

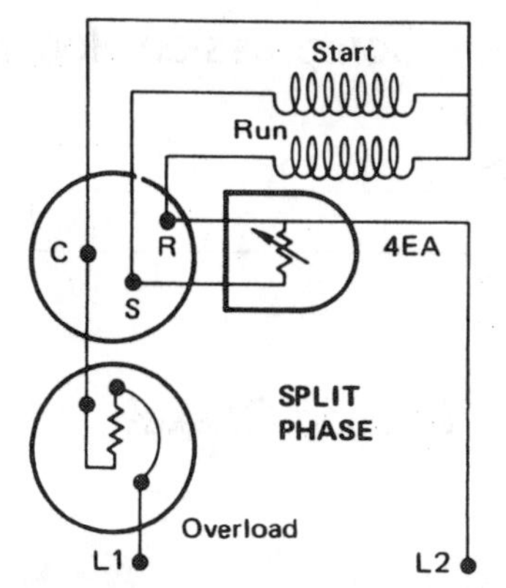

Figure 1–50. Solid-state relay connections. (Courtesy of Klixon Controls Division, Texas Instruments, Inc.)

Figure 1–51. Potential starting relay.

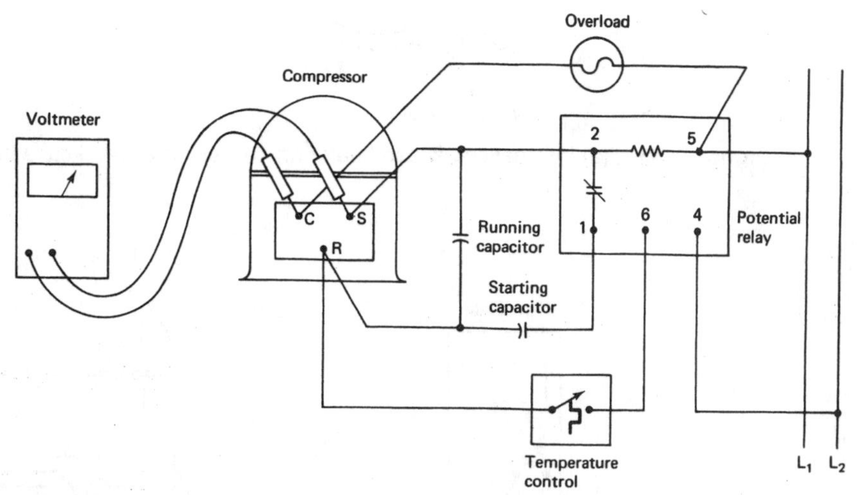

Figure 1–52. Checking voltage between start and run terminals.

ing the process. Be sure to replace the relay with the proper type and size, and if it is a positional relay, it must be mounted to satisfy the manufacturer's recommendations.

15–1 CRANKCASE HEATERS:

Crankcase heaters are electrical resistors that are designed to provide just enough heat to keep any liquid refrigerant that might enter the crankcase boiled off. There are two different designs: (1) externally mounted, and (2) internally mounted. See Figure 1–53. Externally mounted crankcase heaters are manufactured by several manufacturers and are easily installed on most compressors. The internally

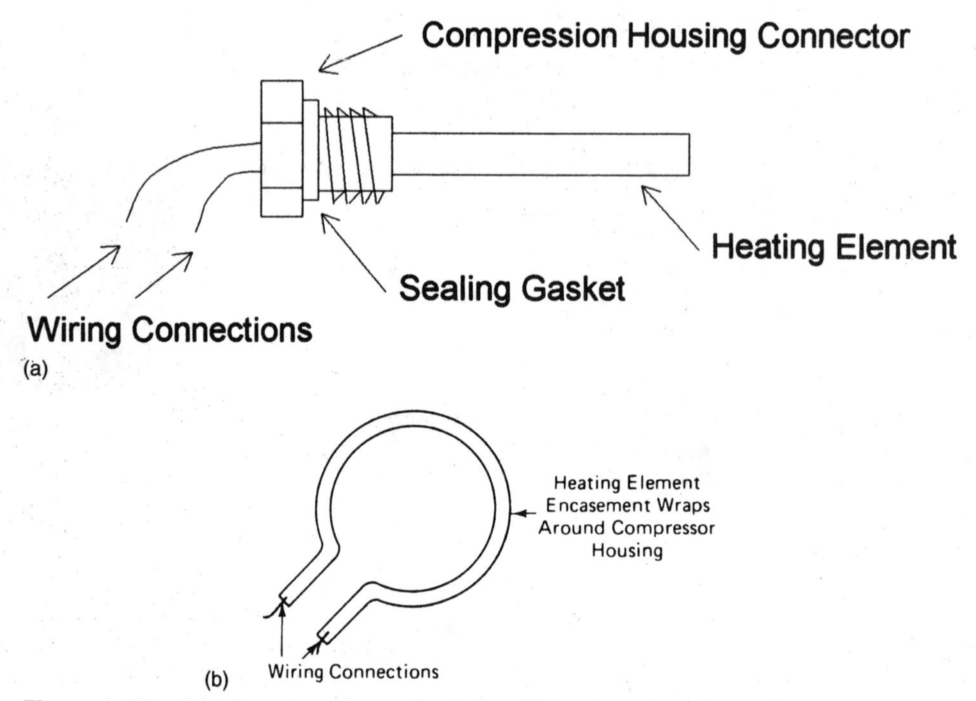

Figure 1–53. (a) Internal crankcase heater and (b) external crankcase heater.

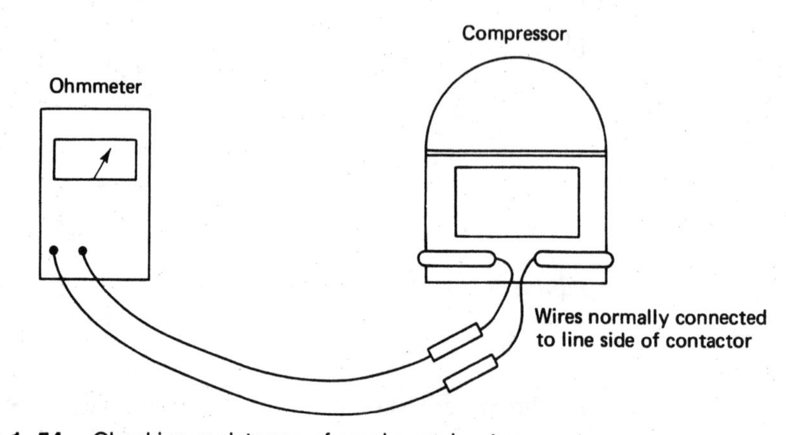

Figure 1–54. Checking resistance of crankcase heater.

mounted heaters are usually designed for a specific model of compressor. To check crankcase heaters, dampen a finger and gently touch the heater. If it is working, it will be hot to the touch. The electrical power must be on for several hours before this test can be performed. Another method of checking is to disconnect both electrical wires and check the resistance of the heater. See Figure 1–54. Crankcase heaters are generally energized continuously and are designed to prevent overheating of the compressor oil.

16–1 UNEQUALIZED PRESSURES:

Unequalized pressures is a term used to indicate that the refrigerant pressures in the high and low sides of the system are not close enough to allow a PSC motor to start the compressor. When there is a great difference between these two pressures, the starting load is too great for the motor to start. Less starting torque is required as the difference between these two pressures becomes less; that is, with zero pressure difference between the high and low side, a minimum of starting torque is required. There are two remedies to this situation. One is to allow the system to sit idle for a longer period of time. The second is to install a hard-start kit. See Section 17-1.

17–1 HARD-START KIT:

Hard-start kits are designed for use when conditions are encountered that prevent the PSC motor from starting normally. Such conditions are when the electrical power fluctuates or is too low to provide the necessary power for the proper starting of a compressor motor. Another such condition occurs when a PSC motor is used on a system that requires rapid cycle operation. Hard-start kits are designed to convert PSC motors to CSCR (capacitor start capacitor run) motors. Hard-start kits consist of the proper starting relay and the proper starting capacitor, along with the necessary wiring to install the kit. The individual components may also be combined to make up a hard-start kit. To install a hard-start kit, turn off the electrical power and complete the connections as shown in Figure 1–55.

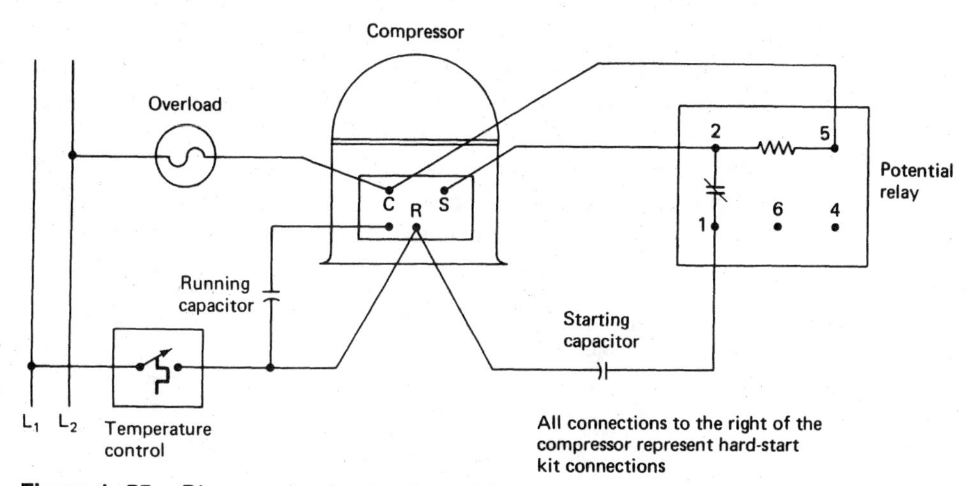

Figure 1–55. Diagram showing hard-start kit connections.

PARALLEL

18–1 HIGH DISCHARGE PRESSURE:

High discharge pressure can be the result of one or a combination of several things. It is a condition that causes an overload on the motor and decreases the efficiency of the compressor and the refrigeration system. The most common causes are: (1) compressor discharge service valve front-seated, (2) lack of cooling air, (3) lack of cooling water, (4) an overcharge of refrigerant, or (5) noncondensable gases.

18–1A

The compressor discharge service valve when front-seated will reduce or completely stop the flow of refrigerant from the compressor. Exercise caution to prevent this condition because damage to the compressor or motor is likely. The rapid build-up of pressure within the compressor cylinder head is tremendous and increases rapidly as the piston completes its upward stroke. Never front-seat the compressor discharge service valve while the compressor is running or start the compressor with the valve front-seated.

18–1B

The lack of cooling air over an air-cooled condenser will cause the discharge pressure to increase. This is because the higher pressure is required to condense the vapor to a liquid. The higher refrigerant temperature also reduces the unit efficiency because of the increased flash-gas, as well as reducing the compressor efficiency. This condition is generally caused by a dirty condenser, loose fan belt, or bad fan motor bearings.

A dirty condenser can be cleaned by using a garden hose with a high-pressure nozzle. The water should be forced through the condenser from both sides. Be sure to prevent water from entering the fan motor, which might cause an electrical short. This can be done by wrapping a piece of sheet plastic around the motor. Be sure to remove the plastic before starting the unit. Turn the unit off during this procedure.

A loose or broken fan belt will prevent the blower from moving air to cool the condenser. This condition is generally obvious and is easily corrected. To adjust the belt tension, turn the adjustment until the belt can be flexed about 1 in. with one finger using moderate pressure. See Figure 1–56. If the belt is frayed, has wear grooves on the sides, or has become hard, it should be replaced. Be sure to use the proper size belt. A belt that is too narrow will ride the bottom of the pulley. See

Figure 1–56. Proper belt tension adjustment.

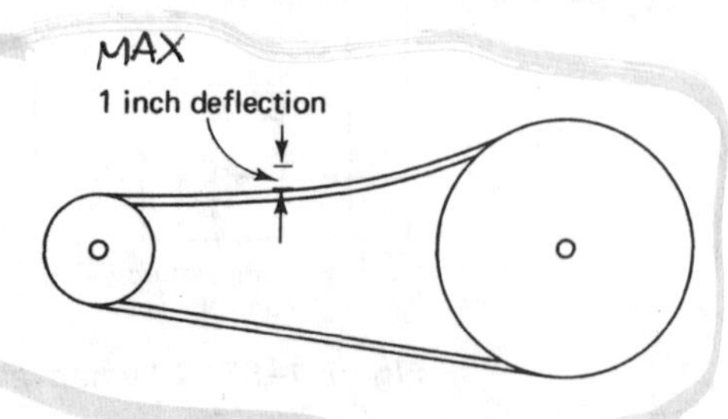

Figure 1–57. It will slip, causing decreased efficiency. A belt that is too wide will ride too high in the pulley and will not maintain the desired efficiency and will possibly overload the fan motor.

Bad fan motor bearings will cause the motor to overheat and cut out on the overload. When the condenser fan motor stops, the condenser overheats and the compressor will cut out on high pressure. To check for bad bearings, stop the unit, remove the belt, and move the motor shaft from side to side. Any movement of the shaft in a sideways direction indicates bad bearings. Replace the bearing or the motor, depending on the motor size.

18–1C

The lack of cooling water on a water-cooled condenser will cause the discharge pressure to increase because the higher pressure is required to condense the vapor to a liquid. The higher refrigerant temperature also reduces the unit efficiency because of the increased flash-gas, as well as reducing the compressor efficiency. This condition is generally caused by plugged strainers, pumps, or spray nozzles. When this condition is suspected, check the temperature rise of the water through the condenser. This rise should not be more than 10°F (5.56°C). If a temperature rise of more than 10°F is found, a strainer, pump, or spray nozzle is stopped up or there is not enough water in the tower water sump. The necessary steps must be taken to relieve this condition.

A temperature rise of the water less than 10°F (5.56°C) indicates that the condenser is scaled and must be cleaned. There are several commercial cleaners available for cleaning (acidizing) water-cooled condensers. The amount used is recommended by the manufacturer of the cleaner. Use caution to prevent damage to the equipment, personnel, and the surrounding vegetation when acidizing a unit. The most common method of acidizing a unit is to be sure the strainers,

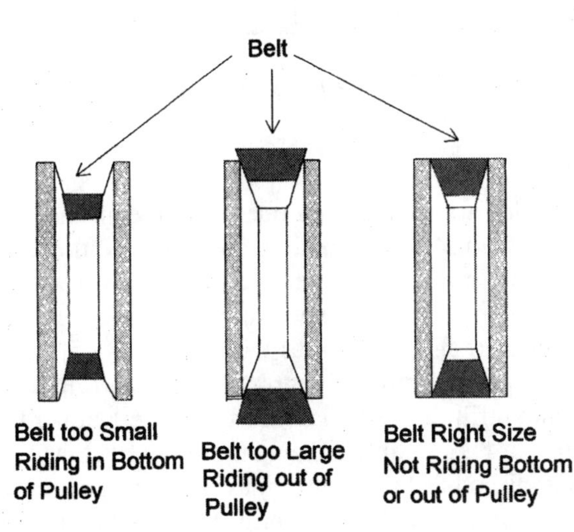

Figure 1–57. Comparison of belt and pulley fitting.

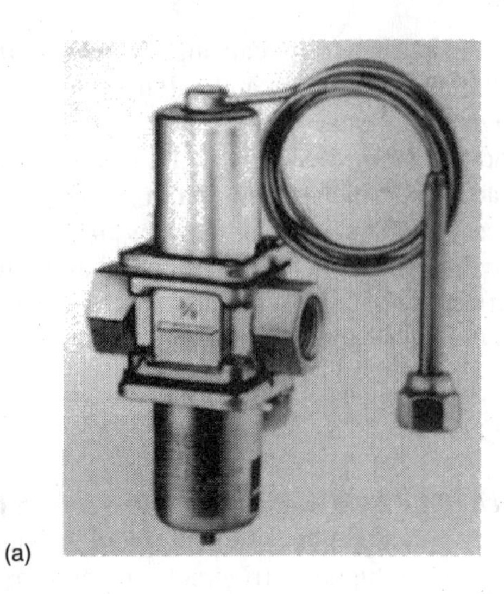

(a)

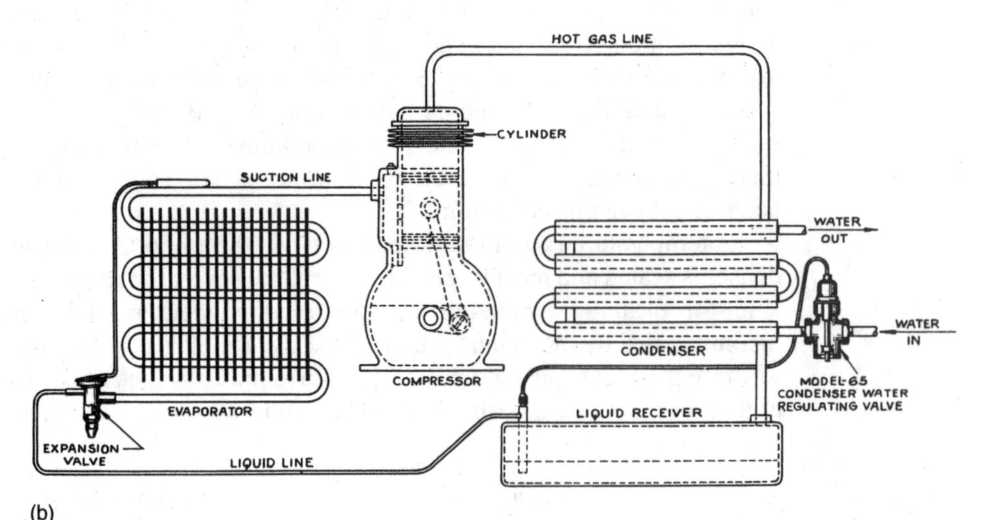

(b)

Figure 1–58. (a) Water flow control valve (Courtesy of Singer Controls Division) and (b) location of water flow control valve.

pump, and spray nozzles are clean. Then dump the cleaner into the tower sump and check the mixture with pH strips, adding cleaner until the desired pH is indicated, and allow the unit to run until all the scale has been removed. The cleaner and sediment must be completely removed from the system. To do this, drain and flush the tower, condenser, and water lines with fresh water. Then add a neutralizer, allowing the unit to run until the neutralizer has had time to neutralize the cleaner; then drain and refill the system with fresh water. Never leave the cleaner in the system because of the possible damage that may be done to the equipment. Another method is to disconnect the water lines from the condenser and circulate the cleaner through the condenser with a pump. This method is usually expensive and therefore not popular.

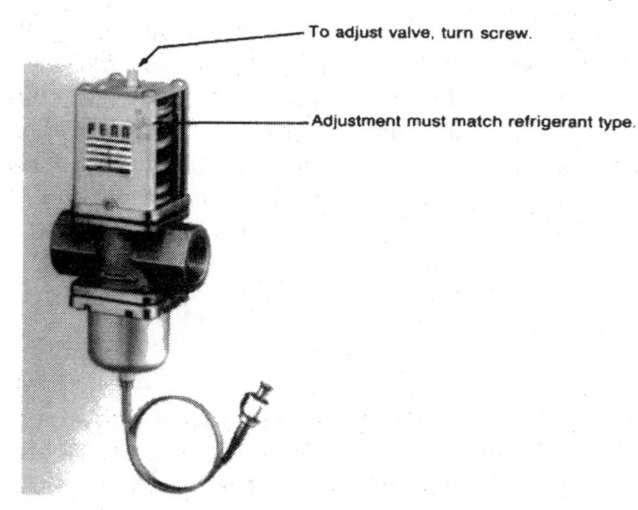

Figure 1–59. Pressure-actuated water regulating valve. (Courtesy of Penn Controls, Inc.)

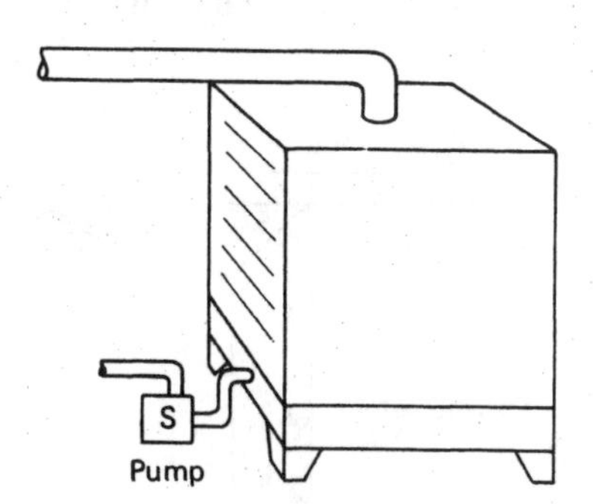

Figure 1–60. Cooling tower.

18–1D

Water-cooled condensers are sometimes equipped with a valve that controls the amount of water passing through the condenser. See Figure 1–58. These valves eventually become corroded with minerals, which causes them to become inoperative. When this occurs, remove the valve and either repair or replace it. Be sure to relieve the pressure in that part of the system where the pressure attachment is connected. When the valve is repaired or replaced, adjust the valve to provide the proper discharge pressure. See Figure 1–59.

18–1E

The water used to cool water-cooled condensers must be of low enough temperature to cause the refrigerant vapor to condense at the pressures encountered in the normal operating cycle. Cooling towers are used to cool the water. See Figure 1–60. The spray nozzles cause the water to atomize and mix with the air. This mixing with the air causes evaporation of the water, which lowers its temperature. If the spray nozzles do not atomize the water, it will not be cooled sufficiently. To determine whether or not the tower is functioning properly, measure the water temperature as it leaves the tower sump and as it leaves the spray nozzle. The difference in these two temperatures should be a minimum of 10° F (5.56° C). See Figure 1–61. Should the spray nozzles become clogged, the required cooling of the water will not be accomplished. Be sure that a full cone of water is coming from each nozzle. It may be necessary to clean the pump and strainers to obtain the desired amount of water to the system.

18–1F

An overcharge of refrigerant will cause a high discharge pressure because the extra refrigerant will take up space in the condenser that is needed to condense the vapor to a liquid. An overcharge of refrigerant will cause at least one-half of the condenser tubes to be cooler than the rest. The cool tubes are the ones that are full of liquid refrigerant. See Figure 1–62. When a capillary tube is used, the suction

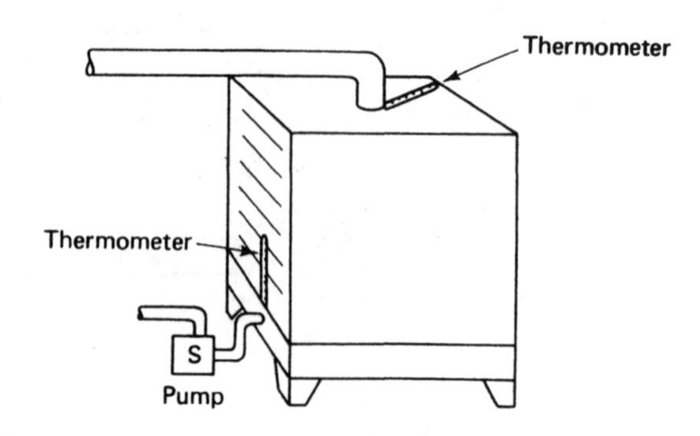

Figure 1–61. Checking cooling tower water temperatures.

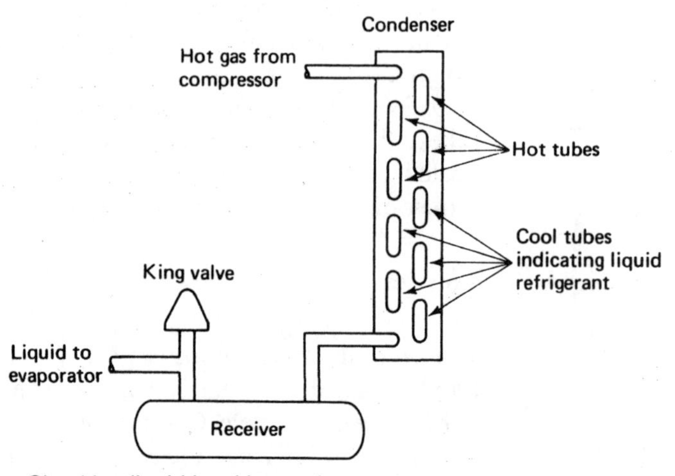

Figure 1–62. Checking liquid level in condenser.

pressure will also be higher than normal. The suction line will also be cooler than normal and may be frosted over, depending on the amount of overcharge.

The excess of refrigerant must be purged from the system. When purging, allow only a small amount of refrigerant to escape at a time. This procedure is recommended to prevent purging too much refrigerant from the system, which would require the addition of refrigerant back into the unit. Also, it prevents the removal of lubricating oil from the compressor crankcase.

18–1G

The presence of noncondensable gases in a system will cause a high discharge pressure because the noncondensables will not condense under the normal pressures encountered in refrigeration systems and will take up space needed by the refrigerant in the condenser. To determine if noncondensables are present in a system, pump the system down; that is, pump all the refrigerant into the receiver or condenser by front-seating the liquid line service (king) valve. See Figure 1–63. Operate the compressor until the low side has been pumped down to approximately 5 psig. Stop the unit and allow it to stand idle until it has cooled to the ambient temperature. Compare the idle discharge pressure to the pressure indicated on a pressure-temperature chart at that temperature. If the discharge pressure is higher than that indicated by the chart, slowly purge the noncondensables from the highest point in the system. Purge the air slowly to prevent purging an excess of refrigerant from the system. If the system contains only a small charge of refrigerant, it will probably be better to purge the entire charge, evacuate the system, and add a complete new charge. Noncondensables should be kept from a system because they not only reduce the system efficiency but they also contain moisture, which is extremely dangerous to a refrigeration system.

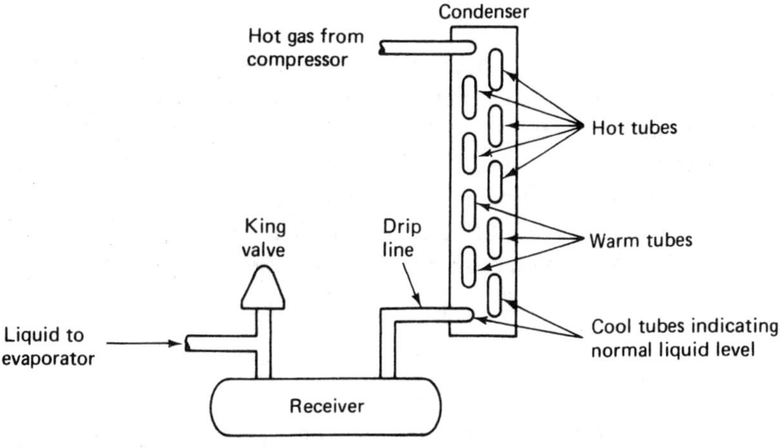

Figure 1–63. Location of king valve.

19–1 LOW SUCTION PRESSURE:

A suction pressure that is lower than normal may be due to any or all of the following reasons: (1) shortage of refrigerant, (2) dirty air filter, (3) dirty evaporator, (4) dirty blower, (5) loose fan belt, (6) iced-over evaporator, (7) TXV (thermostatic expansion valve) superheat set too low, (8) AXV (automatic expansion valve) set too low, or (9) restriction in the refrigerant system. The problem or problems causing a low suction pressure should be found and corrected because the compressor may pump all the lubricating oil out of the crankcase and into the system, resulting in possible damage to the compressor as well as inefficient operation.

19–1A

A shortage of refrigerant is usually due to a leak that has developed in the refrigerant circuit. The leak should be found and repaired and the system recharged with the proper charge of refrigerant. If the leak is not found and repaired, the refrigerant will escape from the system, requiring more service. A refrigerant leak is generally indicated by oil on the place where the leak has occurred. Many times a visual inspection will locate the problem area. See Figure 1–64. If the problem area is not found by visual inspection, an electronic or halide torch leak detector should be used. Either of these will pinpoint a leak under most conditions. Two conditions that make leak detection difficult with these devices are when the wind is blowing outdoors or in an enclosed place where the refrigerant concentration is high. When these conditions are encountered, a soap and water solution or one of the liquid plastic gas leak detectors must be used. The leak may be found by applying the solution to the suspected area and watching for bubbles to appear. See Figure 1–65. The bubbles will appear within 5 seconds. When repairing leaks that require the tubing to be heated, be sure to relieve any pressure from inside the tubing that is being heated. This will prevent a blowout, which can result in damage to the equipment and injury to the service technician.

19–1B

Dirty air filters are a frequent cause of low suction pressure in air conditioning systems. Dirty air filters restrict the flow of air over the evaporator, resulting in a low load on the refrigeration system. The air filter is located on the air inlet side of the blower. See Figure 1–66. The filters should be changed or cleaned on a regular basis to insure peak efficiency from the unit. The throwaway-type filters should not be cleaned. They should be replaced. When cleanable-type filters are used, they should be coated with a filter coater after cleaning to increase the effectiveness of the filter.

19–1C

Dirty evaporator coils are often the cause of low suction pressures on air conditioning and refrigeration systems. A dirty evaporator restricts the air flow through

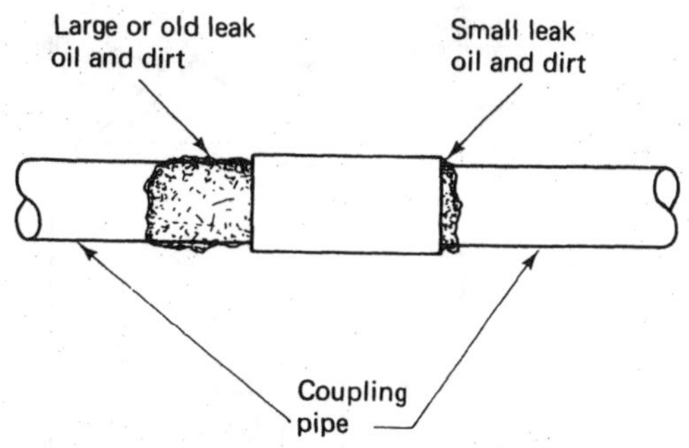

Figure 1–64. Oil around refrigerant leaks.

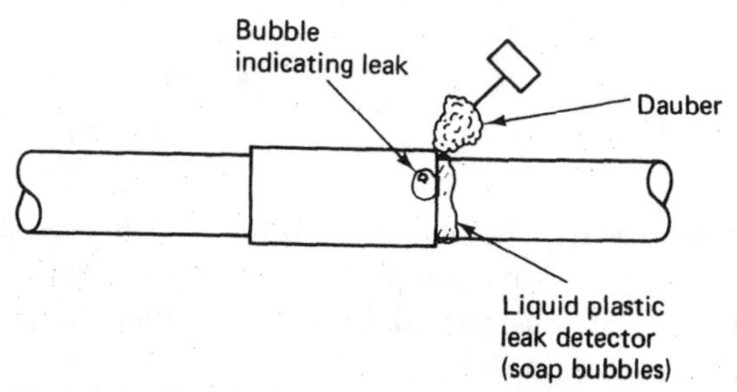

Figure 1–65. Leak testing with liquid plastic leak detector (soap bubbles).

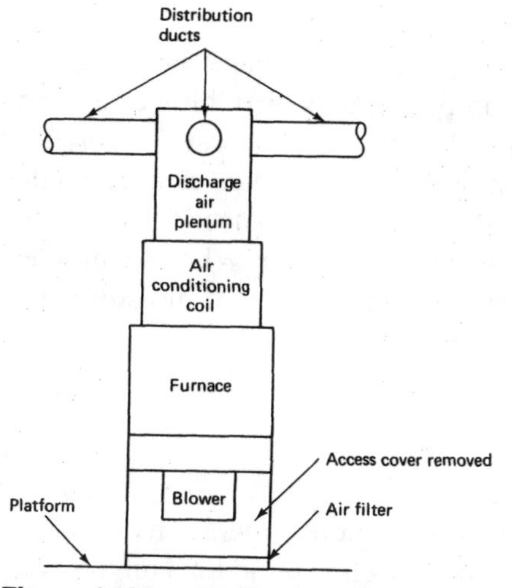

Figure 1–66. Air filter location.

the unit and the insulating value, and the dirt prevents the proper transfer of heat to the refrigerant. A dirty evaporator coil is the result of a dirty air filter or one that does not fit properly. When an air filter does not fit properly, the dust-laden air will bypass and the dust will stick to the evaporator coil fins. See Figure 1–67. Precautions

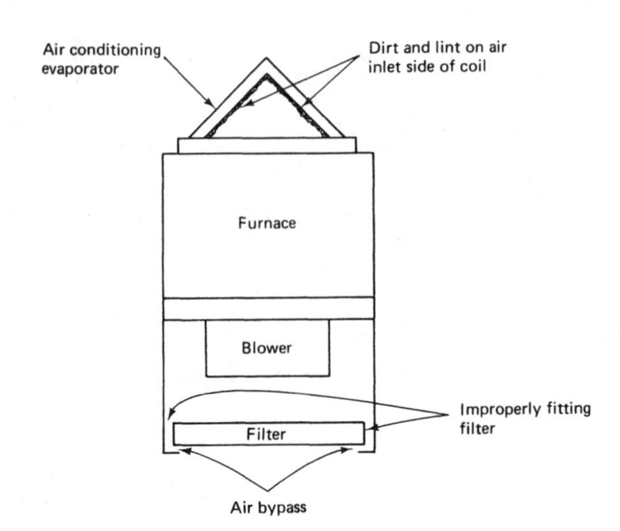

Figure 1–67. Dirty coil because of filter air bypass.

must be taken to insure proper fit of the air filter, and the evaporator must be cleaned before the system will function properly. Many times the evaporator will need to be removed and be steam-cleaned. Be sure to prevent moisture entering the refrigerant system during this process.

19–1D

A dirty blower will not deliver enough air to place the proper load on the evaporator, resulting in a low suction pressure. A dirty blower is the result of improperly fitting air filters or a unit that has been operated with an excessively dirty filter. The blower must be removed and the vanes cleaned so that only the metal can be seen. The filter must be cleaned, or replaced, and any air bypass must be prevented. Otherwise, the problem will reoccur. When steam-cleaning or washing the blower with water, take precautions to prevent moisture entering the electric blower motor.

19–1E

A loose or broken fan belt will prevent proper operation of the blower and the flow of air over the evaporator, resulting in a low or no load on the evaporator and a low suction pressure. A loose belt that is in good condition can be properly adjusted and the unit put back in service. A broken or frayed belt or one that has become hard on the edges must be replaced and properly adjusted. To properly adjust a belt, tighten the adjustment screw until the belt can be flexed about 1 in. with one finger using moderate pressure. See Figure 1–56. Be sure to use the proper sized belt. A belt that is too narrow will ride the bottom of the pulley and will slip, causing decreased efficiency. See Figure 1–57. A belt that is too wide will ride too high in the pulley and will not maintain the desired efficiency and will possibly overload the motor.

19–1F

An iced or frosted evaporator will cause a low suction pressure because the insulating function of the ice or frost will prevent proper heat transfer from the air to the refrigerant. An iced evaporator is usually caused by an insufficient load on the refrigeration system or a shortage of refrigerant. A reduced load can be the result of (1) a dirty filter, (2) a dirty evaporator coil, (3) a dirty blower, or (4) a loose or broken fan belt. A shortage of refrigerant is due to a refrigerant leak. Any of these conditions must be corrected and the coil de-iced before the system will operate satisfactorily. De-icing can be accomplished by turning on the blower and turning off the compressor. Allow the fan to operate until the ice has been melted. A faster method is to apply a small amount of heat to the ice. Caution should be exercised to prevent overheating of the evaporator coil or fins.

20–1 THERMOSTATIC EXPANSION VALVES:

Thermostatic expansion valves are the most common type of flow control devices used on units of 5-ton capacity and above. They operate as a result of pressure and temperature inside the evaporator. The proper superheat adjustment of these valves will have a tremendous effect on the efficiency of the equipment.

A thermostatic expansion valve that is adjusted for too great a superheat or an improperly located feeler bulb will result in a low suction pressure. If the superheat is found improperly adjusted, it may be adjusted by the following procedure:

1. Measure the temperature of the suction line at the point where the bulb is clamped.
2. Obtain the suction pressure that exists in the suction line at the bulb location by either of the following methods:
   a. If the valve is externally equalized, install a pressure gauge in the external equalizer line to read the pressure directly and accurately.
   b. Read the gauge pressure at the suction service valve of the compressor. To the reading add the estimated pressure drop through the suction line between the bulb location and the compressor suction service valve. The sum of the gauge reading and the estimated pressure drop will equal the approximate line pressure at the bulb.
3. Convert the pressure obtained in Step 2(a) or Step 2(b) to the saturated evaporator temperature by using a temperature-pressure chart as shown in Table 1–1.
4. Subtract the two temperatures obtained in Step 1 and Step 3. The difference is the superheat setting of the valve.

Table 1–1. Refrigerant Pressure Temperature Relationship

°F	22	410A	12	134a	401a	409A	502	404a	507	408A	402A
−40	***4.8***	6.9	***14.6***	***17.7***	***17.7***	***16.4***	0.7	0.8	1.7	*1.0*	2.1
−44	***2.0***	9.1	***12.9***	***16.2***	***16.0***	***14.8***	2.3	2.5	3.5	1.1	3.9
−40	0.5	11.6	***11.0***	***14.5***	***14.5***	***13.1***	4.1	5.5	5.5	2.8	5.9
−36	2.2	14.2	***8.9***	***12.8***	***12.5***	***11.2***	6.0	7.5	7.6	4.6	8.0
−32	4.0	17.1	***6.7***	***10.8***	***10.6***	***9.2***	8.1	9.7	9.9	6.6	10.3
−28	5.9	20.1	***4.3***	***8.6***	***8.3***	***6.9***	10.3	12.0	12.4	8.7	12.8
−24	7.9	23.4	***1.6***	***6.2***	***6.0***	***4.5***	12.7	14.5	15.0	11.0	15.5
−20	10.1	26.9	0.6	***3.6***	***3.5***	***1.9***	15.3	17.1	17.8	13.5	18.4
−16	12.5	30.7	2.1	***0.8***	***0.5***	0.5	18.1	20.0	20.9	16.1	21.5
−12	15.1	34.7	3.7	1.1	1.4	2.0	21.0	23.0	24.1	18.9	24.8
−8	17.9	39.0	5.4	2.8	3.1	3.6	24.2	26.3	27.6	21.9	28.3
−4	20.8	43.7	7.2	4.5	4.8	5.3	27.5	29.8	31.3	25.2	32.1
0	24.0	48.6	9.2	6.5	6.7	7.2	31.1	33.5	35.2	28.7	36.1
2	25.6	51.1	10.2	7.5	8.0	8.2	32.9	34.8	37.3	30.5	38.1
4	27.3	53.8	11.2	8.5	8.8	9.2	34.9	37.4	39.4	32.3	40.4
6	29.1	56.7	12.3	9.6	9.9	10.2	36.9	39.4	41.6	34.3	42.6
8	30.9	59.4	13.5	10.8	11.0	11.3	38.9	41.6	43.8	36.3	44.9
10	32.8	62.3	14.6	12.0	12.2	12.5	41.0	43.7	46.2	38.3	47.3
12	34.7	65.4	15.8	13.1	13.4	13.6	43.2	46.0	48.5	40.4	49.7
14	36.7	68.6	17.1	14.4	14.6	14.8	45.4	48.3	51.0	42.6	52.2
16	38.7	71.9	18.4	15.7	15.9	16.1	47.7	50.7	53.5	44.9	54.8
18	40.9	75.2	19.7	17.0	17.2	17.4	50.0	53.1	56.1	47.2	57.5
20	43.0	78.3	21.0	18.4	18.6	18.7	52.5	55.6	58.8	49.6	60.2
22	45.3	82.3	22.4	19.9	20.0	20.1	54.9	58.2	61.5	52.0	63.0
24	47.6	85.9	23.9	21.4	21.5	21.5	57.5	60.9	64.3	54.5	65.9
26	49.9	89.7	25.4	22.9	23.0	22.9	60.1	63.6	67.2	57.1	68.9
28	52.4	93.5	26.9	24.5	24.6	24.4	62.8	66.5	70.2	59.8	72.0
30	54.9	96.8	28.5	26.1	26.2	26.0	65.6	69.4	73.3	62.5	75.1
32	57.5	101.6	30.1	27.8	27.9	27.6	68.4	72.3	76.4	65.3	78.3
34	60.1	105.7	31.7	29.5	29.6	29.2	71.3	75.4	79.6	68.2	81.6

36	62.8	110.0	33.4	31.3	31.3	30.9	74.3	78.5	82.9	71.2	85.0
38	65.6	114.4	35.2	33.2	33.2	32.7	77.4	81.8	86.3	74.2	88.5
40	68.5	118.0	36.9	35.1	35.0	34.5	80.5	85.1	89.8	77.4	92.1
42	71.5	123.6	38.8	37.0	37.0	36.3	83.8	88.5	93.4	80.6	95.7
44	74.5	128.3	40.7	39.1	39.0	38.2	87.0	91.9	97.0	83.9	99.5
46	77.6	133.2	42.7	41.1	41.0	40.2	90.4	95.5	100.8	87.3	103.4
48	80.7	138.2	44.7	43.3	43.1	42.2	93.9	99.2	104.6	90.7	107.3
50	84.0	142.2	46.7	45.5	45.3	44.3	97.4	102.9	108.6	94.3	111.4
52	87.3	148.5	48.8	47.7	**60.0**	**63.6**	101.0	**109.0**	112.6	97.9	**120.0**
56	94.3	159.3	53.2	52.3	**65.0**	**68.9**	108.6	**117.0**	121.0	105.5	**129.0**
60	101.6	169.6	57.7	57.5	**70.0**	**74.5**	116.4	**125.0**	129.7	113.5	**138.0**
64	109.3	182.6	62.5	62.7	**76.0**	**80.3**	124.6	**134.0**	139.0	121.8	**147.0**
68	117.3	195.0	67.6	68.3	**82.0**	**86.3**	133.2	**144.0**	148.6	130.6	**157.0**
72	125.7	208.1	72.9	74.2	**89.0**	**92.8**	142.2	**153.0**	158.7	139.7	**168.0**
76	134.5	221.7	78.4	80.3	**95.0**	**99.4**	151.5	**164.0**	169.3	149.3	**179.0**
80	143.6	235.3	84.2	86.8	**102.0**	**106.4**	161.2	**174.0**	180.3	159.4	**190.0**
84	153.2	250.8	90.2	93.6	**109.0**	**113.7**	171.4	**185.0**	191.9	169.8	**202.0**
88	163.2	266.3	96.5	100.7	**117.0**	**121.2**	181.9	**197.0**	203.9	180.8	**214.0**
92	173.7	282.5	103.1	108.2	**125.0**	**129.1**	192.9	**209.9**	216.6	192.2	**227.0**
96	184.6	299.3	110.0	116.1	**133.0**	**137.4**	204.3	**222.0**	229.8	204.1	**240.0**
100	195.9	317.2	117.2	124.3	**142.0**	**146.0**	216.2	**235.0**	243.5	216.6	**254.0**
104	207.7	335.2	124.7	132.9	**151.0**	**154.9**	228.5	**249.0**	257.9	229.5	**269.0**
108	220.0	354.3	132.4	142.8	**160.0**	**164.2**	241.3	**264.0**	272.9	243.0	**284.0**
112	232.8	374.2	140.5	151.3	**170.0**	**173.9**	254.6	**279.0**	288.6	257.0	**299.0**
116	246.1	394.8	148.9	161.1	**180.0**	**183.9**	268.4	**294.0**	304.9	271.6	**316.0**
120	259.9	417.7	157.7	171.3	**191.0**	**194.4**	282.7	**311.0**	321.9	286.8	**332.0**
124	274.3	438.7	166.7	182.0	**202.0**	**205.2**	297.6	**328.0**	339.7	302.6	**350.0**
128	289.1	462.0	176.2	193.1	**213.0**	**216.5**	312.9	**345.0**	358.2	319.0	**368.0**
132	304.6	486.2	185.9	204.7	**225.0**	**228.2**	328.9	**364.0**	377.6	336.0	**387.0**
136	320.6	511.4	196.1	216.8	**237.0**	**240.3**	345.4	**383.0**	397.7	353.6	**406.0**
140	337.3	539.0	206.6	229.4	**250.0**	**252.9**	362.6	**402.0**	418.7	372.0	**426.0**
144	354.5	564.8	217.5	242.4	**263.0**	**265.9**	380.4	**423.0**	440.6	390.9	**447.0**
148	372.3	593.8	228.8	256.0	**277.0**	**279.5**	398.9	**444.0**	462.0	410.6	**468.0**

BOLD ITALIC = VACUUM

BLACK = VAPOR PSIG (CALCULATING SUPERHEAT)

BLACK BOLD = LIQUID PSIG (CALCULATING SUBCOOLING)

Figure 1–68 illustrates a typical example of superheat measurement on an air conditioning system using Refrigerant 12 as the refrigerant. The temperature of the suction line at the bulb location is read at 51°F (10.6°C). The suction pressure at the compressor suction service valve is 35 psi (241.316 Kp), and the estimated pressure drop is 2 psi (13.790 Kp). The total suction pressure is 35 psi + 2 psi = 37 psi (255.106 Kp). This is equivalent to a 40°F (4.4°C) saturation temperature. Then, 40°F subtracted from 51°F equals 11°F (6.2°C) superheat setting.

Notice that subtracting the difference between the temperature at the evaporator inlet and the temperature at the evaporator outlet is not an accurate measure of superheat. This method is not recommended because any evaporator pressure drop will result in an erroneous superheat indication.

Adjust the valve to provide the desired superheat setting. To determine if the expansion valve is operating, remove the remote bulb and warm it with the hand while observing the suction pressure. If the pressure increases, the valve needs adjusting.

20–1A

The feeler bulb on a thermostatic expansion valve is one of the major forces controlling the valve. If the bulb is located in a cold location, it will cause the valve to starve the evaporator, causing a lower than normal suction pressure. The location of the feeler bulb is extremely important and in some cases determines the success or failure of the refrigeration plant. For satisfactory expansion valve control, good thermal contact between the bulb and the suction line is essential. The bulb should be securely fastened with two bulb straps to a clean, straight section of the suction line.

Installation of the bulb to a horizontal run of suction line is preferred. If a vertical installation cannot be avoided, the bulb should be mounted so that the capillary tubing comes from the top. See Figure 1–69. To install, clean the suction line thoroughly before clamping the remote bulb in place. When a steel suction line is used, it is advisable to paint the line with aluminum paint to minimize future cor-

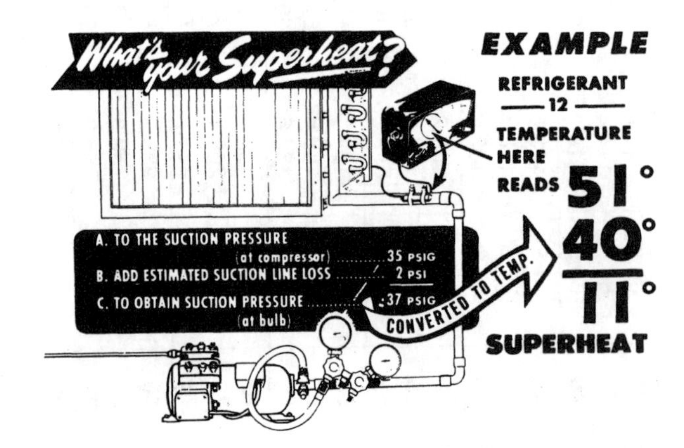

Figure 1–68. Determination of superheat. (Courtesy of Sporlan Valve Co.)

rosion and to eliminate faulty remote bulb contact with the line. On lines under 7/8 in. (22.225 mm) OD (outside diameter), the remote bulb may be installed on top of the line. On 7/8 in. OD and larger, the remote bulb should be installed at about the 4 or 8 o'clock position. See Figure 1–70. If it is necessary to protect the remote bulb from the effects of an air stream after it is clamped to the line, use a material such as sponge rubber that will not absorb water when the evaporator temperatures are above 32°F (0.°C). Below 32°F, cork or similar material sealed against moisture is suggested to prevent ice clogging at the remote bulb location.

20–1B

When a thermostatic expansion valve is stuck open, there will be an excessive amount of sweating on the suction line. The compressor crankcase will also sweat because of the liquid refrigerant being fed back to the compressor. It is best to replace a sticking expansion valve.

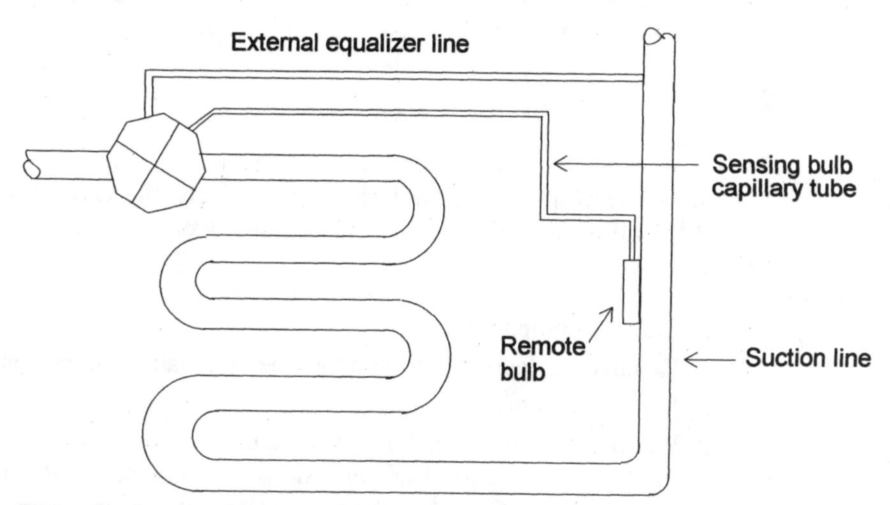

Figure 1–69. Remote bulb installation on a vertical line.

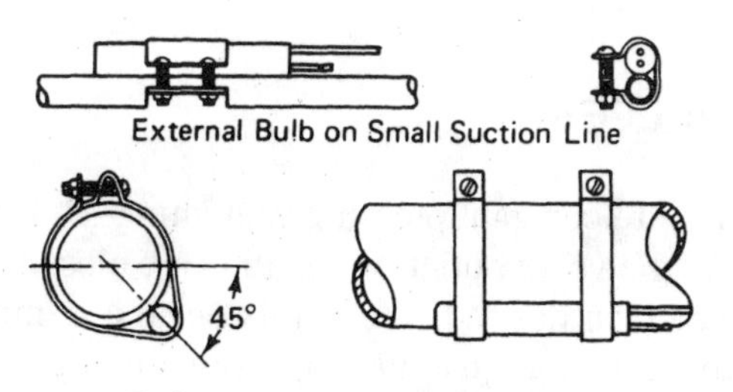

Figure 1–70. Remote bulb installation on horizontal tubing. (Courtesy of Alco Controls Division, Emerson Electric Co.)

20–1C

If the thermostatic expansion valve is too large for the system, it will not maintain a constant suction pressure. The feeler bulb will attempt to control the flow of liquid to maintain the superheat setting, but the oversized valve will admit too much liquid too rapidly. The feeler bulb sensing this liquid will close the valve, and the pressure in the evaporator will drop until the valve opens, admitting more liquid refrigerant. This hunting action will cause the suction pressure to fluctuate, which can be seen on the compound gauge. This variation is usually 10 psi to 15 psi (68.948 Kp to 103.421 Kp). When this condition occurs, either replace the complete valve or replace the valve seat.

20–1D

A thermostatic expansion valve that is too small cannot pass enough liquid refrigerant to properly refrigerate the evaporator. When the unit is heavily loaded, the superheat will be high and the system capacity will be low. An expansion valve that is too small will generally cause a lower than normal suction pressure. Either replace the valve or replace the valve seat with the proper size.

20–1E

The power element of a thermostatic expansion valve contains a vapor charge. If a leak should develop in the power element assembly, the valve will tend to close, stopping the flow of refrigerant. To check a power element, use the following procedure:

1. Stop the compressor.
2. Remove the feeler bulb from the line and place it in ice water.
3. Start the compressor.
4. Remove the bulb from the ice water and warm it with the hand. At the same time feel the suction line for a drop in temperature. If liquid refrigerant floods through the valve, the power element is operating properly. Be sure not to flood liquid back through the suction line for a long period of time because it can cause damage to the compressor.

21–1 AUTOMATIC EXPANSION VALVES:

Automatic expansion valves are constant pressure-regulating devices that respond to the refrigerant pressure at the valve outlet. When they are closed off, the suction pressure may become lower than desired. When this occurs, a minor adjustment will usually correct the problem. To adjust, install a compound gauge on the compressor suction service valve and adjust the valve to obtain the desired pressure. If the valve will adjust, it is operating satisfactorily and does not need to be replaced.

Do not make the final adjustment of the expansion valve until the refrigeration unit has been operating from 24 to 48 hours.

22–1 RESTRICTION IN A REFRIGERANT CIRCUIT:

A restriction in the refrigerant circuit will reduce the flow of refrigerant, resulting in a lower than normal suction pressure. A restriction may be in the form of a kinked line, a plugged strainer, a plugged drier, a plugged capillary tube, or ice in the orifice of a flow control device. The restriction must be removed before so that proper operation can be realized.

22–1A

A kinked line occurs when the line has been bent too far and has resulted in a flattened place. This type of restriction can be located by visual inspection. However, if the restriction is in the liquid line, there will be a temperature difference across the kink. If the tubing is flattened enough, there may be condensation or frost on the outlet side of the kink. See Figure 1–71. In cases when the tube is not flattened excessively, the flattened spot may be straightened by placing a flaring block around the tube, using the proper size hole, and tightening it down. If this fails, the piece of tubing must be removed and a new piece installed. Be sure to relieve all pressure on the tubing before attempting to make repairs. Personal injury may result when repairs are attempted when there is pressure inside the tube that is being repaired.

22–1B

The purpose of a strainer is to trap foreign particles in the refrigeration system. A plugged strainer will reduce the flow of refrigerant and may completely stop any passage of refrigerant. All expansion valves are equipped with strainers as well as suction line filter-driers, liquid line driers, and a compressor suction inlet. All of these features are used to trap foreign materials in the system to prevent damage to the particular component. A plugged strainer will develop a pressure drop on one side and usually a temperature difference on each side that can be felt with the bare

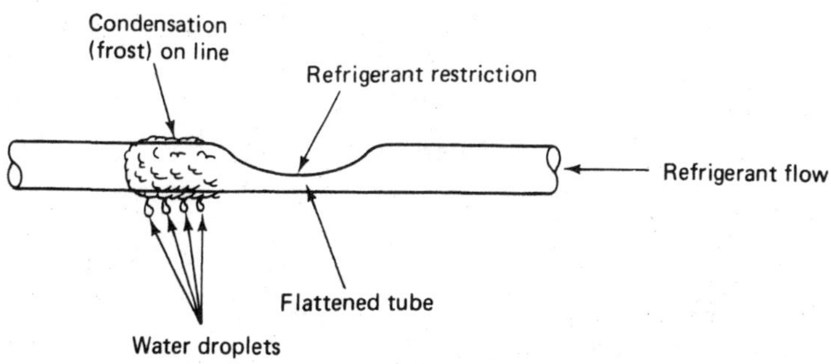

Figure 1–71. Flattened tube causing refrigerant restriction.

hand. When a strainer becomes plugged, it is best to replace it if another one is readily available. If one is not available, thoroughly clean the original and replace it. Do not remove the strainer permanently. To do this will allow foreign particles to enter the protected device, resulting in possible damage.

22–1C

The purpose of a drier is to remove moisture and trap foreign particles in the refrigeration system. A plugged drier will restrict or completely stop the flow of refrigerant through the system. All field built-up systems are equipped with driers to remove any moisture and foreign particles that may accidentally enter the system during the installation process. A plugged drier can be determined by the temperature drop through it. See Figure 1–72. A plugged drier should be replaced. Never permanently remove a drier from the system. A plugged drier is an indication that there are still moisture and foreign particles in the system. When removing a drier be sure to relieve all pressure inside the tube being worked on. To open a refrigeration system while it is pressurized can result in injury.

22–1D

A plugged capillary tube is caused by untrapped foreign particles in a refrigeration system and can result in a reduced or completely stopped refrigerant flow through the system. A plugged capillary tube will be indicated by a longer than usual pressure equalization time, accompanied by a loss of refrigeration. It is recommended

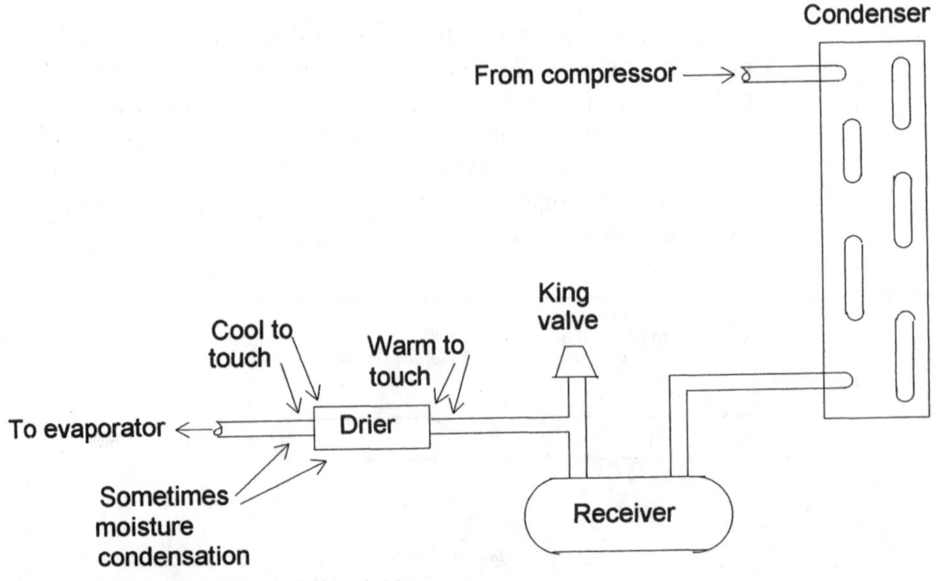

Figure 1–72. Checking a plugged drier.

that a plugged capillary line be replaced rather than cleaned. When replacing it be sure to use the same length and inside diameter tubing as the original tubing. To use a different type will result in an unbalanced refrigeration system and poor refrigeration. Always replace the filter strainer along with the capillary tube. Be sure to relieve the pressure in the low side of the system before attempting repairs or injury may result.

22–1E

Ice in the orifice of a refrigerant flow control device is due to free moisture in the system. This condition generally occurs when the drier has trapped all the moisture it can absorb, and there is surplus free moisture that turns to ice in the orifice of the flow control device. This condition is usually indicated by poor operation, a lower than normal suction, and low discharge pressure. To be sure that moisture is causing the problem, stop the unit and apply a cloth moistened with hot water to the outlet of the flow control device. If after a few minutes a hissing sound is heard and an increase in the low side pressure is detected, moisture is the culprit. To correct this problem a new drier must be installed. If the problem continues, it may be necessary to completely discharge the refrigerant from the system, triple-evacuate the system, install oversize driers, and recharge the system with dry refrigerant.

23–1 EXCESSIVE LOAD:

An excessive load condition normally occurs if the unit was undersized during the initial installation or there has been an addition to the building. The only solution to this problem is to resize the unit to better fit the application. Another condition that could cause excessive loading of the equipment occurs when the evaporator fan speed has been increased to the point of overloading the evaporator. This condition will cause an increase in the suction pressure. The obvious solution here is to reduce the fan speed while measuring the temperature drop of the air flowing through the evaporator. This temperature drop should be maintained at the manufacturer's recommendations to allow the equipment to operate at peak efficiency. Usually 20°F (11.2°C) for air conditioning and 10°F (5.6°C) to 15°F (8.4°C) for refrigeration applications are recommended.

24–1 TRAPS IN LINES:

Incidental traps that are left in the refrigerant lines during the installation process will trap oil and will cause damage to the compressor. These traps are generally in the form of low sags in the lines. The oil will settle out of the refrigerant and collect in these low places. The refrigerant will pass through the pipe over the oil, leaving the oil in the line instead of returning it to the compressor crankcase. See

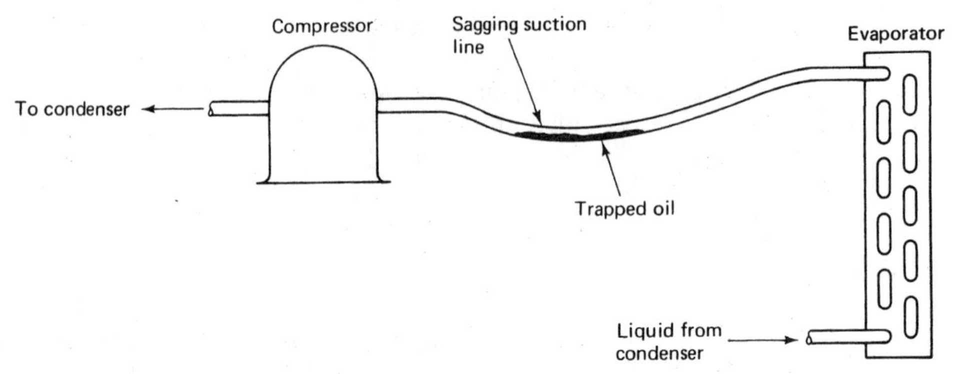

Figure 1–73. Trapped oil in sagging suction line.

Figure 1–73. There are two ways to correct this problem. The easiest and most correct method is to properly install the lines, removing such traps. The second way, which requires a bit more expense, is installing an oil separator.

24–1A

Oil separators are installed in the compressor discharge line to collect the oil and return it to the compressor crankcase. See Figure 1–74. The proper size of oil separator must be used to effect proper oil return. Be sure to locate the oil separator where the temperature will not be lower than the receiver temperature during the "off" cycle. This will prevent liquid refrigerant condensing in the oil separator and being fed into the compressor crankcase.

A leaking valve in the oil separator will allow the discharge pressure to go into the compressor crankcase rather than to the condenser. This causes a lower than normal discharge pressure and a higher than normal suction pressure. The compressor crankcase will also be warmer than normal. The valve or the complete separator must be replaced to correct the problem.

24–1B

Oil return traps that are properly designed will aid in oil return on some installations where oil collects in the evaporator, which frequently occurs in multiple-evaporator systems. The principal function of an oil trap is to start the oil rising in a vertical line. Vertical risers must be correctly sized to carry the oil to the top of the riser.

A trap that is properly constructed will have as short a horizontal run as possible. A good oil trap may be constructed by connecting a street ell into a regular ell. See Figure 1–75. On larger size piping where street ells are not available, use two regular ells with a short piece of pipe. When oil traps are used on systems using

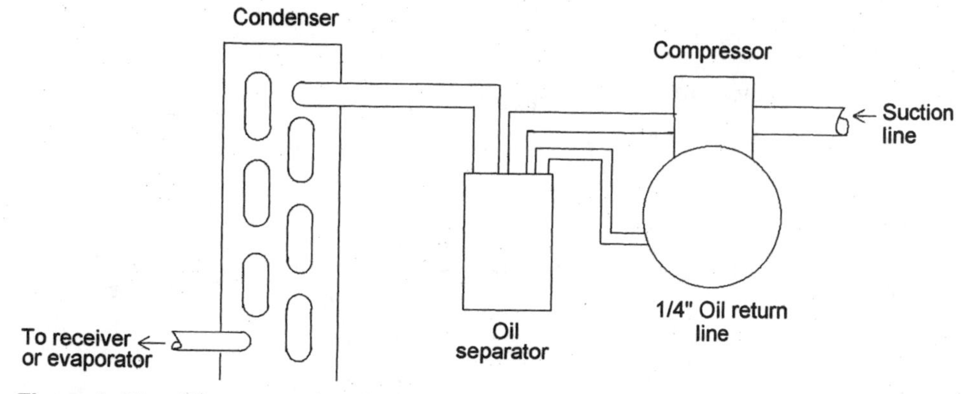

Figure 1–74. Oil separator installation.

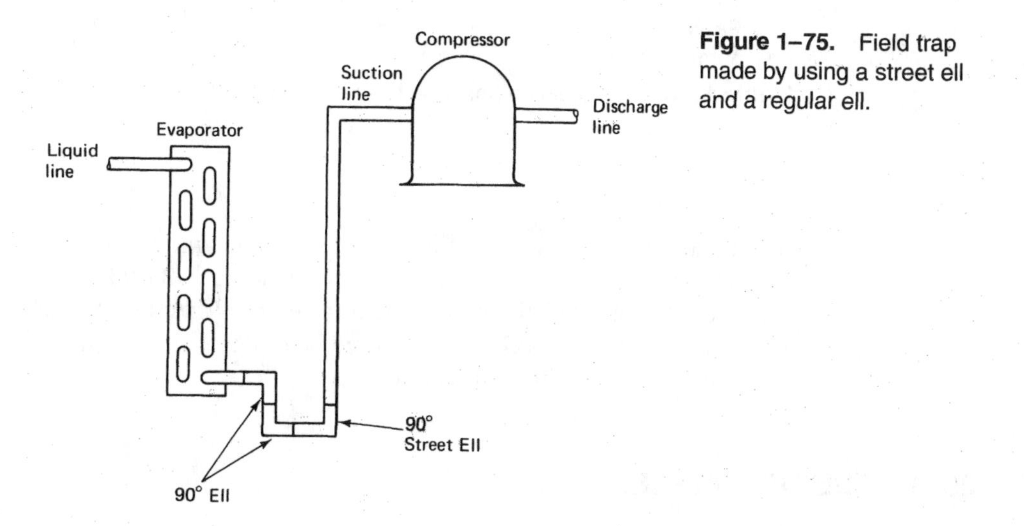

Figure 1–75. Field trap made by using a street ell and a regular ell.

thermostatic expansion valves, place the feeler bulb on the horizontal line before the trap. See Figure 1–76.

24–1C

When the refrigerant velocity is too low in the refrigerant lines, the oil will not be returned to the compressor properly. The efficiency of the system drops, resulting in unsatisfactory operation. When this condition is found, there are two alternatives: (1) Resize the lines to the proper size using tables and charts that are printed by various manufactures; or (2) install an oil return trap as discussed in Section 24-1B.

24–1D

Gas-oil ratio in a system results in oil flowing with the refrigerant throughout the system. The design of reciprocating compressors provides for the correct mixing

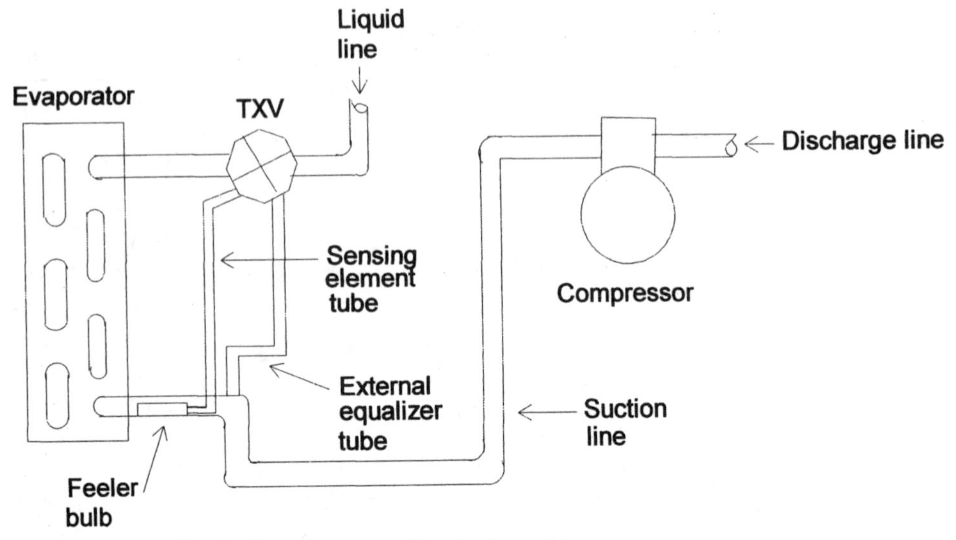

Figure 1–76. TXV bulb on horizontal line before oil trap.

of oil and refrigerant vapor. When a considerable amount of refrigerant has been lost from a system because of leaks, a certain amount of oil has also been lost. The oil should be replaced along with the refrigerant. When the shortage of refrigerant results in a shortage of oil, the oil should be replaced at the ratio of 1 pt (.47321 g) of oil to 10 lbs (4.5359 g) of refrigerant.

25–1 TUBING RATTLE:

A tubing rattle is the result of mishandling or misuse of the refrigeration equipment. Not only is this condition annoying, it is also damaging to the tubing. If let go long enough, a refrigerant leak will result because of a hole being rubbed in the tubing. To eliminate tubing rattle, either slightly bend the tubing apart with the hands or place a cushion between the tubes. Be careful not to apply enough pressure to the lines to break or kink them.

26–1 LOOSE MOUNTING:

When mountings become worn or broken, an annoying noise will result, as well as possible damage to the equipment. The only proper method of correcting this situation is to replace the mounting. In cases where a bolt or nut has become loose, it can be tightened without replacement of the mounting. When springs are used as the mounting, it is usually best to replace all the springs, not just the broken one, because the rest are probably weak and may break in the near future.

27–1 COMPRESSOR DRIVE COUPLINGS:

Compressor drive couplings are used on large-size compressors that are driven directly from the end of the motor shaft. A coupling is used to connect the two shafts. See Figure 1–77. There is a cushion included in the coupling to absorb the shock developed by the compressor pumping. These two shafts must be properly aligned or a vibration will be present that will cause excessive wear on the bearings and shaft seal. To align this coupling, loosen the hold-down bolts on either the motor or the compressor or both. Place a straight edge on both halves of the coupling. See Figure 1–78. This must be a metal straight edge that is in good condition. Move the loosened component around until the straight edge is touching the complete face of both halves of the coupling in at least three different places around the coupling. Tighten the loosened component hold-down bolts and recheck the alignment. Then start the unit. If vibration is still present, repeat the above procedure. If the vibration still persists, replace the coupling and align it.

28–1 EXCESSIVE PRESSURE DROP IN EVAPORATOR:

An excessive pressure drop through the evaporator is generally due to a restriction in the expansion valve or in the external equalizer line. The first step in correcting the problem is to check the operation of the expansion valve as outlined in Section 20-1. If the trouble is not found there, check the external equalizer line to see that it is not restricted. If a restriction is found in the external equalizer line, it may be removed by connecting a drum of the proper refrigerant type and blowing out the restriction. Complete replacement of the equalizer line may be required. The unit should function properly when the restriction is removed.

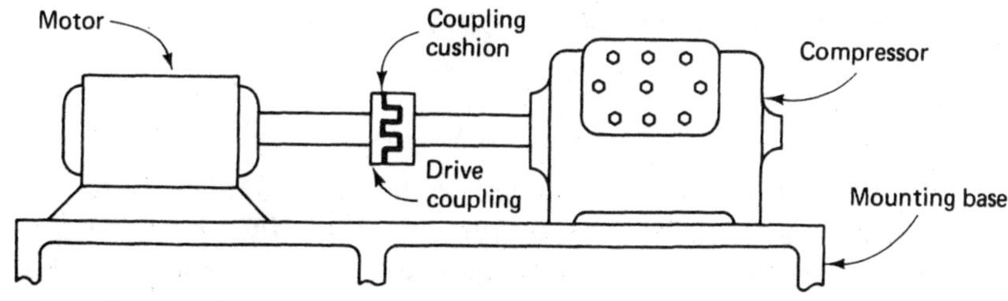

Figure 1–77. Direct drive compressor-motor coupling location.

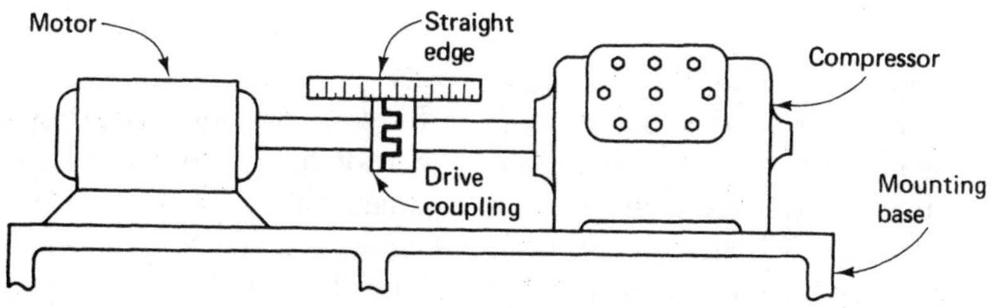

Figure 1–78. Alignment of a drive coupling using a straightedge.

29–1 CONTROL SETTING TOO HIGH:

A too high control setting generally occurs when several people are involved in storing and removing products in a refrigeration unit. Someone will set the control at a higher than desired temperature either accidentally or for defrosting purposes and will forget to set it back to the desired temperature. The control should be returned to the desired temperature setting and the user informed. Do not set the control at a temperature lower than required because this may cause excessive dehydration of the products.

30–1 EVAPORATOR TOO SMALL:

In a properly designed and installed system, the evaporator will be of the correct size. However, sometimes a service technician will find a defective coil and will replace it with one that is too small. The only step required is to replace the evaporator with one of the proper size. An evaporator that is too small will cause a lower than normal suction pressure with probable frosting of the evaporator. As a result, it will be almost impossible to adjust the expansion valve to the desired superheat setting, and the storage area will be at a higher than normal temperature. Therefore, the evaporator should be replaced.

31–1 LIQUID FLASHING IN LIQUID LINE:

Liquid flashing, or turning to vapor, in the liquid line may be due to a shortage of refrigerant, a liquid line that is too small, or a liquid line that has too high a vertical lift. Liquid flashing will be indicated by a hissing or gurgling noise at the expansion valve and a lower than normal suction pressure.

31–1A

When a system is short of refrigerant, the leak should be found and repaired and the system charged to the proper level with the refrigerant and checked for proper operation. Be sure to observe all safety precautions during this process.

31–1B

A liquid line that is too small will offer an excessive amount of resistance to the flow of refrigerant. As a result, the refrigerant pressure will drop near the outlet of the line, resulting in evaporation of a portion of the refrigerant, and picking up heat outside the refrigerated space, thus reducing the effectiveness of the system. The liquid line will also drop in temperature near the evaporator. This drop in temperature can be felt with the hand. See Figure 1–79. Oil return may also be a problem when this situation is experienced. When too small a liquid line is found, replace it

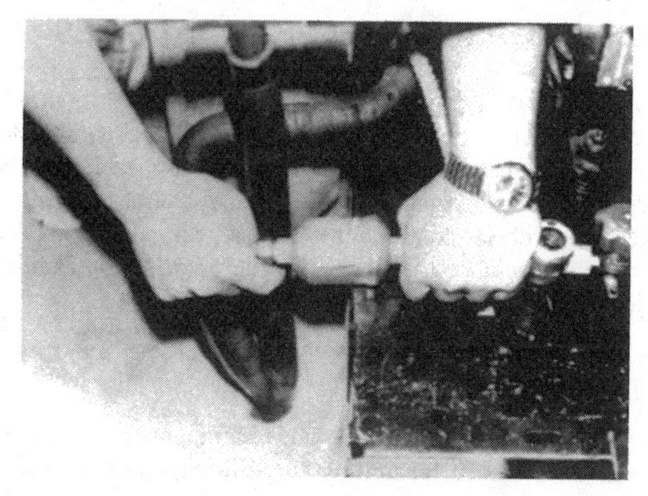

Figure 1–79. Feeling line for temperature drop.

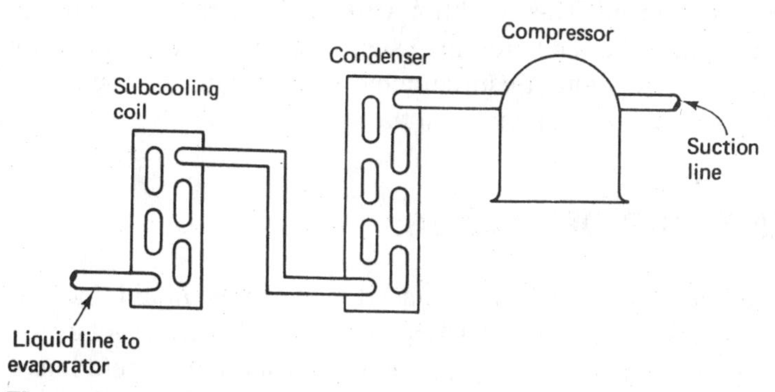

Figure 1–80. Subcooling coil location.

with one of the proper size. Manufacturers provide line sizing charts that can be used to properly size refrigerant lines.

31–1C

When liquid lines extend vertically for more than approximately 20 ft (6.09 m), the weight of the liquid forcing downward will probably cause flash gas. This flash gas is due to the drop in pressure. *Example*: If liquid Refrigerant 22 is forced up 30 ft (9.144 m), the pressure on top of the liquid will be 15 psig (103.42 Kp) lower than that at the bottom of the column. This pressure difference plus the heat in the liquid will cause flash gas. When this condition is encountered, a lower than normal suction pressure and a reduction in temperature of the liquid line near the evaporator will be present. To solve this problem, the liquid refrigerant must be subcooled. This may be accomplished by installing another condenser coil in the liquid line near the condenser outlet, or in severe cases, a suction to liquid heat exchanger should be added to the system. See Figure 1–80. To prevent overloading the compressor, use the proper size heat exchanger.

32–1 RESTRICTED OR UNDERSIZED REFRIGERANT LINES:

Restricted or undersized refrigerant lines can be a source of continual trouble in an otherwise carefully selected unit. Often the system capacity is reduced and oil return problems are frequently experienced with undersized lines. When these two conditions are repeated, a reevaluation of the refrigerant line sizing should be done. There are line sizing charts available for almost every situation encountered. Consult these charts and replace the lines with the proper size, or install an alternate recommended multiple-set of lines.

32–1A

Restricted refrigerant lines are usually due to contaminants or kinks in the system. A restriction in a line can be located by the difference in temperature on each side of the restriction. A restriction may be found in a drier, filter, strainer, or some other system component. This condition will be accompanied by pressures that are lower than normal and a reduction in system capacity. The restriction must be removed before proper operation can be achieved. Be sure to practice safety precautions when removing restrictions from the system.

33–1 LIQUID LINE SHUT-OFF VALVES:

Liquid line shut-off valves are placed between the liquid line and the outlet of the receiver tank or the condenser, depending on whether or not a receiver tank is used. See Figure 1–81. These devices are used in servicing operations. Their most common use is to aid in pumping the system down. When the valve is back-seated (open), the refrigerant will flow through with very little resistance. However, when the valve is front-seated, a complete stoppage of refrigerant occurs. If the valve is partially closed, a restriction in the flow of refrigerant occurs. This restriction allows expansion of the refrigeration outside the conditioned area. This expansion may be found by a temperature difference across the valve, a lower than normal suction pressure, and a lower than normal discharge pressure. The obvious solution is to back-seat the valve.

34–1 COLD UNIT LOCATION:

A cold unit location occurs when a refrigeration unit is operated during the winter season. This cold condition results in an excessively low discharge pressure. An excessively low discharge pressure produces two problems: (1) The pressure required to cause condensation of the refrigerant may not be provided, and (2) the required pressure drop across the flow control device may not be provided. Both problems cause poor refrigeration. There are basically two solutions to this cold unit condition: (1) Provide a warmer condensing medium, and (2) install a head pressure control.

34–1A

When it is desirable to operate small refrigeration units during the winter season, it may be possible to install the condensing unit indoors. This location will provide a warm enough condensing medium to keep the discharge pressure sufficiently high for economical operation.

34–1B

When larger units are used, an indoor location may be impossible. In this situation, other means of controlling the discharge pressure must be used. Two means of controlling the discharge pressure are discussed below.

1. Refrigerant pressure control valves may be used to direct the flow of hot gas around the condenser and into the receiver. See Figure 1–81. During periods of low ambient air temperatures, the discharge pressure drops until the setting of the valve is reached. The flow of refrigerant from the condenser is stopped, allowing a liquid build-up in the condenser, raising the condensing pressure. After a sufficient pressure build-up is reached, the hot gas from the compressor is discharged into the receiver, warming the liquid being discharged from the condenser. Warming the liquid will cause an increase in pressure, which will allow the flow control device to function properly. The manufacturer's recommendation for the particular type of control used should be consulted for adjustment and service procedures.

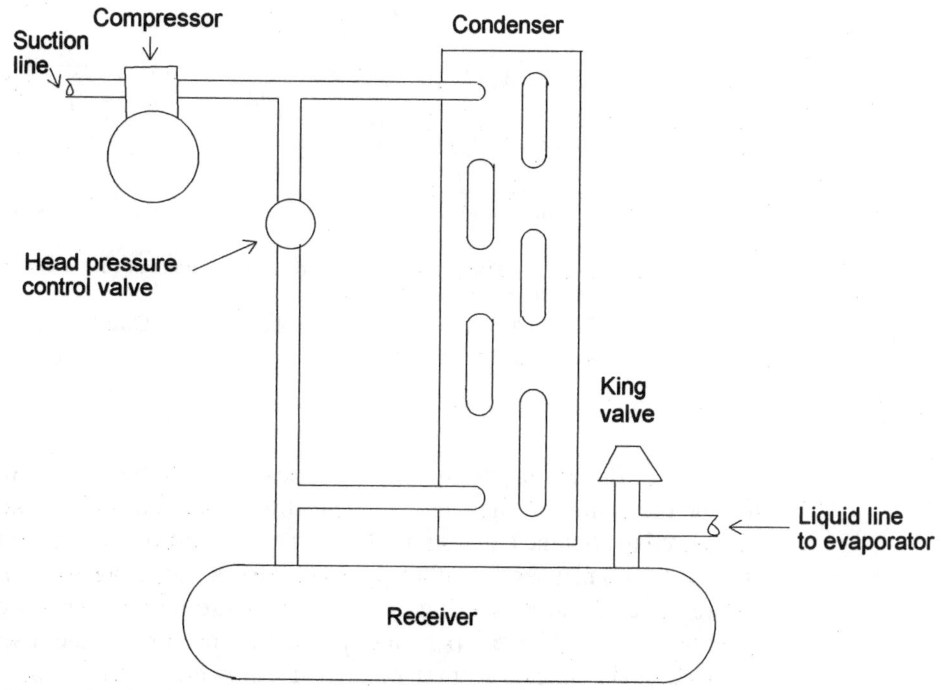

Figure 1–81. Head pressure control valve location.

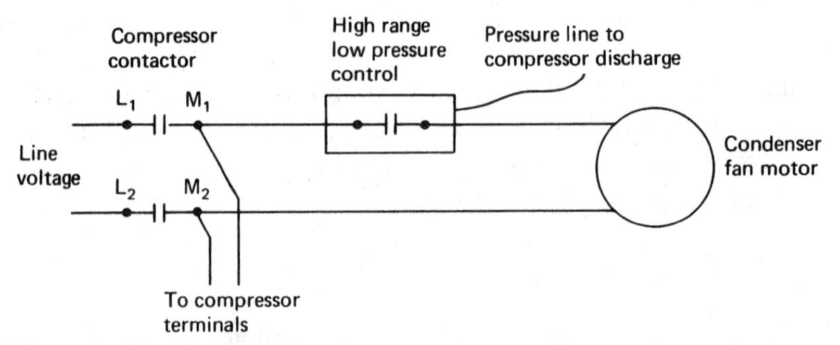

Figure 1–82. Control of head pressure by controlling condenser fan motor.

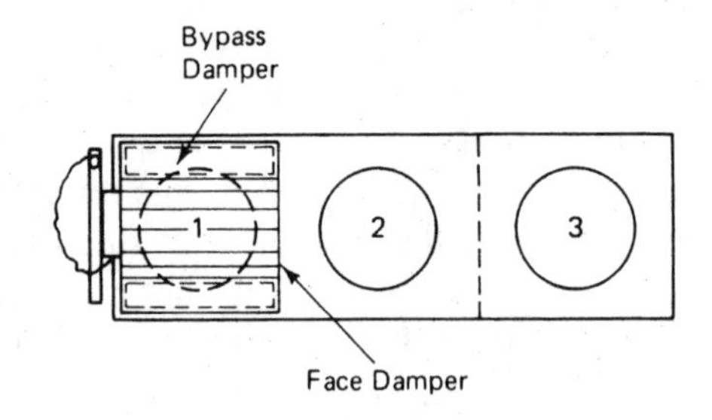

(a) Damper control location

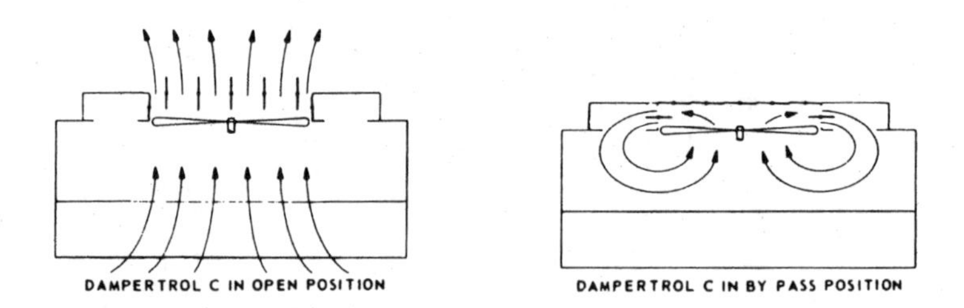

(b and c) Dampertrol C head pressure control selections: air flow in open (normal position) and closed (winter start).

Figure 1–83. Damper control operation. (Courtesy of McQuay Group, McQuay-Perfex, Inc.)

2. The discharge pressure may also be controlled by controlling the amount of cooling medium. This may involve the restriction or complete stoppage of the air or water used to cool the condenser. When an air-cooled condenser is involved, the cycling of the condenser fans in response to the discharge pressure may be used. See Figure 1–82. Or the restricting of the air flow through the condenser may be used. See Figure 1–83. The dampers are operated in response to the discharge pressure. The electrical connections may be made as shown in Figure 1–84. The pump on a water-cooled condenser may be cycled in response to the discharge pressure. See

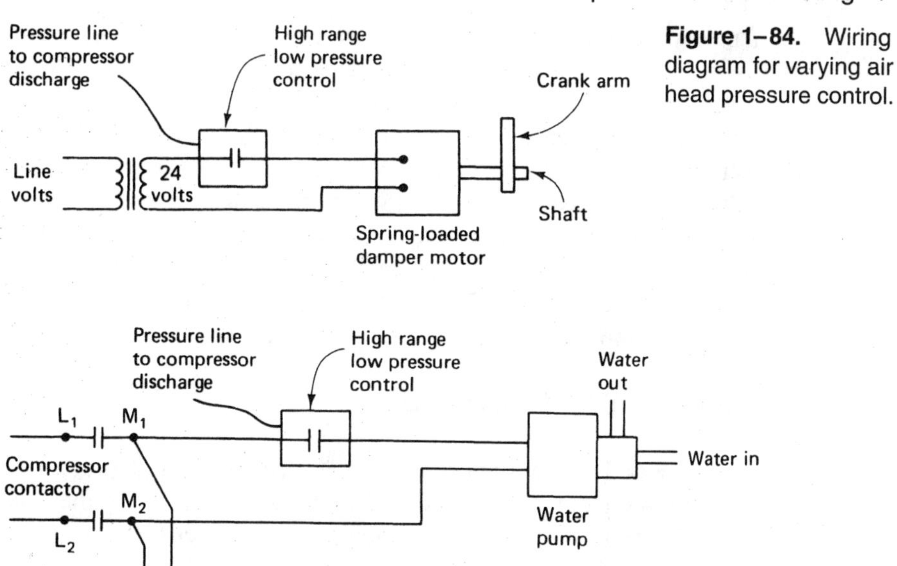

Figure 1–84. Wiring diagram for varying air head pressure control.

Figure 1–85. Head pressure control by cycling water pump.

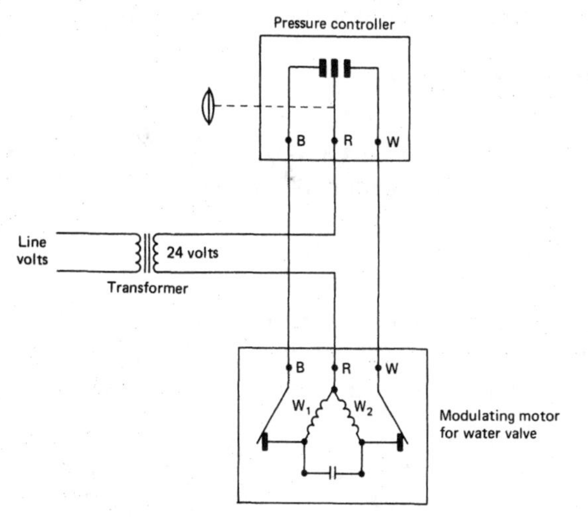

Figure 1–86. Head pressure control by use of modulating motor and water valve.

Figure 1–85. An alternate and more accurate method is to use a modulating valve and control to regulate the flow of water through the condenser. See Figure 1–86.

35–1 EVAPORATOR TOO LARGE:

The purpose of the evaporator is to absorb heat from the space and pass it on to the refrigerant. An evaporator that is improperly sized or one that absorbs too much heat due to an excessive volume of air will cause the compressor to be overloaded.

An evaporator that causes an overloaded condition will cause the suction pressure to be higher than normal. The humidity inside the conditioned space will be higher than normal, and the amperage draw of the compressor will be excessive. The best approach to correcting this problem is to install a properly sized evaporator. An alternate method that may be used with less success is to reduce the air flow over the evaporator coil. Reduce the air flow until the suction pressure is lowered to the desired pressure or until the evaporator starts to collect frost. If frost appears on the evaporator, the air flow should be increased. Use a thermometer to determine what the temperature drop of the air across the coil is. If the desired temperature drop is reached without frost on the evaporator, the system should operate satisfactorily.

36–1 EVAPORATOR UNDERLOADED:

An underloaded evaporator does not absorb the desired amount of heat into the refrigeration system. An underloaded evaporator is generally the result of frost, a dirty coil, or a reduced air flow. When this condition occurs, a lower than normal suction pressure will be noticed. To correct this problem, clean the ice or dirt from the evaporator and from the blower if it is dirty. After all the dirt and ice have been removed, the system should be started and the temperature drop of the air crossing the evaporator checked. If the desired temperature drop and suction pressures are obtained, the system will function satisfactorily. If these two readings are not normal, the situation has not been corrected and further cleaning or inspection is required.

37–1 SUCTION PRESSURE TOO HIGH:

A suction pressure that is too high is generally due to an excessive air flow over the evaporator, an overcharge of refrigerant, or bad suction valves in the compressor.

37–1A

The first step is to check the compressor suction valves. To do this, install the compound gauge on the compressor suction service valve. See Figure 1–87. With the unit running, front-seat the compressor suction service valve. Observe the pressure

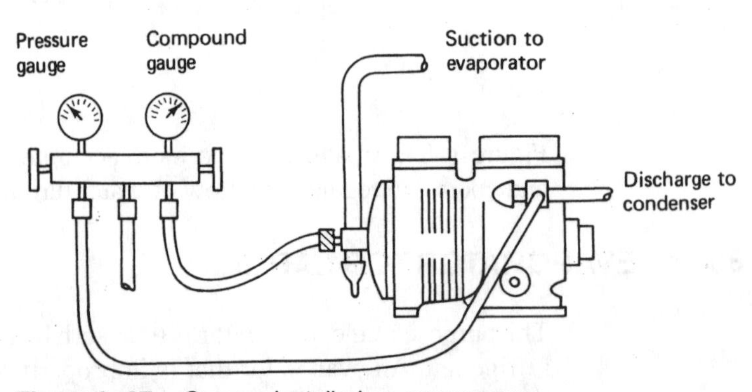

Figure 1–87. Gauges installed on compressor.

on the compound gauge. After the pressure has fallen as low as possible, stop the compressor and observe the compound gauge. The pressure should increase only slightly. If the pressure increases to zero psig (103.421 Kn), start the compressor again and reduce the suction pressure as much as possible. The suction pressure should be a 26-in. vacuum (27.579 Kg/cm^2) or lower. Stop the compressor and observe the compound gauge. The gauge should not indicate an increase in pressure of more than a 5-in. vacuum (.725 Kg/cm^2). If a greater increase than this is shown, replace the compressor valves. It is generally recommended to replace the suction valves, discharge valves, and the valve plate when replacing any one of the three. If the unit is a hermetic compressor, the complete compressor must be replaced.

37–1B

An overcharge of refrigerant will cause both the suction and discharge pressure to be higher than normal. An overcharge of refrigerant will cause approximately one-half of the condenser tubes to be cool to the touch. The cool tubes are full of sub-cooled liquid refrigerant, which must be removed from the system. When removing the excess refrigerant, purge the refrigerant slowly and in small amounts to prevent removing too much refrigerant. When the proper charge of refrigerant is reached, only the bottom two or three rows of tubing will be cool. The others will be warm or hot to the touch. See Figure 1–88.

37–1C

An excessive amount of air flowing over the evaporator will have the same effect as too large an evaporator. The system will be overloaded by the extra amount of heat absorbed into the system. To check for this condition, measure the temperature drop across the evaporator. See Figure 1–89. If the temperature drop is less than that desired, the air flow must be reduced to provide the proper temperature drop. This

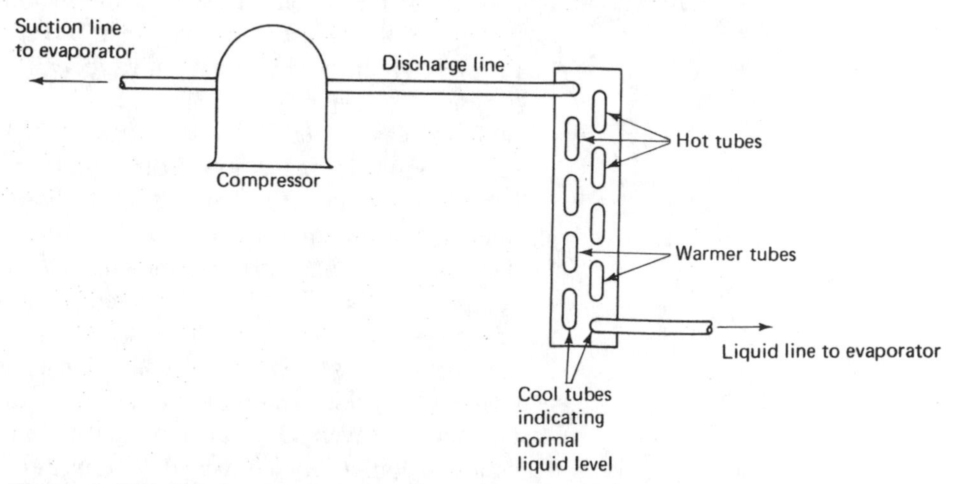

Figure 1–88. Temperatures of normally working condenser.

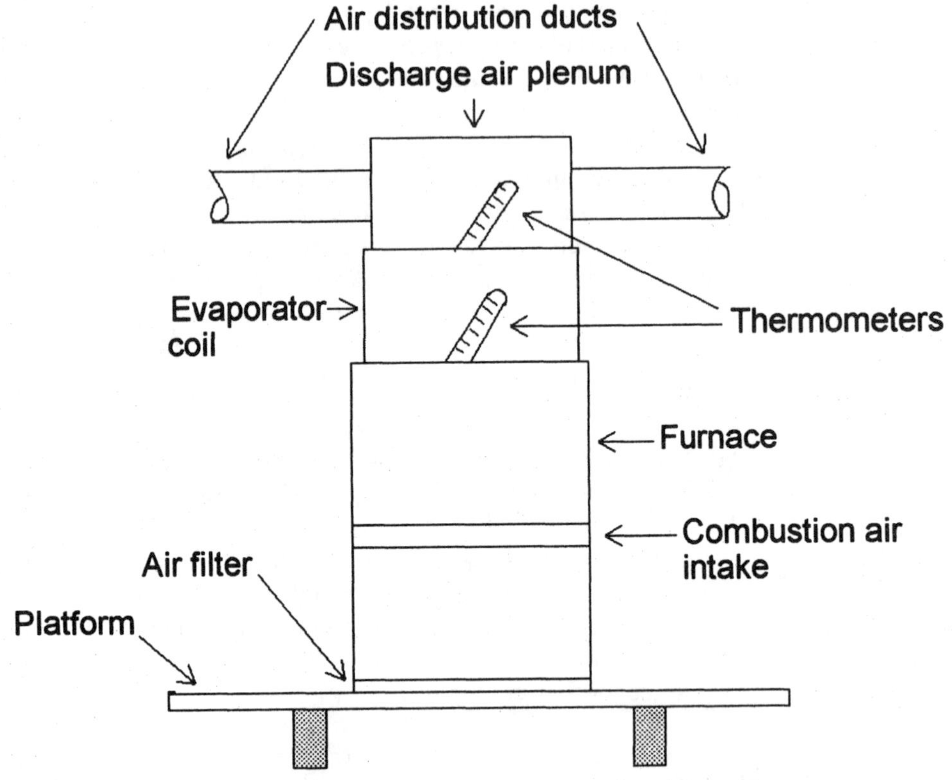

Figure 1–89. Checking temperature drop across air conditioning evaporator.

can be done by reducing the fan speed, replacing the motor with a lower rpm type motor, or in extreme cases, blocking the return air flow will help the situation.

CAUTION: Never heat a flow control device with a welding torch for damage to the flow control may result. If ice is causing the problem, it should melt within a few minutes, allowing the refrigerant to pass through, which will cause a rapid increase in the low side pressure. There may also be a hissing noise in the flow control while the refrigerant is passing through. To correct this situation, install new driers in the system. It is sometimes desirable to oversize the new driers to remove the moisture. In severe cases, complete discharging of the refrigerant, triple-evacuation of the system, and installation of driers in both the liquid and suction lines are required, along with a complete new charge of dry refrigerant. Check the system at least every 24 hours for several days and replace the driers when necessary to completely remove all the moisture.

38–1 MOISTURE IN SYSTEM:

Moisture that has entered the system can cause many serious problems. The most obvious problem is freezing of the moisture in the flow control orifice. Other more serious problems due to moisture in a refrigeration system are: (1) acid forming, and (2) sludging of the compressor lubricating oil.

38–1A

Moisture freezing in the refrigerant flow control orifice will result in poor refrigeration, accompanied by a lower than normal suction pressure and possibly a lower than normal discharge pressure. To determine if moisture is freezing in the flow control device, stop the compressor with the compound gauge connected. Warm the flow control orifice area with a rag wet with warm or hot water while observing the compound gauge.

38–1B

Acid forming in a system due to moisture will cause serious damage to the compressor and expansion valve parts and will attack the insulation on the motor windings in a hermetic or semihermetic compressor. The preliminary indication of acid caused by moisture is copper plating (a copper color on the steel valves and parts) of the components. See Figure 1–90. More advanced stages are indicated by motor burnout or repeated motor burnout. A discoloration of the oil will also be noticed. This condition may be present without freezing of the refrigerant flow control device. To correct this problem, replace the refrigerant driers and check the oil color every 24 hours for several days. Replace the driers when needed until the oil color returns to normal.

38–1C

Sludging of the compressor oil is usually due to contaminants in the system. Oil that has started sludging will not provide proper lubrication to the system. This condition will usually be accompanied by copper plating (a copper color on the steel valves and parts). See Figure 1–90. To correct this situation, remove the oil from the system and charge the proper amount of fresh clean oil into the compressor.

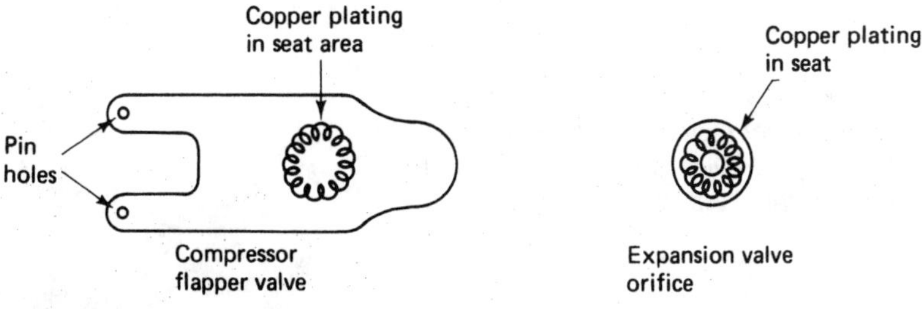

Figure 1–90. Copper plating of steel seats.

Replace all the refrigerant driers and check the system at least every 24 hours for several days. Replace the driers when needed until the oil color returns to normal.

39–1 AUTOMATIC DEFROST CONTROLS:

Automatic defrost controls are timing devices used on refrigeration units that operate with the evaporator temperature below freezing (32°F) (0°C) and on heat pump air conditioning units. The proper operation of these controls is a must for efficient operation of the refrigeration system. See Figure 1–91. There are several methods presently being used for automatic defrosting. The particular method depends on the desired efficiency and the preference of the equipment manufacturer.

39–1A

An erratic or inoperative defrost control will result in poor refrigeration of the space. An inoperative defrost control will allow an excess of frost to build up on the evaporator. This problem is not too difficult to detect. The first check to make is to determine if the timer motor is running. This can usually be determined by a visual check. See Figure 1–92. The motor rotation can be observed through an in-

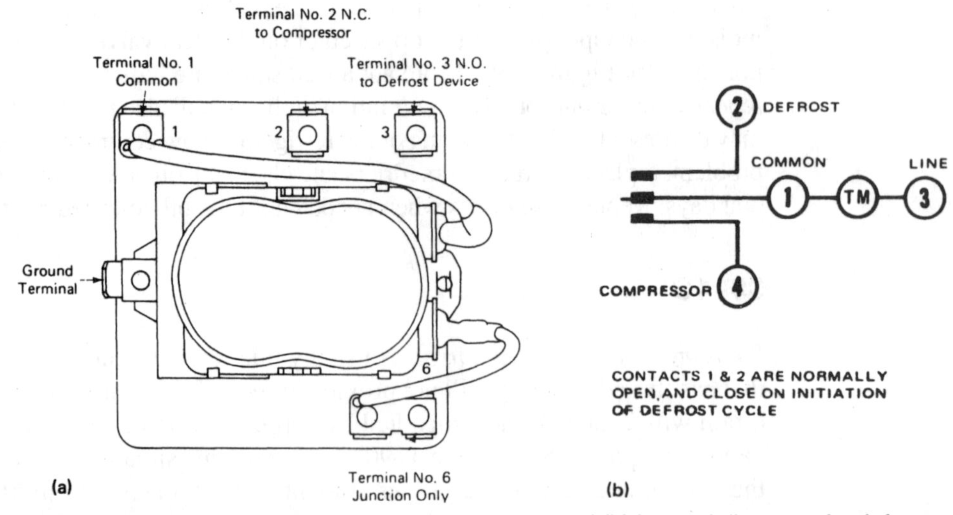

Figure 1–91. (a) Typical defrost timer connections and (b) internal diagram of a defrost timer. (Courtesy of Gem Products, Inc.)

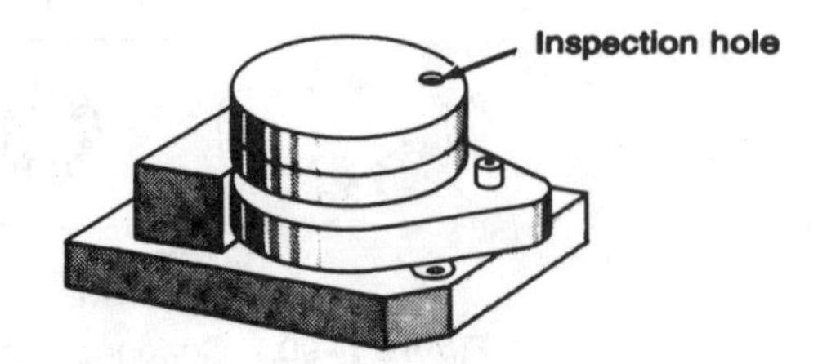

Figure 1–92. Motor inspection hole on defrost timer motor. (Courtesy of Frigidaire Parts and Service Co., Division of White Consolidated Industries, Inc.)

spection window. When the motor is running, manually place the control in defrost by rotating the mechanism to initiate the defrost period. Next, check the voltage drop across the defrost terminals on the control. No voltage reading indicates that the contacts are closed. A voltage reading indicates that the contacts did not close and the control must be replaced. When the motor is not running, check the voltage supply to the clock motor. If a voltage is found, either the motor is bad or a loose connection is indicated, and the control must be replaced.

An erratic defrost control is sometimes difficult to diagnose. The evaporator may have excessive frost on it, or the space temperature may be high because of too many defrost periods. When these conditions exist, a recording voltmeter may be placed in the defrost circuit to record the time and length of each defrost period. The operation should be monitored for at least 24 hours. If erratic operation is indicated by nonscheduled or the lack of defrosts, the control must be replaced.

39–1B

In order for a defrost control to perform properly, it must be correctly wired into the electrical circuit. Consult the wiring diagram of the particular unit to insure that the control is properly wired into the circuit. When faulty wiring practices are found, they must be corrected to gain proper operation.

39–1C

The thermal element on a defrost control is used both to determine when all the frost has melted and to return the system to normal operation. This element has a vapor charge that actuates a set, or sets, of contacts in response to a change in temperature of the evaporator. When this element becomes defective or is improperly installed, the system will not go into the defrost cycle, allowing the evaporator to become excessively frosted. The first step is to determine the condition of the timer motor. See Section 39-1A. If this check proves the timer to be good, the thermal element is defective. The element must be replaced. If the element is not properly installed or is not making good thermal contact with the evaporator, the element must be reinstalled on a clean surface. All paint and corrosion must be removed from the tube.

39–1D

The defrost termination point is the temperature at which the defrost period ends. When this point is too low, a sufficient amount of frost, or ice, will not be removed from the coil to allow proper refrigeration to occur. This temperature is determined by the setting on the defrost control. An accumulation of frost build-up on the coil is an indication of a too low termination point. The termination point should be raised to allow all the frost to be removed from the coil. However, it should not be high enough to allow for too long a defrost period because the temperature inside the cabinet will go too high. In instances when the control cannot be adjusted to allow proper operation, it must be replaced. Consult the manufacturer's information regarding the particular make and model of the control for proper adjustments.

39–1E

When the hot gas type of defrost is involved, a solenoid valve is used to direct the flow of hot gas around the flow control device and into the evaporator. A burned-out defrost valve solenoid coil will prevent operation of the defrost circuit. To check for a defective defrost valve solenoid, turn off all electrical power to the unit and disconnect the electric wiring from the solenoid coil. Using an ohmmeter, check the continuity of the coil. See Figure 1–93. No continuity will be found when the coil is open. A shorted coil will provide less than normal resistance. A visual check of a burned coil will show discoloration of the insulation, and a distinct burnt insulation odor will be present. Replace the coil with one of the proper type, size, and voltage.

39–1F

A stuck-closed defrost valve will not allow the defrost period to occur. All the other components will function normally, but the coil will have a heavy accumulation of frost. To check for a stuck defrost valve, check the voltage to the defrost valve while manually placing the defrost control in and out of defrost. See Figure 1–94. If voltage to the valve coil is indicated and there is no clicking of the valve, the

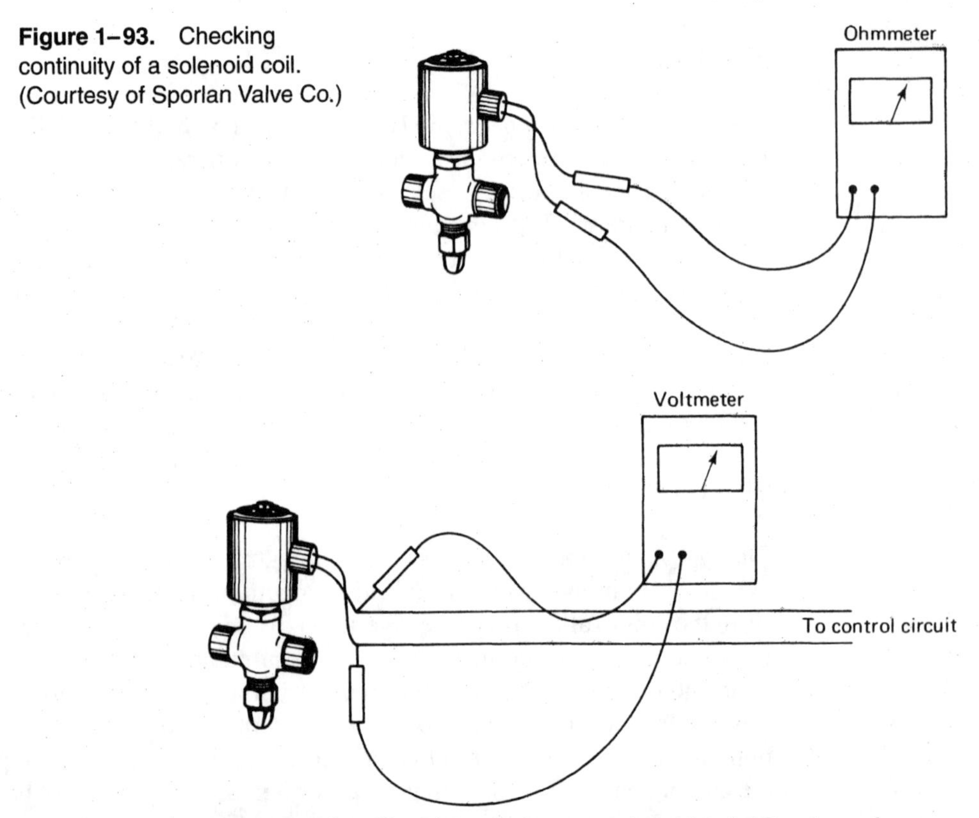

Figure 1–93. Checking continuity of a solenoid coil. (Courtesy of Sporlan Valve Co.)

Figure 1–94. Checking voltage on a solenoid coil. (Courtesy of Sporlan Valve Co.)

valve is faulty and should be replaced. A sticking defrost solenoid valve will cause a great deal of trouble. Therefore, repairing it is not generally recommended. A new one should be installed.

39–1G

A restricted or plugged hot gas bypass line will prevent an otherwise properly operating defrost system from functioning. If a heavy accumulation of frost on the evaporator is experienced and all the defrost components are operating satisfactorily, the hot gas bypass line is likely to be plugged. The bypass line must be cleaned or replaced before the system will operate normally. See Figure 1–95.

39–1H

An inoperative freezer compartment door switch will allow an excessive amount of frost to build up on the evaporator because it will prevent operation of the freezer compartment fan. See Figure 1–96. This will allow the evaporator temperature to become excessively low. To check the door switch, remove the switch and carefully by-pass the fan contacts. If the fan motor operates, the door switch is faulty and should be replaced with the proper style and type.

39–1I

An inoperative freezer-refrigerator compartment fan will allow excessive frost to build up on the evaporator because of the lower evaporator temperature caused by the reduced load. To check for an inoperative fan, first make sure that the fan blades are free to rotate. Then check for voltage to the fan motor while the door switch button is depressed. If voltage is present and the motor still does not run, it is

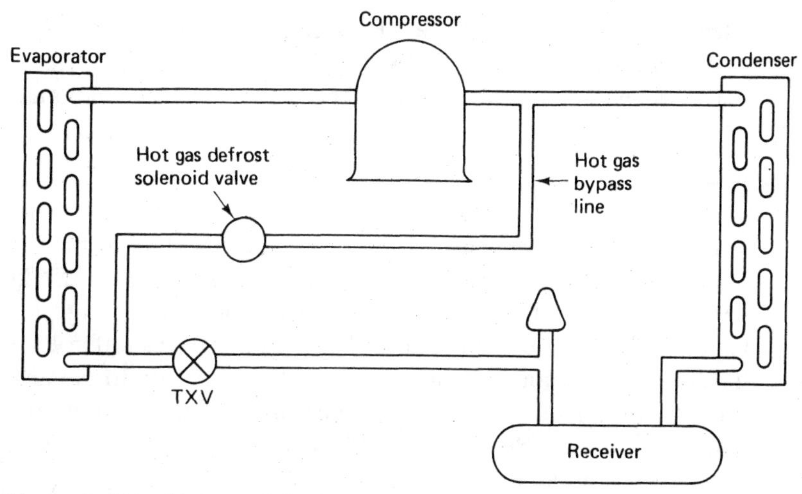

Figure 1–95. Hot gas defrost system.

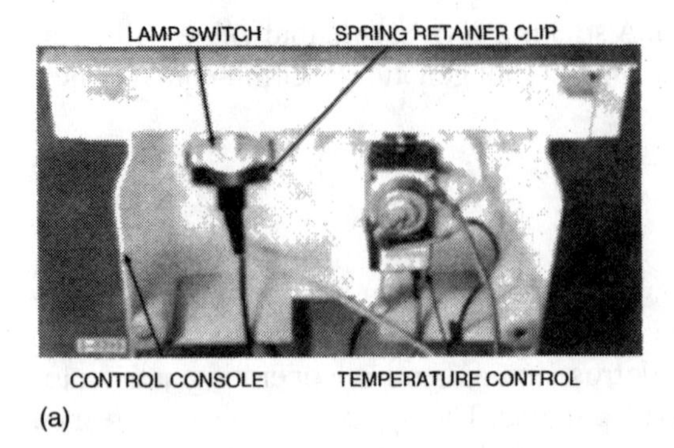

(a)

SPRING CLIP
LAMP SWITCH
LAMP SHIELD

A-6296

(b)

Figure 1–96. Door (lamp) switch location: (a) lamp switch removal—top freezer models; (b) lamp switch removal—model FPC13-165VU. (Courtesy of Frigidaire Parts and Service Co., Division of White Consolidated Industries, Inc.)

defective and must be replaced. Be sure to use the proper replacement and check the rotation of the motor. See Figure 1–97. The fan must have the proper rotation or a sufficient amount of air will not be blown over the evaporator.

39–1J

Electric defrost systems use electric resistance heaters to melt the frost from the evaporator. A faulty defrost element will allow an excessive frost accumulation on the evaporator. To check the defrost element, check for voltage to the element while manually placing the unit in defrost by rotating the timer. If voltage is indicated and the element does not get warm, turn off the electricity to the unit. Disconnect the electric wiring from the defrost element and check the continuity of the element with an ohmmeter. See Figure 1–98. If no continuity is found or the element is found to be shorted, it must be replaced.

Figure 1–97. Location of evaporator fan, fan motor, and defrost limiter switch. (Courtesy of Frigidaire Parts and Service Co., Division of White Consolidated Industries, Inc.)

39–1K

A drain trough or drain line that does not allow proper drainage of the defrost water will cause an excessive amount of frost to build up on the evaporator. This is because the water left in the drain trough will refreeze on the evaporator, thus reducing the evaporator surface area and causing it to reach too low a temperature. If ice or water is found standing in the drain trough, the drain line must be opened, and operation of the drain trough or drain pan heater must be checked. To check the heater, determine whether or not voltage is applied to the heater while the unit is manually placed in defrost. If voltage is indicated, disconnect the electrical power to the unit. Remove the electric wiring from the heater and check for continuity with an ohmmeter. See Figure 1–99. If the heater is found defective, it must be replaced.

39–1L

An incorrectly wired defrost cycle may not allow the unit to terminate the defrost cycle. When proper operation of the control cannot be obtained, check the present wiring against the wiring diagram on the unit. There are several different wiring hookups depending on the type of defrost system, type of control, and the manufacture of the control. To insure proper operation, wire the control according to the equipment manufacturer's recommendations.

39–1M

A defrost control that is set too high will allow the unit to remain in the defrost cycle longer than necessary. This will be indicated by an increase in cabinet temperature during the defrost period and excessive operating costs. The control should be set so as to dry the moisture from the evaporator and return the system to the

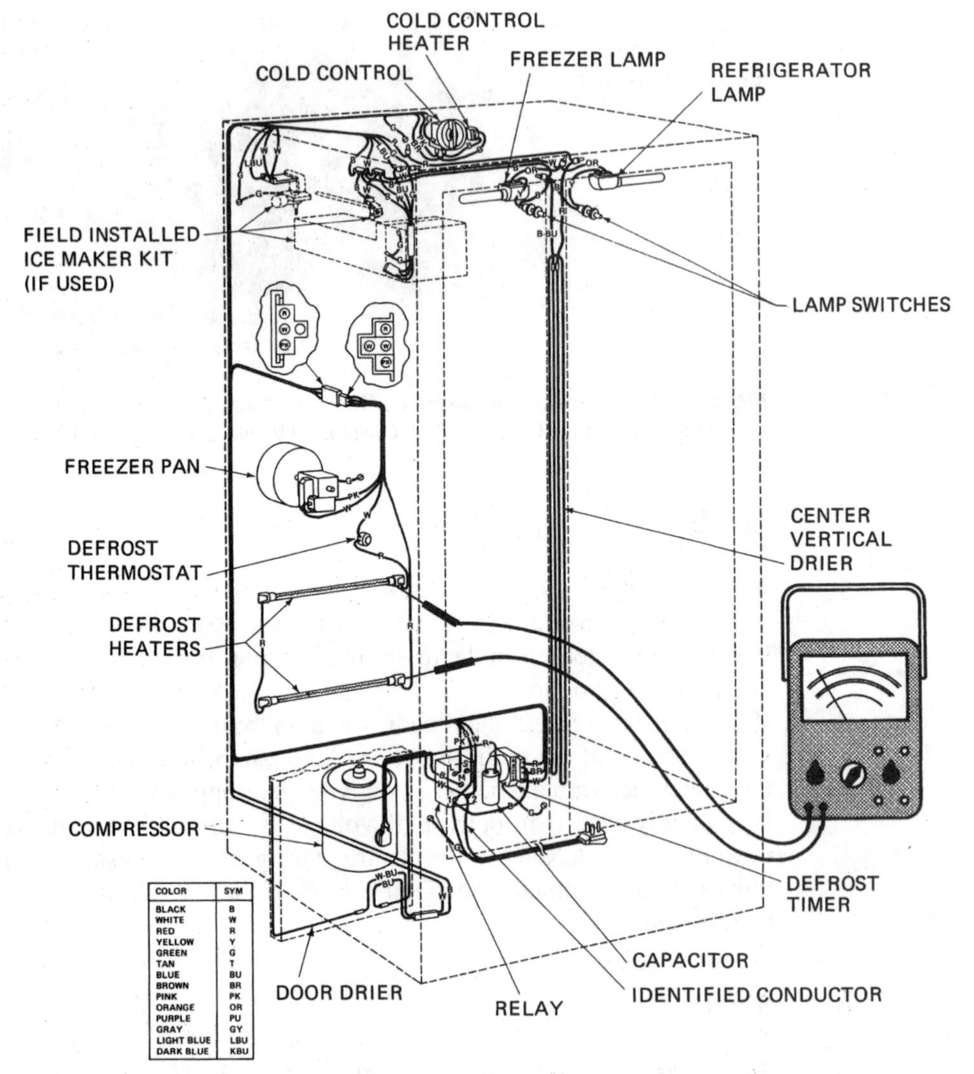

Figure 1–98. Checking continuity of defrost heaters. (Courtesy of Frigidaire Parts and Service Co., Division of White Consolidated Industries, Inc.)

normal operating cycle. If the system remains in defrost after the moisture is removed from the evaporator, the control should be adjusted or replaced. Be sure to use the proper replacement. When adjusting consult the manufacturer's specifications for the particular control in question.

39–1N

A stuck-open defrost solenoid valve will not allow termination of the defrost cycle, and the cabinet temperature will be too high. To check for a stuck defrost valve, check the voltage to the defrost valve while manually placing the defrost control in and out

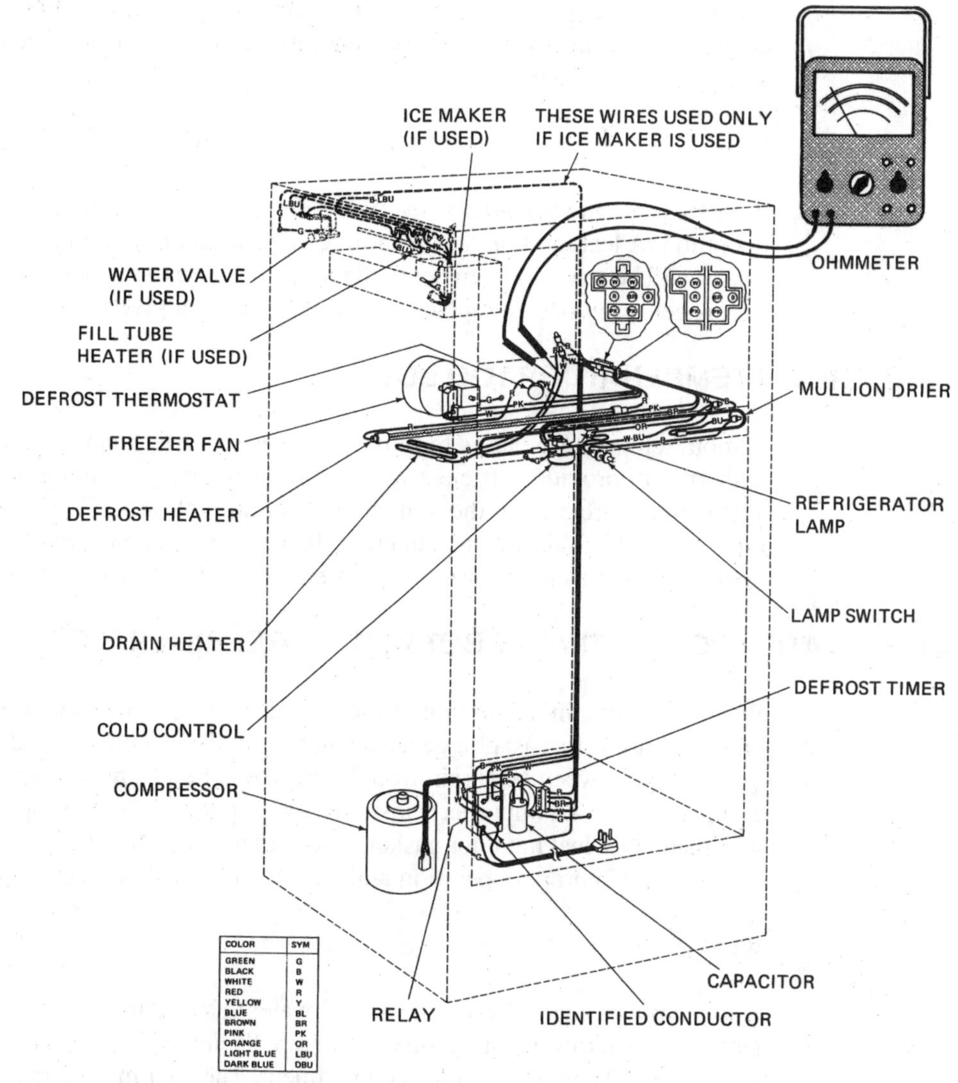

Figure 1–99. Checking continuity of drain heater. (Courtesy of Frigidaire Parts and Service Co., Division of White Consolidated Industries, Inc.)

of defrost. See Figure 1–94. If voltage to the valve is applied and is interrupted as the defrost control is rotated and there is no clicking of the valve, the valve is faulty and should be replaced. A sticking defrost valve will cause a great deal of trouble. Therefore, repairing it is not generally recommended. A new one should be installed.

39–1P

The defrost limiter is a temperature-sensing device that terminates the defrost period when the evaporator temperature reaches approximately 50°F (10°C). A faulty defrost limiter will allow the unit to stay in defrost and will increase the inside cabinet temperature. To check a defrost limiter, remove one of the wires from the control. If the unit returns to the normal refrigeration cycle, replace the limiter with an exact replacement.

39–1Q

A leaking hot gas solenoid valve will cause poor refrigeration and excessive unit operation. A leaking hot gas solenoid can be found by feeling the hot gas bypass line. It will be warm or even hot to the touch because of the hot gas passing through it. It is generally best to replace solenoid valves that give trouble.

40–1 ROOM TEMPERATURE TOO LOW:

A room temperature that is too low will cause a lower than normal head pressure and will not provide sufficient heat to properly operate a hot gas defrost system. The temperature around the unit should not fall below 55°F (12.8°C). If temperatures lower than this are encountered, the unit should be moved to a warmer location, or some means of warming the area around the unit should be provided.

41–1 WATER COLLECTING IN BOTTOM OF REFRIGERATOR:

Water collecting in the bottom of the refrigerator compartment may be due to several conditions such as: plugged, frozen, or split drain tube; a split drain trough; water leaking between the drain trough and the cabinet liners; a warped fresh food compartment liner; improperly installed evaporator baffle; improperly adjusted humidiplate; or a leaking door gasket. These conditions should be found and be corrected to aid in proper operation and usage, and to provide customer satisfaction.

41–1A

A split drain trough will allow the defrost water to leak into the compartment. This water will cause the evaporator to frost more rapidly and will reduce unit efficiency, along with ruining the products in the compartment. The split may be repaired with a nontoxic, nonodorous type of sealer. However, it is usually preferable to replace the trough.

41–1B

Water leaking between the drain trough and liner will allow the defrost water to collect in the bottom of the compartment. This is an undesirable situation. See Section 41-1A. The crack between the drain trough and liner may be repaired with a nontoxic, nonodorous type of sealer.

41–1C

A warped compartment liner will allow the defrost water to drain to the bottom of the compartment. To correct this situation, use a nontoxic, nonodorous type of sealer or replace the fresh food compartment liner.

41–1D

An improperly installed evaporator baffle may be providing an alternate path for the defrost water. If the baffle is touching the evaporator, some of the defrost water may be channeled down the baffle around the drain trough and to the bottom of the compartment. To correct this situation, correctly install the baffle.

41–1E

A humidiplate that is adjusted so that it provides an alternate path around the drain trough for the defrost water should be adjusted to correct the situation. It should be noted that when something is in contact with the evaporator coil above the drain trough a water leak is likely to occur.

42–1 LEAKING DOOR GASKETS:

Door gaskets that do not seal properly will allow warm, moisture-laden air to enter the refrigerated compartment. Moisture due to condensation may collect in the bottom of the compartment. The warm air will cause the temperature inside the cabinet to be high because of the extra load on the unit, causing the unit to run excessively. To check for a leaking door gasket, first visually check for cracks or deterioration of the gasket. Then open the door and insert a piece of paper about the size of a dollar bill between the gasket and the refrigerator. Close the door, then gently pull on the paper. There should be a noticeable drag on the paper. See Figure 1–100. Repeat this procedure until the complete gasket has been checked. If a noticeable drag is not experienced, adjust the refrigerator door until a slight drag on the paper is obtained. If the door cannot be properly adjusted, replace the gasket and then adjust the door to the proper tightness.

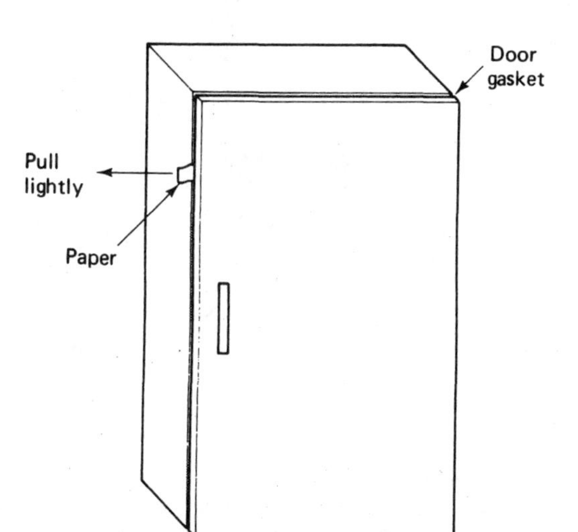

Figure 1–100. Checking a refrigerator door gasket with a piece of paper.

43–1 CONDENSATION ON OUTSIDE OF CABINET:

Moisture condensing on the outside of a refrigerator cabinet can cause rusting as well as becoming an annoyance to the user. This condition can be caused by leaking door gaskets, mullion heater not working, or abnormally high humidity.

43–1A

A mullion heater (drier) is a resistance heater that is placed around the door openings and inside the outer shell to help in preventing condensation of moisture on the cabinet. See Figure 1–101. This type of heater generally is a low-wattage heater and only supplies enough heat to keep the cabinet at room temperature. To check a mullion heater, determine if voltage is present at the mullion heater with a voltage

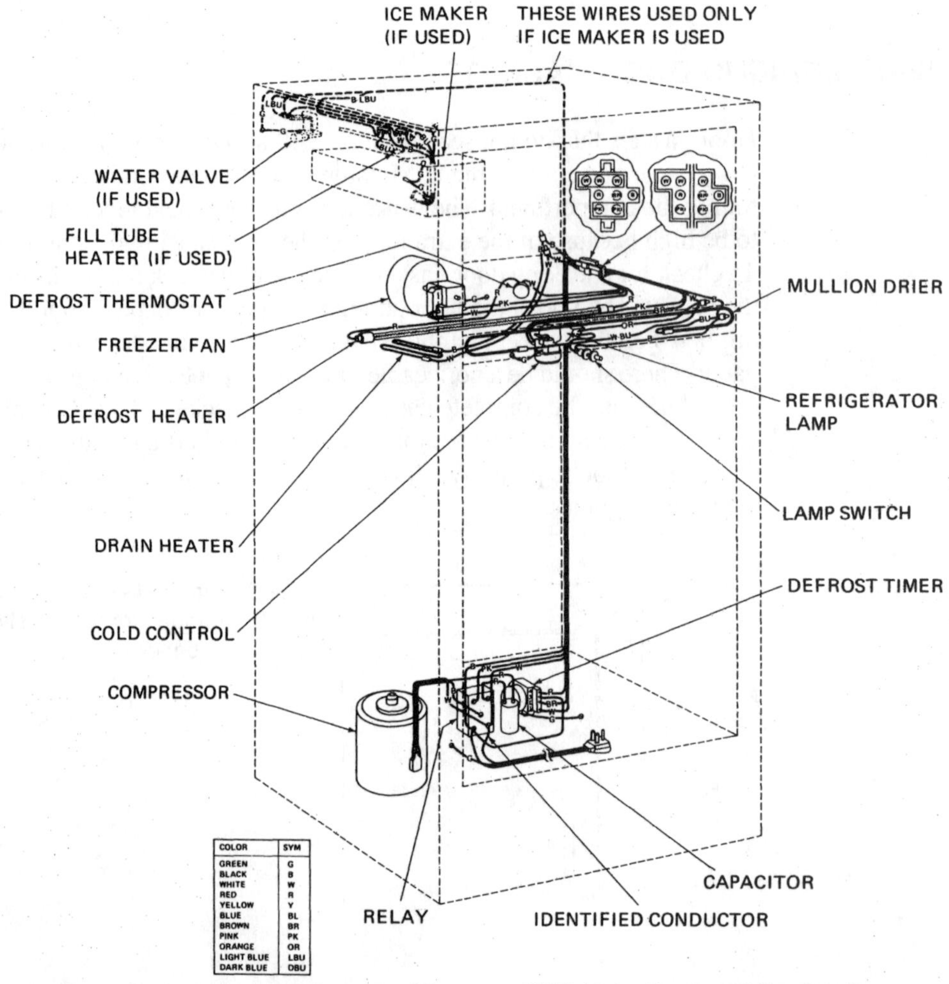

COLOR	SYM
GREEN	G
BLACK	B
WHITE	W
RED	R
YELLOW	Y
BLUE	BL
BROWN	BR
PINK	PK
ORANGE	OR
LIGHT BLUE	LBU
DARK BLUE	DBU

Figure 1–101. Mullion drier location. (Courtesy of Frigidaire Parts and Service Co. Division of White Consolidated Industries, Inc.)

meter while the unit is connected to electrical power. If voltage is found, turn off the electrical power and disconnect the electrical wiring from the mullion heater. Check for continuity of the heater with an ohmmeter. If no continuity, replace the mullion heater. If no voltage is found, check the wiring and repair any loose connections or replace any broken or frayed wiring.

43–1B

Abnormally high humidity will cause condensation on the outside of a refrigerated cabinet. This condensation occurs because the cabinet temperature is below the condensation temperature of the surrounding air. There is very little, if anything, that can be done to stop condensation under these conditions. The most practical step is to educate the user about the problem. When the humidity drops, the condensation will stop.

44–1 POOR AIR DISTRIBUTION TO FOOD COMPARTMENT:

Poor air distribution is the cause of several problems. The most frequent problem is a warm compartment. A warm compartment occurs because the heat is not removed from the contents of the cabinet. To correct this problem, the compartment contents must be placed so that the air flow will not be restricted in any way. The fan intake and outlet must be kept free at all times. The controls on cabinets that are equipped with air controls should be checked to be certain that proper air distribution is obtained.

45–1 COMPARTMENT LIGHT STAYS ON:

The constant burning of interior lights will cause the compartment to be warmer than normal. This condition can be corrected by replacing the door switch or by properly wiring the unit. If the door switch is bad, the light will go off when the switch is disconnected from the electrical wiring. Be sure to use an exact replacement so that mounting difficulties and operating problems will not be encountered. Check to see that the new switch is operating properly by manually depressing the door switch button. The light should go off and the fan come on when the switch button is depressed. When the button is released the light should come on and the fan go off.

46-1 EXCESSIVE COMPARTMENT LOAD:

Hot foods or an excessive amount of food stored in a refrigerated cabinet will place an excessive load on the unit and will cause a higher than normal cabinet temperature. The user should be instructed not to place hot food in the cabinet. Freshly cooked food should be allowed to cool to room temperature before storing in the refrigerator. Foods should not be stored in such a way as to block the air flow through the cabinet. The user should be advised as to the proper use of the equipment.

47–1 ELECTRIC MOTORS:

Electric motors are used for many purposes in refrigeration and air conditioning other than to drive the compressor. The majority of uses for which these motors are used include fan motors and pump motors. There are several different types of motors in use. They are: (1) split-phase (SP), (2) permanent split capacitor (PSC), (3) capacitor start (CS), (4) capacitor start capacitor run (CSCR), and (5) shaded pole. The amount of starting and running torque required to do the job will determine the type of motor used. The steps listed below are used to check electric motors. If any of the following conditions are found, the motor must be either repaired or replaced.

47–1A

Open motor windings occur when the path for electric current is interrupted. This interruption occurs when the insulation on the wire becomes bad and allows the wire to overheat and burn apart. See Figure 1–102. To check for an open winding, remove all external wiring from the motor terminals or connections. Using the ohmmeter, check the continuity from one terminal to another terminal. See Figure 1–103. Be sure to zero the ohmmeter. The open winding will be indicated by an "infinity" resistance reading on the ohmmeter. There should be no continuity from any terminal to the motor case. Repair or replace the motor if found to be faulty.

47–1B

Shorted motor windings occur when the insulation on the winding becomes bad and allows a shorted condition (two wires to touch), which allows the electric current to bypass part of the winding. See Figure 1–104. In some instances, depending on how much of the winding is bypassed, the motor may continue to operate but will draw excessive amperage. To check for a shorted winding, remove all external wiring from the motor terminals. Using the ohmmeter, check the continuity from one terminal to another terminal. See Figure 1–105. Be sure to zero the ohmmeter. The shorted winding will be indicated by a less than normal resistance. In some cases it will be necessary to consult the motor manufacturer's data for the particular motor to determine the correct resistance requirements. There should be no continuity from any terminal to the motor case. Repair or replace the motor if found to be faulty.

Figure 1–102. Open motor winding.

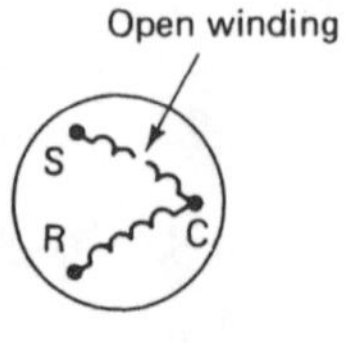

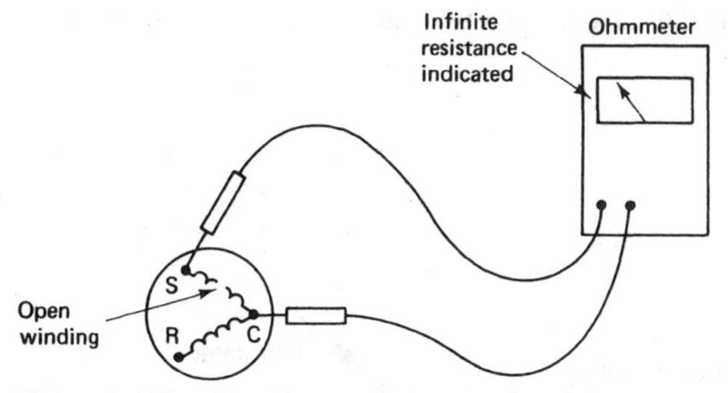

Figure 1–103. Checking continuity of open motor winding.

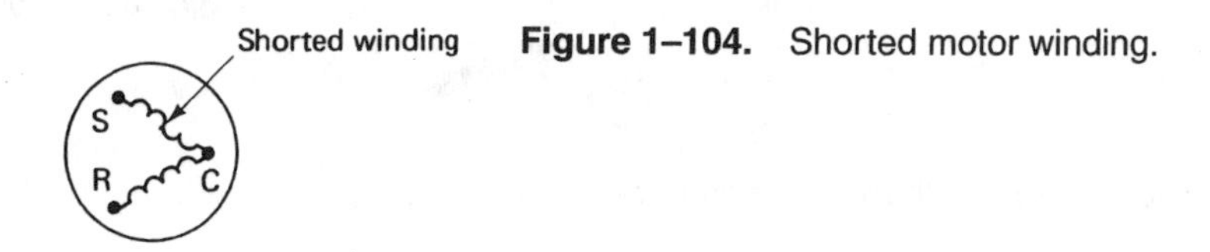

Figure 1–104. Shorted motor winding.

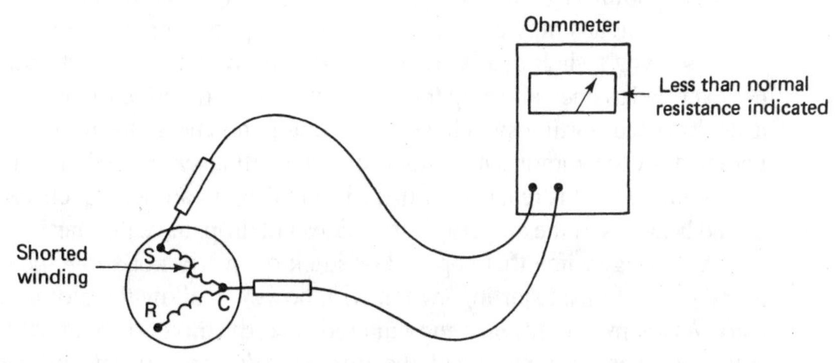

Figure 1–105. Checking continuity of shorted motor winding.

47–1C

Grounded motor windings occur when the insulation on the winding is broken down and the winding becomes shorted to the motor housing. See Figure 1–106. Be sure to zero the ohmmeter. The grounded winding will be indicated by a low resistance reading. See Figure 1–107. It may be necessary to remove some paint or scale from the motor case so that an accurate reading can be obtained. If a reading is obtained, repair or replace the motor.

47–1D

A bad starting switch on a split-phase motor will prevent the proper starting of the motor. This switch provides a path for electrical power to flow to the starting (auxiliary) winding during the starting period, and it interrupts the electrical power

Figure 1–106. Grounded motor winding.

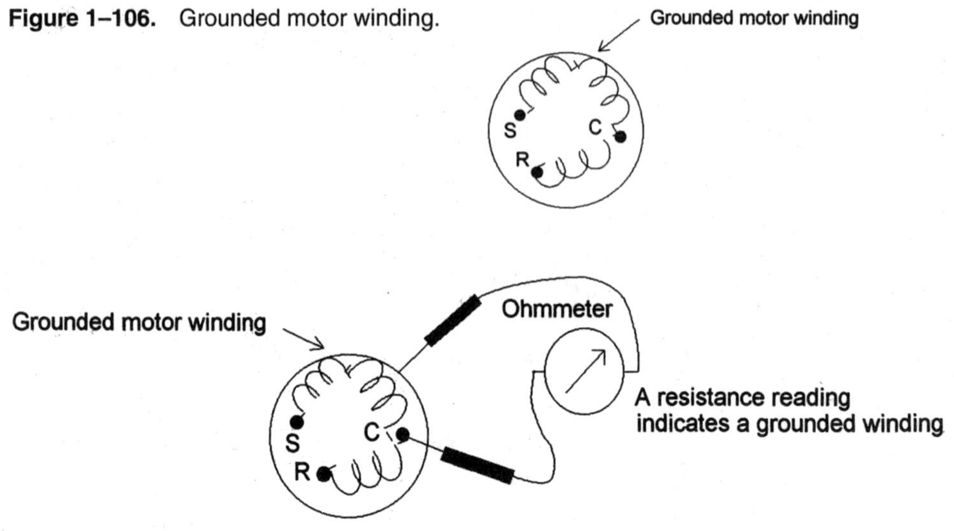

Figure 1–107. Checking for grounded motor winding.

when the motor has reached approximately 75% of its running speed. The contacts on these switches may become stuck closed, pitted, or stuck open.

A starting switch that is stuck closed will allow the motor to start but will cause the overload to open and stop the motor after a short period of operation. To check for a stuck-closed starting switch, start the motor and check the amperage draw. The amperage draw to a motor with a stuck-closed starting switch will not drop when 75% of the running speed is reached. If there is any drop in amperage, check for an overload or bad bearings. If the amperage draw does not drop, have the starting switch replaced.

A starting switch that is pitted or stuck open will not allow the motor to start. A pitted or stuck-open starting switch will, however, allow the motor to hum and try to start. When this condition is encountered, start the motor turning while it is humming. If the motor comes up to speed, the amperage draw is normal, and the motor operates normally, have the starting switch replaced. If the motor sometimes runs in the wrong direction, the starting switch is probably the cause and it should be replaced.

47–1E

Bad bearings in a motor will cause an overloaded condition, which will cause an excessive current draw. The motor may cut off due to the overload after operating for a while. To check for bad bearings, remove the belt if one is used, and free the motor shaft. Move the motor shaft in a sideways movement with the hand. See Figure 1–108. If movement in the shaft is found, either have the bearings replaced or replace the entire motor.

47–1F

Motor replacement should be done with great care. An exact replacement must be found to be certain that efficiency is maintained. Be certain that the motor manufacturer's wiring diagram is followed. After tightening all the motor

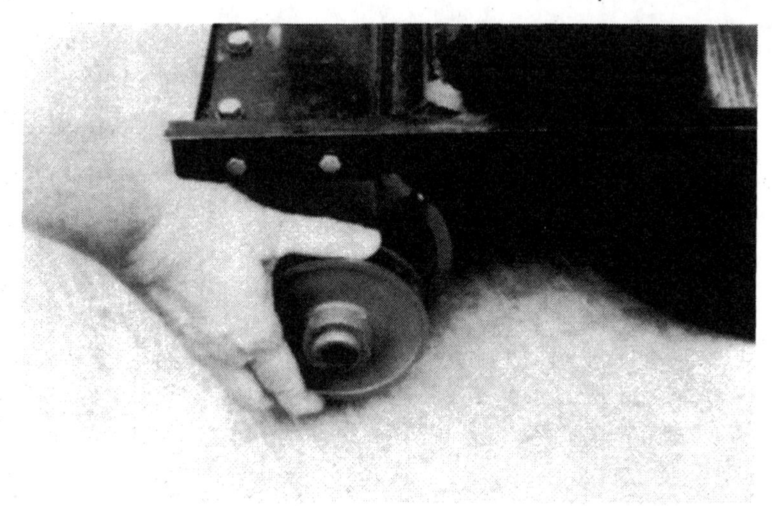

Figure 1–108. Checking motor bearings.

mounts, check to be sure that the shaft is free to rotate. Start the motor momentarily to see that it turns in the right direction and that nothing is dragging. Then start the motor and allow it to operate under normal conditions while checking the amperage draw. If the amperage draw is excessive, the reason should be found and corrected before leaving the job.

48–1 TRANSFORMERS:

Transformers are electrical devices that are used to reduce, or increase, electrical voltage. In refrigeration and air conditioning work, they are used to reduce line voltage to low voltage (24 volts). See Figure 1–109. The 24 volts is used in the control circuit because it is safer, the controls are cheaper to manufacture, and the controls are more responsive to temperature change than line voltage controls. To check a transformer, use the voltmeter and check the input voltage to the primary side of the transformer. See Figure 1–110. If voltage is found, check the output voltage from the secondary side of the transformer. See Figure 1–111. If no secondary voltage is found and primary voltage is found, the transformer is bad and must be replaced. If no voltage is found to the primary side of the transformer, the trouble is elsewhere. An alternate method of checking a transformer is to first disconnect all external wiring to the transformer. Then check the primary and secondary windings, in turn, with an ohmmeter. If either winding is open, the transformer is bad and must be replaced.

Some transformers incorporate a fuse in the secondary winding. If this fuse should blow, the transformer is rendered inoperative and usually must be replaced. Fuses can be replaced in only a few transformers.

48–1A

When replacing a transformer, be sure to use one with at least an equal VA (volt ampere) rating as the one being replaced. Always use a replacement designed for the same electrical characteristics as the one being replaced. Never use one with a

Figure 1–109. Class 2 transformers. (Courtesy of Honeywell, Inc.)

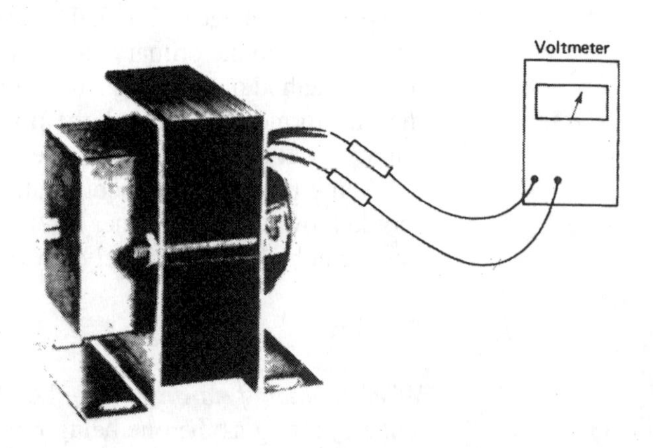

Figure 1–110. Checking voltage on a transformer primary. (Courtesy of Honeywell, Inc.)

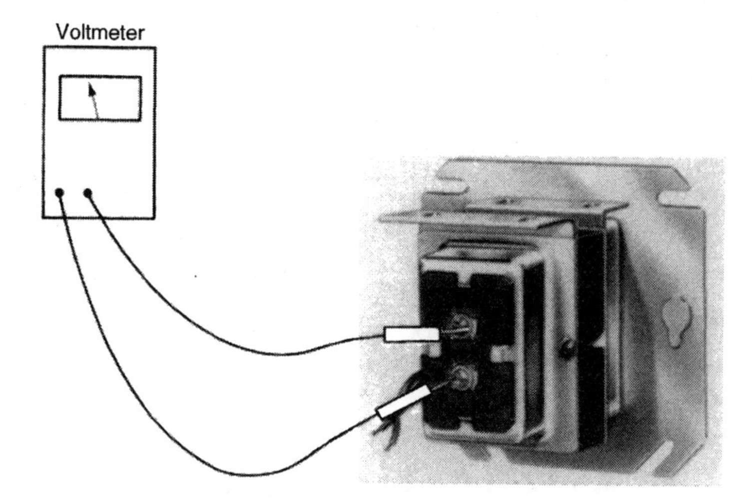

Figure 1–111. Checking voltage on a transformer secondary. (Courtesy of Honeywell, Inc.)

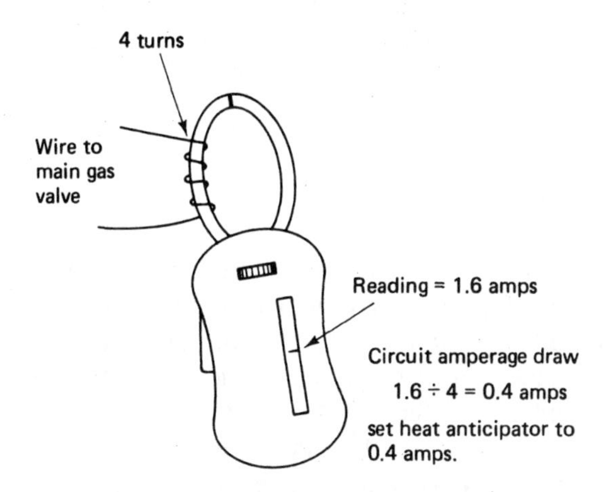

Figure 1–112. Checking amperage in temperature control circuit.

smaller VA rating unless it is known that the replacement will have sufficient capacity. A transformer with too small a VA rating will only burn out because of an overloaded condition. Always check to be sure that there is no short to cause the replacement transformer to burn out. To check for an overload or short, measure the amperage draw in the low voltage circuit. Then multiply the amperage times the voltage to obtain the VA draw of the circuit. The circuit VA must be equal to or smaller than the transformer rating. This check must be made with the unit running. Because of the small current draw in the low voltage circuit, it may be necessary to use a multiplier. These multipliers may be purchased or may be handmade. To make one, simply coil a piece of wire around the tong of an ammeter. See Figure 1–112. Connect the multiplier into the circuit and check the current draw. Then divide the current draw by the number of turns in the handmade coil. *Example:* A coil has ten turns and the current reading is 4.5 amps. Therefore 4.5 ÷ 10 = .45 amps. The VA would be .45 × 24 = 10.8 VA.

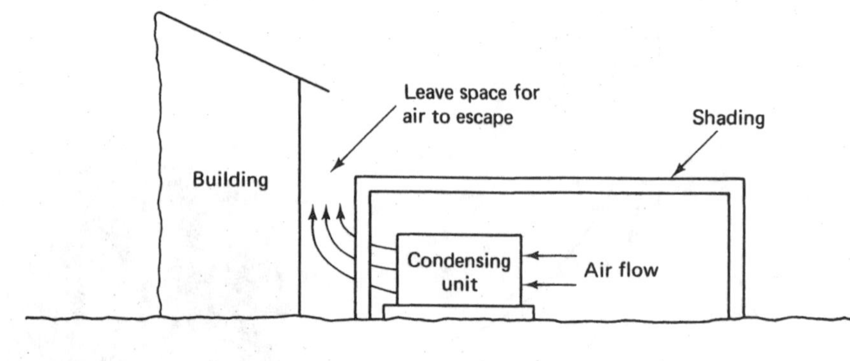

Figure 1–113. Shading for condensing unit.

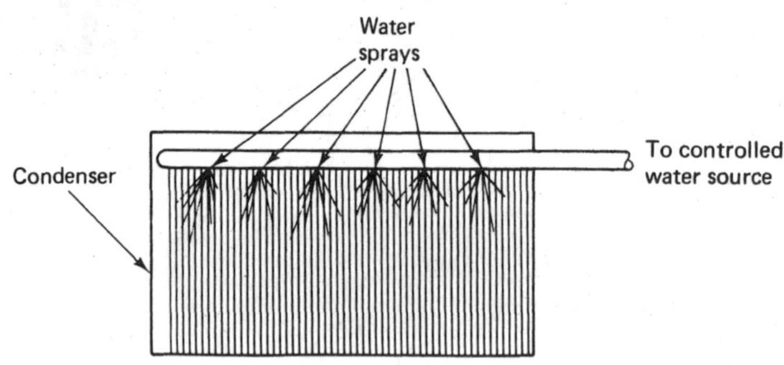

Figure 1–114. Water spray on air-cooled condenser.

49–1 HIGH AMBIENT TEMPERATURES:

High ambient temperatures are usually encountered in the summertime, especially in air conditioning installations. High ambient temperatures result in high discharge pressure and poor operation of the unit. The best method of helping this situation is to provide a shade of some type over the unit. The shade should extend several feet past the unit toward the direction of air flow. See Figure 1–113. An alternate method is to provide a spray of water on the condenser. This can involve an intricately designed unit or a lawn sprinkler, depending on the efficiency required. See Figure 1–114. Care should be taken to protect the electric motors and components from becoming electrically shorted to the unit.

50–1 HIGH RETURN AIR TEMPERATURE:

High return air temperature is usually encountered when a unit has been shut down for awhile and then is restarted. The high temperature of the air flowing over the evaporator will cause a high suction pressure, resulting in a temporary overload of the unit. The only solution is to allow the unit to operate until the return air temperature is lowered to the designed operating temperature. The unit should then operate normally.

51–1 INDOOR FAN RELAY:

The purpose of an indoor fan relay is to allow the user to select continuous or intermittent operation of the indoor fan motor at the thermostat, and to allow the fan to operate on a slow speed during the heating season and on a high speed during the cooling season. This type of relay is normally equipped with one set of open and one set of closed contacts. There are two general types of fan relays. See Figure 1–115. The open type has marked terminals, and the shrouded type has colored wires. To check out a fan relay, first turn the thermostat fan switch to "on." The relay should click. If there is no click, check the voltage on the two coil connections. See Figure 1–116. If voltage is indicated and no click is heard when the relay is energized, the relay is sticking and should be replaced. If a click is heard and the fan still does not start, check the line voltage across the common connection

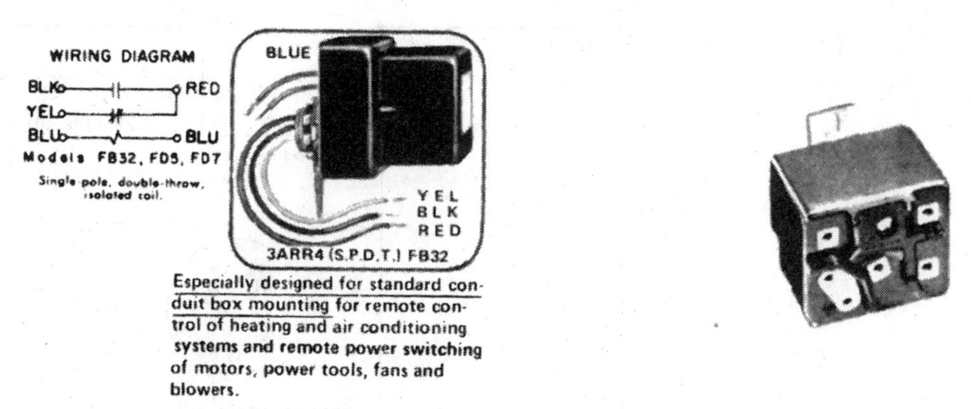

Figure 1–115. Shrouded (a) and open-type (b) fan relays and corresponding schematics.

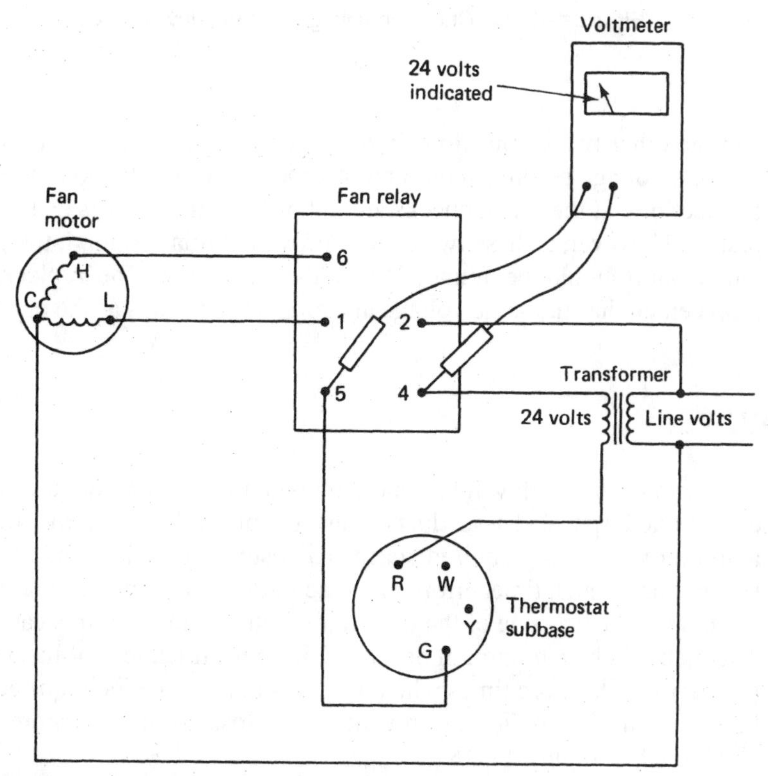

Figure 1–116. Checking voltage on fan relay coil.

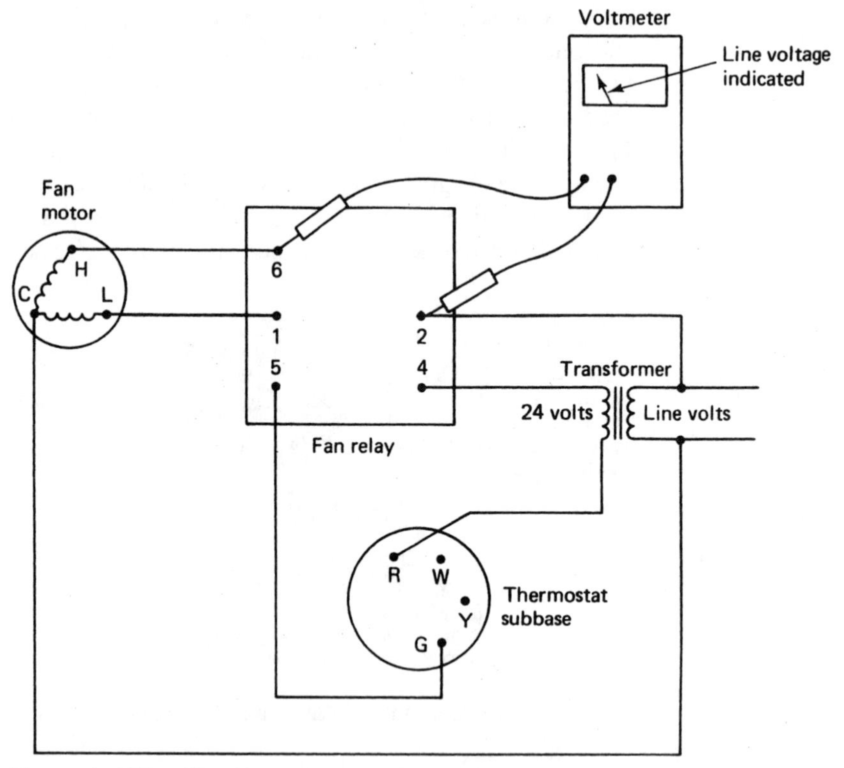

Figure 1–117. Checking voltage on fan relay contacts.

and the other two connections in turn. See Figure 1–117. With the relay deenergized, a voltage reading should be obtained between the common and low speed connections. If these two checks are not proven, the relay is bad and should be replaced. However, if these two checks prove the relay is good, the problem is elsewhere and should be found. When the relay must be replaced, be sure the replacement has the same voltage and amperage ratings as the one being replaced.

52–1 AIR DUCTS:

Air ducts are the hollow tubes that direct the flow of air from the air handler to the conditioned space. These ducts must be properly sized to direct the desired amount of air to the required space. To design a practical and efficient duct system requires much time, effort, and calculations. The proper amount of insulation around the ducts is one of the important features that requires careful calculation. There should be a minimum of 2 in. (25.4 mm) thickness of insulation with a vapor barrier. There are times when more insulation may be required. Enough insulation should be applied to prevent a heat loss or gain of more than about 2°F (1.11°C). See Figure 1–118.

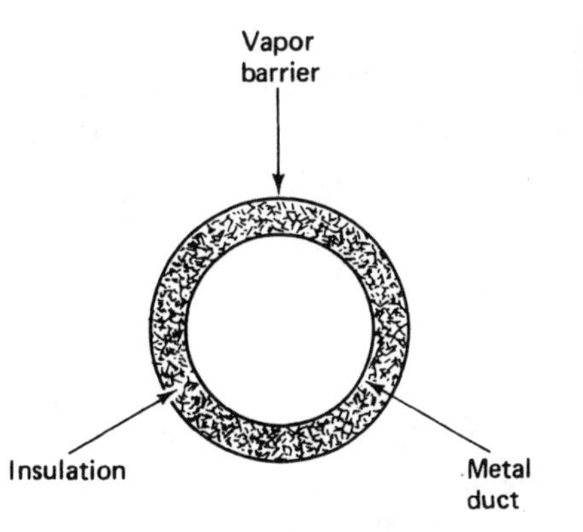

Figure 1–118. Cross section of metal duct with insulation.

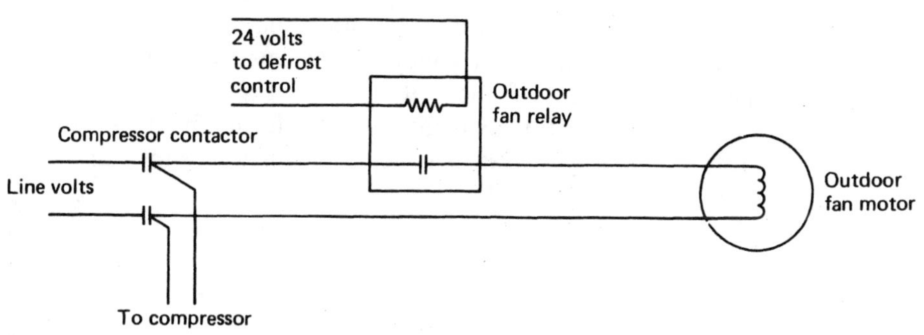

Figure 1–119. Outdoor fan relay wiring connections.

53–1 UNIT TOO SMALL:

Occasionally a designer will miscalculate and the wrong-sized unit will be installed. When this occurs, the only remedy is to install a unit of the proper size. A unit that is too small will operate properly, but the space temperature will be higher than desired on cooling and lower than desired on heating. See Section 61-1.

54–1 OUTDOOR FAN RELAY:

Outdoor fan relays are used on heat pump systems to aid in the starting and stopping of the outdoor fan in the various cycles involved in this type of system. The relays are generally SPST-type (single pole single throw) switches, which are actuated by a 24-volt coil. The contacts are in the line voltage circuit to the fan motor. See Figure 1–119. To check an outdoor fan relay, set the thermostat to a temperature that will cause the unit to run. With a voltmeter, check the voltage across the relay coil. A reading of 24 volts should be indicated. Cycle the unit and listen for the relay to click. No click with voltage present indicates a faulty relay. If a click and voltage to the coil are present, check the contacts. With the relay energized and the system turned on, there should not be a voltage indicated across the contacts. The relay is bad if a voltage is indicated. If no voltage is indicated and the fan does not run, the trouble is not in the relay and further checking is necessary. When replacing a relay be sure that the replacement has the correct voltage and amperage rating.

55–1 HEATING RELAY:

Heating relays are used on electric furnaces to complete the electrical circuit to the heating elements on demand from the thermostat. These relays operate by an electrically operated bimetal warp switch. See Figure 1–120. The bimetal heater is in the 24-volt circuit, and the contacts are in the line voltage to the heating elements. See Figure 1–121. To check the heating relay, use a voltmeter to check the voltage across the contacts. If a voltage is indicated here, the contacts are open. Next, jumper the contacts. If the heating element is energized, the trouble is in the relay. An energized heating element will be indicated by a current flow to the element. Replace the relay with an exact replacement or the unit will not operate as designed.

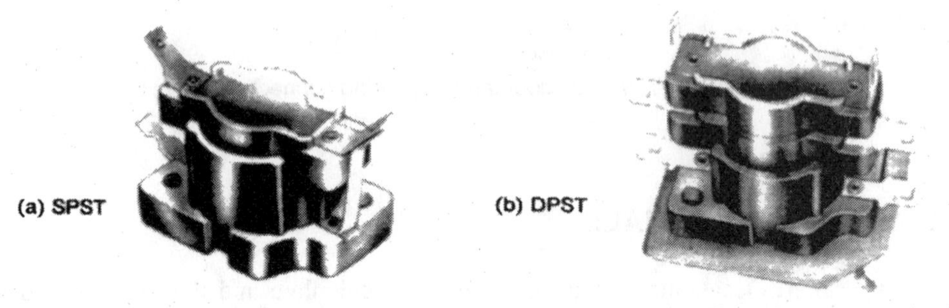

Figure 1–120. Bimetal heating relays. (Courtesy of Klixon Division, Texas Instruments, Inc.)

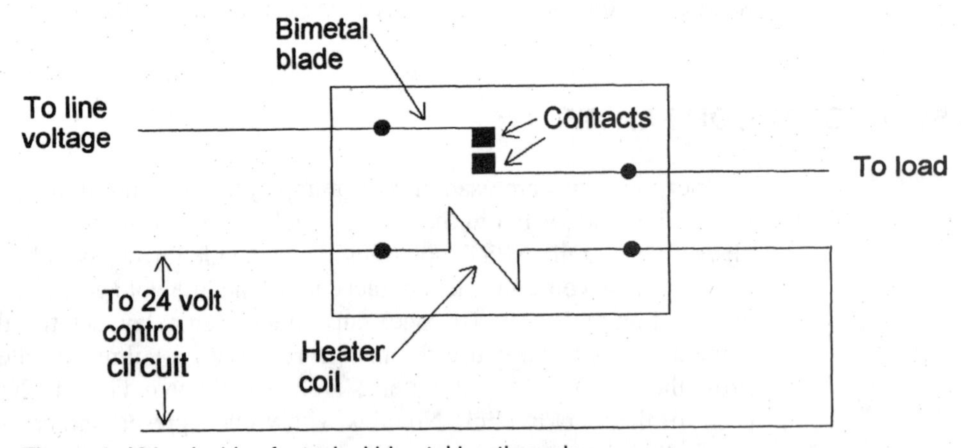

Figure 1–121. Inside of a typical bimetal heating relay.

56–1 FAN CONTROL:

The fan control is used on heating systems to start and stop the fan motor on an increase or decrease in temperature inside the furnace. This control is manufactured in adjustable and nonadjustable types as well as in extended element and bimetal disc types. See Figure 1–122. Some manufacturers are using electric time delay relays for fan controls on some units. The bimetal types are set to bring the fan on when the temperature at the sensing element reaches a temperature of 135°F (57°C) to 150°F (66°C), and to stop the fan when the temperature drops to approximately 100°F (37.8°C). The electric types are preset for a specified time to lapse for the "on" and "off" settings. To check out a fan control, insert a thermometer in the unit as close as possible to the sensing element. See Figure 1–123. Start the main burner and observe the thermometer. When the desired temperature is reached, or the specified time has lapsed, the fan should start. If not, adjust the control if it is of the adjustable type. Next, stop the main burner and observe the thermometer. When approximately 100°F (37.8°C) is reached, the fan should stop. If it does not stop, adjust the control if it is of the adjustable type. If the control is not adjustable, or more than about 15°F (8.33°C) of adjustment is needed, replace the control. Be sure

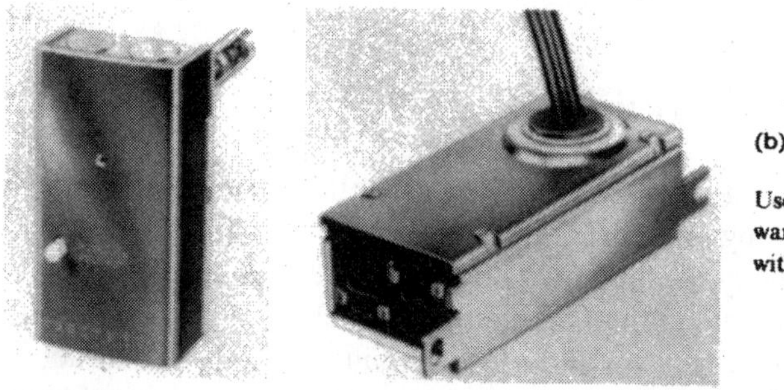

(b) Electric

Used for timed fan operation for forced warm air furnaces when wired in parallel with low voltage gas valve.

(a) Standard

Figure 1–122. Fan controls. (Courtesy of Honeywell, Inc.)

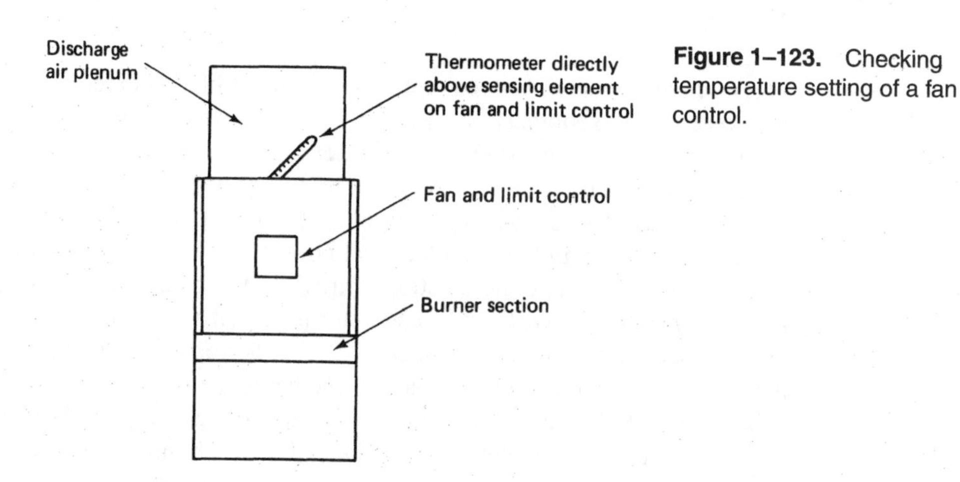

Figure 1–123. Checking temperature setting of a fan control.

to use an exact replacement or the unit will not operate as designed. Never attempt to make internal adjustments on these controls because of the dangers of overheating and permanently damaging the equipment, possible fire, and injury. (See Standard Service Procedure 4-1.)

56–1A

If the fan should cycle when the main burner is off, check to see that the air flow is not restricted by a dirty filter or dirty cooling coil. When no air restriction is found, widening the differential between the fan "on" and "off" settings will usually solve the problem.

56–1B

If the fan cycles on and off when the main burner remains on, the air temperature may be too low or too much air may be blown through the furnace. Allow the room temperature to increase, and reduce the air flow to provide the desired temperature rise, as indicated on the furnace nameplate. Sometimes widening the differential between the fan "on" and "off" settings will solve the problem.

57–1 HEAT SEQUENCING RELAYS:

Heat sequencing relays are used on electric furnaces to prevent all of the heating elements coming on or off at the same time. Therefore, these relays may be used to vary the capacity of the unit. There are three basic types of relays used to accomplish sequencing: (1) thermal element switches, also known as bimetal time-delay switches, which heat up and close circuits, (2) relays that activate another relay along with the heating element, and (3) modulating motors that rotate to close contacts at various points in the rotation.

57–1A

Thermal element switches make use of a bimetal blade heated by an electric current to actuate the contacts. See Figure 1–120. These switches are generally termed *stack relays.* They have one heated bimetal that closes more than one set of contacts. The use of the different contacts will vary depending on the equipment design. One set of contacts may complete the electrical circuit to the fan motor; another set may complete the electrical circuit to the heating element; and yet another set may complete the electrical circuit to the heating element of another relay. The specific use will need to be determined on each unit.

To check out these switches, first set the thermostat to demand heating. Then listen to the switch. If a click is heard, the heating element is working. If no click is heard, check the voltage across the heating coil. See Figure 1–124. If a voltage is indicated, the heating element is bad and the switch must be replaced. If a click is

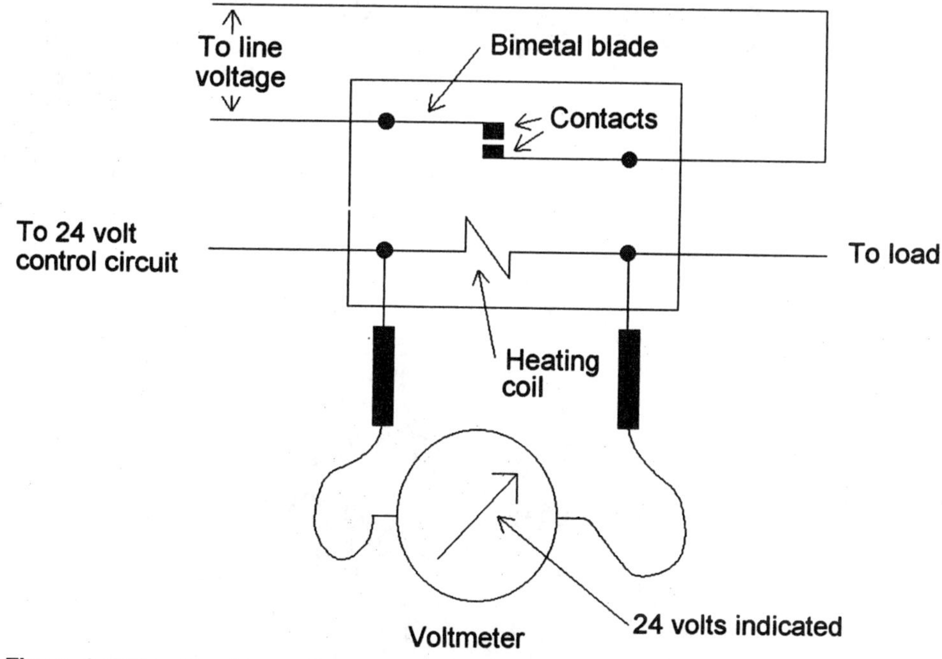

Figure 1–124. Checking voltage across thermal element on a time delay relay.

heard, indicating the heating element is working, check the voltage across each set of contacts. If a voltage is indicated, the contacts are bad and the switch must be replaced. Be sure that full electrical power is supplied to the unit. The replacement relay must be an exact replacement or the unit will not function as designed.

57–1B

Time delay relays that actuate relays along with the heating element are also sometimes used. These are single-contact relays. See Figure 1–125. When these relays are used, the relays energized by these devices must be equipped with heating elements of the same voltage as the heating element energized by the original relays. The checkout procedure is the same as that described in Section 57-1A.

57–1C

Modulating motors are generally used on more sophisticated units requiring more exact control than other units. See Figure 1–126. These motors will operate in any position within the number of degrees of rotation for which they are designed. The motor is controlled by use of a Wheatstone bridge. See Figure 1–127. The motor shaft is extended through the housing.

The switches are placed on the shaft to complete the desired circuit at a given degree of rotation. The thermostat controls the motor by the use of potentiometers,

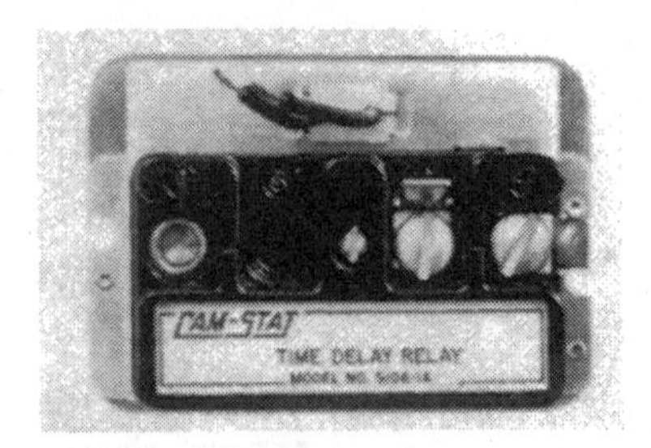

Figure 1–125. Single-contact time delay relay. (Courtesy of Cam-Stat, Inc.)

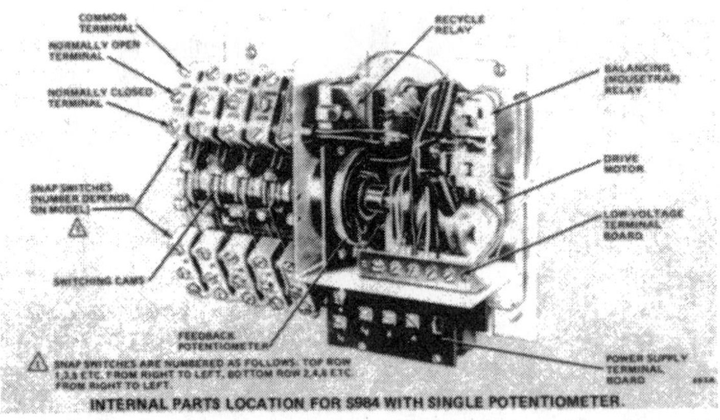

Figure 1–126. Modulating motor with end switches. (Courtesy of Honeywell, Inc.)

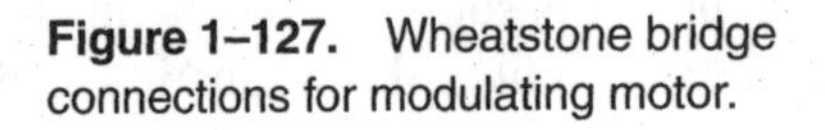
Figure 1–127. Wheatstone bridge connections for modulating motor.

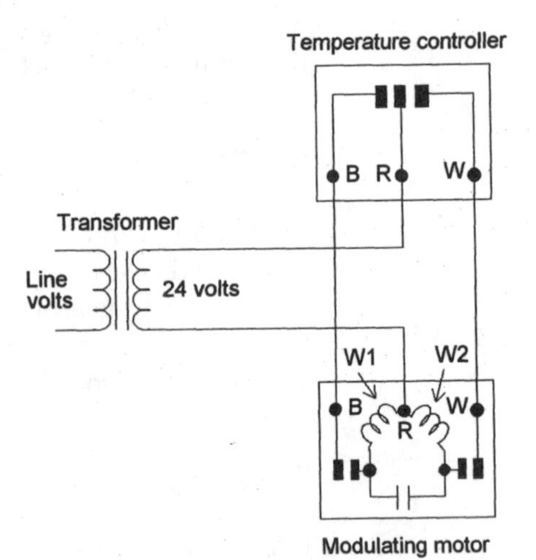

which produce a balanced voltage effect and control the balancing relay in the motor. To check out these motors, first be sure that the required voltage is being supplied to the motor and controls. Next, set the thermostat to demand heating. The motor should start turning. If not, clean the potentiometers and the balancing relay contacts with a cleaning solvent. It is not generally recommended to burnish a potentiometer because of possible damage. If the motor still does not operate, check to see that voltage is supplied to the motor. The manufacturer's wiring diagram may be required to determine the motor terminals. If voltage is indicated at these terminals and the motor still does not operate, replace the motor. If the motor operates but the heating elements do not function, check the voltage across each switch after making certain that the correct voltage is available. If voltage is indicated across the contacts, the switch is defective and must be replaced. Be sure to install the replacement switch in the exact position as the replaced switch to maintain equipment efficiency.

58–1 HEATING ELEMENTS:

There are several types of heating elements available. However, the open wire element is the most popular. Heating elements generally are equipped with protective devices in case of excessive current or overheating of the element. See Figure 1–128. These heating elements produce heat when electricity is applied to them. Sometimes they will develop hot spots and the wire will separate, causing an open circuit, and the heating will stop. See Figure 1–129. A visual inspection of the element will show this condition. A separated heating element must be replaced before normal operation can be resumed.

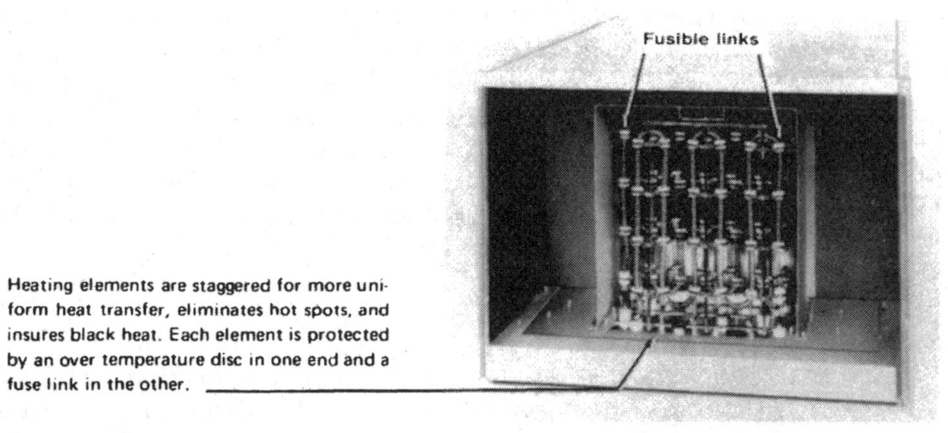

Fusible links
Heating elements are staggered for more uniform heat transfer, eliminates hot spots, and insures black heat. Each element is protected by an over temperature disc in one end and a fuse link in the other.

Figure 1–128. Open wire type heating element. (Courtesy of Square D Company)

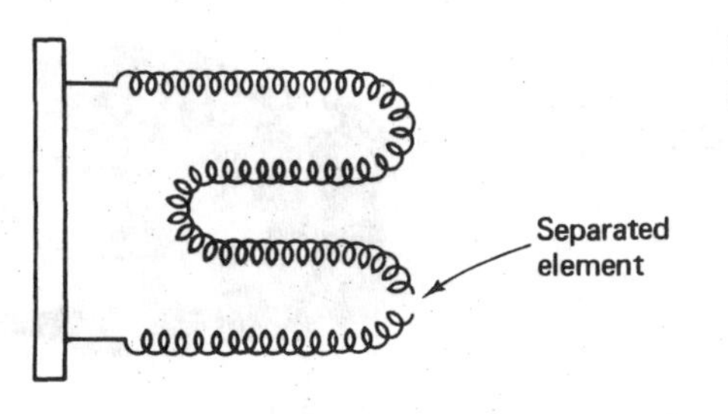

Figure 1–129. Separated electric heating element.

58–1A

Over temperature protective devices are required by both UL (Underwriters Laboratories) and NEC (National Electrical Code). There are two types of protection required by both agencies. The primary system is automatically reset and is designed to interrupt the flow of electrical current to the heater if insufficient airflow, air blockage, or other causes result in an overheating condition. The secondary or backup system is designed to operate at a higher temperature and interrupt the flow of electrical current to the heater in the event of failure of the primary system or continued operation under unsafe conditions.

When standard slip-in heaters are used, either a disc type or a bulb and capillary type thermal cutout may be used for both the primary and secondary system of temperature protection. See Figure 1–130. While this type of cutout is standard in the field for primary protection, some manufacturers also use them for control of the secondary system. The fusible links require field replacement, but the manual reset controls can immediately be put back in operation by simply pressing the reset button located in the heater terminal box. Be sure to determine and correct the cause of the trouble before leaving the job.

58–1B

Overcurrent protection is required on all electrical devices by UL and NEC. These agencies require that an electric heater drawing more than 48 amps total must be subdivided into circuits that draw no more than 48 amps each. Each circuit must be protected by fuses or circuit breakers. Because electric heaters are regarded as a continuous load, overcurrent protective devices must be rated for at least 125% of the circuit load. Thus, while the circuits are limited to 48 amps, the fuses are limited to 60 amps. These fuses must be applied by the heater manufacturer. When a fuse is blown there is an electrical overload, which must be found and removed before the service job is complete.

58–1C

Loss of air flow protection in an electric heater is required by both UL and NEC. Both agencies require that each heater be electrically interlocked with the fan so that the heater cannot operate unless the fan circuit is energized and air is therefore flowing. See Figure 1–131. Most slip-in heaters include a built-in differential pres-

Figure 1–130. Element protective device. (Courtesy of Industrial Engineering and Equipment Co.)

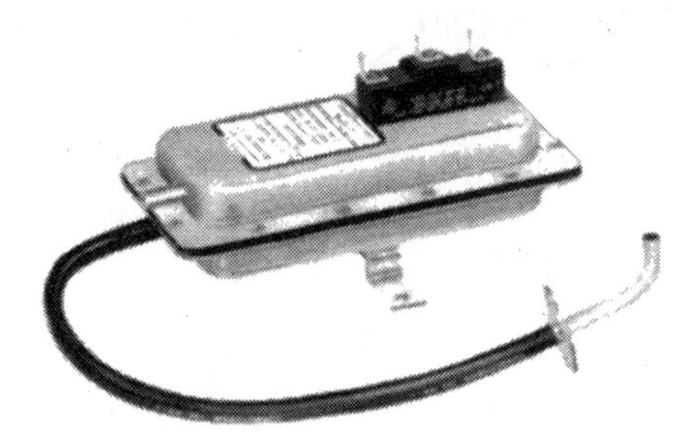

Figure 1–131. Air flow protective device. (Courtesy of Industrial and Engineering Equipment Co.)

sure, diaphram operated, air flow switch to meet this requirement. A loss of air flow may be caused by dirty air filters, dirty blower, an inoperative blower, or air outlets closed off. The problem must be corrected before the service job is complete.

59–1 TEMPERATURE RISE THROUGH A FURNACE:

The amount of temperature rise through a furnace is recommended by the furnace manufacturer. The minimum and maximum limits are designated on the furnace nameplate. When these limits are exceeded, the unit will not perform as it was designed. The temperature rise may be determined by subtracting the inlet dry bulb temperature from the outlet dry bulb temperature. To obtain these temperatures, insert a thermometer as close to the air inlet and outlet of the furnace as possible. See Figure 1–132. When the maximum temperature rise is reached, the minimum of air is flowing through the furnace. If this limit is exceeded, damage to the heater may occur. When the minimum temperature rise is reached, the maximum amount of air is flowing through the furnace. If this limit is exceeded, cold drafts may be experienced. When operating on a maximum temperature rise, the relative humidity will be lowered. When operating on a minimum temperature rise, the relative humidity will be raised. To change the temperature rise, adjust the fan speed to deliver the amount of air required.

60–1 BLOWERS:

Blowers are the devices used to force air through the equipment and into the conditioned space. See Figure 1–133. Blowers are generally located at the air inlet to the furnace or to the air distribution system. A blower will not move a sufficient amount of air if the vanes become clogged with dirt, if the belt becomes loose and slipping, if the bearings become worn or out of lubricant, or if the motor develops a problem.

60–1A

The blower should be removed from the equipment before it can be effectively cleaned. Sometimes a brush and soapy water are required to clean the blower. In such cases take the blower out of the building and protect the motor from the water spray.

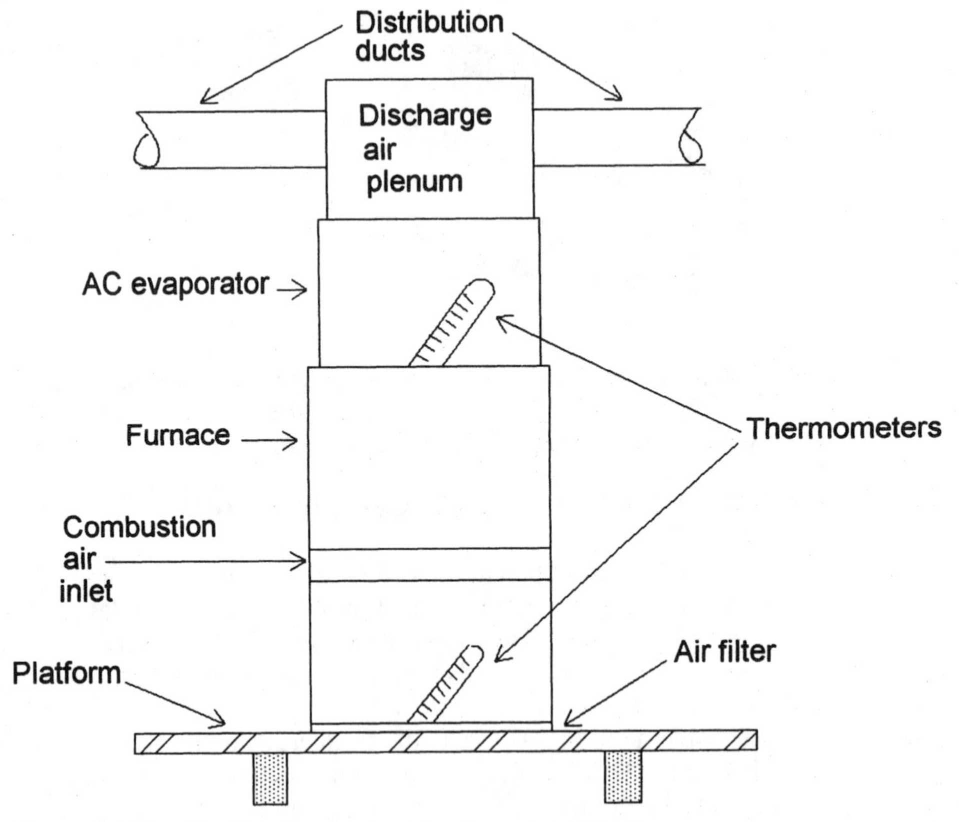

Figure 1–132. Checking temperature rise through a furnace.

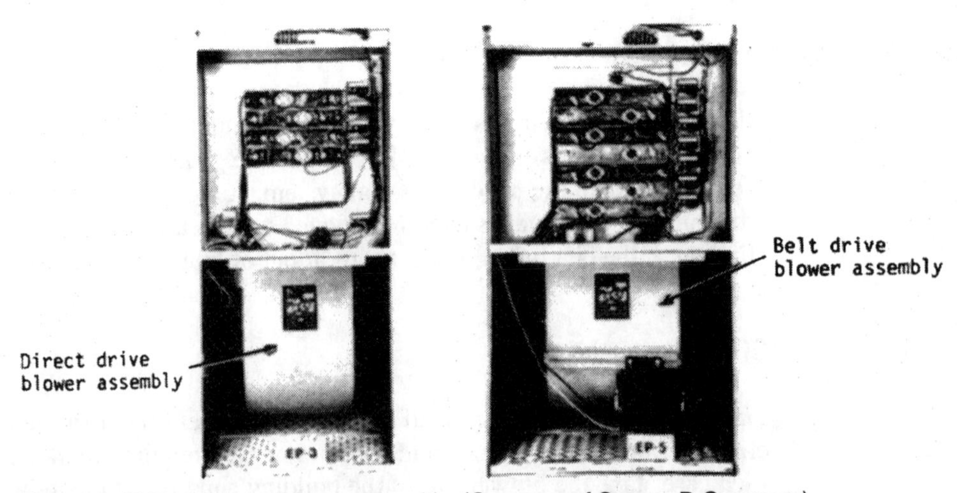

Figure 1–133. Furnace blower assembly. (Courtesy of Square D Company)

Figure 1–134. Checking fan bearings.

60–1B

Defective bearings will overload the motor and will cause the blower to turn more slowly than is required. Therefore, the air flow will be reduced, affecting the equipment operation. To check for defective bearings, remove the fan belt and grasp the blower shaft in the hand. Move the shaft from side to side. See Figure 1–134. If sideways movement is detected, replace the bearings. Be sure to check the blower shaft for wear or scoring. If wear or scoring is indicated replace the shaft also. If no sideways movement is detected, spin the blower and observe if the blower turns free or if it is stiff. If a stiffness is indicated, lubricate the bearings. If lubrication does not solve the problem, replace the bearings as described above. Do not overtighten the belt or excessive bearing wear will result. Belt adjustment is discussed in Section 18-1. Motor service is discussed in Section 47-1.

61–1 BTU INPUT:

The Btu input to an electric heating unit can be changed in the field without much problem. When more or less heat is required, change the number of heating elements that are in operation. If more heat is required, add enough heating elements to supply the demand. When less heat is required, remove the extra heating elements. The Btu input of a gas furnace cannot be increased above the Btu input rating of the furnace. To do so would cause overheating of the heat exchanger, and this would upset the vent system. The Btu input to a gas furnace can be reduced by only 20% of the input rating. To reduce the input Btu more would upset the vent system. An improperly operating vent system is very dangerous and must be avoided. When changing the Btu input of any furnace, be sure to check the temperature rise through the furnace as described in Section 59-1. Make any adjustments on the blower speed to maintain the recommended temperature rise. In some cases a larger blower motor will be required.

62–1 HUMIDITY:

During the winter months a desired relative humidity inside a conditioned space will be between 40% and 60%. However, when the outdoor temperature is very low, condensation on doors, walls, and windows may be experienced. This is an indication of excessive relative humidity, which must be lowered to prevent damage. This humidity may be the result of cooking with open pots and pans or bathing. When cooking with open pots and pans, the pans should be covered to prevent the escape of moisture into the conditioned space. If covering the pots is not desirable, a vent system may be installed above the stove. When bathing causes the problem, install a vent system in the bathroom.

63–1 PILOT SAFETY CIRCUIT:

The pilot safety circuit is the device that makes fuel burning equipment safe. The components that make up the pilot safety circuit are the thermocouple, the pilot burner, and the pilot safety device. These devices are designed to stop the flow of gas to the burners when the pilot flame is not sufficient to safely ignite the main burner gas.

63–1A

Thermocouples are devices that when heated produce a small electrical current. See Figure 1–135. This small current (approximately 30 mV) is used to energize an electromagnetic coil in the pilot safety device. The heat is provided by the pilot burner flame. The cartridge end of the thermocouple is mounted in the pilot burner so that the cartridge is in the pilot flame from 3/8 in. (9.525 mm) to 1/2 in. (12.7 mm). See Figure 1–136. When the cartridge is heated to a dull red, almost black color,

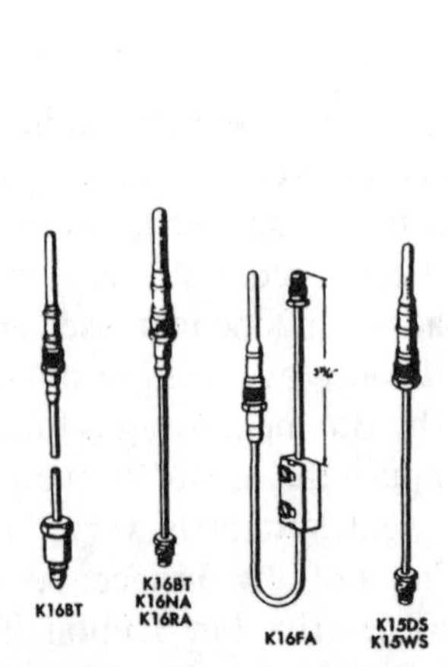

Figure 1–135. Thermocouples. (Courtesy of Penn-Baso Controls)

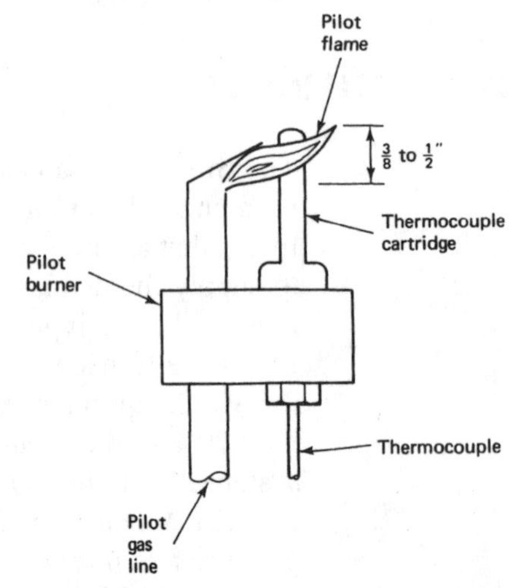

Figure 1–136. Thermocouple properly mounted on pilot burner.

the maximum output is gained from the thermocouple. There are three checks that can be made on a thermocouple: (1) the open circuit test, (2) the closed circuit test, and (3) the drop-out test. All three checks are made with a dc millivolt meter. (See Standard Service Procedure 3-1.)

The open circuit test is made while the thermocouple is being heated with the pilot flame. The first step is to keep the pilot burner burning by depressing the bypass knob on the pilot safety device. See Figure 1–137. Then disconnect the thermocouple from the pilot safety device and check the thermocouple output. See Figure 1–138. Be sure to connect the meter leads properly or a reading may not be obtained. If the meter does not indicate a voltage or the needle moves off the scale, reverse the meter leads. The voltage reading obtained in this test should be between 25 and 35 mV. If the reading is as low as 20 mV, replace the thermocouple. Be sure that all connections are clean.

The closed circuit test is made with the thermocouple connected to the pilot safety device with an adapter while the cartridge end is in the pilot flame. See Figure 1–139. This test is used to check the thermocouple while being used. The

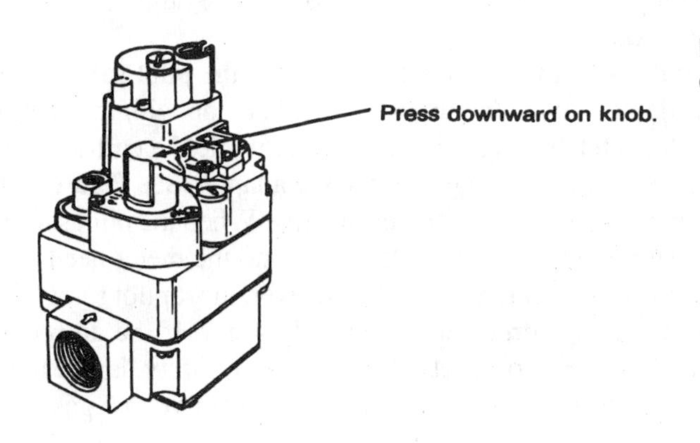

Figure 1–137. Pressing knob on gas valve to light pilot.

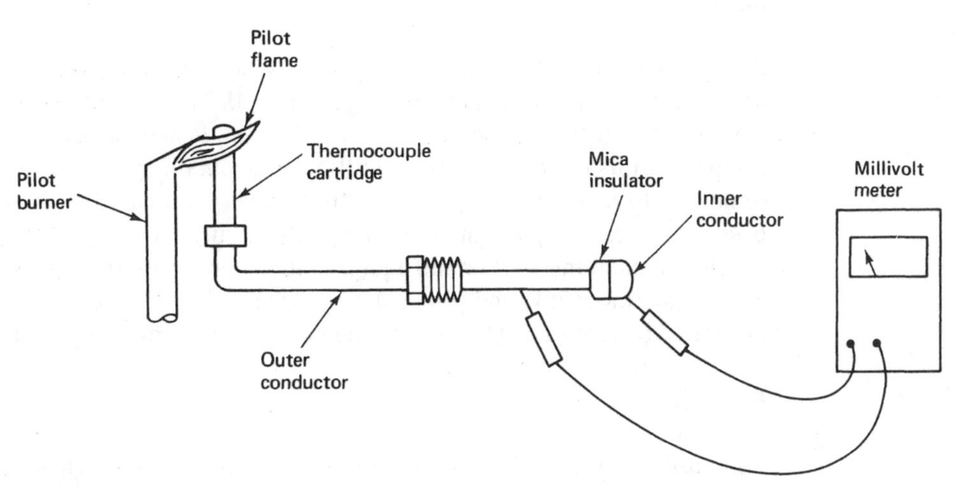

Figure 1–138. Open circuit thermocouple test.

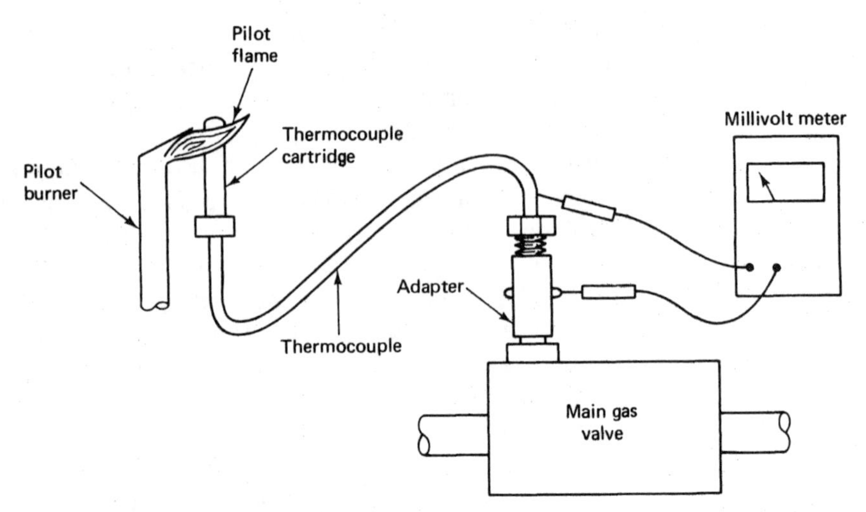

Figure 1–139. Closed circuit thermocouple test.

thermocouple output should be 17 mV or more. If the output is less than 17 mV, replace the thermocouple.

The drop-out test is made to determine when the pilot safety device will close when the pilot flame is out. The drop-out test is made with the thermocouple connected to the pilot safety device with an adapter while the cartridge end is in the pilot flame. A millivolt meter is connected to the adapter. See Figure 1–139. Next turn off the pilot flame and observe the meter reading. When the pilot safety device drops out or closes, there will be a click in the valve and the meter needle will jump. The recommended drop-out voltage is 8 mV. However, most pilot safety devices will not drop out until the voltage drops approximately 4 mV. If the drop-out voltage is as low as 2 mV, replace the pilot safety device. A pilot safety device that will not drop out above 2 mV is unsafe because it will allow excessive raw gas to escape.

63–1B

A pilot burner has two functions: (1) to provide heat for the thermocouple, and (2) to ignite the main burner gas. See Figure 1–140. The only thing that can happen to a pilot burner is the orifice may become clogged. When this happens, the pilot cannot perform its duties. The flame will become small, or yellow, or both. The only remedy is to remove the pilot burner from the equipment, dismantle, and clean the orifice. Use care to prevent enlarging or damaging the orifice. Reassemble the pilot burner and reinstall it on the equipment. When a pilot flame repeatedly goes out and the thermocouple and pilot safety device is found to be good, shielding of the pilot may be required. The pilot flame is possibly being blown out by drafts.

63–1C

The pilot safety device is used to prevent raw gas escaping when the pilot flame is not functioning properly. There are two checks to make on a pilot safety device: (1)

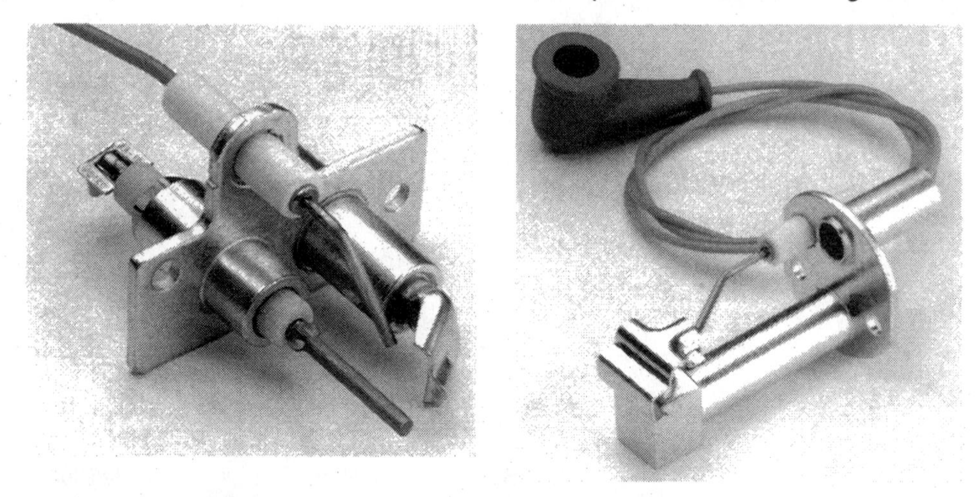

Figure 1–140. Pilot burner installation. (Courtesy of Penn-Baso Controls)

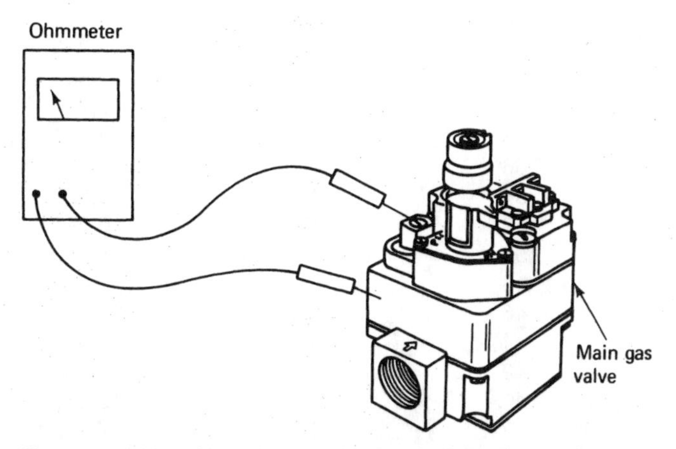

Figure 1–141. Checking continuity of pilot safety coil.

the drop-out test, and (2) the continuity test. The drop-out test is outlined in Section 63-1A. Refer to that information for the testing. The continuity test is made with an ohmmeter. First remove the thermocouple from the pilot safety device; then with an ohmmeter check the continuity of the electrical coil. See Figure 1–141. Be sure to zero the ohmmeter. If no continuity is indicated or the drop-out test fails, replace the pilot safety device.

64–1 ORIFICES:

The purpose of an orifice is to regulate the flow of gas to the burner. The orifice is sized according to the type of gas, the manifold pressure, and the Btu rating of the burner and gas. See Figure 1–142. Usually, orifices are trouble-free and need little attention. However, if someone tried cleaning an orifice and ruined it, the orifice must be replaced. When replacing an orifice be sure to use an exact replacement.

Figure 1–142. Burner orifices.

Table 1–2. Orifice Capacities for Natural Gas, 1000 Btu/ft^3, Manifold Pressure 3½ In. Water Columns.

Wire Gauge Drill Size	Rate ft^1/hr	Rate Btu/hr
70	1.34	2,340
68	1.65	1,650
66	1.80	1,870
64	2.22	2,250
62	2.45	2,540
60	2.75	2,750
58	3.50	3,050
56	3.69	3,695
54	5.13	5,125
52	6.92	6,925
50	8.35	8,350
48	9.87	9,875
46	11.25	11,250
44	12.62	12,625
42	15.00	15,000
40	16.55	16,550
38	17.70	17,700
36	19.50	19,500
34	21.05	21,050
32	23.70	23,075
30	28.50	28,500
28	34.12	34,125
26	37.25	37,250
24	38.75	39,750
22	42.50	42,500
20	44.75	44,750

Often the furnace and the type of gas used are different. When this happens, the orifice must be sized and must be replaced for the type of gas used. There are tables for every type of gas, gas pressure, and Btu rating thinkable. See Table 1–2. The local supplier will usually have this information and can drill the orifice to the proper size. Drill the orifice from the rear so as to center the orifice in the spud. Use care not to elongate the orifice and ruin the spud.

64–1A

Orifices that are too large will admit excess gas to the burner. This excess gas will not burn properly and will cause carbon monoxide, a deadly gas, to be produced. A large orifice will cause a larger than desired flame, which will float off the main burner. See Figure 1–143. Sometimes the flame will be all blue and sometimes it will be partially yellow.

64–1B

Misaligned orifices will misdirect the gas flow down the burner. The gas stream will often strike the burner, slowing down the velocity. See Figure 1–144. This reduction in velocity will reduce the amount of primary air drawn into the burner and will prevent the proper mixing of the gas and air. This condition is accompanied by a yellow flame. The orifice must be replaced or the manifold redrilled to correct the problem. Dirty orifices will often misdirect the gas stream. The remedy here is to clean the orifice.

CAUTION: Be careful not to damage the orifice when cleaning.

64–1C

Orifices that are too small will cause smaller than normal flame. See Figure 1–145. The flame will be all blue but will not provide the required amount of heat. Orifices that are too small may be redrilled to the proper size. When drilling an orifice, insert the drill in the rear of the orifice and use extreme care to obtain a round hole. See Figure 1–146. Use an orifice chart to determine the correct size.

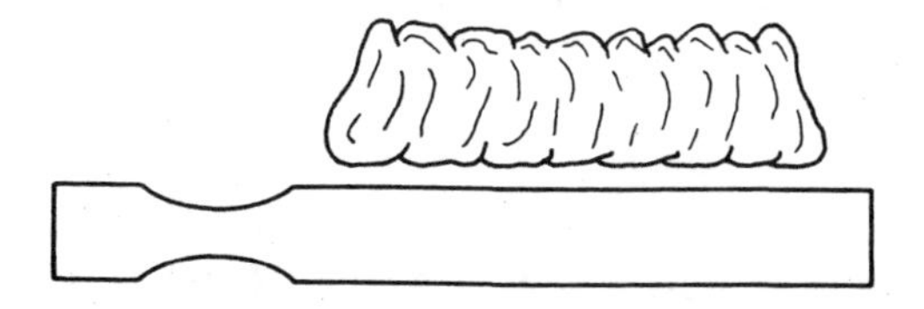

Figure 1–143. Floating main burner flame.

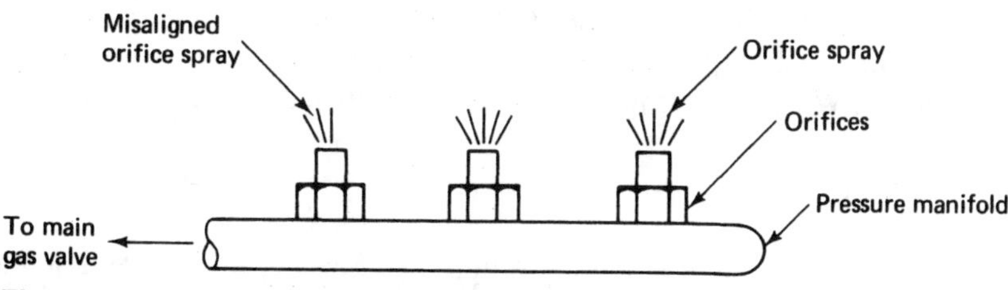

Figure 1–144. Misaligned orifice gas stream.

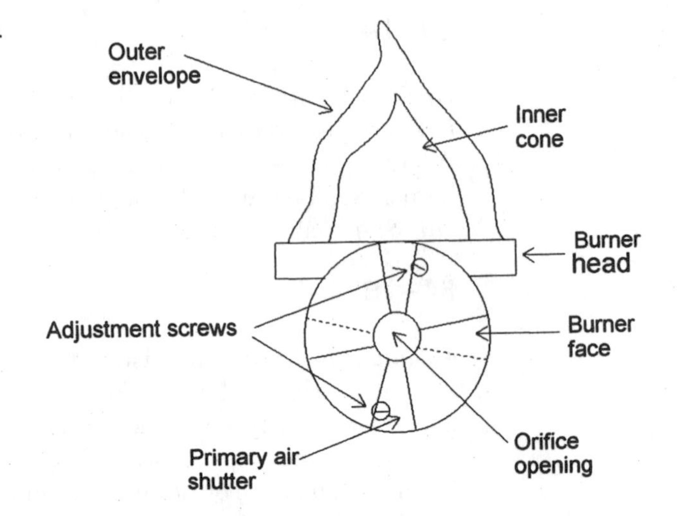

Figure 1–145. Small flame.

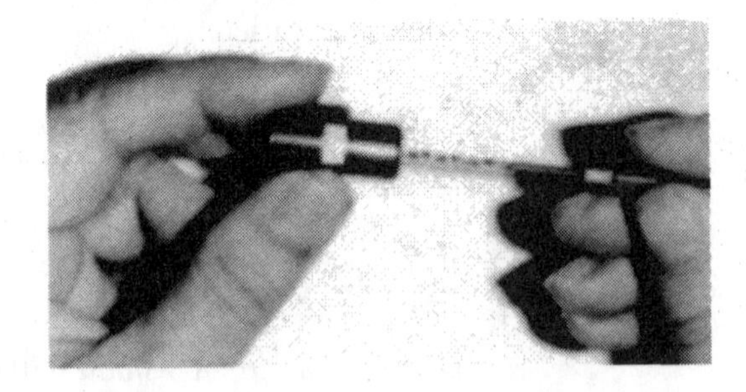

Figure 1–146. Drilling an orifice.

65–1 MAIN BURNERS:

The purpose of the main burner is to provide a place for the mixing of gas and air and a place for the burning of the mixture. See Figure 1–147. There are several conditions that can affect the operation of a burner, such as: distorted carry-over wing slots, dirty burner, primary air adjustments, poor flame travel to the burner, and poor flame distribution.

65–1A

The carry-over wing on a main burner provides a path for the pilot flame to travel from burner to burner to provide smooth gas ignition. See Figure 1–148. If the slots in the carry-over wing should become distorted, the ignition flame will not travel properly and flashback (gas burning at the orifice) will occur. The solution to this problem is to repair the slots. This is done by restoring them to their original condition. When repairing is not practical, replace the burners.

65–1B

Gas burners become dirty through use. See Figure 1–149. Dust and lint particles from the air, and rust particles from the heat exchanger, collect in the burner and upset the flow of the gas and air mixture, which upsets the normal combustion process.

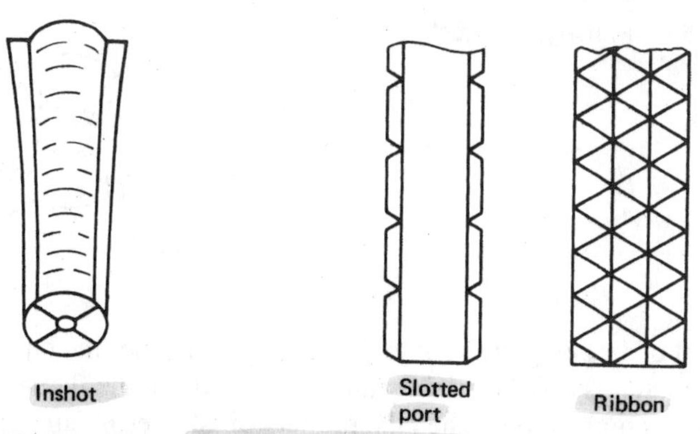

Figure 1–147. Types of main burners.

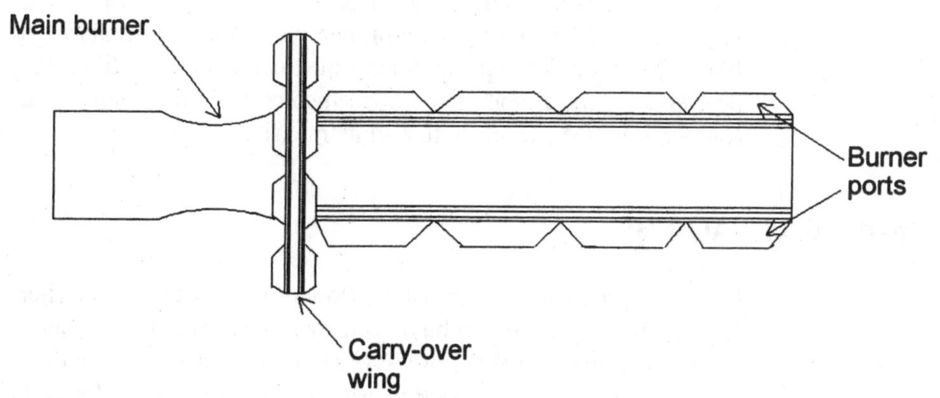

Figure 1–148. Main burner carry-over wing.

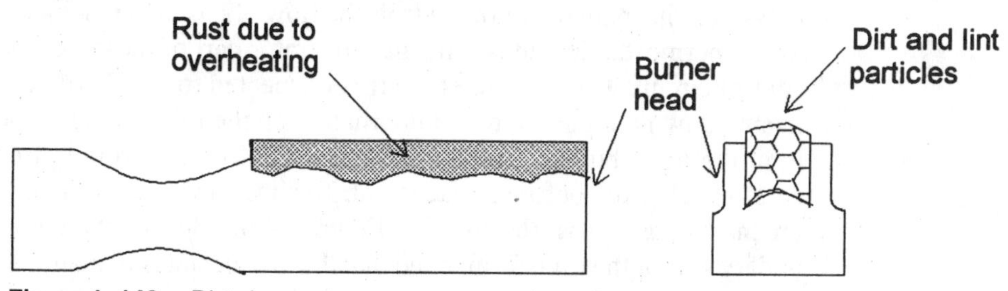

Figure 1–149. Dirty burner.

Dirty burners are the cause of flashback (gas burning at the orifice) and possible delayed ignition. To correct this problem, remove the burners and clean both inside and outside. The burner ports must be cleaned with extreme care to prevent damage. Air pressure may be used to blow out the inside of the burner to remove any collected dirt and lint. The burners should be removed from the furnace to aid in cleaning.

65–1C

The primary air adjustment (shutter) is located on the face of the burner and is used to regulate the amount of primary air being drawn into the burner. See Figure 1–150.

Figure 1–150. Primary air adjustment.

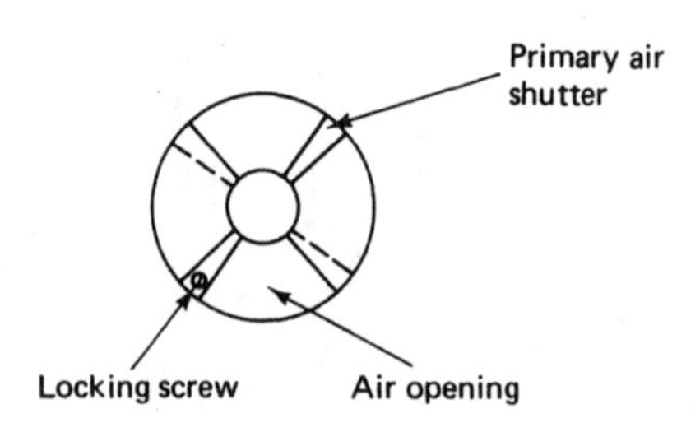

The amount of air being drawn into the burner determines the type of flame present. When the shutter is off, the primary air is restricted and the flame will be yellow, indicating poor combustion. When the primary air shutter is fully open, too much air will be drawn into the burner, resulting in a hard, blowing flame, also indicating improper combustion. The ideal flame will be neither hard and blowing nor contain yellow tinges. To adjust a main burner, close off the primary air and obtain a yellow flame. Next, open the primary air adjustment shutter until all the yellow has disappeared from the flame. Then open the shutter adjustment about one-eighth more (one-eighth of the total shutter adjustment).

66–1 GAS VALVES:

Gas valves are used to control the flow of gas to the main burner. Usually there are four main parts, each part having its own function: (1) the manual gas cock, (2) the pilot safety device, (3) the pressure regulator, and (4) the main operator. See Figure 1–151. The manual gas cock is used to shut down the pilot and main burner manually. The pressure regulator is used to supply gas to the main burner at a constant pressure. The main operator controls the flow of gas to the main burners in response to the thermostat demands. The main operator part of these valves is in the 24-volt control circuit. The pilot safety part is connected to the thermocouple and is used to shut down the pilot and main burner when the pilot flame is too small to safely light the main burner. These valves are usually very trouble-free and should be carefully checked before replacement. To check a valve, use a voltmeter and measure the voltage across the coil. See Figure 1–152. Be sure there is power on the secondary side of the transformer and that the thermostat is demanding heat. If 24 volts are present and the pilot safety circuit is functioning, the gas valve should be replaced. See Section 63-1. Also, if the gas valve operates properly at times and not at other times, the valve should be replaced. If a valve of the same physical size is used as the replacement, there will be no plumbing changes needed. Do not attempt to repair gas valves. Replace components or the complete valve as needed.

66–1A

Quick-opening main gas valves cause rough ignition and flame roll out. To prevent this, install a surge arrestor in the vent line. Do not attempt to repair any gas valve. Replace components or the complete valve as needed.

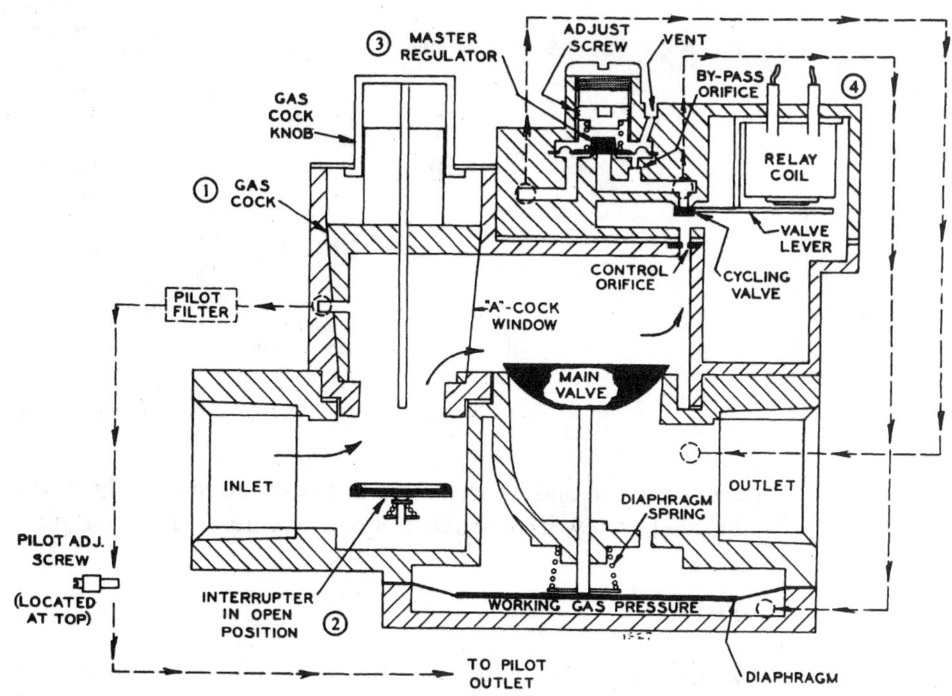

Figure 1–151. Main gas valve cutaway. (Courtesy of White Rodgers Division, Emerson Electric Co.)

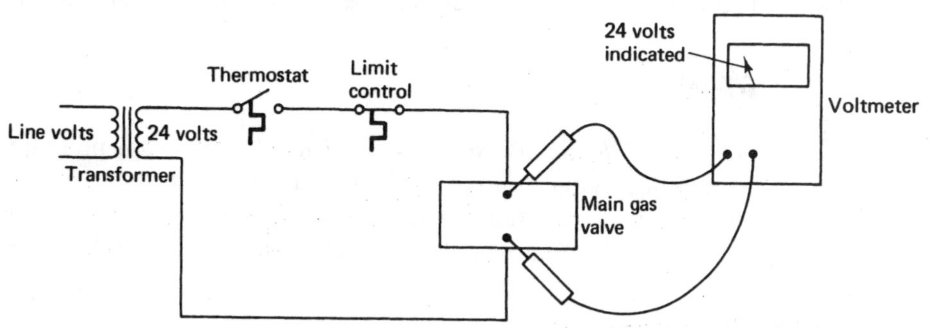

Figure 1–152. Checking voltage on a main gas valve coil.

67–1 PRESSURE REGULATOR:

Natural gas furnaces are designed to operate with 3½ to 4 in. water-column pressure and LP (liquified petroleum gas) furnaces with 11 in. water-column pressure in the furnace manifold. The purpose of the pressure regulator is to maintain these pressures. Pressure regulators are in the same housing as the main gas valve. See Figure 1–151. Except on LP gas furnaces, the regulator is located at the storage tank. To check the outlet pressure of the regulator, connect a manometer or a manifold pressure gauge to the manifold. See Figure 1–153. To make any adjustments,

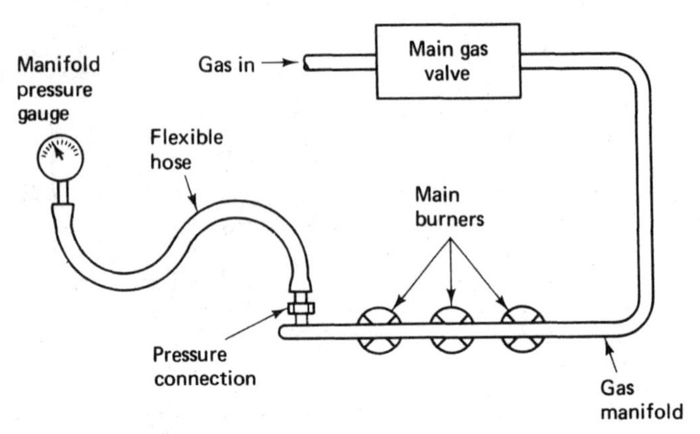

Figure 1–153. Checking manifold pressure on a gas furnace.

remove the cap and insert a screwdriver inside to the adjustment screw. Turn the adjustment screw in to increase or out to decrease the outlet pressure.

67–1A

On some large-capacity units, a step-opening regulator is used. Step-opening regulators are used to prevent large amounts of gas input to the combustion area all at once. Initially these regulators open to about 80% of their full capacity, then open fully after the ignition stage has been accomplished. When erratic operation is experienced with these valves, replace the valve rather than make repairs.

67–1B

When an unstable gas supply is experienced, a two-stage pressure regulator should be installed in place of the original regulator. An unstable gas supply may be detected by wide variations in the gas manifold on the furnace. A two-stage pressure regulator will smooth out these pulsations and will permit a more even flame.

68–1 HEAT EXCHANGER:

The heat exchanger in a gas-fired furnace acts as the combustion area, the heat transfer medium, and the flue passages. See Figure 1–154. The heat exchanger is sometimes called the "heart" of the heating system. The most common problems that occur with a heat exchanger are: (1) restrictions, and (2) cracks or openings in the metal.

68–1A

Restrictions in the flue passages of a heat exchanger upset the draft through the furnace and interfere with the combustion process. These restrictions must be re-

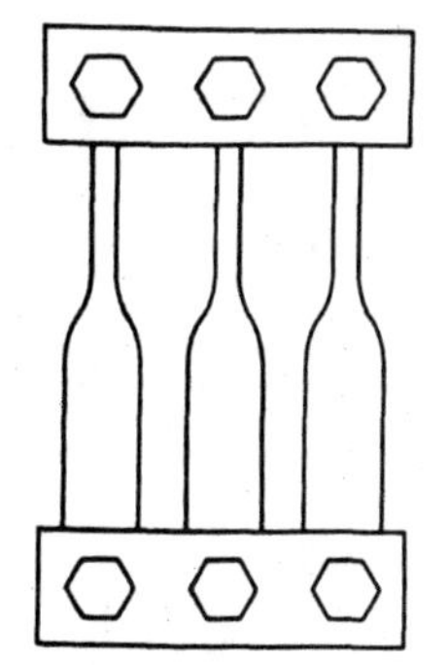

Figure 1–154. Gas furnace heat exchanger.

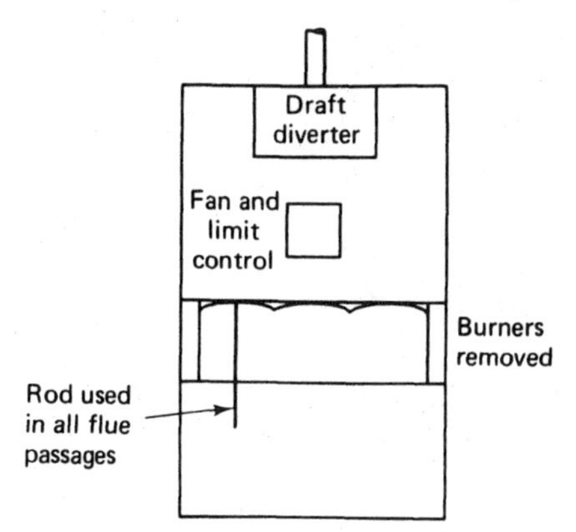

Figure 1–155. Cleaning a plugged heat exchanger.

moved to restore normal operation. To remove the restrictions the first step is to remove the main burners and the draft diverter. The top and bottom openings are now clear to insert a straightened-out coat hanger. See Figure 1–155. Scrape each side of the flue passage from front to back several times to loosen any rust particles and soot. The loosened rust and soot will fall to the bottom of the combustion zone and must be removed. After the rust and soot are removed from the heat exchanger, clean the burners and reinstall them. Light the furnace and adjust the main burners to obtain good combustion. Next, check the heat exchanger for cracks. See Section 68-1B.

68–1B

Heat exchangers that are cracked are hazardous because of possible fire and carbon monoxide poisoning. A cracked heat exchanger allows the recirculating air and the products of combustion to mix. When the fan is running, a pressure is built up in the duct system that forces air into the combustion zone and causes the flame to be blown around. When the crack is large enough, the flame will be blown out of the front of the combustion zone. To make a more positive check, turn off the gas to the furnace and place a smoke bomb in the combustion zone. Place the bomb first in one section and then in another while the fan is running. The smoke will be blown out of the burner opening. A cracked heat exchanger should be replaced, never repaired.

69–1 GAS VENTING:

The purpose of a gas venting system is to provide an escape of the products of combustion to the atmosphere. As the products of combustion leave the furnace heat exchanger, they enter the draft diverter. See Figure 1–156. In the draft diverter the products of combustion are mixed with air from the room. The mixture then passes through the vent pipe to the outside atmosphere. When the vent pipe fails to remove

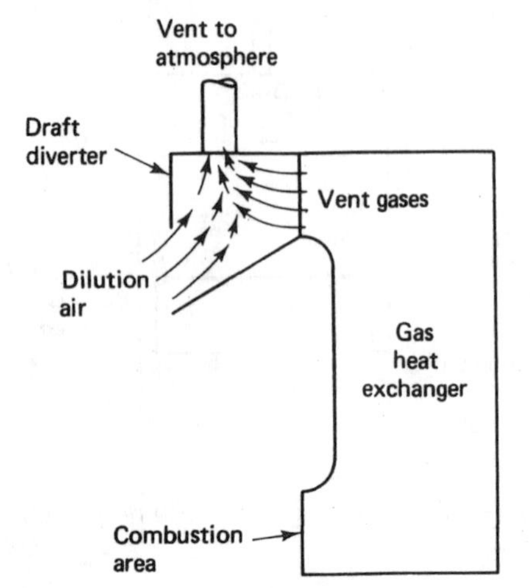

Figure 1–156. Vent gases and dilution air mixing in draft diverter.

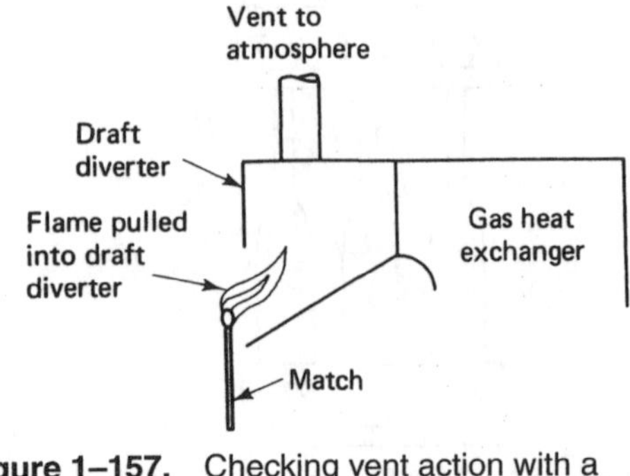

Figure 1–157. Checking vent action with a wooden match.

the products of combustion, they are directed from the draft diverter into the building, thus creating odors in the building and upsetting the combustion process. This condition can be caused when an exhaust fan is installed in the equipment room and is not supplied with enough make-up air. To check the vent action, hold a lighted match close to the draft diverter opening while the furnace is operating. See Figure 1–157. If the vent is operating properly, the match flame will be drawn toward the draft diverter. If the vent is not operating properly, the match flame will be blown away from the draft diverter and will be snuffed out. To correct, remove the obstruction and stop the down draft condition by installing a new vent cap or by locating the vent termination point out of the high static pressure area. When an exhaust fan is used be sure that enough make-up air is provided.

70–1 LIMIT CONTROL:

The limit control is a safety device used to interrupt electrical power to the main gas valve when an overheated condition occurs. It is a bimetal temperature-sensing switch with normally closed contacts that open at approximately 200°F (93°C). See Figure 1–158. To check a limit control, insert a thermometer in the recirculating air stream as close to the sensing element as possible. Reduce the air flow through the furnace and observe the thermometer when the main burner flame goes off. If this temperature is about 200°F (93°C), the limit control is functioning normally. The trouble is something else. If the temperature indicated is not quite 200°F (93°C), make only minor adjustments on the limit control. If the temperature is off by as much as 20°F (9.4°C), replace the limit control. Be sure that the replacement control has the same length sensing element. (See Standard Service Procedure 4-1.)

Figure 1–158. Limit control. (Courtesy of Honeywell, Inc.)

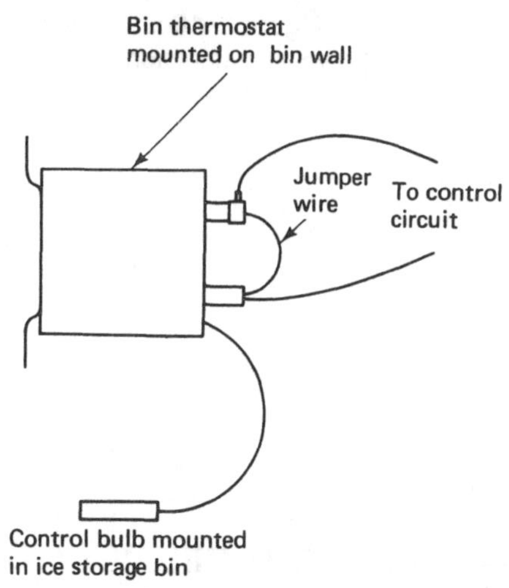

Figure 1–159. Ice machine bin thermostat.

71–1 ICE STORAGE BIN THERMOSTAT:

The ice storage bin thermostat is used to maintain a predetermined level of ice in the storage bin. The sensing element is usually clamped in brackets at the desired ice level. Testing a thermostat is relatively simple. If voltage is supplied to the unit and the ice level in the storage bin is below the sensing bulb on the thermostat with a bin temperature of 45°F (7.2°C) or higher, the unit should operate. If the unit does not operate, place a jumper across the two terminals on the thermostat. See Figure 1–159. If the unit begins operating, the thermostat is bad. Replace the thermostat with an exact replacement. If the unit does not begin operating, the trouble is elsewhere and further checking is necessary. Be sure to check the thermostat setting and adjust before replacing the thermostat.

72–1 POWER RELAY:

The power relay is used in conjunction with the lock-out relay and is energized when an overloaded condition exists when a circuit is completed through terminals 2 and 3 of the lock-out relay. This energizes the power relay coil and opens the contacts on one side of the electrical line. At the same time, another set of contacts are closed, which provides a self-energizing circuit to the power relay coil. This condition will remain until the circuit is interrupted by turning off the electrical power and then turning it back on.

If the condensing unit does not start after interrupting the electrical power, and all other conditions are normal, check between terminals "A" and "B" with a test lamp. A circuit indicates that the self-energizing contacts are stuck closed. Replace the control with an exact replacement.

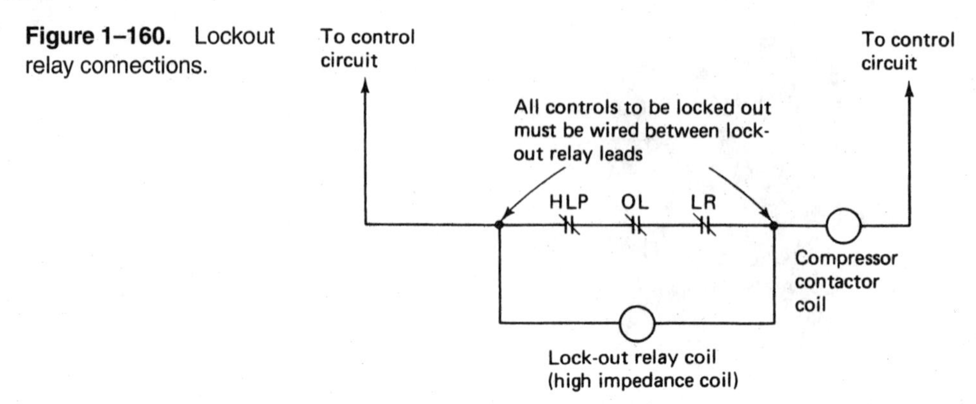

Figure 1–160. Lockout relay connections.

73–1 LOCKOUT RELAY:

Lockout relays are used to interrupt the system electrical circuit if an overload occurs in the auger motor circuit. To reset a lockout relay, manually interrupt the electrical circuit, and then turn it back on. If the condensing unit does not come on after interrupting the electrical circuit, and all other conditions are normal, check for continuity between terminals 2 and 3 with an ohmmeter. Be sure to turn off the electrical power and remove the wires from the terminals. Continuity will indicate that the contacts are stuck closed. Replace the relay with an exact replacement. See Figure 1–160.

74–1 "O" RING SEAL:

The purpose of the evaporator "O" ring is to prevent water leakage where the water line enters the evaporator shell. A pinched or torn "O" ring will allow water to leak, affecting the capacity of the unit and causing a mess on the floor. The manufacturer's recommended procedures should be checked and followed when making repairs involving the "O" ring.

75–1 FLOAT ASSEMBLY:

The float assembly is used to maintain a constant level of water in the reservoir. The level is regulated by a float ball that allows water to enter the reservoir when the level is low and stops the flow of water when the level has reached the desired level. A float ball that becomes partially filled with water will not float on the water and will not stop the flow of water. To check a float ball, remove it from the valve and shake it. If movement is detected inside the ball, replace it with a new one.

75–1A

Water contains minerals that will become solid when the moisture has been evaporated. These minerals will collect in the form of scale on the outside and inside of the valve and will make it inoperative. When this condition occurs, the float valve may be cleaned to resume normal operation. At times the corrosion may be so bad that replacement of the valve is required. Be sure to adjust the float valve assembly to maintain the desired water level in the reservoir. Adjustment can be accomplished by loosening the adjustment clamp screw, making the adjustment, and tightening the screw.

76–1 DRAIN LINE:

The purpose of a drain line is to carry away water that has collected due to condensation or the melting of ice or frost. Because water contains minerals and because water will attract solid particles from the air, a sludge will usually form in the drain pan. This sludge will eventually stop the flow of water through the drain line. When this occurs, the stoppage must be removed. Usually blowing through the drain line with compressed air will remove the stoppage. The drain line should then be flushed with clear water to remove any remaining sludge. The unit may then be put back in operation.

77–1 HOLD-DOWN CLAMPS:

Hold-down clamps are used to hold the auger gear motor rigid and in line with the other components. The vibration set up by the gear motor will sometimes cause these clamps to become loose, and the gear motor will make a chattering noise due to the vibration of the clamps. This condition is both annoying to the user and damaging to the bearings in the auger and motor. To correct, be sure that lock washers are under the nuts; then tighten sufficiently to hold the gear motor rigid. If the clamps are worn they should be replaced.

78–1 END PLAY IN MOTOR:

Electric motors are equipped with fibre washers, or spacers, on the shaft so that alignment of the rotor and the magnetic field may be possible. See Figure 1–161.

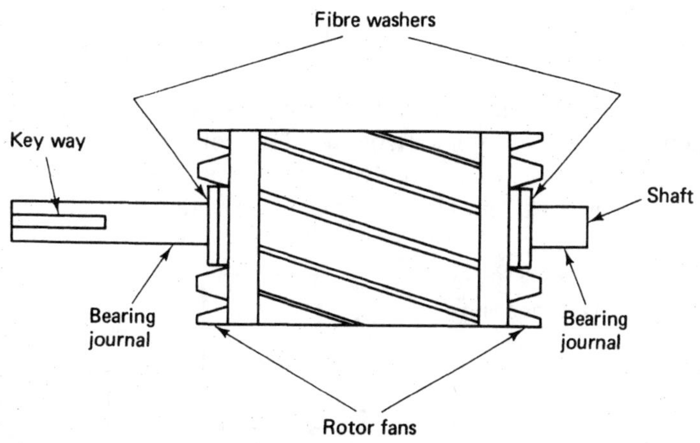

Figure 1–161. Fibre washers on motor shaft.

These washers or spacers become worn and will allow the rotor to float back and forth in the bearings. This floating movement will cause a noise indicating that repair is needed. This repair is usually done in an electric motor repair shop. Therefore, the motor must be removed and delivered. When replacing the motor, follow the manufacturer's recommendations for the proper procedures.

79–1 AUGER GEAR BEARINGS:

Auger gear bearings that become worn must be replaced to keep the unit operating smoothly and economically. When replacing auger bearings, be sure to use the recommended bearings, replacement procedure, and lubrication practices. These procedures will insure the best possible repair and operation.

80–1 WATER LINE AIR LOCK:

A water line air lock occurs only on units that use a gravity-fed water system, such as some flake ice machines. An air lock will stop the flow of water into the evaporator. To remove the air lock, loosen the water line at the lowest point in the system and allow water to flow until a clear, solid stream comes from the tube. Tighten the connection while the water is flowing. The unit should now operate satisfactorily.

81–1 CLEANING EVAPORATOR PLATES:

Evaporator plates will occasionally become covered with scale deposits caused by minerals in the water. This scale acts as an insulator, slowing down the freezing process and preventing a proper harvest of ice. Scale deposits cause the plate to become rough to the touch and may be detected by sliding the fingers across the plate. There are scale removers that may be purchased at most refrigeration supply houses. Follow the instructions on the bottle to obtain the best results. Be sure to use ice machine scale remover. Other scale removers are too dangerous to use on ice machines.

81–1A

When scaling of the evaporator plates becomes excessive due to the high mineral content of the water, a water softener will help solve the problem. Most water softners are installed in the water supply line to the unit. See Figure 1–162. They should be given serious consideration when excessive scaling becomes a problem.

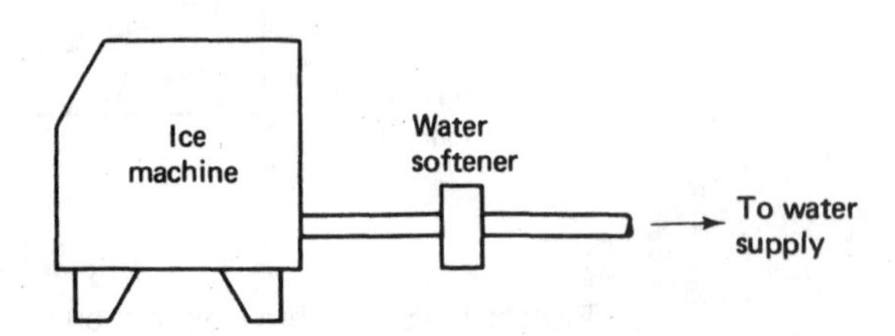

Figure 1–162. Water softener installation on an ice machine.

82–1 ICE BIN SWITCH:

The bin switch is a lever-operated control used to stop unit operation when the ice level reaches a predetermined level in the storage bin. If the ice level is not at the desired level, it should be adjusted. Follow the manufacturer's adjustment procedures.

83–1 CHECK VALVES:

Check valves are used on heat pump systems to help in directing the flow of refrigerant through the proper path for the desired cycle—heating, cooling or defrosting. See Figure 1–163. The check valve will allow the passage of refrigerant either around the flow control device or through the flow control device. Use the following suggestions to check the operation of a check valve:

1. A check valve sticking closed in the outdoor section during the cooling cycle will cause the suction pressure to be low. The superheat will be high on the indoor coil.
2. A check valve sticking open in the indoor section during the cooling cycle will cause high suction pressure, flooding of the compressor, and low superheat on the indoor coil.
3. A check valve sticking closed in the indoor section during the heating cycle will cause the suction pressure to be low, and the superheat to be high on the outdoor coil.
4. A check valve sticking open in the outdoor section during the heating cycle will cause high suction pressure, flooding of the compressor, and a low superheat on the outdoor coil.

If any of the above conditions are found, the check valve should be replaced. Be sure to purge the refrigerant charge or pump the system down to prevent personal

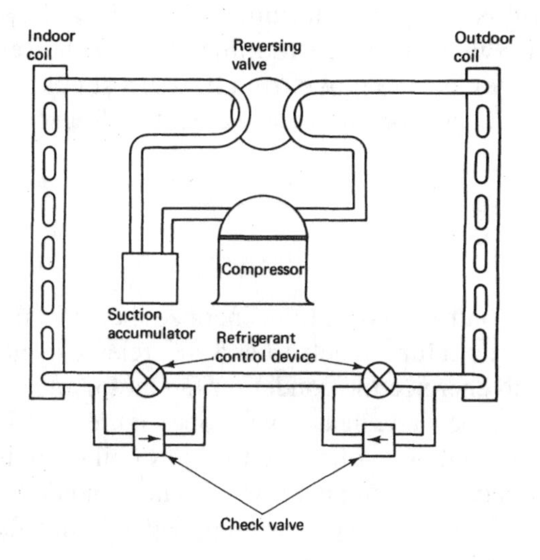

Figure 1–163. Check valve installation on a heat pump system.

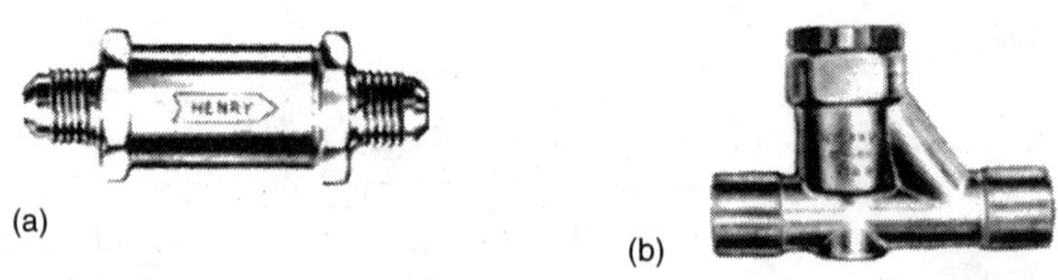

Figure 1–164. Check valves: (a) ball check and (b) swing check. (Courtesy of Henry Valve Co.)

injury. Be sure to install the check valve with the arrow pointing in the direction of refrigerant flow at that point. See Figure 1–164.

84–1 ICE THICKNESS SWITCH:

The purpose of the ice thickness switch is to initiate the harvest cycle when the slab thickness is reached. In the freezing cycle the contacts are open. When the slab becomes thick enough, the contacts close, placing the unit in the harvest cycle. To change the slab thickness, adjust the thumb screw on top of the sensing element. Lowering the sensing element reduces the slab thickness, raising the element increases slab thickness. See Figure 1–165(a) and (b). To check the thickness switch, place a jumper across the switch terminals. The unit should go into the harvest cycle with the terminals jumpered.

85–1 FREEZE RELAY:

The freeze relay is used to control the electrical power to the slab completion relay and the water pump motor. See Figure 1–166. To test the relay, with no power to terminals F_7 and F_8, continuity should exist between terminals F_1 and F_2. When 14 volts of electrical power are supplied to terminals F_7 and F_8, continuity should be found between terminals F_1 and F_3, also between terminals F_4 and F_6.

If the freeze relay coil is energized (terminals F_7 and F_8) beyond three to five minutes, the bimetal switch between terminal F_7 and the relay coil will open, permitting the relay contacts to move from the defrost position to the ice-making position. Any other conditions indicate a defective freeze relay. Replace the relay.

86–1 DEFROST CONTROL RELAY:

The purpose of the defrost control relay is to deenergize the condenser fan motor and energize the hot gas solenoid for the defrost and slab release cycle, on demand from the freeze relay, and to energize the condenser fan motor and deenergize the hot gas solenoid at the end of the slab release cycle. See Figure 1–167. Test the defrost control relay as follows: with no voltage to the relay coil (terminals 7 and 8), there should be continuity between terminals 1 and 2. There should be no continuity between terminals 1 and 3. When 14 volts are supplied to terminals 7 and 8, the

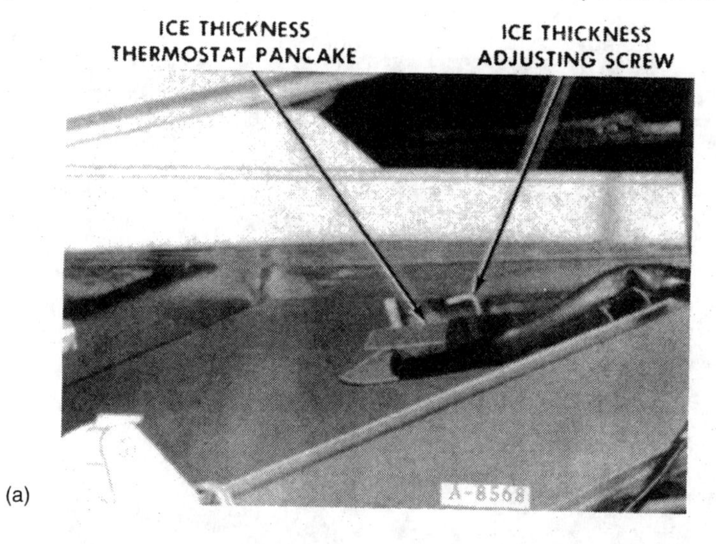

(a)

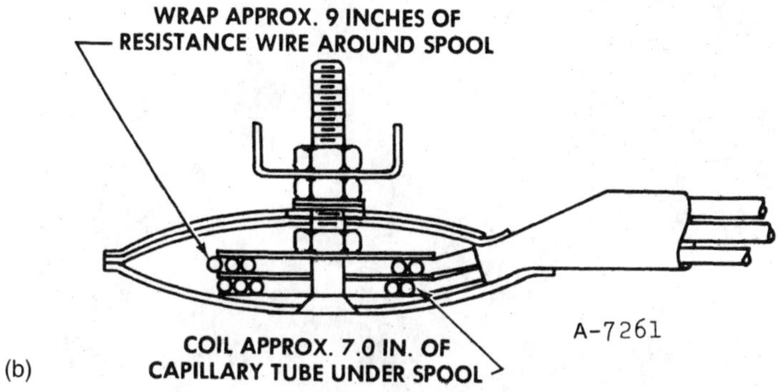

(b)

Figure 1–165. Ice thickness thermostat. (Courtesy of Frigidaire Parts and Service Co., Division of White Consolidated Industries, Inc.)

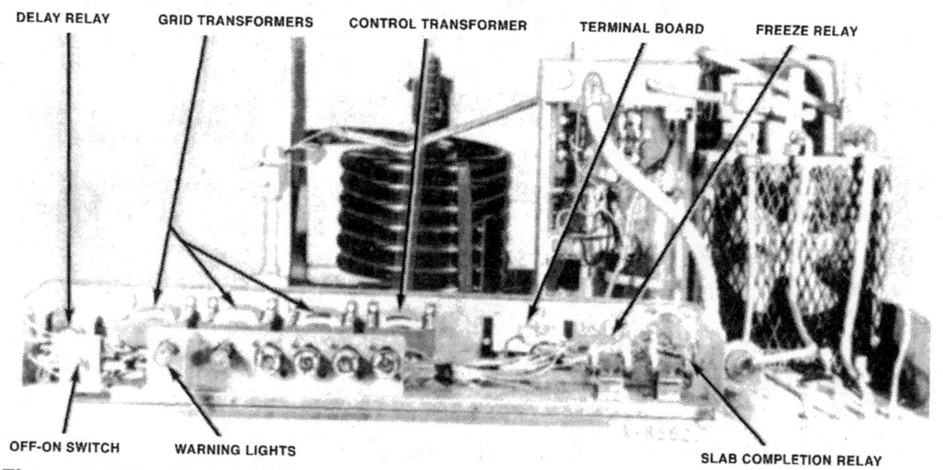

Figure 1–166. Freeze relay location. (Courtesy of Frigidaire Parts and Service Co., Division of White Consolidated Industries, Inc.)

Figure 1–167. Defrost control relay location. (Courtesy of Frigidaire Parts and Service Co., Division of White Consolidated Industries, Inc.)

relay coil will be energized and continuity should exist between terminals 1 and 3. There should be no continuity between terminals 1 and 2 when the relay coil is energized. Any conditions other than these indicate a defective defrost control relay. Replace the relay.

87–1 OPERATION CONTROL SWITCH:

The purpose of the operation control switch is to interrupt electrical power to the unit. To test the operation control switch, jumper the terminals of the switch. If the unit runs, replace the operation control switch.

88–1 EVAPORATOR PLATE:

The purpose of the evaporator plate is to provide a freezing surface for the water on the outside and an evaporating space for the refrigerant on the inside. Occasionally these plates will become worn and rough or warped, preventing removal of the ice slab. In such cases the evaporator plate will need to be replaced. To replace the evaporator plate, the refrigerant must be purged or the unit pumped down, the tubing disconnected and the evaporator removed. When the new evaporator is installed be sure to leak test the system and repair all leaks. Evacuate the refrigerant circuit and install new filter-driers. Recharge the system with the proper amount of refrigerant if the refrigerant was lost. Place the unit back in operation. Check the operation of the unit after 24 hours of operation. Make any necessary adjustments. Sometimes the evaporator plate will become covered with lime from the water and will need cleaning. To clean, see Sections 81-1 and 81-1A.

89–1 WATER DISTRIBUTOR:

The purpose of the water distributor is to distribute the water over the evaporator plate to insure an even ice slab. The water is transferred from the sump to the distributor by the water pump. See Figure 1–168. Occasionally the outlet holes in the header will become clogged with lime or scale, causing uneven distribution of water. This uneven water distribution will result in an uneven slab and will cause slab hang-ups. Physically remove as much of the scale as possible without damaging the distributor. Then apply a cleaner as outlined in Sections 81-1 and 81-1A.

90–1 SYPHONING CYCLE:

The purpose of the syphoning cycle is to drain all the water from the reservoir during the "off" or defrost cycle. As the ice is frozen, some of the minerals return to the reservoir. These minerals must not be allowed to remain in the water or cloudy ice will result. Therefore, the reservoir is emptied and refilled after each freezing cycle. If the syphoning tube becomes clogged or out of adjustment, the minerals

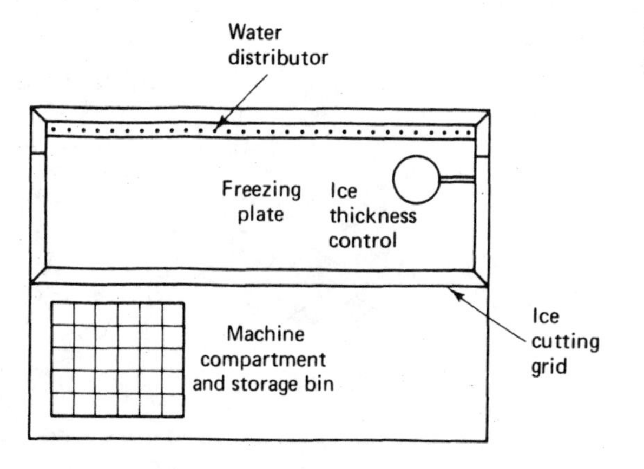

Figure 1–168. Ice machine water distributor location.

will not be removed. The tube must be either cleaned or adjusted to clear up the ice slab. Follow the manufacturer's recommendations for the proper adjustment.

91–1 ICE-CUTTING GRID:

The purpose of the ice-cutting grid is to cut the slab into the desired size cubes. This is done by applying electricity to the high-resistance grid wires. See Figure 1–169. This causes the grid wires to heat slightly and melt thin lines through the ice. Should one of the wires become broken, corroded, or loose, the grid will not operate properly.

To service the ice-cutting grid, use the following procedures:

1. Disconnect the unit from the source of power. Open the front panel to gain access to the grid. Remove cube deflectors, water return troughs, and any other components required.
2. Remove the grid lead-wire plug from its receptical and remove the mercury switch from the switch arm.
3. Remove the grid assembly. Take precautions not to damage the assembly or other components. Ice-cutting grid servicing tools are shown in Figure 1–170.
4. Mark the location of the terminals where the grid wire harness connects. Also, mark which set of wires the silver-plated pins connect. The fastest way to remove the old wires will be to cut each one individually.
5. Insert the new wires from the side of the grid having insulators only and locate the silver-plated connecting pins. After all of the wires are in position, clamp the tool over these wires to keep them securely in their proper location. See Figure 1–171.
6. Turn the grid assembly over to the opposite side and place the insulators and spring clips in position. See Figure 1–171. This tool keeps the spring clip and insulator in position while depressing the clip to insert the pin through the loop of the grid wire.

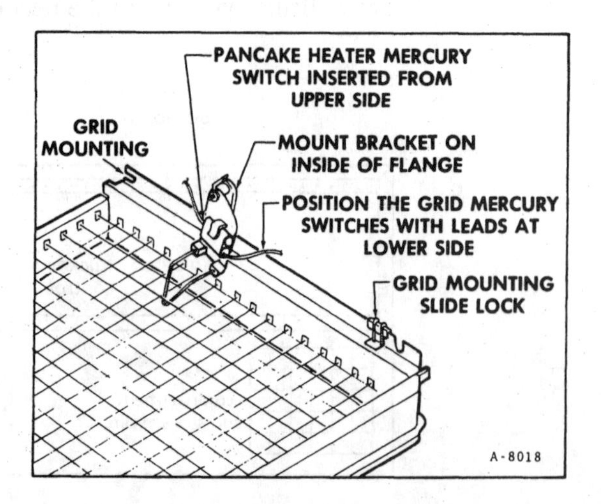

Figure 1–169. Ice-cutting grid. (Courtesy of Frigidaire Parts and Service Co., Division of White Consolidated Industries, Inc.)

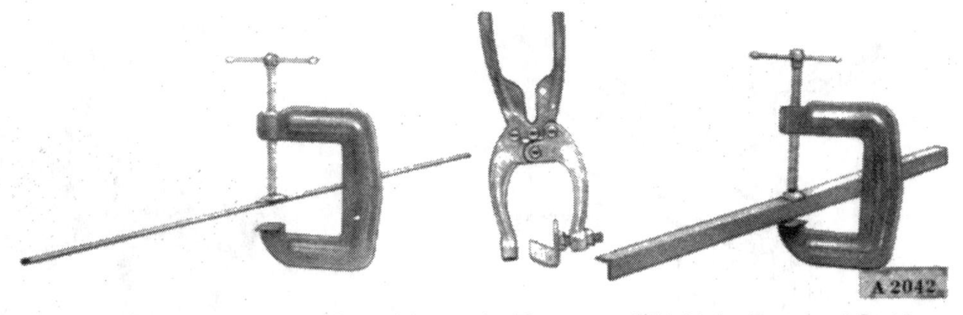

Figure 1–170. Ice cutting grid servicing tools. (Courtesy of Frigidaire Parts and Service Co., Division of White Consolidated Industries, Inc.)

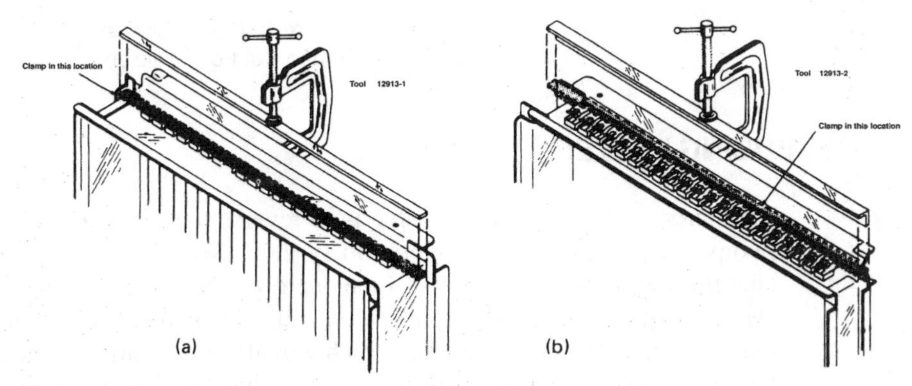

Figure 1–171. Grid wires clamped in place. (Courtesy of Frigidaire Parts and Service Co., Division of White Consolidated Industries, Inc.)

7. Depress two spring clips at a time and insert the connecting pins through the loops at the end of the grid wires. See Figure 1–172. At the same time reconnect the grid wire harness to the proper terminals.
8. Be sure that all wires have uniform tension when the clips are released. A loose wire will give a poor electrical circuit.
9. When old connecting pins are used, be sure that the silver-plating is in good condition and not corroded. Corrosion also can cause a poor electrical circuit, heating of the connection, and melting of the insulators.

91–1A

The purpose of the ice grid fuse is to protect the grid electrical circuit in case of a short. If the fuse is blown, check the grid wires and circuit and remove any shorts that are present. Sometimes a check of the wires to the frame resistance with an ohmmeter is required. Also, check the condition of the fuse holder. If the fuse is not held tightly in the holder, heat will cause the fuse to blow. A faulty fuse holder should be replaced.

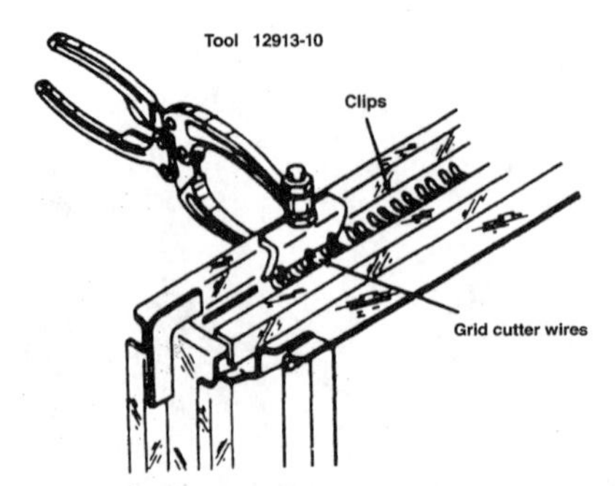

Figure 1–172. Ice cutting grid wire or insulator replacement. (Courtesy of Frigidaire Parts and Service Co., Division of White Consolidated Industries, Inc.)

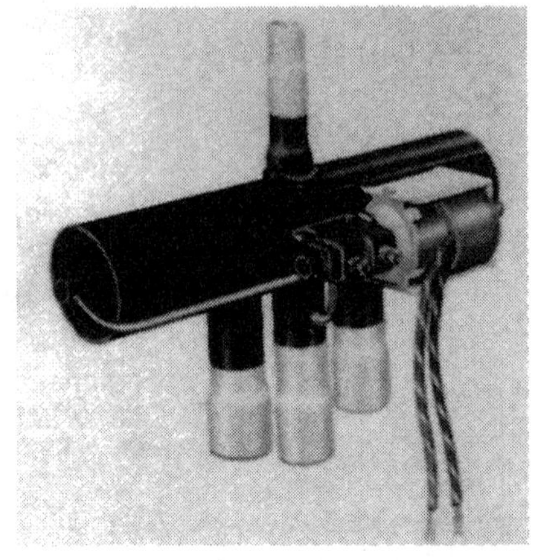

Figure 1–173. Reversing valve. (Courtesy of Ranco Controls Division)

92–1 REVERSING VALVE:

The purpose of the reversing valve is to reverse the refrigerant direction on heat pump, and reverse cycle refrigeration systems. See Figure 1–173. Some systems energize the coil on heating and some energize the coil on cooling. Therefore, the system in question will determine the energized condition of the coil. Any troubles that occur in a heat pump system, which will affect the normal operating pressures, could possibly affect proper shifting of the valve. Some examples of problems are: a refrigerant leak in the system resulting in a shortage, a compressor that is not operating at full capacity, a defective check valve, a damaged valve, or a defective electrical system. Any of these conditions will indicate an apparent malfunctioning valve. The following checks should be made on the system and its components before attempting to diagnose any valve troubles by using the "touch test" method.

1. Inspect the electrical system. This is best done by having the system in operation so that the reversing valve solenoid is energized. While the unit is operating, remove the lock nut from the solenoid cover. Pull the solenoid coil part way off the stem. There should be a magnetic force trying to hold the coil on the stem. Be sure that electric power is applied to the solenoid. If the coil is moved off the stem, the valve will return to the deenergized position, which will be indicated by a clicking sound. A clicking sound will also be heard when the coil is replaced on the stem.
2. Inspect the reversing valve for physical damage. Check the solenoid coil for deep scratches, dents, and cracks.
3. Check the operation of the system against the equipment manufacturer's recommendations. Make any corrections indicated.

If all of these checks prove that the system is operating properly, perform the "touch test" on the reversing valve, using Table 1–3. The "touch test" is performed

Table 1–3. Touch Test Chart

Valve Operating Condition	DISCHARGE TUBE from Compressor 1	SUCTION TUBE to Compressor 2	Tube to INSIDE COIL 3	Tube to OUTSIDE COIL 4	LEFT Pilot Back Capillary Tube 5	RIGHT Pilot Front Capillary Tube 6	Possible Causes	Corrections
NORMAL OPERATION OF VALVE								
Normal Cooling	Hot	Cool	Cool, as (2)	Hot, as (1)	*TVB	*TVB		
Normal Heating	Hot	Cool	Hot, as (1)	Cool, as (2)	*TVB	*TVB		
MALFUNCTION OF VALVE								
Valve will not shift from cool to heat	Check electrical circuit and coil						No voltage to coil. Defective coil.	Repair electrical circuit. Replace coil.
	Check refrigeration charge						Low charge. Pressure differential too high.	Repair leak, recharge system. Recheck system.
	Hot	Cool	Cool, as (2)	Hot, as (1)	*TVB	Hot	Pilot valve okay. Dirt in one bleeder hole.	Deenergize solenoid, raise head pressure, reenergize solenoid to break dirt loose. If unsuccessful, remove valve, wash out. Check on air before installing. If no movement, replace valve, add strainer to discharge tube, mount valve horizontally.
							Piston cup leak	Stop unit. After pressures equalize, restart with solenoid energized. If valve shifts, reattempt with compressor running. If still no shift, replace valve.
	Hot	Cool	Cool, as (2)	Hot, as (1)	*TVB	*TVB	Clogged pilot tubes.	Raise head pressure, operate solenoid to free. If still no shift, replace valve.
	Hot	Cool	Cool, as (2)	Hot, as (1)	Hot	Hot	Both ports of pilot open. (Back seat most port did not close.)	Raise head pressure, operate solenoid to free partially clogged port. If still no shift, replace valve.
	Warm	Cool	Cool, as (2)	Warm, as (1)	*TVB	Warm	Defective Compressor.	
Start to shift but does not complete reversal	Hot	Warm	Warm	Hot	*TVB	Hot	Not enough pressure differential at start of stroke or not enough flow to maintain pressure differential.	Check unit for correct operating pressures and charge. Raise head pressure. If no shift, use valve with smaller ports.
							Body damage.	Replace valve.
	Hot	Warm	Warm	Hot	Hot	Hot	Both ports of pilot open.	Raise head pressure, operate solenoid. If no shift, replace valve.

NOTES:
*Temperature of Valve Body.
**Warmer than Valve Body.

(continued)

Table 1–3. (cont.)

Valve Operating Condition	DISCHARGE TUBE from Compressor 1	SUCTION TUBE to Compressor 2	Tube to INSIDE COIL 3	Tube to OUTSIDE COIL 4	LEFT Pilot Back Capillary Tube 5	RIGHT Pilot Front Capillary Tube 6	Possible Causes	Corrections
Start to shift but does not complete reversal	Hot	Hot	Hot	Hot	*TVB	Hot	Body damage.	Replace valve.
							Valve hung up at mid-stroke. Pumping volume of compressor not sufficient to maintain reversal.	Raise head pressure, operate solenoid. If no shift, use valve with smaller ports.
	Hot	Hot	Hot	Hot	Hot	Hot	Both ports of pilot open.	Raise head pressure, operate solenoid. If no shift, replace valve.
Apparent leak in heating	Hot	Cool	Hot, as (1)	Cool, as (2)	*TVB	**WVB	Piston needle on end of slide leaking.	Operate valve several times then recheck. If excessive look, replace valve.
	Hot	Cool	Hot, as (1)	Cool, as (2)	**WVB	**WVB	Pilot needle and piston needle leaking.	Operate valve several times then recheck. If excessive leak, replace valve.
Will not shift from heat to cool	Hot	Cool	Hot, as (1)	Cool, as (2)	*TVB	*TVB	Pressure differential too high.	Stop unit. Will reverse during equalization period. Recheck system.
							Clogged pilot tube.	Raise head pressure, operate solenoid to free dirt. If still no shift, replace valve.
	Hot	Cool	Hot, as (1)	Cool, as (2)	Hot	*TVB	Dirt in bleeder hole.	Raise head pressure, operate solenoid. Remove valve and wash out. Check on air before reinstalling, if no movement, replace valve. Add strainer to discharge tube. Mount valve horizontally.
	Hot	Cool	Hot, as (1)	Cool, as (2)	Hot	*TVB	Piston cup leak.	Stop unit, after pressures equalize, restart with solenoid deenergized. If valve shifts, reattempt with compressor running. If it still will not reverse while running, replace valve.
	Hot	Cool	Hot, as (1)	Cool, as (2)	Hot	Hot	Defective pilot.	Replace valve.
	Warm	Cool	Warm, as (1)	Cool, as (2)	Warm	*TVB	Defector compressor.	

NOTES:
*Temperature of Valve Body.
**Warmer than Valve Body.
VALVE OPERATED SATISFACTORILY PRIOR TO COMPRESSOR MOTOR BURN OUT—caused by dirt and small greasy particles inside the valve. To CORRECT: Remove valve, thoroughly wash it out. Check on air before reinstalling, or replace valve. Add strainer and filter-dryer to discharge tube between valve and compressor.

simply by feeling the temperature differences on the six tubes on the valve and comparing these differences. Refer to the table for possible causes and corrections.

When replacing the reversing valve, the refrigerant will need to be purged. Follow the manufacturer's recommendations when installing a new valve. Evacuate the system, install new filter-driers, and recharge the system according to the manufacturer's recommendations.

93–1 DEFROST CONTROLS:

Heat pump defrost controls are used to detect ice and frost on the outdoor coil during the heating cycle. The most popular type of defrost control is the time temperature method. See Figure 1–174. This method means that both the time and temperature sections must demand defrost before the defrost cycle can be initiated. When the defrost cycle is initiated, the reversing valve changes and the outdoor fan motor stops running. The clock motor operates only when the compressor is running. The timer can be set to initiate the defrost cycle on 30, 60, and 90-minute intervals. If it is found that the 90-minute cycle allows too much frost on the outdoor coil, then the timer can be adjusted to provide the proper amount of defrost time. The defrost can be initiated only by demand from both the timer and the temperature sections of the defrost control. The defrost control temperature will initiate the defrost cycle at a temperature of approximately 30°F (21.1°C) and will terminate the defrost cycle at approximately 60°F (15.6°C).

The timer motor is operated by 230 volts and operates only when the compressor is operating. The temperature portion of the control is attached to a return bend on the outdoor coil. See Figure 1–175. The bulb is insulated to insure that the ambient air does not affect the bulb. To check a time-temperature defrost control, be sure that the outdoor coil is cold enough to need defrosting; then use the following steps:

1. Place a jumper between terminals 2 and 3 on the clock timer. See Figure 1–176. If the reversing valve changes, the clock is not operating. Check to be certain

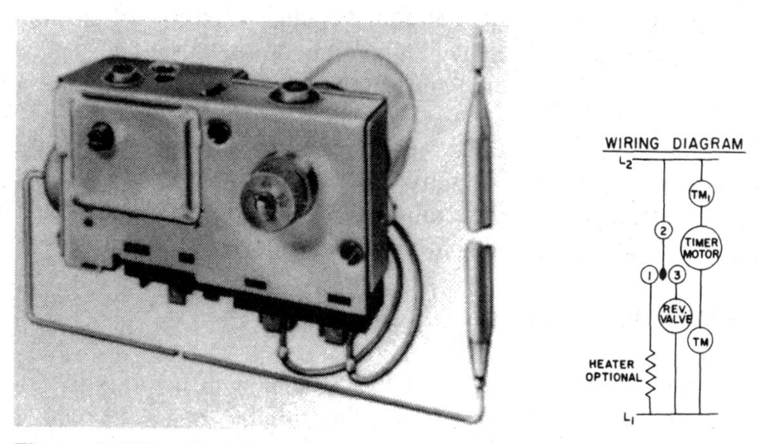

Figure 1–174. Time-temperature defrost control and schematic. (Courtesy of Ranco Controls Division)

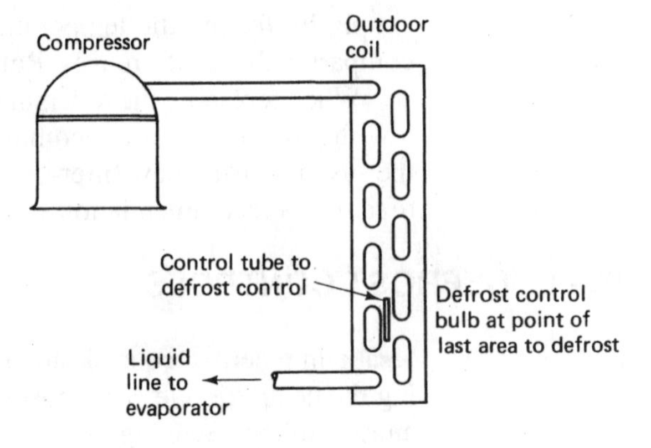

Figure 1–175. Installation of temperature bulb for a defrost control.

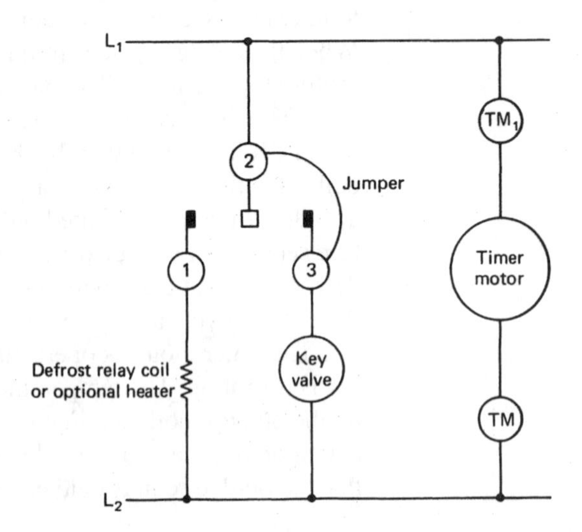

Figure 1–176. Jumper between terminals 2 and 3 on defrost timer.

that power is supplied to the clock motor terminals. If power is found, the motor is defective and must be replaced.

2. If the reversing valve does not change, jump the terminals of the defrost thermostat while the jumper remains on terminals 3 and 4. If the reversing valve changes, the defrost overload is defective and must be replaced.

If the additional jumper does not cause the reversing valve to change, the trouble is not in the defrost control. The reversing valve solenoid coil, the reversing valve, and the defrost relay must be checked to find the trouble. Be sure that the specified voltage is applied to these controls.

94–1 AIR BYPASSING COIL:

Air bypassing a coil will reduce the efficiency of the unit. See Figure 1–177. The greater the amount of air bypassing the coil, the greater the reduction in efficiency. The opening through which the air bypasses must be closed so that peak efficiency can be obtained. Air bypassing an evaporator causes a low suction pressure. Air bypassing a condenser causes a high discharge pressure.

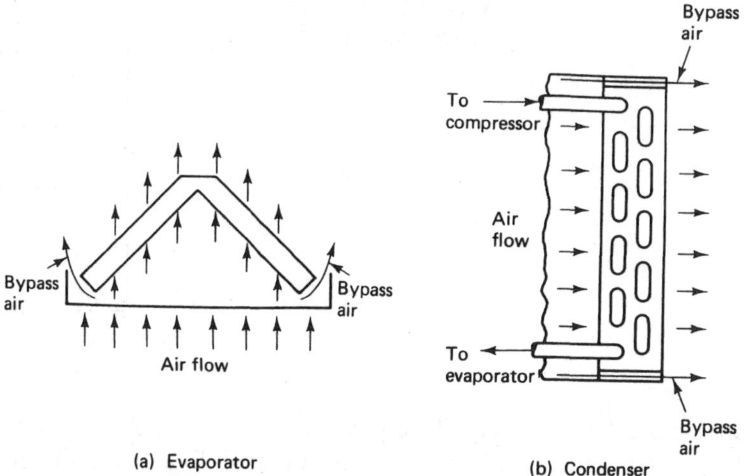

Figure 1–177. Air bypassing coils.

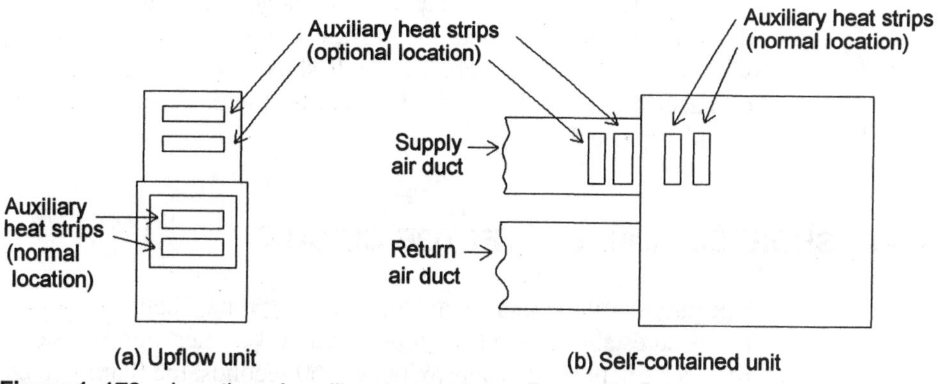

Figure 1–178. Location of auxiliary heat strips on heat pump systems.

95–1 AUXILIARY HEAT STRIP LOCATION:

The auxiliary heat strips must be installed downstream of the indoor coil. See Figure 1–178. If the heat strips are installed upstream of the indoor coil, the heat pump will act as a reheat system. This hot air entering the indoor coil during the heating cycle will cause the discharge pressure and temperature to be excessively high, thus causing a high compression ratio on the compressor, one that will probably cause permanent damage and will require compressor replacement.

96–1 DEFROST RELAY:

The purpose of the defrost relay is to cause the reversing valve to change to the cooling position, stop the outdoor fan motor, and energize the auxiliary heaters

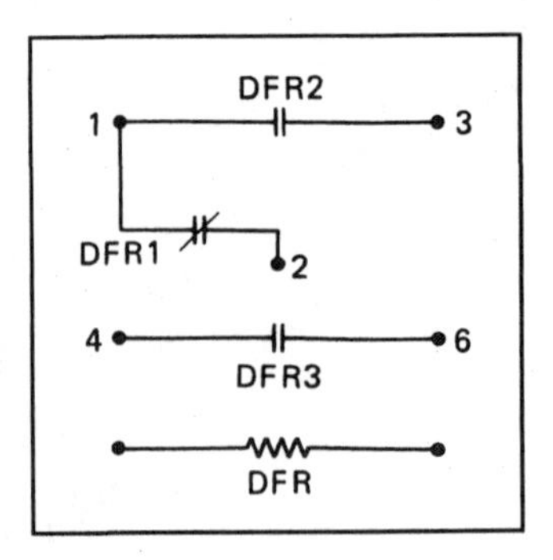

Figure 1–179. Internal schematic of defrost relay.

during the defrost cycle. The coil voltage is usually 230 volts. See Figure 1–179. This relay is energized by the defrost control. Because some manufacturers prefer to energize the reversing valve during the heating cycle and some during the cooling cycle, the wiring diagram for the particular unit in question should be used.

If the relay is energized and a click is heard when the defrost control demands defrost but the system does not shift, check the voltage across each set of contacts with a voltmeter. Open contacts will show full line voltage across them. Closed contacts will show no line voltage across them. If conditions other than these are found, replace the relay.

97–1 SHORTED FLAME DETECTOR CIRCUIT:

The purpose of the flame detector is to determine whether or not a satisfactory flame has been established in the firebox of an oil burner unit. If a satisfactory flame has not been established in approximately 60 seconds, the flame detector will shut down the oil burner. See Figure 1–180. To check to determine whether the flame detector is defective, improperly positioned, or dirty, use the following procedure:

1. Remove the cad cell from the flame detector socket assembly. To remove cell, push in and turn counterclockwise. If there is any soot accumulation on the face, remove it with a soft cloth. Replace the cell in the socket assembly and start the burner. If the burner runs and does not lock out, the cad cell was dirty.
2. If the primary control is locked out on safety or if the cad cell face was found clean (in step 1), install a new cad cell and start the burner. If the burner runs and does not lock out, the cad cell was defective.
3. If the primary control locks out on safety (in step 2), the flame detector is not properly positioned, or there is an open circuit between the contacts on the cad cell and the socket assembly. Follow the manufacturer's recommendations in repositioning the flame detector.

4. If the burner will not start, the flame detector leads or the cad cell may be shorted, or there could be a false light affecting the cad cell. Use the following procedure for testing:
   a. Remove one of the leads from FD-FD terminals.
   b. Turn on the electricity. If the burner starts, the flame detector is defective and must be replaced.
   c. If the burner does not start, the trouble is elsewhere and other checks must be made.

The proper application, location, and mounting of flame detector units are determined by the equipment manufacturer. Every service technician should be familiar with the following factors, which are the basis for flame detector location:

1. The flame detector must be located so that it's face temperature never exceeds 140°F (60°C). Therefore, it must be mounted in a cool location.
2. The flame detector must have a direct view of the burner flame. On some installations, it might be necessary to drill holes in static discs to provide a clear view for the flame detector.

The flame detector should not be adjusted to react to reflected light because the amount of reflection changes after the burner is in operation a short period of time. Some equipment manufacturers blacken the inside of the firebox to simulate actual field conditions when designing new installations.

98–1 SHORTED FLAME DETECTOR LEADS:

Shorted leads will bypass the flame detector unit and will prevent the burner from starting. The shorted leads must be found, and must be separated and insulated. Precautions should be made to prevent the return of this problem. It may be necessary to install new wiring, using rubber grommets and other protection at points where wear of the wiring may occur.

99–1 FRICTION CLUTCH:

The purpose of the friction clutch in a flame detector is to prevent damage to the detector due to the stack temperature. When a flame has been established, the rise in stack temperature causes the thermal element on the stack control to expand, moving the stack switch arm down, closing the stack switch, and shunting out the safety switch heater. The stack switch arm moves on down to the lower (hot) stop. The additional expansion of the thermal element is absorbed by the friction clutch. After the thermostat stops the burner, the drop in stack temperature causes the thermal element to contract, opening the stack switch and returning the stack arm to the

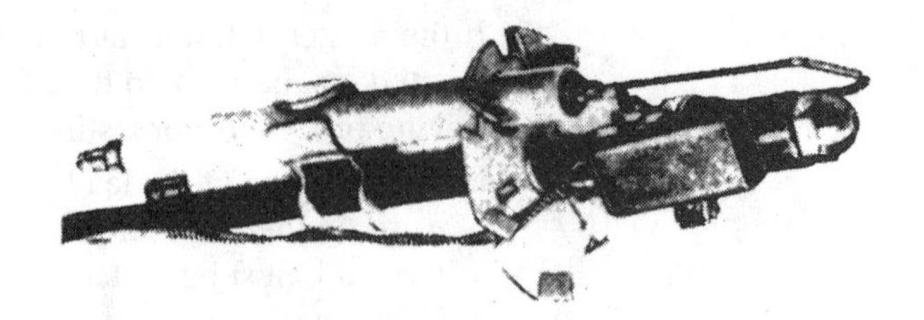

Figure 1–180. Flame detector mounted on nozzle, General Controls CT-99 Perfxray. (Courtesy of ITT General Controls Division)

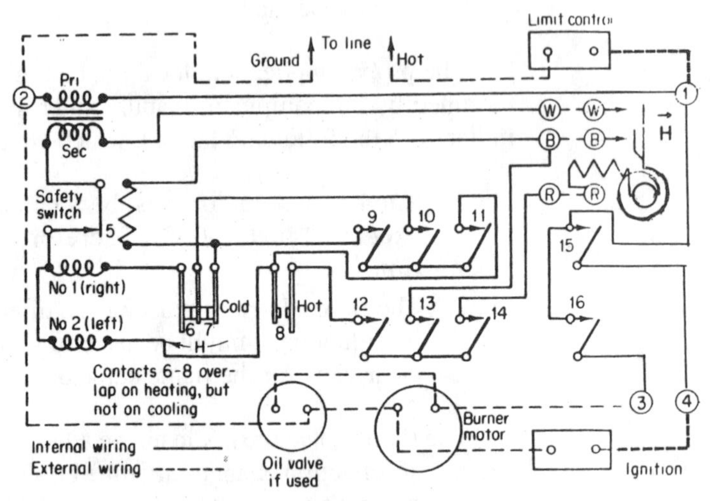

Figure 1–181. Schematic circuit of Honeywell RA117A. (Courtesy of Honeywell, Inc.)

starting position. Any further contracting of the thermal element is absorbed by the friction clutch. A faulty friction clutch will prevent both the normal operation of the flame detector and starting of the oil burner. The detector, or the element, must be replaced to correct the problem.

100–1 HOT AND COLD FLAME DETECTOR CONTACTS:

The hot flame detector contacts are used to shunt out the heater element in the safety switch, and the cold contacts are used to control the ignition process. In operation, when the thermostat calls for heat, the relay coils are energized and the right-hand relay pulls in. See Figure 1–181. The ignition is started when contact 15 closes. The left-hand relay is energized when contacts 9, 10, and 11 are energized. Contact 16 starts the oil burner motor. The hot contact 8 is closed, shunting out the heater element in the safety switch as the stack temperature rises. A further increase in the stack temperature causes contacts 7 and 6 to break, in that order, thus interrupting the circuit to the right-hand relay, dropping it out, and shutting off the ignition. The burner is now operating normally.

If any one of these contacts becomes dirty, pitted, stuck, or inoperative, it must be serviced before normal operations will resume. Dirty contacts may be cleaned by placing a business card between them and pulling it back and forth until the contacts are clean. Pitted contacts may be filed lightly with a contact file. If the con-

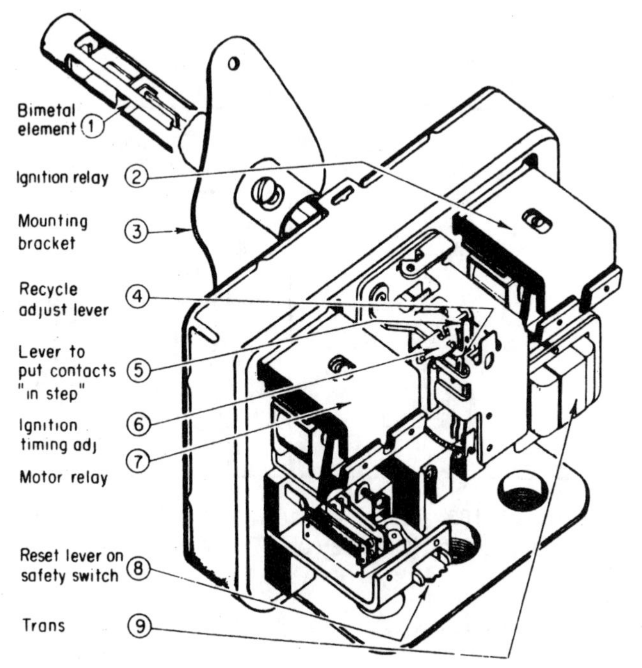

Figure 1–182. Honeywell RA117A stack-mounted Protectorelay. (Courtesy of Honeywell, Inc.)

tacts are severely pitted, the control should be replaced or the trouble will reoccur. Stuck or inoperative contacts require that the control be replaced.

101–1 FLAME DETECTOR BIMETAL:

The purpose of the bimetal in a flame detector is to sense the stack temperature and cause the proper relays and contacts to act accordingly. See Figure 1–182. Should this bimetal become covered with soot or warped, the flame detector will not respond properly. If the bimetal is covered with soot, the safety switch will lock the burner out before the detector has time to switch into the normal operating position. And if the bimetal is warped, it will not permit proper operation of the flame detector and must be replaced. It is generally best to replace the complete control rather than simply replacing the bimetal. A dirty bimetal must be cleaned in order to resume normal operation. A good cleaning solvent such as kerosene or gasoline may be used for cleaning. Use caution not to twist or warp the bimetal.

102–1 PRIMARY CONTROL:

The purpose of the primary control is to control the operation of the oil burner and the ignition procedure. See Figure 1–183. After the control has been properly installed and wired, the burner should operate as follows: on demand from the thermostat, the combustion relay is closed, starting the oil burner motor and the ignition process. The safety-switch heater is also energized at the same time.

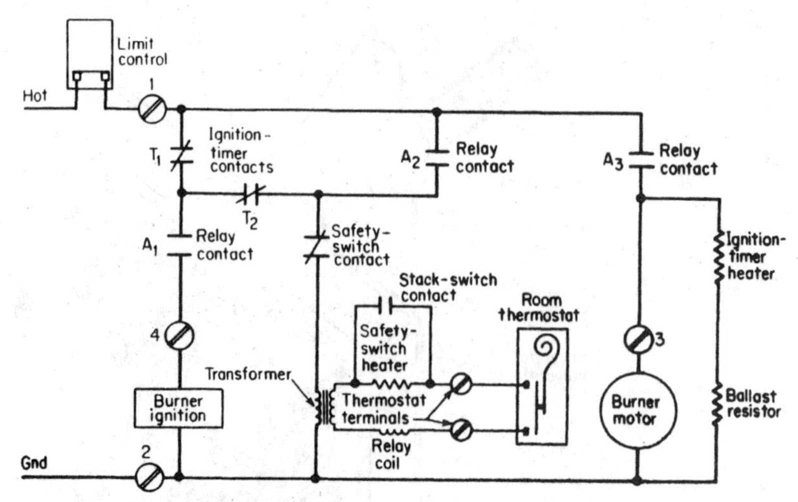

Figure 1–183. Schematic wiring diagram of General Controls 5520-D21 primary control. (Courtesy of ITT General Controls Division)

Figure 1–184. Air leak in vent pipe.

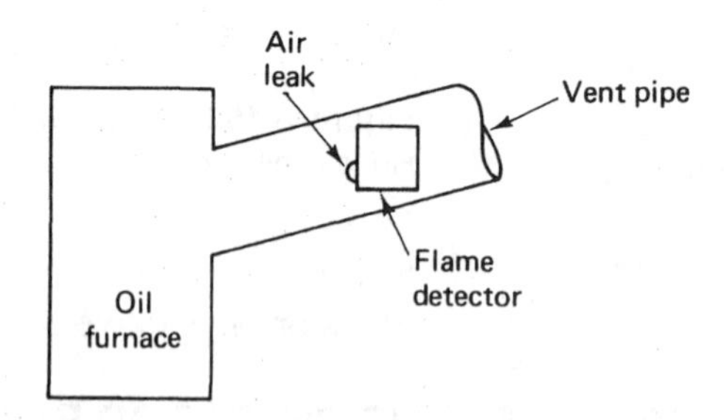

When a predetermined period of time has passed, the ignition timer stops the ignition process. If the safety-switch heater is not shunted out by the stack switch, a safety shutdown will occur. To restart the unit, reset, the safety switch.

Any malfunction in the primary control could be due to a defective internal circuit or dirty combustion relay contacts. A defective internal circuit will require a new primary control. Dirty combustion relay contacts may be cleaned by moving a business card back and forth between them. File pitted contacts lightly or the contacts will be ruined, requiring replacement of the control.

Air leaking into the vent pipe around the flame detector mount will keep the bimetal cool and will cause the control to lock out on safety. Sealing any leaks here should be done with asbestos-type gaskets because of the high temperatures encountered. See Figure 1–184.

103–1 OIL BURNER BLOWER WHEEL:

The purpose of the blower on an oil burner is to furnish the combustion air to the flame. See Figure 1–185. A binding blower wheel will reduce or completely stop the combustion air. This binding could be due to a loose set screw, allowing the

Figure 1–185. Location of oil burner blower wheel. (Courtesy of Lennox Industries, Inc.)

blower to get out of alignment, or it may be due to dry or worn bearings. To test for a binding blower wheel, simply rotate the blower by hand. If the blower rubs on the fan scroll, it needs adjustment to clear all components. If the blower comes to a sudden stop after manual rotation, rather than coasting to a stop, the trouble is in the bearings. The bearing should be lubricated and tested for wear. If worn, they should be replaced.

104–1 FUEL PUMP:

The purpose of the fuel pump is to supply the fuel oil to the burner with sufficient pressure to atomize the fuel for combustion. There are basically two different pressures used to atomize the fuel: low pressure (100–200 psi) and high pressure (200–300 psi). A fuel pump is adjustable, within these limits, to provide the prescribed pressure for the burner. See Figure 1–186. The method and location of the pressure adjustment will vary with the type and make of the pump. Refer to the manufacturer's specifications for the pump in question. These pumps are usually very trouble-free. However, occasionally a bearing will seize, due to contaminants rendering a pump inoperative. In such cases, the pump must be replaced. Be sure to follow the manufacturer's installation and adjustment recommendation. Also, occasionally a drive belt or a drive coupling will break and need to be replaced. Fuel pumps are usually equipped with strainers to prevent the entrance of foreign matter into the pump. These strainers will occasionally become plugged and will not allow the pump to produce the desired pressure. Plugged strainers require cleaning. Cleaning the strainers is a relatively simple task. Stop the pump, close the fuel line valve, disconnect the inlet fuel line, and remove the strainer. Clean the strainer with gasoline or some other suitable solvent. Reinstall the strainer, reconnect the fuel line, and bleed any air from the line by opening the bleeding valve and allowing a small amount of fuel and air to escape before tightening the bleeding valve. Place the unit back in operation.

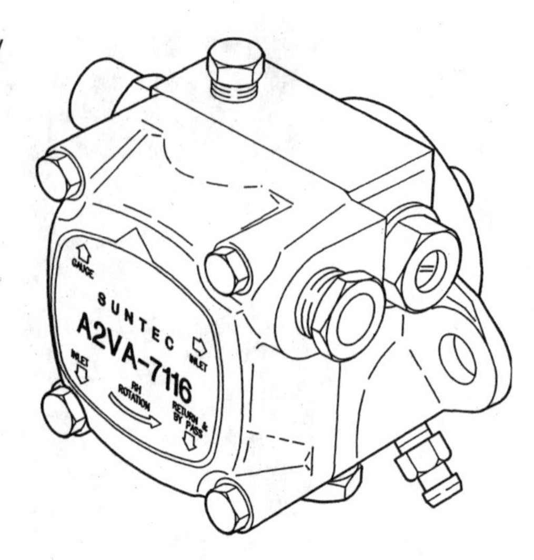

Figure 1–186. Oil burner pump. (Courtesy of Suntec Industries Incorporated)

105–1 OIL BURNER ADJUSTMENTS:

To properly set an oil burner, a CO_2 analyzer and a smoke gun must be used. Do not set oil burners by eye or estimation. When instruments are not used, the burner will be set with too high a CO_2, and smoke will result, which is usually accompanied by a noisy fire and carbon build-up in the heat exchanger. A proper setting with instruments will result in a quiet and clean flame at 8 to 10% CO_2, with zero to a slight trace of smoke.

Before making any final burner adjustments, allow the burner to operate continuously for five to ten minutes. This operation will purge the fuel lines and will level out the combustion process. Take the readings and make the adjustments as follows.

1. Make sure that the inspection door is closed tightly and that any fitting joints between the furnace and the point where the CO_2 and smoke readings are taken are tightly sealed or taped. An air leak at the inspection door or fitting will cause a false CO_2 reading because of diluting the flue gases with air.
2. Punch a 5/16 in. (7.94 mm) diameter service hole in the flue outlet between the furnace and the draft control. The draft readings, the CO_2, and the smoke test should be taken at this point.
3. Adjust the barometric draft control in the stack for the correct draft. See Figure 1–187. The draft should be measured with a draft gauge and set for .03 in. to .035 in. water column. See Figure 1–188.
4. Loosen the air control locking screw, and rotate the air control cover until a clean flame is produced. See Figure 1–189.
5. Take a CO_2 reading at the service opening in the stack using a CO_2 indicator. See Figure 1–190. Follow the instructions with the CO_2 indicator. If the CO_2 reading is between 8 and 10%, the setting is correct. If not, rotate

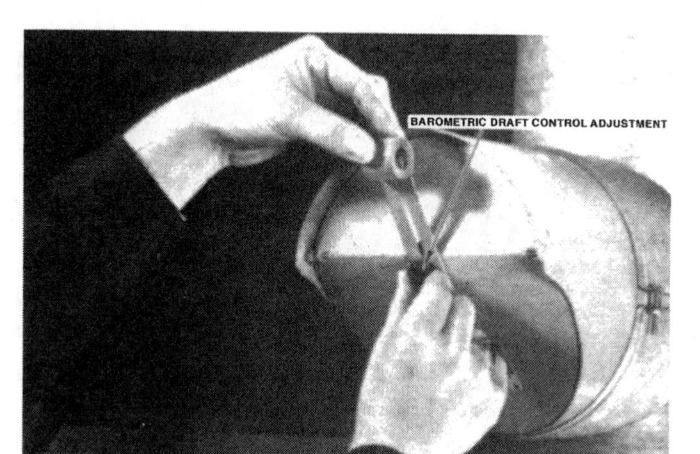

Figure 1–187. Barometer draft control. (Courtesy of Lennox Industries, Inc.)

Figure 1–188. Measuring flue draft on oil burner. (Courtesy of Lennox Industries, Inc.)

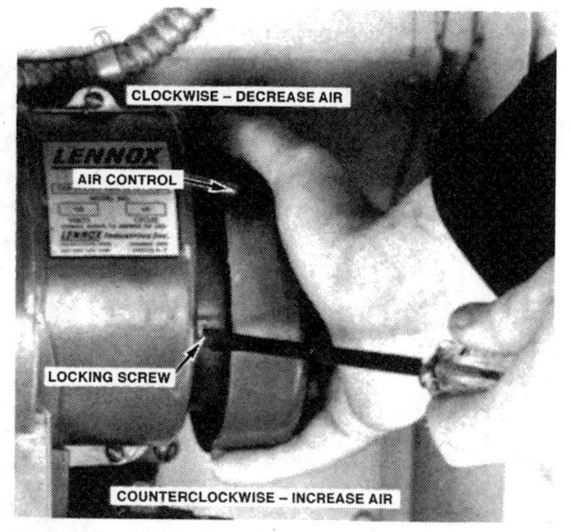

Figure 1–189. Adjusting an oil burner air control cover. (Courtesy of Lennox Industries, Inc.)

the air control cover and recheck until the CO_2 readings fall within the 8 to 10% range. Tighten the air control locking screw.

6. Take a smoke reading in the same service hole used for taking the CO_2 reading. See Figure 1–191. Use a standard smoke tester such as the Bacharach True Spot Tester. The smoke reading obtained at an 8 to 10% CO_2 reading will generally be between zero and a number one spot. The smoke reading should never be more than a number one spot. If the smoke test provides greater than a number one spot, it could be caused by a poor nozzle or by air leakage at the inspection door or fitting. Occasionally it could be caused by a difference in oil or by some unusual condition created during the original installation. Rotate the combustion air control cover until a number one spot or less is obtained; then recheck the CO_2 to

Figure 1–190. Checking stack CO_2 content. (Courtesy of Lennox Industries, Inc.)

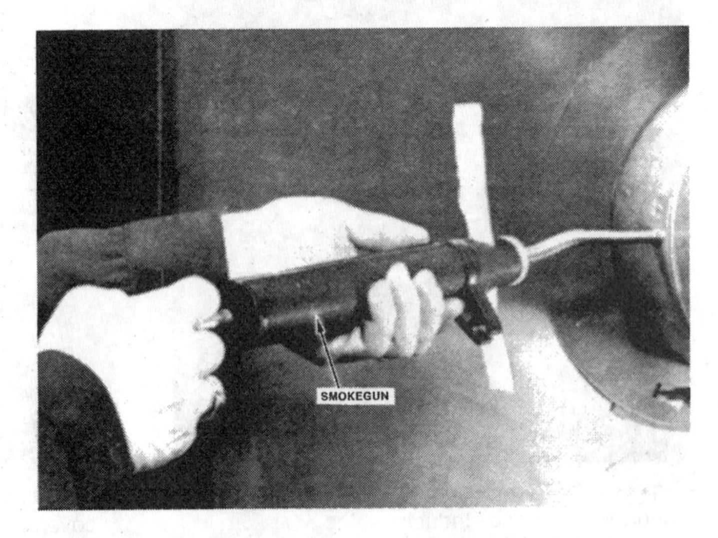

Figure 1–191. Taking a smoke reading of an oil burner. (Courtesy of Lennox Industries, Inc.)

make sure it is 8 to 10%. If the CO_2 is less than 8%, look for air leakage, a bad nozzle, or improper settings of the burner gun assembly. Air leakage can often be determined by taking a CO_2 reading both at the stack and over the flame.

If the stack CO_2 is more than 1/2% below the CO_2 reading taken over the flame, it indicates an excess air leakage. Any air leakage will lower the furnace efficiency and should be found and corrected.

106–1 OIL TANKS:

Oil tanks are used to store the fuel oil until it is needed. These tanks may be installed inside, outside above ground, or outside below ground level. When fuel oil has been stored for sometime, condensation will occur. This condensation will form in the bottom of the tank in the form of water. This water may be drawn into

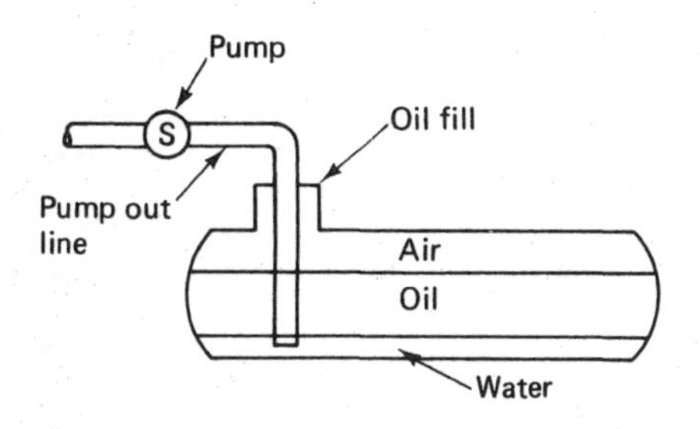

Figure 1–192. Pumping water out of an oil storage tank.

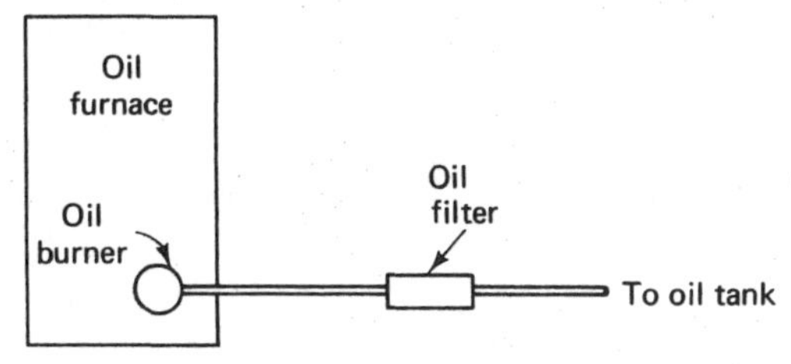

Figure 1–193. Oil filter location.

the oil burner and can upset the combustion process and at times cause a flameout. When this condition exists, the water must be removed from the tank. This may be accomplished by lowering a line into the tank through the oil and into the water collected at the bottom of the tank. See Figure 1–192. A pump is then connected to the line and the water pumped out. Sometimes it may be necessary to completely drain the tank to remove all the water.

107–1 FUEL OIL LINES:

Fuel oil lines are used to move the oil from the tank to the oil burner. There are both one-pipe and two-pipe systems in use. The number of pipes is dependent upon the type of fuel unit used. There is usually a filter installed in the supply line to prevent any foreign matter entering the fuel unit. See Figure 1–193. These fuel lines should be protected to prevent damage such as a kinked line, a dented line, or a leak.

The fuel supply line is used to move the oil from the tank to the burner and usually has a reduced pressure inside it. Should this line or filter become kinked or clogged, the fuel flow will be reduced, and this will affect the operation of the burner. The kink must be removed or a new filter installed. A leak in the supply line will allow air to enter the system and will cause air blockage of the oil pump. The leak must be found and repaired. Usually a leak will be indicated by a slight trace of oil around the leak. The air must be purged from the system before normal operation can be resumed.

The fuel return line carries any excess oil back to the oil tank. This line is under pump pressure. Therefore, a leak in this line will allow the fuel oil to escape, causing a mess. A restriction will cause a reduced flow to the tank, thereby putting the rest of the system under excessive pressure and affecting normal operation. Any leaks or restrictions in this line must also be removed.

108–1 OIL BURNER NOZZLES:

One of the functions of a nozzle is to atomize the fuel to break it up into tiny droplets that can be vaporized in a short period of time. The atomizing nozzle performs three basic and vital functions for an oil burner: atomizing, metering, and patterning. See Figure 1–194.

Figure 1–194. Typical high pressure nozzle. (Courtesy of Delavan, Inc.)

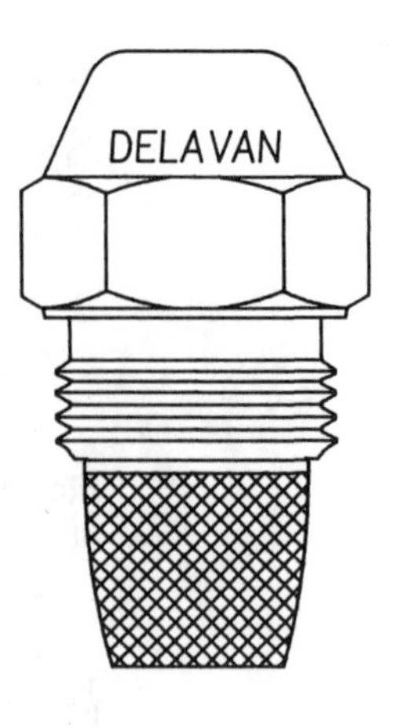

To accomplish proper atomization of the fuel, the nozzle must be designed for the pressure being pumped by the oil pump. If proper atomization is not being obtained and the pump is delivering the prescribed pressure, the trouble is in the nozzle.

The size of the orifice determines the amount of fuel being delivered to a heating unit. Too small an orifice will not deliver enough fuel, whereas too large an orifice will deliver too much fuel. A clogged orifice will not deliver enough fuel and the flame will be one-sided. To correct this problem, remove the orifice and clean it with a suitable solution. Use caution to prevent damage to the orifice.

The nozzle is expected to deliver atomized oil to the combustion chamber in a uniform spray pattern and at a spray angle best suited to the requirements of a specific installation. If the combustion chamber is almost but not completely filled with flame, the patterning of the nozzle may be wrong. In this case the nozzle must be replaced with one having the correct patterning characteristics. When replacing a nozzle, consult the orifice manufacturer's chart to obtain the correct nozzle for the installation.

109–1 NOZZLE FILTER:

The nozzle filter or strainer is designed to prevent dirt or other foreign matter from getting into the nozzle and clogging its passages. Nozzle manufacturers specify the type and size based on protecting the smallest passage in the nozzle slots. See Figure 1–194. When the nozzle filter becomes clogged, the flow of the fuel is reduced, and underfiring of the unit results. A clogged nozzle filter requires that the nozzle be replaced with one having the proper characteristics, as outlined in Section 108-1.

110–1 IGNITION TRANSFORMER:

The ignition transformer, commonly known as a step-up transformer, is used to step up the voltage from line voltage to 5,000 volts or more to jump the gap at the end of the ignition electrode. See Figure 1–195. Ignition transformers are quite trouble-free and require only that the primary voltage correspond to that of the transformer, that the electrical connections be clean and tight, and that the secondary terminals be kept clean.

If the secondary terminal post becomes fouled with soot, dust, or other foreign matter, a high-tension spark can easily short-circuit to the transformer case, resulting in no spark at the ignition electrode. A transformer in good condition should cause a spark to jump a gap of not less than 1/4 in. (6.4 mm). The transformer case must be securely grounded to provide a path for the current to flow. The spark should be a bright blue color.

To check an ignition transformer, first check to see that voltage is supplied to the primary side. If voltage is found on the primary side, check to see if there is voltage at the secondary terminals.

CAUTION: Be sure that the meter used is positioned on a high enough scale to prevent meter damage. If no voltage is found on the secondary terminals, and there is voltage to the primary side, the transformer is bad and must be replaced.

If secondary voltage is found and there is no sparking at the electrode ends, check the transformer ground wire and make certain that it is clean and tight. Next check the electrodes for cleanliness, gap, and loose connections or broken insulators. See Figure 1–196. Clean and gap the ignition electrodes and replace any defective electrode leads or defective insulators. The electrode gap should be set according to the manufacturer's specifications. Clean and tighten all electrical connections in the ignition circuit.

111–1 FUEL OIL GRADE:

Fuel oil is refined in several different grades, ranging from Number 1 to Number 6 and omitting Number 3. The higher the number, the thicker the fuel oil. Therefore, Numbers 1 and 2 are the most popular grades for domestic and light commercial use. Occasionally in cold weather where the tank is not protected from the low temperatures, Number 2 fuel oil will become too thick for the pump to move properly. When this situation occurs, drain the tank and refill it with Number 1 fuel oil. If only a small amount of Number 2 is in the tank, it may not be necessary to drain the tank because the Number 1 oil will reduce the viscosity, allowing the Number 2 oil to flow properly. In such cases it becomes necessary to drain only the fuel supply line.

Figure 1–195. Component location of oil burner. (Courtesy of Lennox Industries, Inc.)

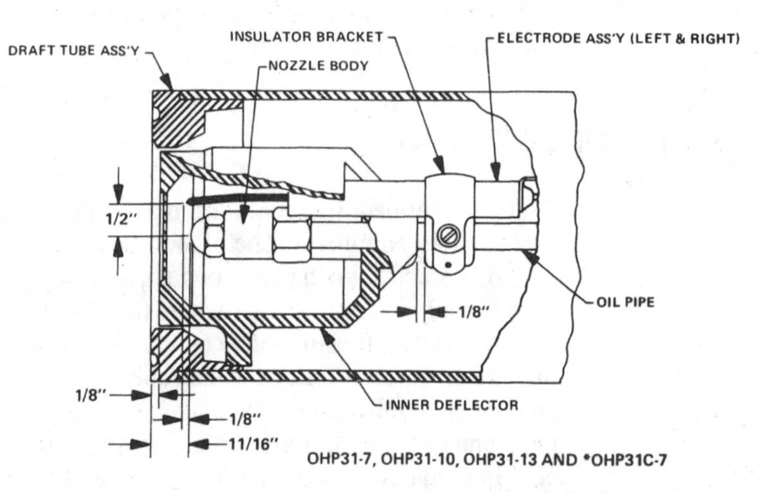

Figure 1–196. Location of ignition electrode. (Courtesy of Lennox Industries, Inc.)

2

Electronic Controls

112–1 CHRONOTHERM III THERMOSTATS:

The purpose of the thermostat is to automatically control the equipment to maintain the desired temperature inside the conditioned space. Any problem that will prevent the thermostat from doing its job must be found and corrected.

113–1 ON-OFF SWITCH:

The ON-OFF switch is commonly called the system switch. See Figure 2–1.

The ON-OFF switch must be in the correct position for the type of conditioning wanted. It usually has three positions: ON, OFF, COOL, and on some thermostats an automatic position is provided. When placed in any one of these positions the

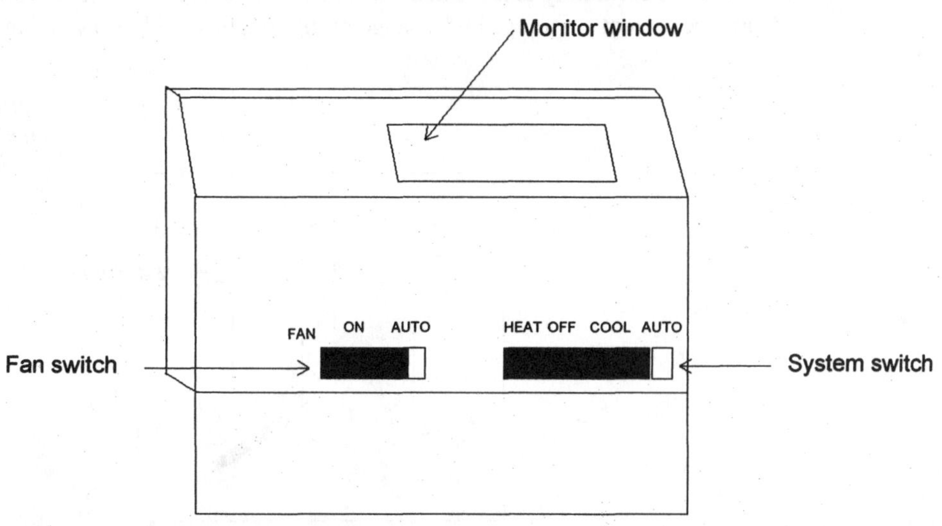

Figure 2–1. System switch positions.

system will operate as demanded by the thermostat. When in the auto position, the thermostat will demand operation from either the heating or cooling system as needed to satisfy the thermostat requirements.

114–1 POWER:

Check the electric power to the furnace, air conditioning or heat pump unit to make certain that power is available. If no power is available at these places, check the fuses, and/or circuit breakers.

115–1 FUSES:

Fuses are used to protect the circuit and its components from damage in case of overload. To test cartridge type fuses, leave the voltage on.

CAUTION: Be careful not to cause a shorted condition or personally come into contact with any of the bare wiring to prevent being shocked. Some electrical shocks are fatal. Then check the voltage from the bottom of one fuse to the top of another phase, then alternate between fuses. See Figure 2–2.

Do this until all the fuses have been checked. A bad fuse will not show line voltage across the phases. It must be replaced before the system will operate again. Be sure to use one with the correct amperage and voltage rating.

Another method of checking a fuse is to remove it from the fuse box and use an ohmmeter to check the continuity through it. See Figure 2–3.

If no continuity is indicated the fuse is bad. Replace the fuse with a good one having the correct amperage and voltage rating. If a fuse with too small an amperage rating

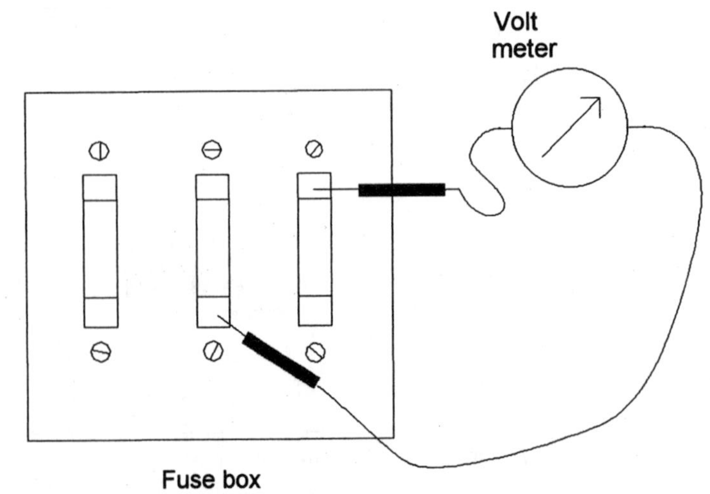

Figure 2–2. Checking fuses.

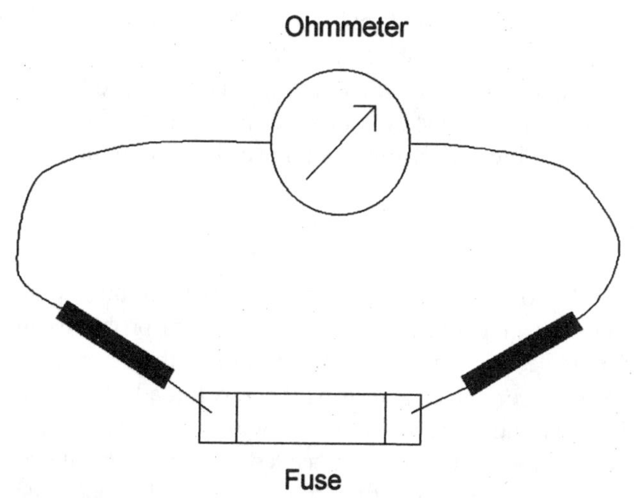

Figure 2–3. Alternate method of checking fuses.

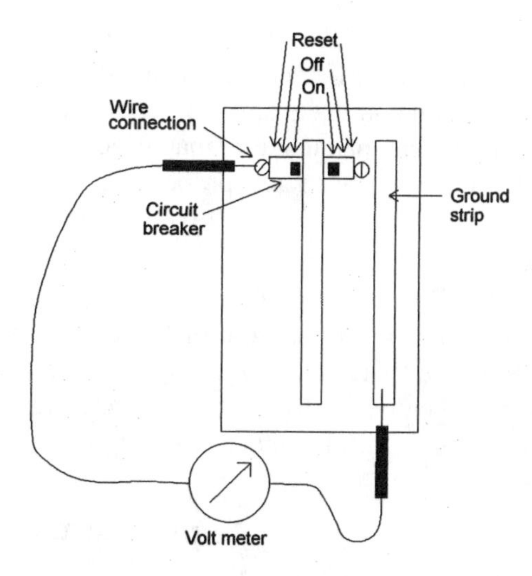

Figure 2–4. Checking circuit breaker.

is used it will blow again very soon and require replacement. If one with too large an amperage rating is used, the circuit and its components will not be correctly protected.

Next test the power at the unit to make certain that it is available there. If not, check further so the problem can be found and repaired.

116–1 CIRCUIT BREAKERS:

The purpose of a circuit breaker is to protect the circuit and its components from being damaged by an overload. Circuit breakers are designed to trip the handle to either the off position or to the tripped position. However, sometimes the handle does not move. To reset the breaker push the handle all the way to the off position and apply a little pressure to be sure that the breaker is reset. Then push the handle to the on position. Check the voltage between the wire connection and the ground, or to the other phase for 240 volts, to see if voltage is present. See Figure 2–4.

If no voltage is present, the breaker is bad and must be replaced. The correct make and amperage rating must be used. If the wrong make is used it will not fit into the breaker panel. If the wrong amperage rating is used it will soon trip again or the circuit will not be properly protected. Next check for voltage at the unit. If none is found the problem is elsewhere and further checking is necessary.

117–1 LOW VOLTAGE:

Check for loose 24 voltage connections in the 24 volt circuit. Do not take for granted that the difficult ones are good. This could lead to many problems and unnecessary replacement of parts. It is a good idea to try to tighten the screwed connections with a screwdriver. Do not over tighten and strip the screw threads. When twisted connections are used it is best to remove the insulator and twist the wire to make certain that it is not corroded. Make any repairs that may be needed to the wiring connections. Also, check for broken wiring or cracked insulation. If any are found, replace the wiring.

118–1 INCORRECT WIRING:

This could happen if it is a new system, or if it has been worked on recently. Check the wiring with the unit wiring diagram and make any corrections required. If this does not correct the problem, the equipment manufacturer may need to be contacted for assistance with the problem. The unit will not operate properly until it is wired correctly.

119–1 HIGH LIMIT OPERATION:

The purpose of the high limit control is to prevent the unit from overheating. If it is in the line voltage supply to the transformer primary winding, low voltage will not be provided to the low voltage circuit. To determine if the high limit control is open, check the voltage across the control contacts. See Figure 2–5.

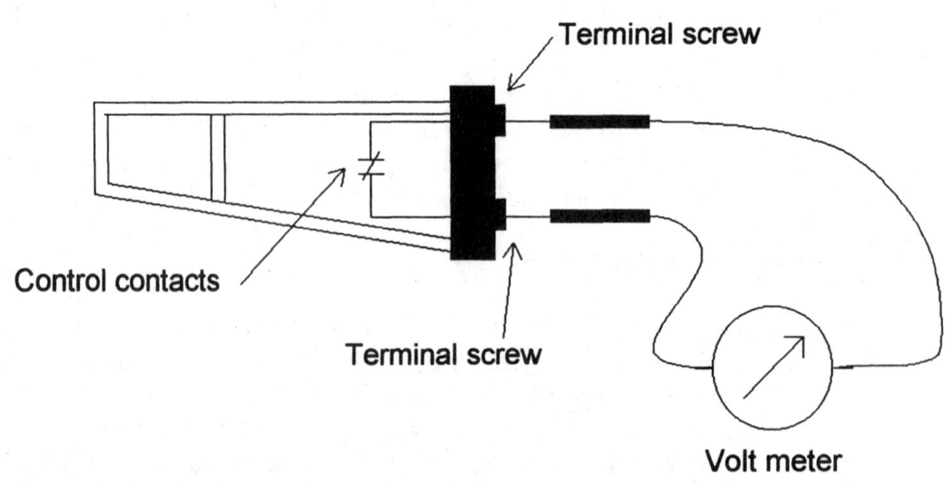

Figure 2–5. Checking continuity of high limit control contacts.

If voltage is indicated, the contacts are open. If no voltage is found, either the circuit is open elsewhere or the contacts are closed. To check for voltage at the control, check the electrical connections to ground. If voltage is found at one of them, there is voltage to the control. Usually in this case the control contacts are open. There are several things that could cause the high limit control to open such as: dirty filter, dirty blower, or restricted air flow through the unit. These conditions must be checked and if found to be the problem they must be repaired.

120–1 DIRTY FILTER:

The purpose of the filter is to remove dirt and dust from the air before it enters the unit. In the unit it can clog up the evaporator coil, the blower wheel, the blower motor, and/or the air distribution system.

Instead of removing the filter when it becomes clogged with dirt, replace it. The filter should be changed or cleaned at least every month or more often when dirty or dusty conditions are present. When the air flow is restricted by a dirty filter during the heating season, the air flow will be reduced enough to cause the unit to operate at higher than normal temperatures. These higher temperatures will cause the high limit control to open and stop the heating process. However, the air flow will be reduced and the efficiency of the unit will drop long before the system becomes completely shut down. Thus, the increased cost of operation is so slow that the user does not usually notice it until the unit completely stops. Always replace the filter with the correct size and make certain that all air passes through the filter media. Any air bypassing the filter media takes dirt into the system. This is problematic because the system will eventually become clogged with dirt and the system will need to be cleaned.

During the cooling season, a clogged filter will slow the air flow, lowering the suction pressure and reducing the cooling process. The unit will eventually begin operating so poorly that the customer will realize that something is wrong with it. When air has been by-passing the filter and collecting on the blower, evaporator coil, blower motor, and/or the system, causing the system to operate at a reduced capacity and increased cost, the system must be cleaned and a new, tight fitting filter installed. This will allow the correct amount of air to flow through the unit. On some units the low-pressure switch will stop the compressor and all cooling, warning the customer that something is wrong.

121–1 POOR AIR FLOW:

Proper air flow is needed for any type of air conditioning, heating, or refrigeration equipment to operate properly. Without proper air flow, the correct number of Btu's will not be transferred to or from the conditioned space.

Poor air flow can be caused by clogged filters, a dirty blower wheel, closed air vents, or a restricted return air system. All of these systems must work properly or the complete system will not function as designed.

The filters were discussed earlier. A dirty blower wheel must be removed and cleaned either with a water hose using a high pressure nozzle or with a steam cleaner. Be careful when cleaning the blower wheel to prevent moving or losing any of the balance weights. When the balance weights are moved or lost, the blower wheel will cause vibration and other problems associated with an unbalanced rotating object.

122–1 CLOSED DISCHARGE AIR VENTS:

The purpose of the discharge air vents is to direct the air into the conditioned space at the proper velocity and direction. When either of these is changed, the air will not do its intended job.

Many people close the discharge air vents to stop the air going into an unused room or space. This is done so that the room will not be conditioned. The thinking is that closing the vents will reduce the cost of operation. Even some technicians share this thought. However, closing the vents to an unused part of the building generally will not reduce the cost of operation. It will probably increase the cost. The cost may be increased because the heat transfer through the uninsulated wall will be greater than through the outside insulated wall. The unit will not have the proper air flow through it to keep the operating characteristics normal. The closing of discharge air vents will have the same effect as a dirty air filter. Be sure that all the discharge air vents are properly adjusted for the proper air to flow through the unit for the best efficiency and operation.

123–1 RESTRICTED RETURN AIR SYSTEM:

The purpose of the return air system is to provide a path for the air to return from the space to the conditioning equipment. It must be properly sized and unrestricted for proper operation.

The most common cause of a restricted return air system is the return air grills being blocked by drapes or furniture. Another cause may be that something large enough to cause the problem has fallen into the return air system. In refrigeration systems the product may have been stacked too high to allow for proper air to return to the evaporator. This restriction must be removed before the system will operate properly. The user should be instructed not to place any furniture, drapes, or product where they will interfere with the flow of return air.

124–1 FURNACE POWER:

There must be continuous power to the heating unit even when the system is in the cooling mode. When there is no power to the heating unit during the cooling mode there will probably be no power to the thermostat. Remember that some units interrupt the power to the heating unit when the system is in the cooling mode. This causes the system to be incompatible with the thermostat. The thermostat must be replaced with the proper type. Be sure to wire the thermostat into the system using the proper wiring diagram.

125–1 COMPATIBILITY:

Make certain that the current draw through the control circuit is more than 0.08 amps. To check the current draw through the control circuit see Figure 2–6.

There are different ways to check this amperage. The easiest way is to have an amperage multiplier for your ammeter. Another way is to wrap the voltage (usually the red) wire around one tong of your ammeter and read the current flow, then divide the current draw by the number of turns made on the ammeter tong. Example: We take ten turns of the voltage wire on the ammeter tong. Measure the current flow. Then divide the current flow by ten, using the following formula. If the meter shows that 8 amperes are flowing through the circuit, what is the actual current draw of the circuit?

$$\text{Current Flow} = 8 \text{ amps} \div 10$$
$$= 0.08 \text{ amps}$$

125–1A

Use an Isolation Relay. The addition of an isolation relay separates the control circuit from the thermostat circuit. Remember that the isolation relay must have a current draw of more that 0.08 amperes, or 80 milliamperes. To wire the isolation relay into the circuit see Figure 2–7.

The relay must draw 0.08 amperes through the complete on cycle of the equipment. Be sure to that this current draw is present or the problem probably will not be solved.

125–1B

Add A Shunt Resistor. A shunt resistor may be installed to help in solving a low amperage draw in the thermostat circuit. See Figure 2–8.

This must be a 250 ohms, 10 watt resistor.

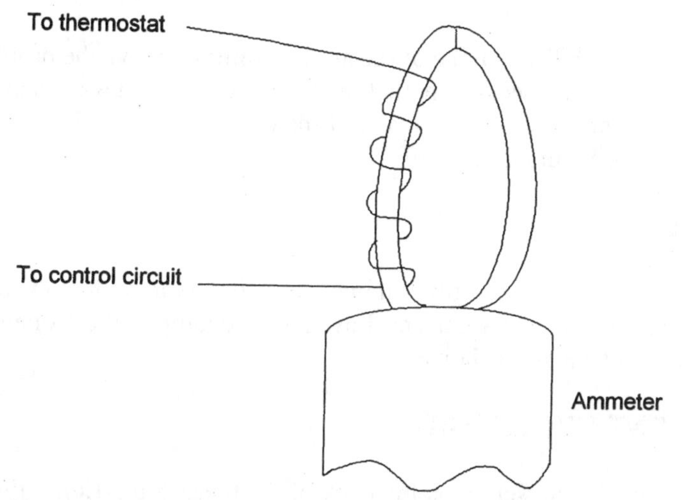

Figure 2–6. Checking current in a control circuit.

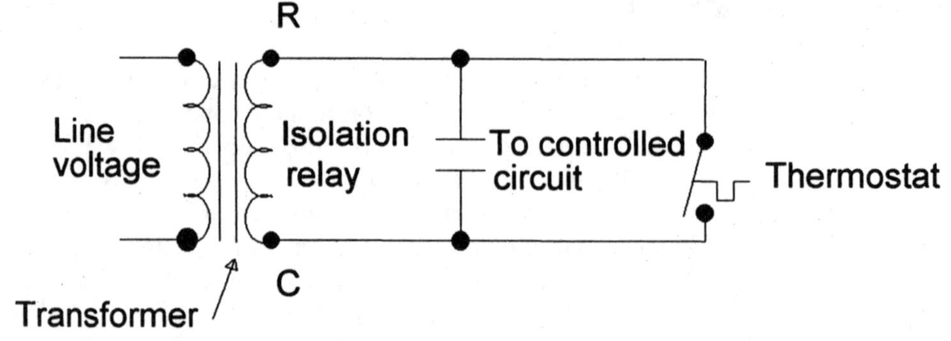

Figure 2–7. Wiring for isolation relay.

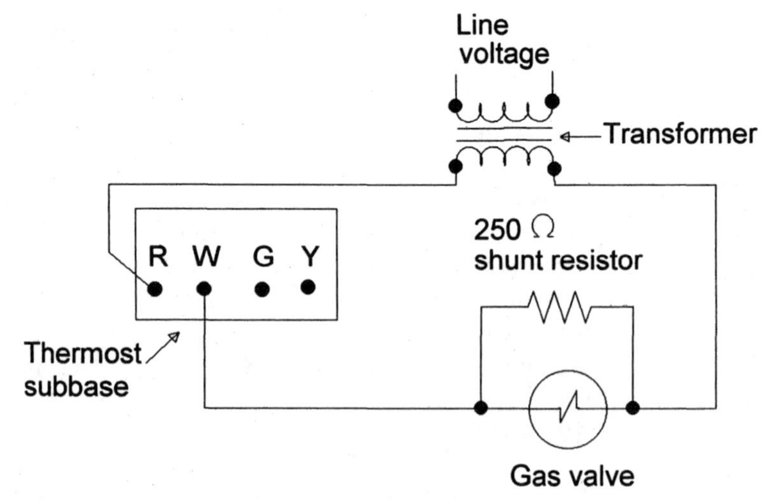

Figure 2–8. Wiring a shunt resistor into the circuit.

125–1C

Add a Fan Center. When none of the above examples solve the problem, a fan center may be installed. Make sure that the fan center coil draws at least 0.08 amperes during the complete operating cycle of the equipment. To wire the fan center into the circuit, see Figure 2–9.

125–1D

Replace Thermostat. Sometimes it may be necessary to replace the thermostat with one that requires a lower current draw. As an example, the Honeywell T8602 or T8603 would be a good choice.

126–1 INTERMITTENT CYCLE RATE:

Check the cycle rate screw on the back of the thermostat. Either tighten or loosen the screw to change the equipment cycle rate for proper operation. See Figure 2–10.

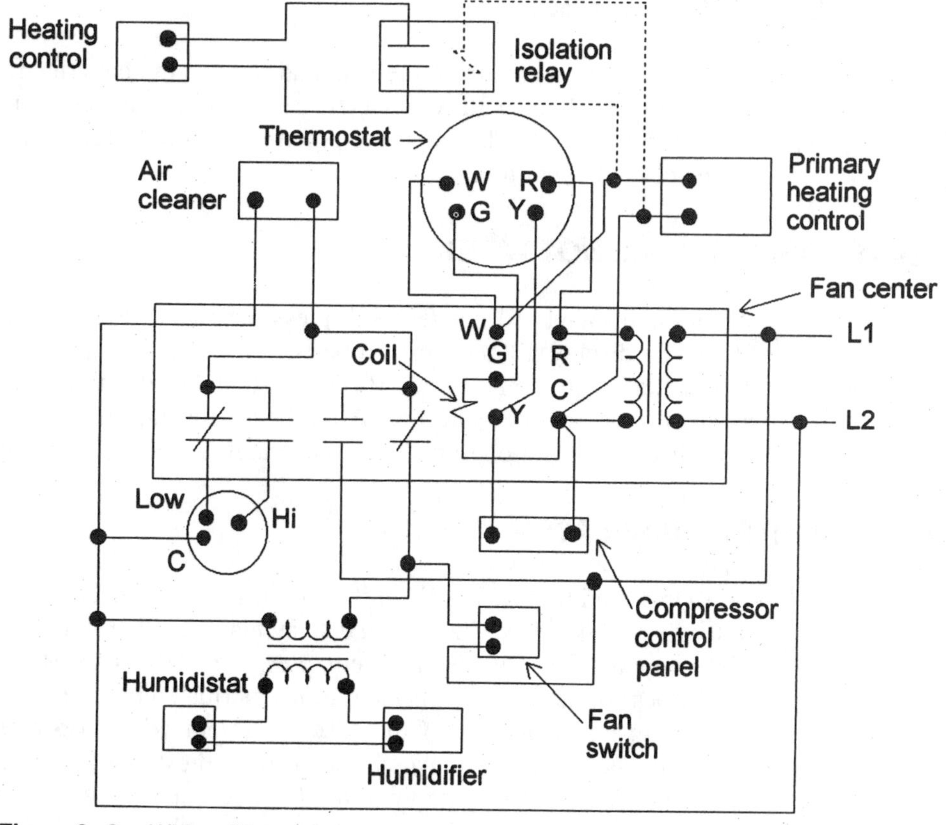

Figure 2–9. Wiring diagram using a fan center.

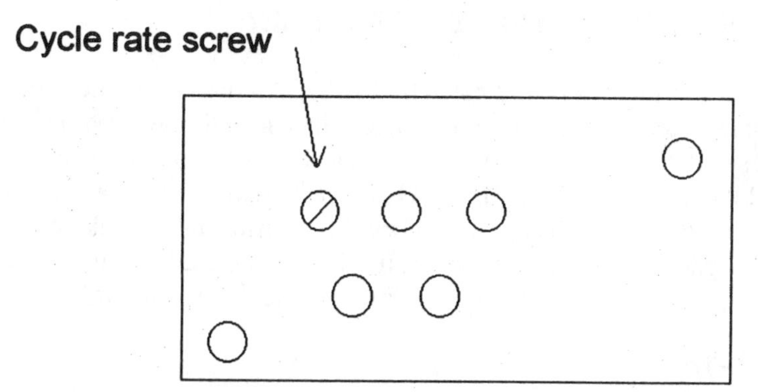

Figure 2–10. Cycle rate screw. Check each make and model for exact location.

127–1 POWER:

Remove the thermostat and then remove the batteries from the thermostat. Leave the batteries out of the thermostat for at least five minutes. Then replace the batteries and reprogram the thermostat. If the problem recurs, install an isolation relay.

128–1 LOOSE BATTERIES:

Remove the thermostat from the wall and remove the batteries from their holder. Make sure that all the battery connections in the holder are clean and tight. Replace the batteries in the holder in proper polarity. Be sure to reprogram the thermostat for the desired operation.

129–1 DISPLAY WILL NOT WORK:

When the display will not respond, press the RUN PROGRAM key two times. Should the display still not work properly, replace the thermostat. Program the thermostat and install it on the wall. Be sure that all the adjustments are made to make the thermostat operate as desired. After the thermostat has reached room temperature, check its operation to make certain that it is working properly.

130–1 PARTIAL DISPLAY:

When this condition occurs either the thermostat is not in the "Run Program" mode or the display is faulty. The partial display may be either in the top or bottom half of the window. To solve this problem, press the RUN PROGRAM key two times. If this does not cause the display to return to normal, replace the thermostat. Be sure to use an exact replacement. Otherwise the system will not operate as desired. Program the thermostat and make all the needed adjustments so that the system will operate as desired. After the thermostat has reached room temperature, check its operation to make certain that it is working properly.

131–1 BATTERIES NOT INSTALLED PROPERLY:

This condition will be indicated by REPL BAT flashing in the display. To solve this problem, remove the thermostat and either install new batteries or install the old ones with the correct popularity as indicated by the markings on the battery holder and the batteries. The REPL BAT indicator will flash for about 15 minutes. After this period of time has passed, check the thermostat to make certain that it is off. Reprogram the thermostat and reinstall it on the wall. After the thermostat has returned to room temperature, check it to make certain that it is working as desired.

131–1A

Batteries Are Discharged. This condition is indicated by the REPL BAT indicator flashing. This usually occurs when the batteries are discharged below a usable voltage. The thermostat is not getting enough voltage to operate properly. To solve this problem, remove the thermostat from the wall and replace the batteries with the correct size and the correct polarity as indicated on the battery holder and the batteries. The REPL BAT indicator light will flash for about 15 minutes. After this time, check to make certain that the thermostat is not flashing. If not, reinstall it on the wall. After the thermostat has reached room temperature, make certain that it operating as desired.

132–1 DISPLAY FLASHES WHILE PROGRAMMING:

This condition can occur when a wrong programming procedure is used or when the automatic changeover is set too close. Use the following procedures to solve this problem.

Programing: Check the homeowners manual to make certain that the correct programming procedures are being used. Be sure to use these procedures. It is not usually necessary to remove the thermostat from the wall during this procedure. However, removal may make the reprogramming easier.

132–1A

Automatic Changeover. The automatic changeover must not occur unless there is at least a three degree Fahrenheit difference between the heat and cool settings. To solve this problem, set the temperature setting with the heating setting three degree Fahrenheit above the cooling setting.

133–1 PROGRAMMING HAS BEEN LOST:

This condition usually occurs when there has been a loss of power. To solve this problem, install new batteries of the correct size and polarity as marked on the thermostat battery holder. Reprogram the thermostat. Use the homeowners manual if needed. The thermostat may be reprogrammed on the wall. However, in some installations it may be easier to remove the thermostat from the wall, reprogram and then reinstall it.

134–1 UNIT WILL NOT OPERATE:

There are several factors that could cause this problem. Sometimes all of them may need to be checked and adjusted or repaired before normal operation can be realized.

134–1A

No AC power to the thermostat. Check the electric power to the unit. When no power is found here, check the ON-OFF switch, the fuses, circuit breaker, and the electrical connections in both the line voltage and the low voltage circuits. Also, check for incorrect wiring according to the unit wiring diagram.

134–1B

The ON-OFF switch is also known as the system switch. It must be in the ON position before any part of the unit, except the fan, will work. Move the selector switch to the position, heating, or cooling, as desired. The unit should start operating if this was the problem. Be sure to check the system for proper operation.

134–1C

To find out if incorrect wiring is the problem, check the wiring to see if it is in accordance with the unit wiring diagram. If it is not, rewire the unit to match the unit diagram. For some thermostats the "C" terminal must be connected to the common terminal of the transformer.

134–1D

On Honeywell T8603 Thermostats the C and B/C terminals must be connected to both terminals of the clock transformer on a two transformer system. Alternatively, the R terminal can be jumpered to the C and B/C terminals connected to the common terminal of the transformer when two transformers are used. When only one transformer is used, the R, RC, and C terminals are connected to the common terminal of the transformer. Other manufacturer's thermostats have similar recommended wiring. Be sure to use the specific manufacturer's wiring diagram for the thermostat being used or satisfactory operation will not be realized.

134–1E

On Honeywell T8621 Thermostats when two-transformer systems are used, the C terminal must be connected to the common terminal of the heating transformer. When a single transformer, 24 VAC system is used, R and RC must be jumpered and the C terminal connected to the common terminal of the transformer. When wiring any unit or unit component be sure to use the specific manufacturer's wiring diagram for the component being installed so that the correct operation will be realized.

134–1F

To check the system for proper operation, jumper terminals R to W or R to H. The heating unit should come on. If it does not, check the system wiring against the unit wiring diagram. Jumper R or RC to Y, or R or RC to C and the cooling system should come on. If it does not, check the unit wiring against the unit wiring diagram.

CAUTION: Do not jumper terminal Y to R or RC unless the outdoor ambient temperature is high enough for safe compressor operation.

134–1G

When the temperature setting is too high/low, temporarily adjust the temperature setting by pressing the Warmer/Cooler keys. The unit should come on. Reprogram the thermostat to maintain the desired temperatures.

134–1H

When the thermostat system switch is in the wrong position, move the system switch to the desired position. The unit should start operating if there is a demand for operation.

134–1I

To solve minimum off-times in heating, wait about 5 to 10 minutes for the system to reset itself. It is also possible to use the thermostat self-test to override the thermostat time delay. In some cases it may be necessary to check the installation instructions for the make and model thermostat involved. When the Honeywell T8611 is used, this delayed off time may also happen in the cooling mode.

134–1J

When the thermostat will not operate, complete the thermostat self-test procedure. If necessary see the installation instructions for the particular make and model thermostat involved.

134–1K

When the unit will not operate, check the unit to determine what the problem is. It may be in the line voltage circuit, the low voltage circuit, or some mechanical problem. Repair the problem and place the system back in operation.

135–1 THERMOSTAT WILL NOT HOLD THE SETTING:

This condition is usually caused by intermittent connections to the adjustment screws on the back of the thermostat. The cycle rate screw on the Honeywell thermostat is usually the culprit. Other makes and models of thermostats have a similar adjustment. Check the particular thermostat manufacturer's make and model to solve the problem. Check the position of the screw and either tighten or loosen as required for proper operation. See Figure 2–10 to locate the position of this adjustment.

136–1 PRESENT THERMOSTAT SETTING SEEMS WRONG:

This condition usually indicates that the system is in the recovery mode and is showing the ramp temperature. The ramp temperature is the temperature shown on the display when the thermostat is returning to the normal thermostat setting from the energy conservation setting. This is normal operation and there is no corrective action to take.

137–1 ROOM TEMPERATURE SHOWN SEEMS WRONG:

This condition is usually caused when thermostat senses both the wall and the air, and averages these temperatures. To solve this problem, move the thermostat to a location where these temperatures will not affect the thermostat.

The thermostat must not be installed on a wall that will sense the temperature either in or on the other side of the wall. The thermostat should be installed where wall vibration will not affect its operation. It should be located about five feet from the floor to sense the temperature where the majority of people will feel it.

138–1 ROOM TEMPERATURE SEEMS HOTTER THAN THE THERMOSTAT SETTING:

This condition is usually caused by the thermostat being cooled by a temperature other than room temperature, when in the heating mode. There are generally two reasons for this condition. The thermostat is mounted on a cold wall or there is an opening where the wiring comes through the wall. Also, the thermostat could be located close to a window or something that causes the thermostat to sense a temperature cooler than the room temperature.

138–1A

When the thermostat is mounted on a cold wall, check to see if it is mounted on an outside wall. If it is, mount it on a wall that does not have a large temperature difference between the two sides, or where cold air temperatures will not affect its operation. Use the normal installation procedures when relocating the thermostat.

138–1B

Air coming through the wall is usually caused by air passing down or up through the wall and entering the rear of the thermostat where the wiring enters the back of the thermostat. To solve this problem, remove the thermostat from the wall or the subbase and seal the hole where the wiring comes through the wall with putty or some other type of sealer.

138–1C

Solar Affect is usually caused by the thermostat being installed too close to a window, door, or some other opening that could allow the thermostat to sense the cooler air caused by the opening. To solve this problem, relocate the thermostat to a place where these conditions do not exist. Be sure to use the correct procedure during the installation so that the thermostat can do its job. It may also be possible to just close the window blinds or some other type of closure that will reduce the heat loss from the thermostat.

139–1 FURNACE STARTS TOO EARLY IN THE MORNING:

This conditions occurs because of the intelligent (automatic) recovery system built into the thermostat. This is normal operation for this type of system. The intelligent recovery system is designed to bring the space up to temperature without overshooting while conserving energy. The thermostat computer calculates the amount of time needed to bring the space temperature up to the temperature setting at three degree intervals. This is usually done by the number of running periods required during the reset time period. The thermostat then gradually brings the space temperature up to the temperature setting.

This condition can be changed by removing the thermostat from the wall and changing it to a conventional recovery method. See Figure 2–11.

Be sure to use the directions for the particular thermostat being serviced. Another method is to press the WARMER/COOLER keys to stop the recovery for that period. However, any energy savings usually realized will be lost and the thermostat will not be serving its purpose in the system.

140–1 THERMOSTAT DOES NOT PROPERLY CONTROL TEMPERATURE ON WEEKENDS:

This is generally caused by the thermostat not being properly programmed. This condition can be solved by reprogramming the thermostat for the temperature desired on the weekends. Be sure to use the correct procedure for the make and model being programmed. Otherwise the reprogramming will not be properly completed.

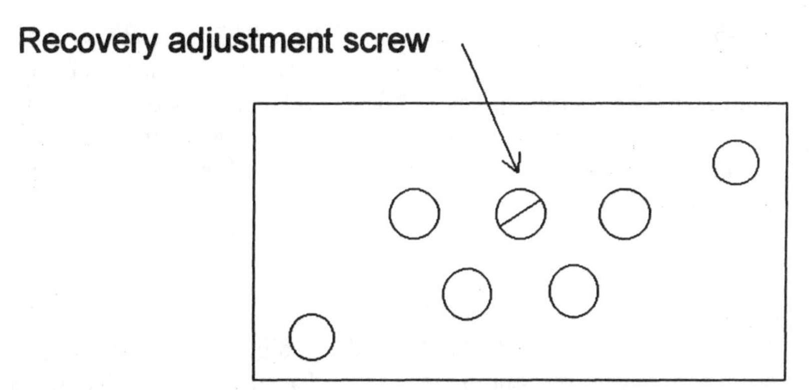

Figure 2–11. Recovery adjustment screw location. Check each make and model for exact location.

141–1 THERMOSTAT TEMPERATURE SETTING NEEDS TO BE ADJUSTED OFTEN:

This problem is usually caused by the WARMER/COOLER keys being used to control the space temperature. This condition can usually be solved by reprogramming the thermostat so that it will automatically maintain the space at the desired temperature. Instruct the user that the WARMER/COOLER keys are to be used only as a temporary thermostat setting unless a period key is pressed first.

142–1 THERMOSTAT PROGRAM NOT OPERATING:

There are several reasons for this problem. Their cause and solutions follow.

142–1A

When the program times are the same for heating and cooling occurs when the thermostat is programmed for heating and cooling temperatures that are too close together. To solve this problem, reprogram the thermostat when the season changes or to have a three degree temperature difference between the heating and cooling settings. Be sure to use the proper procedure for the make and model when reprogramming.

142–1B

Saturday and Sunday not Programmed: The weekends and weekdays are programmed separately. To solve this problem, program the thermostat for weekend operation as well as weekday operation. Be sure to use the proper procedure for the make and model thermostat in question.

142–1C

When temperature adjustment by the WARMER/COOLER keys is required; the thermostat is not properly programmed. To solve the problem, reprogram the thermostat for proper operation. The WARMER/COOLER keys are used only for temporary changes. The thermostat must be reprogrammed to make the change permanent. Be sure to use the proper procedure for the make and model thermostat being serviced.

142–1D

Periods can be canceled by holding down a period key until the time and temperature no longer is present in the display. To solve this problem the thermostat must be reprogrammed. Be sure to use the proper procedure for the make and model thermostat being serviced.

142–1E

A program change generally occurs when the thermostat batteries are bad or low on charge and there is a power failure. To solve this problem, remove the thermostat from the wall. Remove the old batteries and install new ones using the proper size and polarity as indicated on the thermostat battery holder. Reprogram the thermostat. Be sure to use the proper procedure for the make and model thermostat being serviced.

143–1 SHORT BATTERY LIFE:

This problem can be caused by two situations: incorrect wiring or line transient, otherwise know as static discharge.

143–1A

Incorrect Wiring: When Honeywell thermostats T8601, T8603, T8611, or T8621 are used, connect the "C" terminal on the thermostat to the "C" terminal on the transformer. Other make and model thermostats will have a similar solution for this problem. Be sure to use the proper procedure for the make and model being serviced.

143–1B

Line transient or static discharge can be easily solved by replacing the batteries. Use the correct size polarity as indicated on the battery holder. If the problem still occurs, install an isolation relay to give the correct current draw through the thermostat circuit. Be sure to use the proper procedures for the make and model being serviced.

144–1 MELTED PLASTIC:

This is usually caused by the control circuit through the thermostat being higher than 2.25 amps. Sometimes it is caused when line voltage is connected to the control circuit. To solve this problem, check all the relays in the system for proper operation. Replace any that are found bad or defective in any way.

145–1 DOOR WILL NOT STAY CLOSED:

This is usually caused when the door is not properly mounted on the hinge, or the door or hinge is broken. To solve this problem, make sure that the door is hinged properly. If the hinges or the door is broken, replace the door. Be sure to use the door for the thermostat being serviced because they are a slightly different color for the same model.

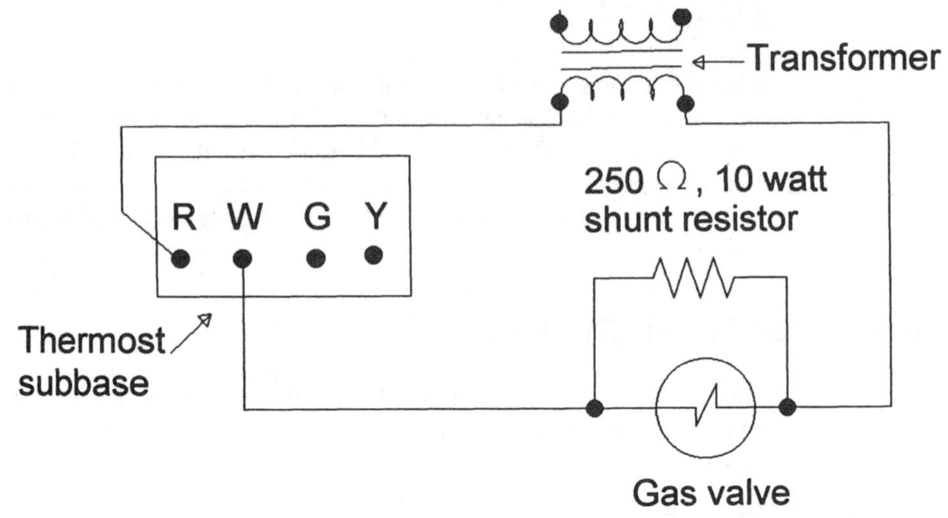

Figure 2–12. Installing a shunt resistor.

146–1 THERMOSTAT LOSES POWER WITH SYSTEM SWITCH IN OFF OR HEAT POSITION:

This problem can be caused when there is no load on the heating system control circuit. Some makes of thermostats, usually the system-powered type, require a load on the heating control circuit for the thermostat to operate in the OFF position. This problem can sometimes be solved by installing a 250 ohm, 10 watt shunt resistor in the heating circuit. Probably the easiest place to install the shunt resistor is around the gas valve. See Figure 2–12.

However, the thermostat or the equipment manufacturer's recommendations must be followed for proper operation of the system.

147–1 ELECTRONIC FAN SPEED CONTROL:

The purpose of the fan speed control is to change the speed of the condenser fan when the ambient temperature changes. When the temperature drops, the fan speed automatically slows to blow less air through the condenser. When the ambient temperature rises, the fan control automatically increases the fan speed to blow more air to help cool the condenser.

The fan speed control is a pressure actuated electronic motor speed controller. As the outdoor air temperature drops, the refrigerant discharge pressure of the compressor drops and causes the fan control to slow the motor. As the outdoor air temperature rises, the compressor refrigerant discharge pressure rises and the fan control speeds the condenser fan to blow more air and keep the condenser cool. See Figure 2–13.

This control can be used with a single phase permanent split capacitor and shaded pole motors that are approved by both the motor and the equipment manufacturer for fan speed control use. These controls directly sense the discharge refrigerant pressure and electronically varies the speed of the fan motor.

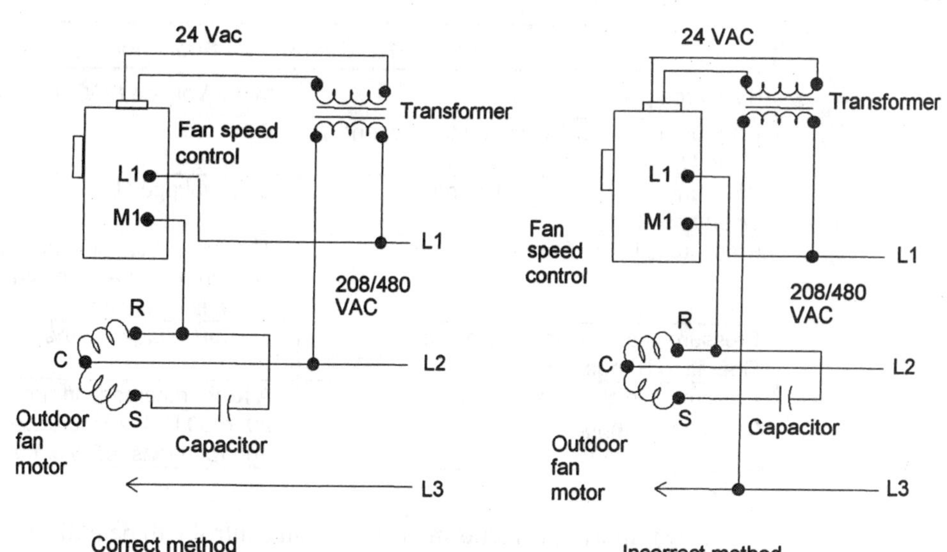

Figure 2–13. Permanent split-capacitor single phase motor connections to an electronic motor speed control used on a three-phase system.

147–1A

Operation of the condensing unit will determine the overall capability of maintaining a satisfactory refrigerant pressure by controlling the discharge pressure. This is done by using a fan speed control. Within the operating range of the control, it will provide air delivery in direct proportion to the heat being rejected by the condenser. The proper heat rejection allows the refrigeration system to operate efficiently in very low ambient outdoor air temperatures.

The pressure sensor(s) respond to changes in the condenser pressure, regardless of the variations in fan delivery curves. The dual pressure input type selects the pressure input with the highest pressure. See Table 2–1.

147–1B

Control Adjustments: The throttling range of most of these controls is internally fixed and cannot be adjusted in the field. However, the operating range can be adjusted in the field; it can be increased or decreased within the control's pressure range. The pressure range is usually from 80 to 200 psig for low pressure models and 140 to 350 psig for high pressure models.

The operating range can be adjusted as follows:

1. Locate the adjustment screw on the pressure transducer. (The second transducer, if supplied, is usually in another location on the control's base.)
2. Turn the adjustment screw clockwise to increase or counterclockwise to decrease the operating range. On low pressure models, one turn equals approximately 17 psig. On high pressure models one turn equals approximately 35 psig.

Table 2–1. Operational Sequence

Discharge pressure input	Motor Voltage (VAC, true RMS)
Pressure between 0 psig and the low end of the Operating Range	0 to 5 Volts
Pressure is at the low end of the Operating Range	Start Voltage (10% to 40% of line voltage)
Pressure is in the Operating Range	The motor voltage varies directly with the system pressure from start voltage to 90% of the line voltage
Pressure is at the high end of the Operating Range	Output voltage is 90% of the line voltage
Pressure is higher than the normal Operating Range	A further increase in normal pressure of 20 to 30 PSI will increase the motor voltage to at least 97% of the applied voltage

Any range adjustments made to the fan control will be indicated in the pressure at which the control applies minimum start voltage to the motor. Do not make any adjustments other than those listed above. If the control is found nonresponsive, replace it.

147–1C

Check the control for proper operation through the desired pressure range before leaving the job. Refer to Table 2–1 for a typical operational sequence.

148–1 NO FAN OPERATION:

When the fan does not operate and the compressor discharge pressure is below the normal operating range of the control there is no problem. This is normal operation.

However, no fan operation with a high discharge pressure could also be caused by no low (24) voltage to the control. Check and find where the problem lies. It could be a bad transformer, broken wire, loose connection, the power to the unit is turned off, a blown fuse, or a tripped circuit breaker.

148–1A

When there is no input pressure to the control, check to make certain that the schrader valve is depressing the valve needle enough to allow pressure to the control. If not, find and correct the reason.

148–1B

To check for a bad fan motor, turn off the power to the unit. Connect a jumper from L_1 to M_1 and turn the power back on. If the fan does not start, the motor is bad and must be replaced. If the motor starts and runs, the problem is either in the control or the wiring to the control.

148–1C

To check for a pressure transducer problem, disconnect the 6 pin connector from the right side of the control. Place a jumper wire between the third pin from the top and the bottom pin on the control, not the cable. If the fan goes full speed, check for input pressure (above). If it has been determined that there is enough pressure, the transducer is bad and the control must be replaced.

149–1 FAN STOPS WHEN THE PRESSURE REACHES THE HIGH END OF THE OPERATING RANGE:

This condition usually indicates that the control is wired incorrectly. When the fan is found to be operating in this manner, check the unit wiring diagram to make certain that it matches the wiring on the unit. If not, rewire the unit to match the diagram.

150–1 NO FAN MODULATION (ON-OFF OPERATION):

This condition usually occurs when the control is wired incorrectly. When this occurs, check the unit wiring diagram to make certain that it matches the wiring on the unit. If not, rewire the unit to match the diagram.

151–1 FAN STARTS AT FULL SPEED:

When the fan starts at full speed, an incorrectly wired unit is indicated. The unit must be wired correctly for proper operation. Check the wiring diagram to make certain that it matches the wiring on the unit. If not, rewire the unit to match the diagram.

152–1 ERRATIC FAN OPERATION:

This condition usually indicates that the control is not wired correctly. Check to make certain that the control (24) volts are on the same phase as the motor being controlled. If not rewire the unit to make certain that they are. This generally occurs when the unit is operating on three-phase voltage. See Figure 2–14.

Some suggested wiring diagrams are shown in Figures 2–15 and 2–16.

152–1A

A dirty or blocked condenser coil will cause erratic condenser fan operation. When the condenser is found dirty, it must be cleaned before the unit will operate properly. Wash the coil in the opposite direction of air flow through the finns. This will usually remove the debris from between the finns. Before cleaning the coil, turn off the electricity to the unit to protect yourself from injury and the unit from damage. Be sure to protect all electrical equipment when cleaning the condenser coil. If spraying the coil with a high pressure nozzle on a garden hose does not clean the coil, some type of noncorrosive cleaner must be used. If the coil is not clean, the problem will only occur again in a short period of time.

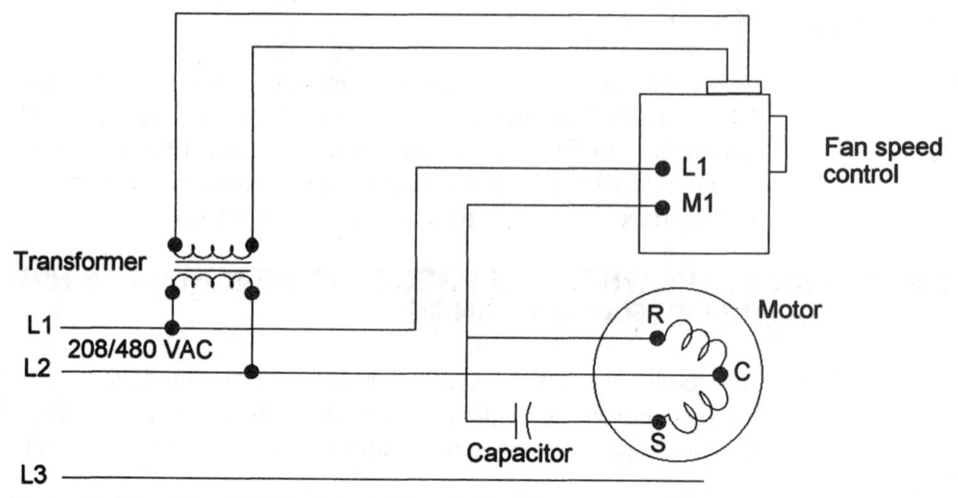

Figure 2–14. Permanent split-capacitor, single-phase motor connections to a fan speed control when used on three-phase systems. Note: the 24 volt control circuit and the motor must be on the same phase.

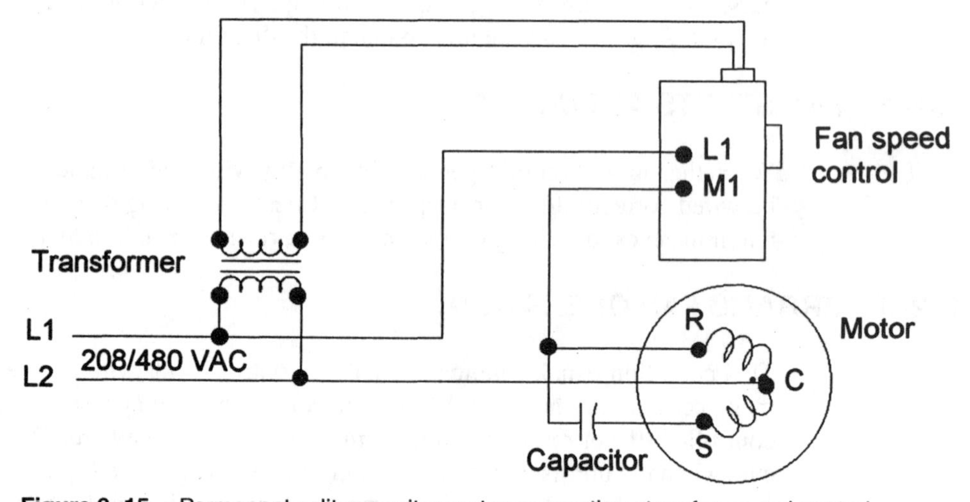

Figure 2–15. Permanent split-capacitor motor connections to a fan speed control.

153–1 FAN MOTOR IS CYCLING ON THERMAL OVERLOAD:

This condition could be caused by a dirty or blocked condenser coil or the wrong motor being used for fan speed control.

153–1A

A dirty or blocked condenser coil will cause the condenser fan motor to cycle on the thermal overload. When the condenser coil is dirty or blocked, clean it. Refer to previous instructions for cleaning the condenser coil for the correct procedure.

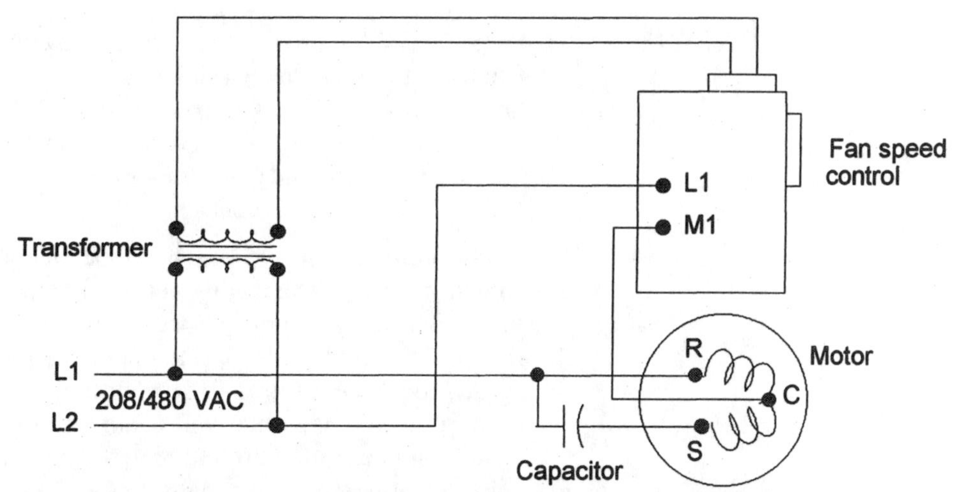

Figure 2–16. Optional wiring diagram for connecting a permanent split-capacitor motor to a fan speed control.

153–1B

When the wrong motor for fan speed control application is used, the motor must be replaced with one that is approved for this type of application. If the proper motor cannot be found immediately, rewire the unit to leave out the fan speed control until one can be located and installed. Be sure to install one that is suited for fan speed control or the unit will not operate as effectively or economically as it was designed.

154–1 ELECTRONIC LUBE OIL CONTROL (WITHOUT R10A SERIES RELAY):

The purpose of the lube oil control is to protect the compressor should there be a drop in, or insufficient, lubricating oil to properly lubricate the compressor. There are several indicator lights on the electronic lube oil control that will be an indication of where to start looking for the problem. See Figure 2–17.

154–1A

The operating status of the LED's is as follows;

GREEN LED ONLY: This indicates that the contactor is energized, and the net oil pressure of the system is at or above the control set point.

GREEN AND YELLOW LED's: The green LED indicates that the the compressor contactor is energized while the yellow indicates that the net oil pressure is below the control set point and the lube oil pressure timing delay is energized.

NOTE: Some lube oil controls are available with either accumulating or nonacummulative lube oil timing delays. Both timers start when the net oil pressure drops below the control set point. If the pressure does not rise above the control set point before the end of the timing cycle, the control will lockout the compressor contactor.

Accumulative Timing: If the lube oil pressure should rise above the control set point value before the time delay is completed, the timer will stop and run back down towards 0 psi at one-half its forward rate. If low lube oil pressure is detected before the timer reaches 0, the timer will again run forward at its normal rate, without resetting to 0. The timer will automatically reset to 0 seconds if power is removed from the control.

Nonaccumulative Timing: Each time the pressure reaches the control set point, the timer stops and resets to 0 seconds.

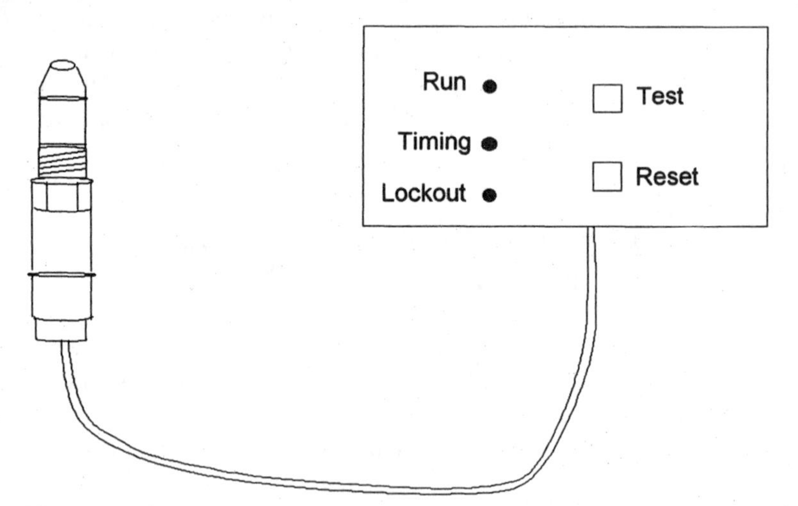

Figure 2–17. Typical electronic lube oil control.

YELLOW LED: Power to the control has been interrupted, and the unit anti-short-cycling timer has started. The compressor contactor will remain de-energized for the length of the time delay, and it will reset automatically after the time delay period has passed.

RED LED: The lube oil pressure delay cycle is completed, and the control has locked out the compressor.

154-1B

The red LED light being on is an indication that either the control is locked out or the control erroneously indicates low oil pressure. The following steps may be used to locate the problem.

154-1C

A control lock out may occur because of failure of the oil pump. After checking to make certain that there is sufficient lube oil in the compressor crankcase, check the oil pump. To check the oil pump, install a set of service gauges on the unit. The suction pressure gauge is connected to the suction pressure connection on the unit. The other gauge is connected on the oil pump pressure connection. The difference between the lube oil pressure and the suction pressure is the net pump pressure. The limits should be checked with the equipment manufacturer's literature. Generally, the required net oil pressure is between 20 psi and 40 psi. However, if at all possible learn what the manufacturer's recommendations are for the compressor. When the oil pump pressure is below these limits the pump is defective and should be replaced. It may be possible to repair the pump if the parts are available.

154-1D

The internal compressor motor overloads have overheated stopping the compressor. There are several problems that could cause this problem. To check for an open internal overload, first turn off all electricity to the unit. Remove both wires from the terminals on the overload circuit. Then with the ohmmeter, check for continuity between the two terminals. If no continuity is found, the overload is open. When continuity is found, the problem is elsewhere and must be located and repaired.

154-1E

When the compressor is hot, wait for it to cool and the overloads to reset. This is usually a long process that can normally be shortened by temporarily bypassing the overloads so that the compressor will run. The suction gas should cool the compressor much more rapidly. Be sure to keep a close watch on the amperage, voltage, refrigerant pressures, the oil pressure, and the temperature of the compressor during this process. If either rises above the normal operating limits, stop the compressor. There is another problem that must be found and repaired. If possible, the compressor could be cooled by flowing a small amount of water over the compressor to cool it. Be sure to keep the water away from the electrical components. In units where the compressor is installed inside the building, using water may not be possible. Another method is to leave all electric power turned off to the compressor until it cools to the operating temperature. Then start the compressor and determine the cause of overheating.

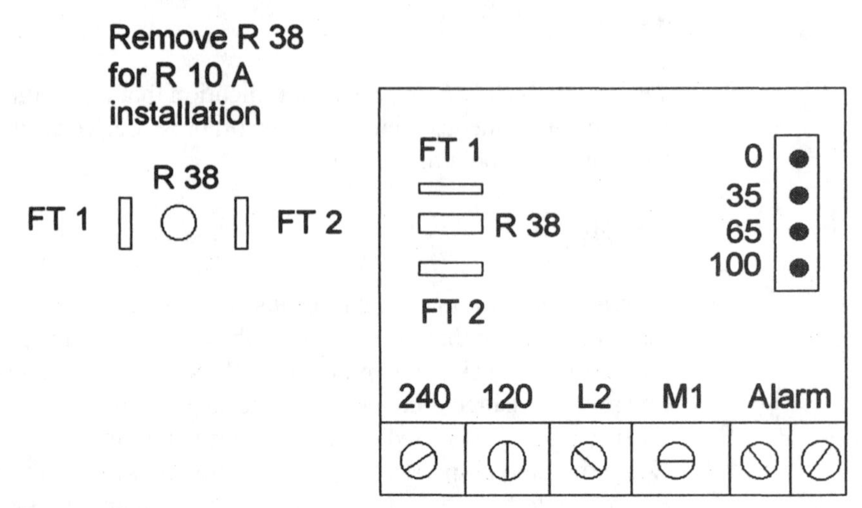

Figure 2–18. Wiring terminal designations.

154-1F

When the unit is equipped with a *reset button,* resetting it may cause the compressor to restart and operate. Be sure to check the operating pressures and amperage of the compressor. If any of them are too high, this will cause the compressor to overheat. Find and repair the problem causing the overheating, otherwise the problem will only recur.

154-1G

To install a wide range current sensing relay, cut and discard resistor R38. See Figure 2–18. Connect the relay to the two blade terminals, FT1 and FT2.

Remember that the relay will not work when the anti-short-cycling delay timer is set at 0 seconds. Set the timer to 35, 65, or 100 seconds.

155–1 YELLOW LED ON:

The yellow LED on indicates that the compressor is not running. When this occurs, check to see if the control is set in 35, 65, or 100 second anti-short-cycling delay mode. Then wait the amount of time indicated by the placement of the control anti-short-cycling timer. After this amount of time has passed, the control will automatically restart.

156–1 GREEN LED ON AND YELLOW LED FLICKERING:

This is the fluctuating oil pressure signal. It could be caused by bad bearings or other compressor components. The oil pump could be starting to give trouble by not pumping enough oil to the compressor parts. It could also be caused by a faulty oil control sensor.

Check the compressor component operation by checking the valves. Refer to Section 4-1 in Chapter 1. Check the oil pump by installing the gauge manifold with the low pressure gauge connected to the low pressure connection on the unit. Connect the other gauge to the outlet of the oil pump. The oil pressure is the difference between the two gauge readings. Make any repairs that are indicated during the above tests.

156–2 CHECKING OUT THE LUBE OIL SENSOR:

Use the following procedure to test for correct operation during the initial installation and maintenance operations: We will use the Johnson P345 Series Electronic Lube Oil Control for illustration purposes. Be sure to use the correct instructions for the make and model control being used.

1. Energize the supply voltage to the control and the compressor circuit.

WARNING: To avoid the risk of electrical shock or damage to the equipment, determine if more than one power supply voltage is present and disconnect all of them before proceeding.

2. Disconnect the wire leads between the contactor and compressor motor ("T" terminals as illustrated in most wiring diagrams). For examples, see Figures 2–19 through 2–24.
3. Re-energize the supply voltage to the control. Check to make certain that all operating and limit controls are closed to ensure that power is being supplied to the control.
4. The compressor contactor will immediately be energized and the yellow and green LED's will be lit. The green LED indicates that the compressor contactor is energized, the yellow LED indicates that the oil pressure differential is low and the timing circuit is energized.
5. In 45 seconds or 120 seconds the control will lockout the contactor. The red LED will illuminate while the yellow and green LED's will turn off. If an alarm is installed, the control's alarm contacts will close and an alarm will sound.
6. Press the RESET button. The red LED will turn off and the green and yellow LED's will turn on. The contactor is now energized.

NOTE: The control will remain locked out until the RESET button is pressed even if power is removed from the control. The control cannot be reset without power.

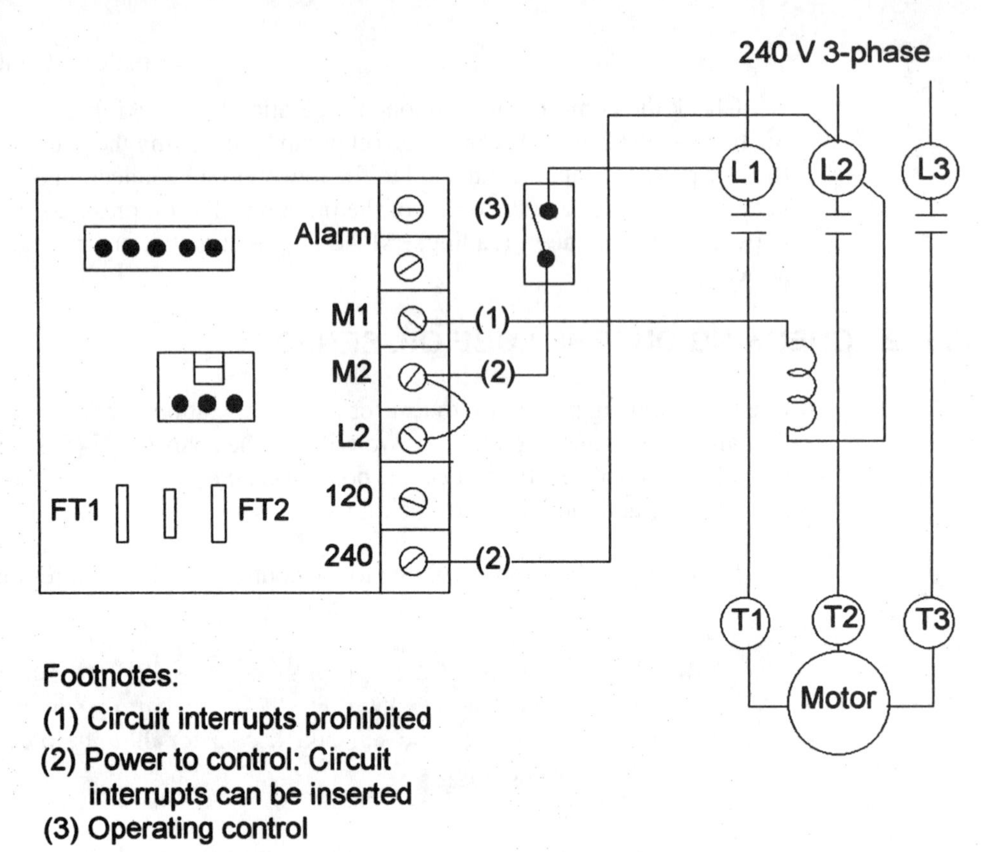

Figure 2–19. Oil failure control in 3-wire application.

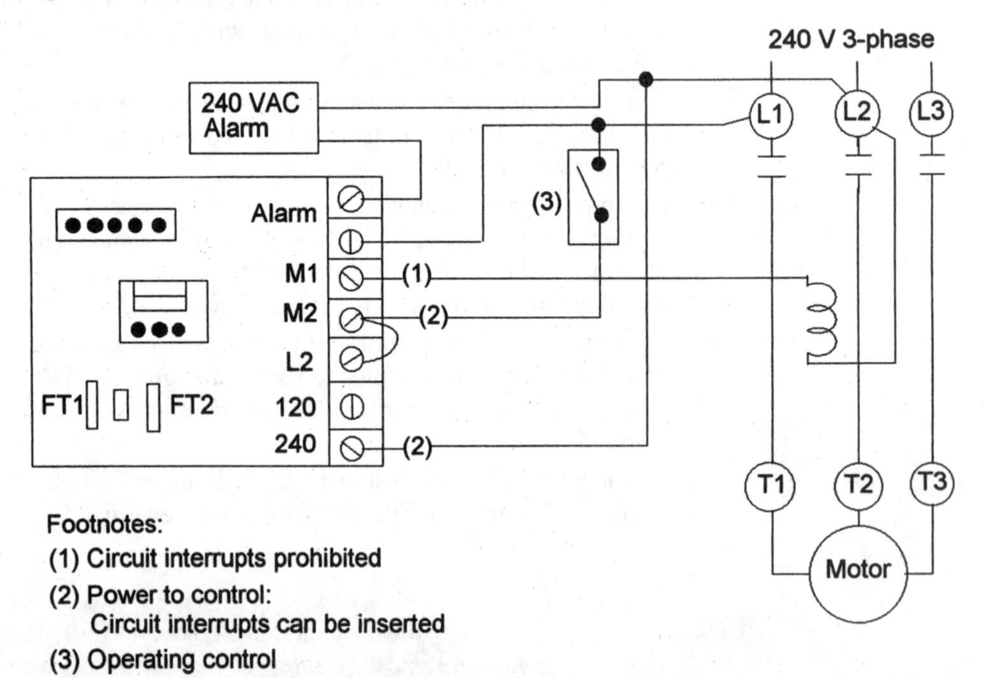

Figure 2–20. Oil failure control and an alarm circuit powered by the same voltage supply as the compressor motor.

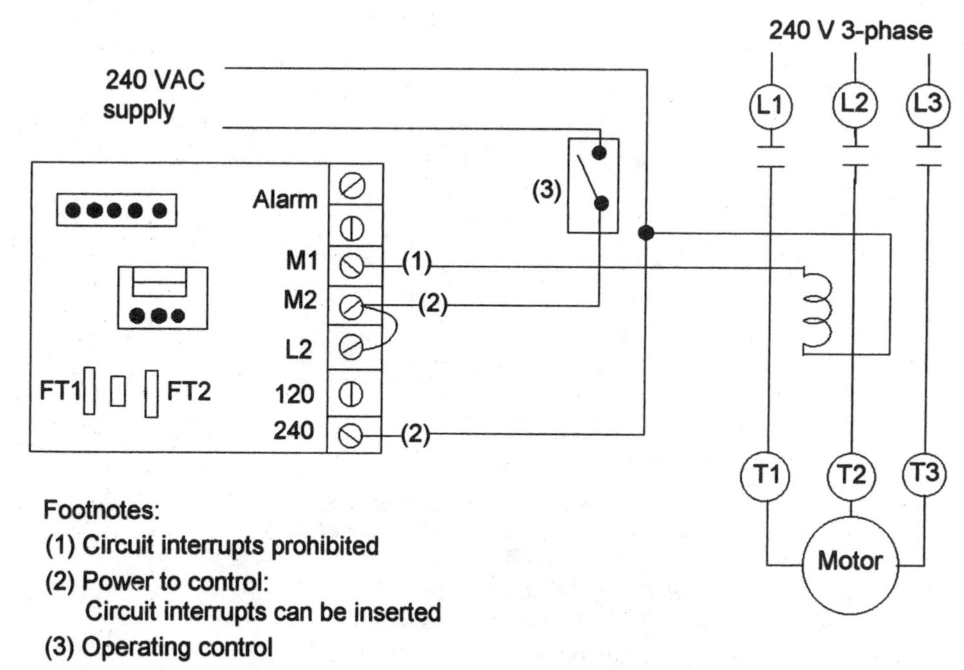

Figure 2–21. Oil failure control with separate voltage supply for the control circuit.

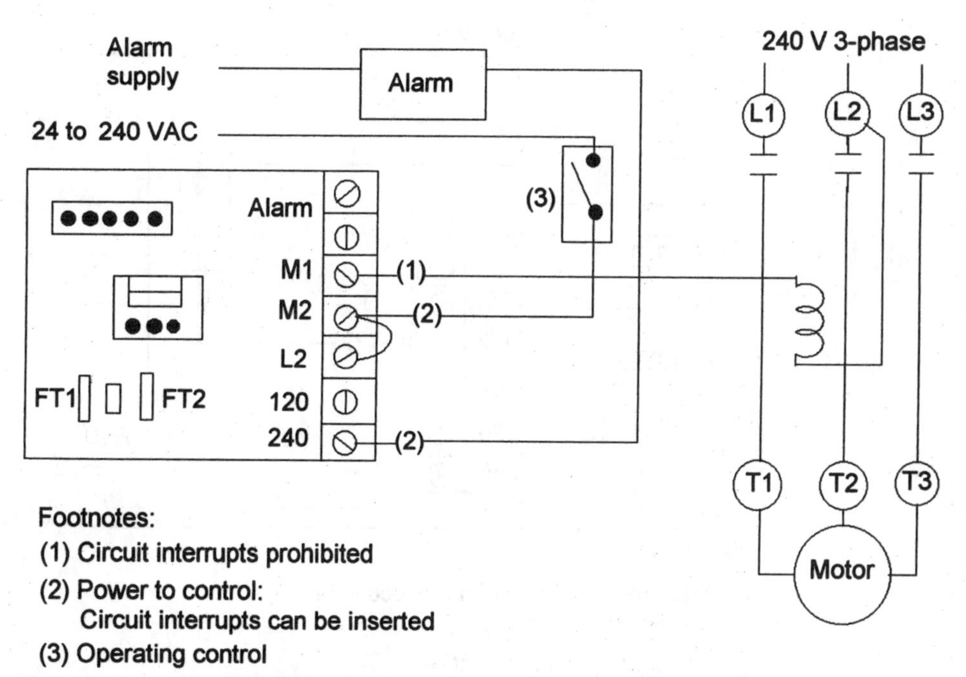

Figure 2–22. Oil failure control with the alarm circuit powered by a separate voltage supply.

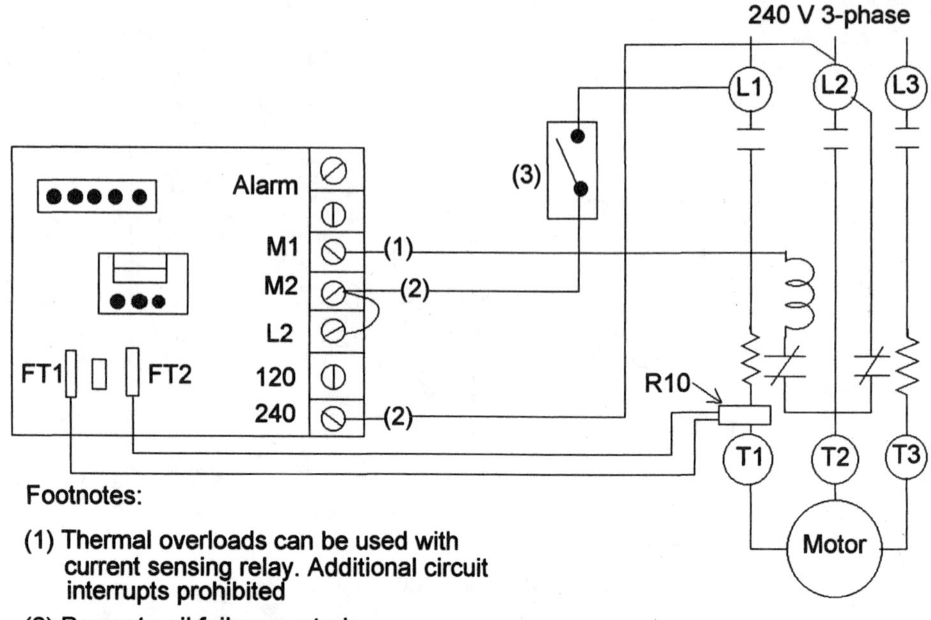

Figure 2–23. Oil failure control operating a motor starter with thermal overloads and an R10 series current sensing relay.

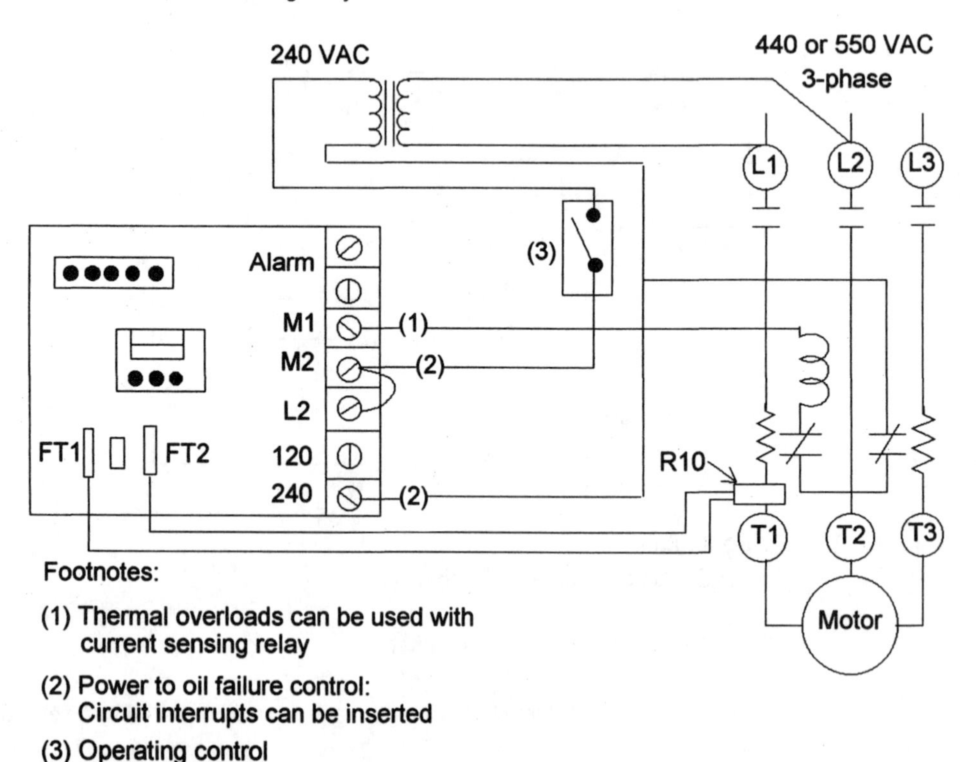

Figure 2–24. Oil failure control using 440 or 550 VAC supply and a 240 VAC starter holding coil and transformer.

7. De-energize the supply voltage. Reconnect the compressor leads to the contactor. If an R10A Series Relay is used, remove the jumper and reconnect the leads to the control.
8. Re-energize the supply voltage. If the operating and limit controls are closed, the compressor will start and both the green and yellow LED's will be lit. The yellow LED will turn off when the pressure level reaches the control set point, generally within seconds of starting the compressor.

156–3 CONTROL OPERATIONAL TEST:

Use this test to check to see if the P345 control is operating correctly. This test simulates a low oil pressure condition and initiates an abbreviated (eight seconds) timing cycle followed by a lockout of the compressor contactor.

1. With the power to the control and the contactor energized (only the green LED is on), press and hold down the test button.
2. The yellow LED (low pressure warning stage) will light for approximately eight seconds before the red LED (lockout stage) illuminates, and the control locks out the compressor contactor. The control alarm circuit is also energized.
3. To restart the contactor, push the RESET button.

157–1 GREEN LED ON:

This condition indicates that the compressor is not running. This could be caused by two problems with the unit: a faulty contactor or wiring or control failure.

157–1A

Faulty Contactor or Wiring: If the unit has been operating satisfactorily for some time before this problem occurred, the most likely place to start looking is at the contactor. Check the contacts to see if they are burned or worn so that they will not mate or move properly. If they are, replace the contacts or the contactor. Then check the wiring to make certain that it has not been changed during some previous service or maintenance procedure. If it has, rewire the unit according to the manufacturer's wiring diagram. To prevent this problem from recurring, when replacing the contactor, be sure to use one that will at least carry the load required of it.

157–1B

Control Failure: To check for proper control operation, use the Operational test described in Section 156–3. When it has been determined that the control has failed, it must be replaced. Never attempt to repair an oil failure control in the field. When installing the new control, make certain to take all of the safety precautions and follow the proper procedure to prevent damage to the control. After the installation is done, make a complete check of the system operation to make certain that the unit is operating satisfactorily.

158–1 NO LED'S ARE ON:

When this condition occurs the compressor will probably not be running. This condition could be caused by either no power to the control or a control failure.

158–1A

When there is no power to the control, check for proper voltage starting at the control and following the circuit back until power is found. Then repair the cause of the problem at this point. When all the problems are corrected the LED's should come on. Be sure to check the unit to determine what caused the power interruption.

158–1B

To check for *control failure* use the Operational Test described in Section 156–3. When it is determined that the oil failure control is defective it must be replaced. Never attempt to make repairs in the field. After completing the installation, check the unit operation to make certain that the unit is operating properly.

159–1 NORMALLY OPERATING LED'S:

Normally Operating LED's will sometimes occur and the compressor will still short cycle. This is usually an indication that the anti-short-cycling timer is set to "0" seconds. To solve this problem, move the anti-short-cycling block to some other timed position. See Figure 2–18. This will generally allow the compressor to run when it is supposed to. It may be necessary to make this selection more than one time to get it set accurately.

160–1 ERRATIC OPERATION, NUISANCE LOCKOUTS, DIMLY LIT LED'S IN THE OFF CYCLE:

This could be caused by the solid state time delay installed in the L_2 circuit as part of the unit control system. This problem can usually be solved by either removing the time delay or changing the time delay to one that operates on a 5 mA minimum current.

Another cause of this condition is that there is an interface with a computerized refrigeration rack control that may include a snubber circuit across its outputs (relay, triac, etc.). In this case install a 0.1 mfd 1000 volt ceramic disc capacitor across L_2 and the 120 or 240 VAC terminals on the control. See Figure 2–25.

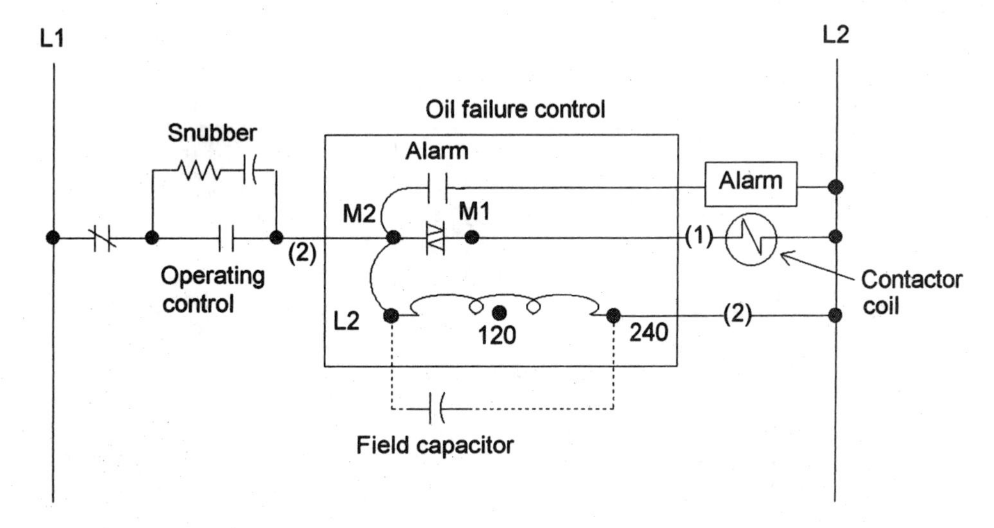

Footnotes:
(1) Circuit interrupts prohibited
(2) Power to oil failure control:
Circuit interrupts can be inserted
* Capacitor should be rated at 10 times the snubber capacitor.

Figure 2–25. Oil failure control ladder diagram showing operation with a field control containing a snubber circuit.*

161–1 ELECTRONIC LUBE OIL CONTROL (USING AN R10A SERIES RELAY):

The major cause of erratic operation is that the control does not respond to the R10A relay or the contactor short-cycles. To solve this problem, determine which is the culprit and make the necessary repairs.

161–1A

There are three things that could cause the *control not to respond to the R10A relay.* Either the resistor on the PC board is not cut, the anti-short-cycle is set at +0 seconds, or there is a faulty R10A Series Relay.

161–1B

Resistor on PC Board not Cut: Cut and discard the resistor. To install the R10A Series Relay into the circuit, cut and discard the R38 resistor. Connect the R10A relay to the two male blade terminals, FT1 and FT2. See Figure 2–26.

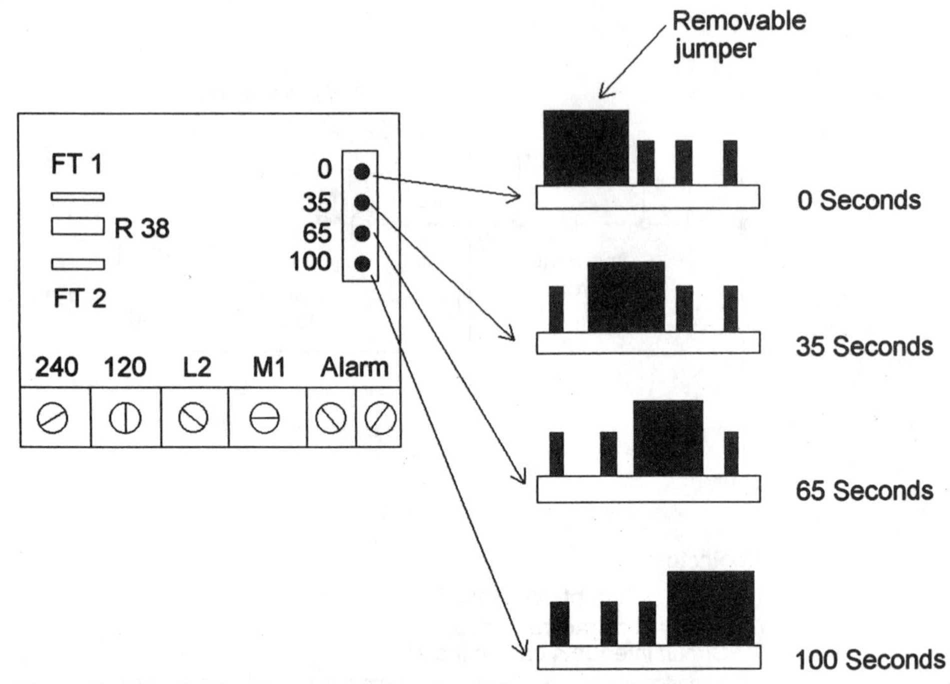

Figure 2–26. Setting the anti-short-cycling delay timer.

161–1C

When the *anti-short-cycle is set at 0 seconds,* change the setting by moving the selector block to a different setting to allow more time delay.

161–1D

When the other tests have been made and nothing has been found to cause the problem, *check the R10A Series Relay.* If any of its operations are found to be erratic or do not operate, replace the relay. Be sure to check the operation of the unit before leaving to make certain that it is operating correctly.

162–1 CONTRACTOR ENERGIZES FOR 3 OR 4 SECONDS, REMAINS OFF FOR ANTI-SHORT-CYCLING TIME DELAY, AND THEN REPEATS (COMPRESSOR IS UNABLE TO START DURING THE 3 TO 4 SECOND PERIOD):

This condition is usually caused by insufficient current to the R10A Current Sensing Relay. There are three possible causes for this problem: the compressor internal overloads, bad R10A relay connections or bad R10A relay, or not enough current passing through the current sensing loop.

162–1A

To *check the compressor internal overloads,* make certain that all electricity to the unit is turned off. Remove the wiring from each of the overload terminals. Be sure to mark the wires so that they can be placed back in their proper places. With an ohmmeter, check the continuity of each of the overloads. If either of the overloads does not have continuity, allow the compressor to cool to the reset temperature of the overloads and then recheck the continuity. If there is still no continuity, either replace the compressor or if the overload is replaceable, replace it. When the compressor must be replaced, be sure to use the correct refrigerant removal and recharging techniques. Put the unit back in operation and completely check it to make certain that it is operating correctly before leaving the job.

162–1B

Check all the *connections to the R10A relay* and if any are found bad, make the connection correct. If any wires are burnt, they should be replaced, or if long enough they could be cut off until good insulation will reach the terminal.

162–1C

Check the R10A relay and *determine if any part of it is not functioning properly.* If improper operation is found, replace the relay. Be sure to check the unit operation before leaving the job to make certain that it is repaired.

162–1D

If it is found that *insufficient current is flowing through the sensing loop,* the best thing to do is run another loop of wire through the sensing loop. This will provide enough current flow to energize the R10A relay.

163–1 ELECTRONIC OUTDOOR THERMOSTAT:

Electronic outdoor thermostats are made up of a control board that mounts on or near the indoor blower coil and a remote temperature sensor that mounts on the outdoor unit. See Figure 2–27.

The control board has a terminal strip for connecting the wiring and indicator lights to indicate which stage is being used.

Depending on the model, there are either 2 or 4 stages with an overall range of −10°F to 50°F. The individual range of each stage and the factory setting is shown in Table 2–2.

When the 24 VAC common wire from the indoor blower is connected to terminal “B” and not one of the staging terminals, the electric heat will come on when the second stage of the indoor thermostat calls for heat. Some utility companies will not like this because it places an extra load on the electric distribution system.

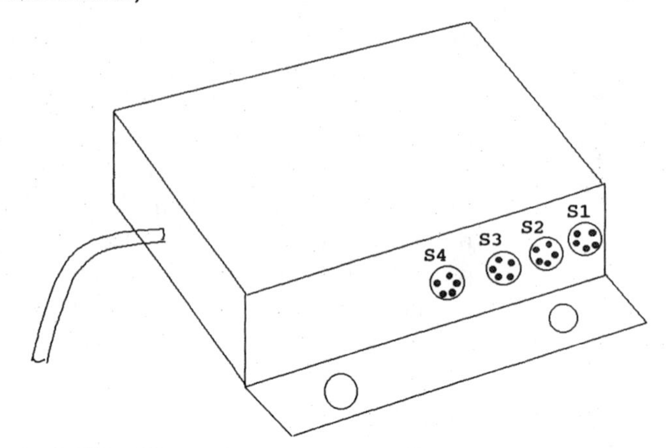

Figure 2–27. Electronic outdoor auxiliary heat thermostat.

Table 2–2. Typical Heil Time Delay Stages

Model	Stage	Degree Range	Factory Setting	Timing Delay* Between Stages
NAMA0010T	1	15 to 50	50	5 to 30 secs.
	2	−10 to 15	15	5 to 10 secs.
NAMA0020T	1	30 to 50	50	5 to 30 secs.
	2	15 to 30	30	5 to 10 secs.
	3	0 to 15	15	5 to 10 secs.
	4	−10 to 0	0	5 to 10 secs.

*Actual time for the heat to come on will vary because of the different energizing times of the heater sequencers.

When the "B" wire from the indoor blower is connected to one of the staging terminals, it can only come on when the actual outdoor air temperature is below the thermostat setting. See Figure 2–28.

Determine which stages to connect to the "B" wire based on the temperature range and the balance point of the conditioned space.

Next set the desired ON temperature for the required stages.

When replacing the board, set the new board at the same temperature range(s) as the one being replaced. Make the electrical connections just like those on the old board.

Each stage is factory set at the maximum ON temperature. This is the full clockwise position.

When adjustment is needed, insert a small screwdriver through the hole to reach the adjustment screw. See Figure 2–29.

Make a mark on the screwdriver blade at one of the reference points on the scale markings around the hole. This will show a reference point if needed in future adjustments. Turn the adjustment screw counterclockwise to the desired temperature setting. Each mark on the scale represents a 2°F lower temperature setting.

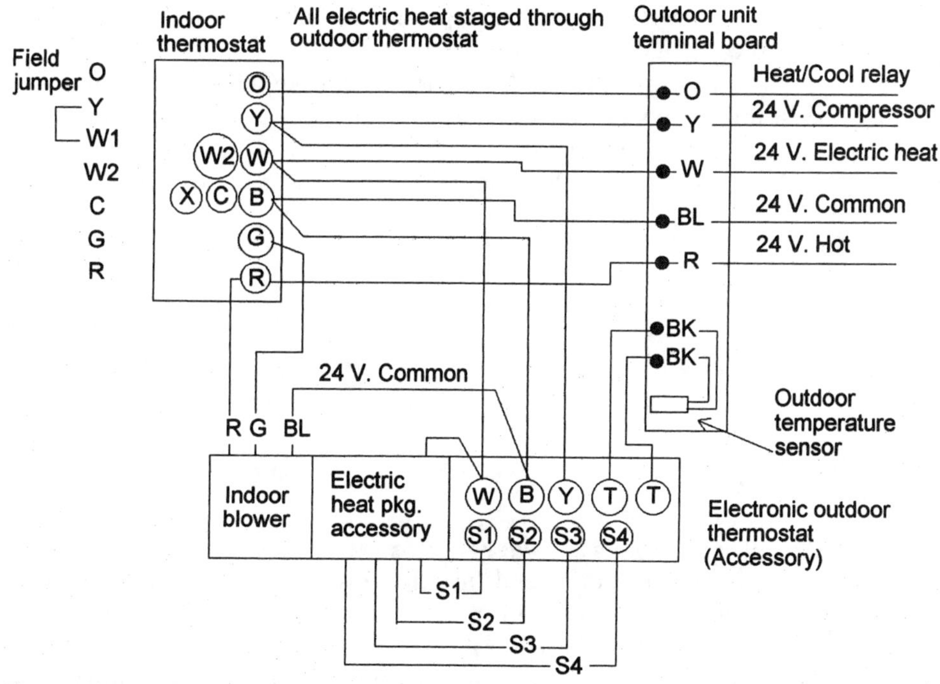

Figure 2–28. Typical heat pump/electric heat circuit.

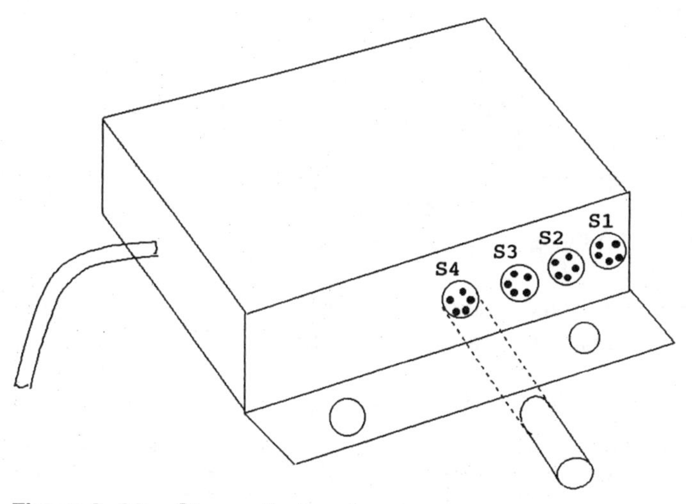

Figure 2–29. Stage adjustment.

163–1A

To complete the *operational check*, switch the indoor thermostat system switch to the heating position and move the temperature selector setting above room temperature so that the second stage (aux.) is calling for heat. At this position there should be 24 VAC at the W terminal on the outdoor thermostat terminal board.

Unless the outdoor temperature is below this stage setting, there will be no indicator lights on.

Disconnect one of the wire leads from the outdoor temperature sensor at terminal T on the board. All of the LED lights should come on.

Reconnect the sensor lead to T. All the LED lights should go off.

Switch the indoor thermostat to the EMER. HEAT position. All the LED lights should come on.

163–1B

There are several steps involved in the *troubleshooting* of an outdoor thermostat. These steps follow.

163–1C

When checking the stages and temperature sensors, in most cases there will be a 5 to 10 second or longer delay after power is applied to the board before the first LED light will come on. This time delay will also occur between each of the stages.

The timing between the stages and the switching can be seen by watching the LED lights by cooling the outdoor temperature sensor with ice and water.

However the sensor must be disconnected at a given temperature. Otherwise the indoor heat strip will come on.

Table 2–3 can be used to check against at least two different temperatures.

The actuating points for each stage can be checked against approximate temperatures by placing a known resistance across terminals "T" and "T" in the board. A variable resistor with a range of 20K to 125K ohms can be used to set the resistance to correspond to the temperature listed in the table. This allows more flexibility than using a single value resistor.

Use an ohmmeter to set the resistance to correspond to the temperature listed in the table.

Connect the resistor across the T and T terminals on the board.

Adjust each stage in this manner until the LED light is on. See Figure 2–30.

Table 2–3. Resistance Values/VS Approximate Temperature Rise

Resistance In Ohms	Approx. Temp. Fahrenheit	Resistance In Ohms	Approx. Temp. Fahrenheit
20K	50	46K	20
23K	45	54K	15
26K	40	63K	10
30K	35	74K	5
35K	30	87K	0
40K	25	100K	−5
		119K	−10

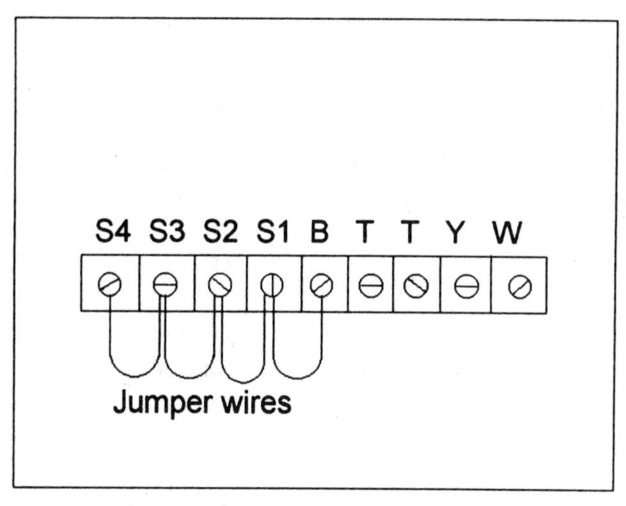

Figure 2–30. Outdoor thermostat control board.

163–1D

When *all the electric heat is on at the same time, this could be an indication* of three things. A failed open outdoor sensor, bad electrical connections, or a broken wire to the sensor.

163–1E

An outdoor sensor can fail in several different ways. They can physically open. In this case there will be no continuity through the sensor. Replace the sensor with the correct replacement. Check to determine what caused the sensor to fail. Make any repairs before installing a new sensor to prevent damage to the new one.

163–1F

Bad electrical connections will reduce the amount of current flowing through the circuit. A low current flow will cause several intermittent problems with the sensor. Check all terminals in the sensor circuit and make any necessary repairs. If the terminals are corroded, remove and clean them, then replace and tighten them to the terminal board. If a wire is burnt, either replace the wire or if it is long enough to allow cutting it off back to good insulation, then cut the wire. Install a new connector and tighten the wire onto the terminal.

163–1G

When a *broken wire to the sensor* is found it must be either repaired or replaced. If the break is close enough so that the wire can be cut at the break and still connect to the sensor terminal, it may be cut. If it is not long enough to cut and still make an easy connection, replace the complete wire and connectors.

164–1 NO ELECTRIC HEAT:

This problem could be caused by a failed (shorted) sensor. When this condition occurs, remove one of the leads at terminal T. The LED lights should come on. Replace the sensor with the proper replacement. Be sure to check all the circuits to make certain that there is not a problem that would cause the sensor to short again.

165–1 NOT ENOUGH HEAT ON TO MAINTAIN THE COMFORT SETTING:

This problem could be caused because either the stages are set for too low a temperature, or a failed stage has occurred on the control board.

165–1A

When *the stages are set for too low a temperature setting,* set the temperature settings on the outdoor thermostat higher. They should not be set to come on before the balance point of the heat pump is reached, except under severe circumstances. This would do away with any savings realized by the heat pump system.

165–1B

To check for *a failed stage on the control board,* remove one of the wires from the outdoor sensor terminal T. All the LED sensor leads lights should come on. This condition requires replacement of the control board before the system will operate as it was designed.

166–1 POSITIVE TEMPERATURE COEFFICIENT RELAY (PTCR):

The PTCR solid state motor starting device operates similar to the motor-starting capacitor and the potential starting relay. See Figure 2–31.

The solid state device improves the starting ability of a PSC motor by temporarily causing an increase in the current supplied to the motor start winding. See Figure 2–32.

When the voltage is initially applied to the motor start winding, the PTCR immediately starts heating up because of the current draw through the solid state medium inside the PTCR Because of this heat build-up, the solid state material

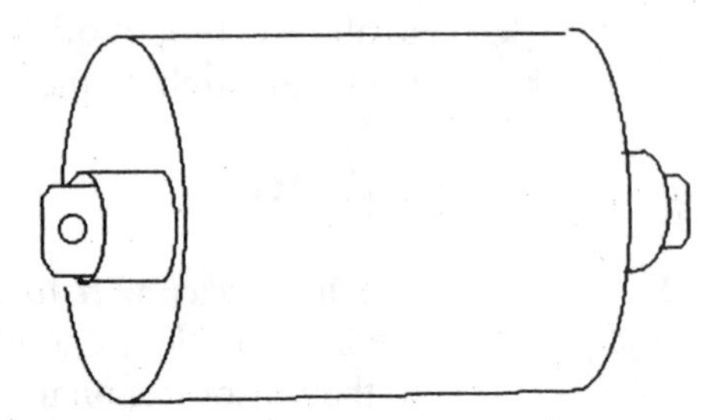

Figure 2–31. PTCR motor starting relay.

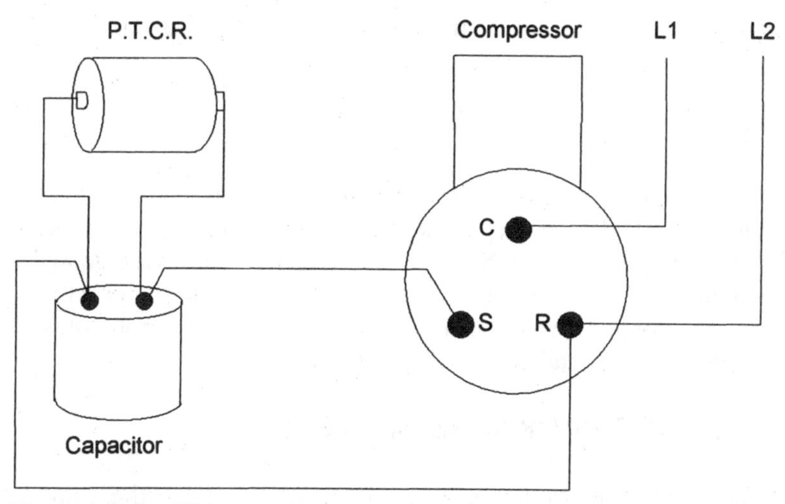

Figure 2–32. Wiring connections for a PTCR device.

changes from a very low resistance to a very high resistance material. When in the heated condition, it essentially becomes a non-conductive control. This allows the motor to return to the PSC running mode.

The PTCR motor starting device may be used on single phase PSC type compressors when some type of starting assistance is needed because of temporary low voltage, or with new or high efficiency compressors.

WARNING: Danger of bodily injury or death from electrical shock if the PTCR is connected to high voltage. Disconnect the electrical power to the unit before working on or removing the PTCR device.

166–1A

When *the compressor hums but will not start* there is usually a problem caused by a defective PTCR device. Using the proper safety precautions, replace the PTCR with the proper replacement. Be sure to check the unit voltage and amperage before leaving the job to make certain that the system is working properly.

166–1B

When *the compressor will sometimes start and sometimes will not,* it is usually an indication of a weak PTCR device. Using the proper precautions, remove the device and install a correct replacement. Be sure to check the voltage, amperage, and run the unit through several starting cycles to make certain that it is operating properly before leaving the job. If the problem is still present, look elsewhere for the problem. It must be found and repaired. Otherwise a call back will result with a disgruntled customer.

166–1C

If *the compressor starts but will stop after a few seconds,* this usually indicates a failed PTCR device. Using the proper precautions, remove the device and install one of the proper type. Start the unit and check the voltage and amperage before leaving the job. It would also be a good idea to run the unit through several starting and stopping cycles before leaving the job. If the problem is still present check elsewhere. Find and correct the problem.

167–1 SOLID STATE CRANKCASE HEATER:

The solid state crankcase heater is energized when the compressor is not running. This is possible because of the change in resistance through the solid state material. The material used in solid state crankcase heaters attempts to maintain a constant temperature. When the compressor is not running, the compressor oil sump cools down and heat is conducted away from the solid state material. See Figure 2–33.

Since the material tries to maintain a constant temperature, it lowers its resistance and allows more current to flow through the heater. This current flow replaces the amount of heat being conducted away from the heater by the cooler crankcase. Thus, the cooler the crankcase is, the more heat is produced by the crankcase heater because of the higher wattage flow through the solid state material.

During the running cycle, heat is generated by the compressor motor. This heat is conducted to the oil sump, which in turn heats the oil. As the oil sump becomes warmer, less heat is conducted away from the solid state material. This in

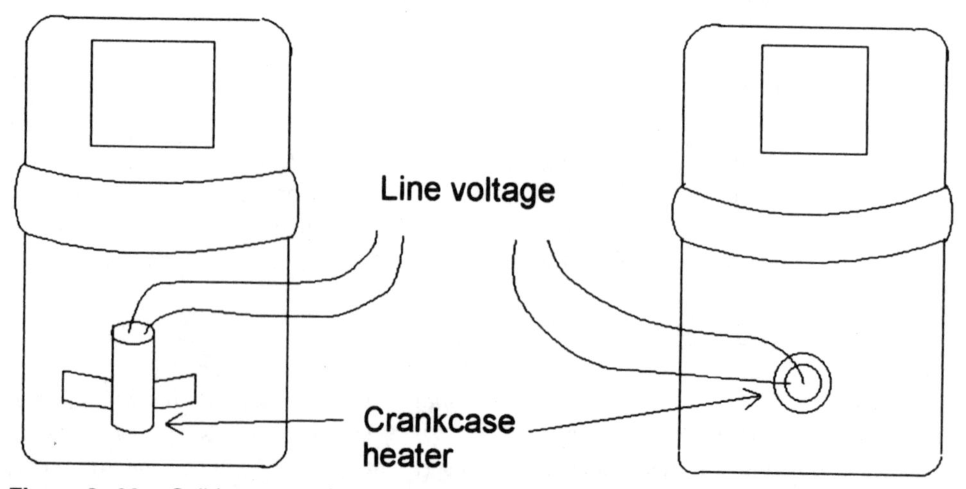

Figure 2–33. Solid state crankcase heaters.

turn causes its resistance to current flow to increase. Less heat is produced by the crankcase heater. As the compressor oil sump reaches about 100°F, the current flow through the heater is very small and for all significant purposes, the heater is turned off.

167–1A

One *possible reason for a noisy compressor when starting* is that the crankcase heater is not working. It is letting the compressor slug liquid refrigerant and oil. Another reason could be that there are bad electrical connections to the heater. Also there could be a broken wire to the heater circuit.

167–1B

To see if *the crankcase heater is not working,* feel the heater with the back of the fingers to see if it is warm or hot to the touch. See Figure 2–34.

If it is not working, check to make certain that there is power to the heater terminals. If not, check the voltage back to the power panel for the problem. Make any repairs needed to make the heater work again. To further check the heater, touch the heater sheath on the protrusion made by the solid state assembly. When the compressor is off and the oil sump is cooled to ambient temperature, the assembly should feel warm to the touch.

If power is found at the heater terminals, turn off the electrical power to the unit. Replace the heater. Turn the power to the unit back on and check for voltage at the heater terminals. If there is voltage, start the unit and allow it to run for about 24 hours. This will allow everything to stabilize. Then check the heater again. The problem must be repaired or the compressor will soon lose its valves because of liquid slugging. This will require that a new compressor be installed.

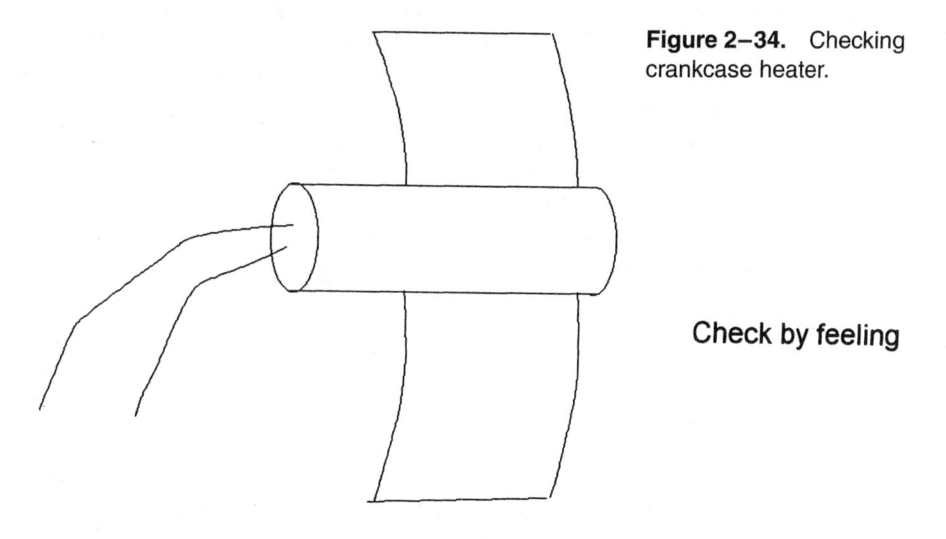

Figure 2–34. Checking crankcase heater.

168–1 ELECTRONIC DEFROST SYSTEM:

Electronic defrost controls are electronic timing devices that are used on refrigeration systems that operate with the evaporator temperature below freezing and on heat pump systems. The correct operation of these controls is necessary for the proper operation of these types of systems.

168–1A

Defrost Control Operation: To initiate the defrost cycle, the temperature sensor, mounted on the outdoor coil, must be closed. The sensor closes at 28 degrees ± 5 degrees and the unit must have run an accumulated total time of 30, 60, or 90 minutes (as selected).

To advance the accumulated run time, the HLD terminal is energized with 24 volts when the unit is running and the sensor is closed.

The accumulated run time is not lost when the unit cycles off, the memory is maintained by the 24 volts supplied by the 24V terminal from the R side of the transformer.

168–1B

When *the unit is not wired properly,* there is only one solution; rewire the unit correctly. Wire the unit according to the equipment manufacturer's wiring diagram for the particular make and model being serviced. Then start and check the system for proper operation.

168–1C

24 volts not available at the control board could be caused by a bad transformer, a loose electrical connection, a burned electrical terminal, or a broken wire. Start at the control board and check for 24 volts back toward the transformer. When 24 volts are found, the problem has also been located and must be repaired. If it is a loose connection it must be tightened. Corroded terminals must be removed and cleaned before tightening. Replace a bad transformer with one of the correct VA rating. Find the cause of the bad transformer. It will usually be an overloaded secondary circuit. This overload may be caused by a wire that has had the insulation rubbed off, a pinched wire, or a shorted component in the secondary circuit. Any overload must be repaired or the transformer will only burn out again. Before leaving the job, start the unit and check the amperage in the secondary circuit to make certain that the overload has been removed and that the system operates properly.

168–1D

The *control switching and timing not functioning properly* could be due to the adjustable defrost time interval. The defrost control board has a jumper wire connected from W_1 to T_1, T_2, or T_3. When the defrost timing interval needs to be changed, remove the jumper from T_1, T_2, or T_3 and connect it to the desired terminal. T_1 = 30 minutes, T_2 = 60 minutes, and T_3 = 90 minutes.

When this problem occurs check as outlined below. To check out the timing and switching functions of the control panel use the following steps:

1. Turn off all power to the unit.
2. Place a jumper wire across the two pins labeled TST. See Figure 2–35. This will speed up the timing cycle as follows: 90 minutes = 21 seconds, 60 minutes = 14 seconds, and 30 minutes = 7 seconds.
3. Connect a jumper from the COM terminal on the defrost control board to the common terminal (#3) of the heat/cool relay. This eliminates the defrost sensor and will simulate a cold coil.
4. Make sure that the jumpers mentioned above are connected properly, touching only the terminals listed in step #2 and #3.

NOTE: Shorting of any of the terminals could possibly cause control board failure.

5. Turn the electrical power on to the unit and adjust the thermostat to demand heating. The unit should start and go into a defrost cycle within the accelerated timing limits outlined in step #2. The unit will cycle in and out of defrost until the jumpers are removed. Prolonged operation in the test mode is not recommended.
6. If the defrost control performs the above tests, remove the test jumpers and place the unit back in operation.

NOTE: This is a timer/switching test and not a complete system check, such as the proper charge, reversing valve operation, etc.

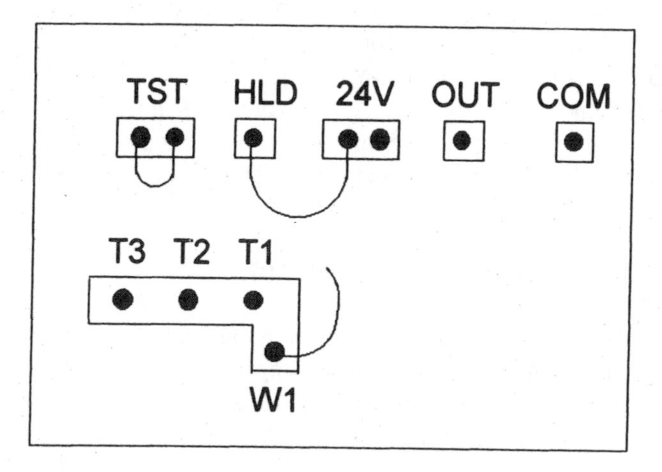

Figure 2–35. Defrost control board.

If the defrost control board failed to perform any of the steps as described on the previous page, refer to the defrost timer troubleshooting section.

168–1E

Cycling of the *defrost relay* is controlled through the defrost control board.

If the defrost relay does not cycle, troubleshoot the control board following the procedure given in the defrost system-defrost control section.

During the defrost cycle the defrost relay coil is energized through the defrost control. If the defrost relay does not cycle, check the defrost control.

Electrical power is supplied from the contactor to terminal 4 and then to 6 through the contacts.

NOTE: Terminal 4 to 5 is now open. The heat/cool relay is not denergized and the fan will not run.

The circuit continues from the 6 terminal of the defrost relay to the 2 terminal of the heat/cool relay. See Figure 2–37.

From terminal 2 to the reversing valve coil. The reversing valve will shift.

The low volt Y to W circuit is also now completed through the defrost relay-contacts 3 to 1, to bring on the electric heat during the defrost cycle. If any of these conditions are not met, check for a bad defrost relay.

168–1F

The purpose of the *temperature defrost sensor* is to determine when a defrost is needed and when the unit should go out of defrost. To field test a sensor, turn off the power to the unit, disconnect the sensor wiring and remove the sensor from the unit. Fill a small container about 75% full of ice and add water to the level of the ice. Place an accurate thermometer in the ice/water solution and add salt until the temperature is in the 20 to 25 degree range. Fill another glass with water and chill or warm it so that its temperature is in the 70 to 75 degree range. Connect an ohmmeter across the sensor leads and submerge the sensor into the salt brine solution. Within approximately 60 to 90 seconds, the sensor switch should close and continuity will be indicated on the ohmmeter. Remove the sensor from the solution and place it in the 70 to 75 degree water. The sensor switch should open (approximately 90 seconds) when warmed. If the sensor fails to close or open during the above test, replace it with a correct replacement.

168–1G Troubleshooting (No Defrost)

Troubleshooting the electronic defrost system can sometimes be a problem if the procedures are not familiar to the person doing the troubleshooting. The following steps should help provide this information. Be sure to use the procedures recommended by the equipment manufacturer for the make and model being serviced.

1. Check the wiring to make certain that the system is wired according to the unit wiring diagram. Refer to Figure 2–36.
2. Check between the terminals marked 24V on the control board and terminal 3 (common) on the heat/cool relay. Make this check with a voltmeter to prevent damage to the components. There should be 24 volts between these two points. If not, check the power source and/or for loose wiring connections.
3. Turn the power to the unit back on and place the thermostat system switch to demand heat. (With the test jumpers in place as described in the section *Check Out Of Timing And Switching Functions.)* Using a voltmeter check between the terminals labeled HLD and COM. There should be 24 VAC between these two terminals. If not, check for loose wiring connections.

   If 24 VAC is indicated between these two terminals, check the voltage between the HLD terminal and the OUT terminal for a minimum of 30 seconds. Since the switching function of the board happens between the COM and OUT terminals, the voltage fluctuates between 0 and 24 volts. The voltage will fluctuate at the accelerated timing rate listed in the section *Check Of Timing and Switching Function.* For example, a 90 minute setting will cause the voltage to fluctuate every 21 seconds. If the voltage fluctuation is not seen, replace the defrost control board.
4. If the voltage fluctuation is seen, the defrost relay should be heard energizing and deenergizing. If not, check the voltage at the low voltage terminals of the relay. If the fluctuating voltage is indicated, replace the defrost relay.

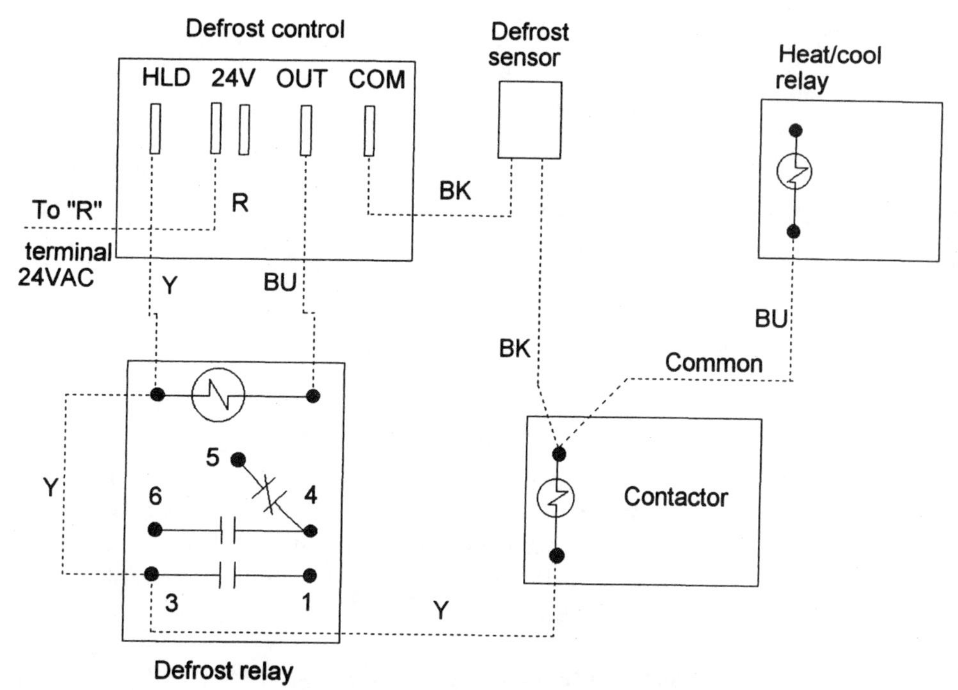

Figure 2–36. Typical electronic defrost system schematic wiring diagram.

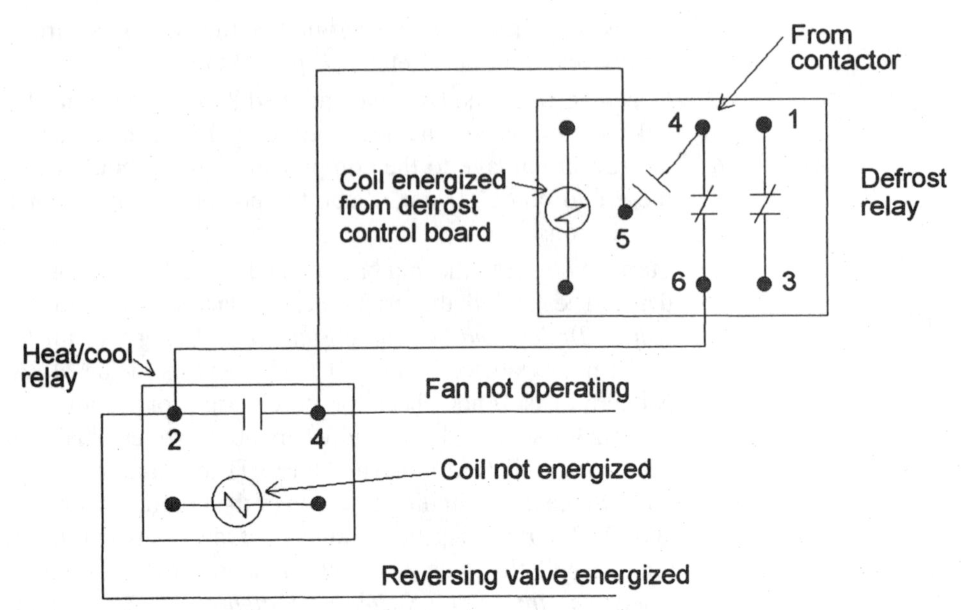

Figure 2–37. Typical defrost relay and cycle operation.

Figure 2–38. Location of defrost sensor.

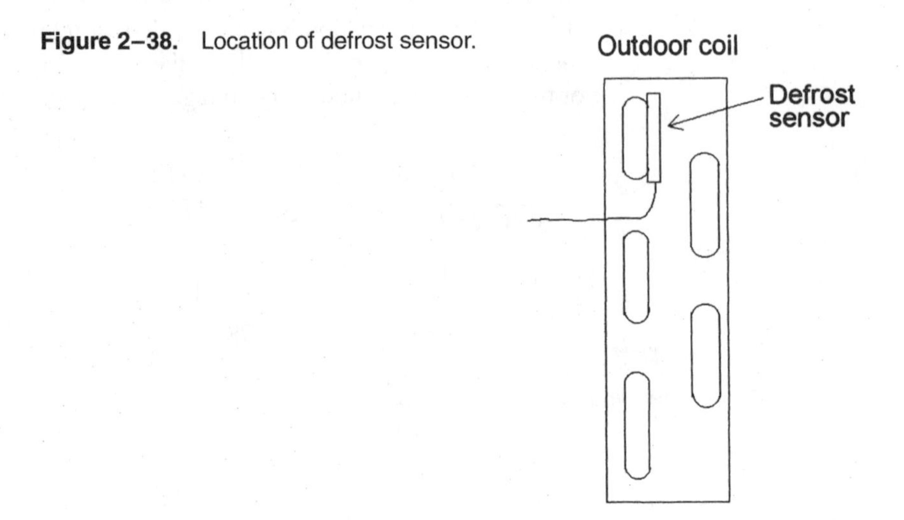

5. The defrost temperature sensor closes at 28 degrees ±5 degrees and opens at 68 degrees ±5 degrees. When the sensor switch is closed, it completes the common circuit to the timing board to start the timing sequence. The sensor is attached to the outdoor coil with a stainless steel clip. The sensor should fit snugly on the tube. The clip can be bent to adjust for a good fit. See Figure 2–38.

Be sure to use the sensor location recommended by the equipment manufacturer for the make and model being serviced.

169–1 ELECTRONIC IGNITION:

The following is a listing of some of the more possible problems that may be encountered in the field when working on electronic ignition systems. These problems may vary depending on the make and model of ignition system being repaired.

CHECKING OUT DIRECT IGNITION SYSTEMS: This is a summary of the check out procedure for the Honeywell direct ignition system. Be sure to use the procedure recommended by the system manufacturer for the make and model being serviced.

Step 1: Perform a Visual Inspection.

1. With the power off, make sure that all wiring connections are clean and tight.
2. Turn on the power to the unit and the ignition module.
3. Open the manual shutoff valves in the gas line to the unit.
4. Complete a gas leak test ahead of the gas control if the piping has been disturbed.

Step 2: Review the Normal Operating Sequence and Timing Summary. See Table 2–4.

Step 3: Reset the Module.

1. Turn the thermostat to its lowest setting.
2. Wait one minute.

Table 2–4. Typical Honeywell Direct Ignition Module Timing Summary

Function	S825	S87A-D	S87J, K	S89A, E	S89B, F	S89C, D	S89G, H	S890C, D	S890G, H
Prepurge	—	—	30 Sec. min.	—	30 Sec. min. 45 Sec. Max.	—	—	30 Sec. min. 37 Sec. Max.	30 Sec. min. 37 Sec. Max.
No. of ignition trials	1	1	1	1	1	1	3	1	3
Safety lockout timing (nom.)[a]	6, 11 or 21 Sec.	4, 6, 11 or 21 Sec.	4, 6, 11 or 21 Sec.	4, 6, 11 15[b] or 21 Sec.	4, 6, 11, 15[b] or 21 Sec.	4, 6, 11 15 or 21[c] Sec.	4, 6, 11, or 15 Sec.	4, 6, 11 or 15[b] Sec.	4, 6, 11 or 15 Sec.
Flame failure response	0.8 Sec. Max.	0.8 Sec. Max.	0.8 Sec. Max.	0.8 Sec. Max.	0.8 Sec.	2 Sec. Max.[d]	2 Sec. Max.[d]	2 Sec. Max.[d]	2 Sec. Max.[d]
Ignitor warmup time.	—	—	—	—	—	34 Sec.	34 Sec.	34 Sec.	34 Sec.

[a]*Ignition continues until burner lights or system locks out.*
[b]S89E, F only.
[c]S89C before 9/87 only.
[d]with 2.5 microamp flame signal.
[e]S89C before 9/87 warmup period was 45 sec. With 2.5 microamp flame signal.

As steps 4 and 5 are completed, watch for points where operation deviates from normal. Refer to the troubleshooting chart to correct the problem.

Step 4: Check Safety Lockout Operation.

1. Turn off the gas supply.
2. Set the thermostat above room temperature to demand heat.
3. Watch for sparking at the ignitor either immediately or following a prepurge period. See Table 2–4.
4. Check the lockout timing.
5. After the lockout, open the manual gas cock and make sure that no gas is flowing into the main burner.
6. Set the thermostat below room temperature and wait one minute before continuing.

Step 5: Check the Normal Operation.

1. Set the thermostat above room temperature to demand heat.
2. Make sure that the main burner lights smoothly and without flashback. Several attempts may be necessary to clear the gas line of air.
3. Make sure that the main burner lights smoothly without floating, lifting, or flame roll-out to the furnace vestibule or heat buildup in the vestibule.
4. If the gas line has been disturbed, complete a gas leak test. Gas Leak Test: Paint the gas control gasket edges and all piping connections downstream of the gas control, including the enrichment tube connections, with a rich soap and water solution. Bubbles indicate gas leaks. Tighten any leaking joints and screws or replace the component to stop the gas leak.
5. Turn the thermostat below room temperature. Make sure that the main burner flame goes out.

169–1A No Heat. Red LED Light not on

There are three things that could cause this problem: no line voltage, a bad door switch, and no 24 VAC at sec-1 and sec-2 of the transformer.

When no line voltage is found between L_1 and L_2, check for a blown fuse or a tripped circuit breaker. Replace the fuse or reset the circuit breaker to get voltage. If voltage at these points is found, check on down the line toward the unit for other problems. Other problems could be a broken wire, bad terminal, or a set of bad contacts in one of the switching relays. Repair the wiring and connections or replace the relays that are causing the problem. Be sure to check the access door switch for continuity. When the door is removed, the switch must be activated manually. Push the switch closed and check for voltage between the two terminals. If voltage is found, the door switch is bad and must be replaced. If voltage is not found, make certain that the door is pushing the switch all the way closed. If not, make the necessary adjustments to make certain that the door causes the switch to close.

169–1B No 24 VAC At Sec-1 and Sec-2 on the Transformer

Check the secondary side of the transformer to see if there is 24 VAC between the two terminals. If none is found, check the primary voltage to the transformer. If voltage is found, the transformer is bad and must be replaced. Be sure to use an exact replacement or one that has enough capacity to handle the secondary load.

To replace the transformer, turn off the electricity to the unit. Disconnect the primary and secondary terminals and remove the transformer from its mountings. Install the new transformer and reconnect the wiring to the correct terminals. Turn on the electricity to the unit and see if it will run. If it does, check the current draw through the secondary wiring of the transformer.

If the unit does not run, check the voltage on both the primary and secondary terminals. If voltage is found here, the problem is elsewhere and it must be found and repaired before the system will operate again. Be sure to check the current draw through the secondary circuit of the transformer to make certain that it is not overloaded. If it is overloaded, some wiring changes must be made to take part of the load off the transformer. Otherwise it will only burn out again.

169–1C 24 VAC at Sec-1 and Sec-2

Usually when there is voltage at these points and the system will not operate, the trouble is in the unit control board. To make the repair, turn off all electricity to the unit and replace the board. Make sure to use the correct board or the unit will not operate properly. These boards are so similar it is usually best to check the board number to make certain that the correct one is being used. Turn the electricity to the unit back on. If the thermostat is demanding heat, the unit should start and operate properly.

170–1 NO HEAT. RED LED LIGHT ON AND BLINKING RAPIDLY WITHOUT PAUSE:

This problem could be caused by the wrong power supply polarity or the wrong control voltage.

170–1A

When *the power supply has the wrong polarity* it may be corrected by simply changing the wiring on the two power inlet terminals. This will change the polarity. If the LED light stops blinking the problem is solved. Turn the thermostat to demand heat. The system should go through its starting cycle. This indicates that the system has been repaired.

170–1B

If the light does not stop blinking in the previous example, the *wrong control voltage* is supplied to the board. Check the transformer to make certain that it is producing the correct voltage required for the control board. When the wrong voltage is supplied, change the transformer to the correct type. When the wiring has been changed, in step 170-1B, it may be necessary to change the wiring back to the original connections. Turn the thermostat to demand heat. The system should go through the complete lighting sequence.

171–1 NO HEAT. RED LED NOT BLINKING:

There are generally two conditions that could cause this problem: the power supply is the wrong voltage or the wrong control voltage is supplied to the unit.

171–1A

To check the power supply for the wrong polarity, Turn off the electric power to the unit. Change the supply voltage wiring to get the correct polarity. Turn the power back on. Set the thermostat to demand heating. The unit should go through the complete lighting sequence.

171–1B

To determine if the *wrong control voltage* is being used, check the secondary side of the transformer to see if the correct voltage is being supplied by the transformer. If not, the transformer must be replaced. Turn off the power supply to the unit. Replace the transformer with the correct secondary voltage. Make the correct wiring connections. Turn the power back on to the unit and test the secondary voltage; if it is correct, connect the wiring to the control board. Set the thermostat to demand heat. The unit should come on and go through the complete lighting sequence. If the supply voltage polarity was changed as in step 170-1A it may need to be changed back to the original connections.

172–1 NO HEAT. RED LED LIGHT BLINKING ON/OFF SLOWLY WITH A COMBINATION OF SHORT AND LONG FLASHES:

The only thing that would cause this condition is that the unit is off on a status code. To repair the unit, determine what the status code is and make the necessary repairs or adjustments. If the unit has been off for 48 hours or the power supply is interrupted to the board the unit should operate again. The status codes are erased after 48 hours or whenever the 115 VAC or 24 VAC power is interrupted.

173–1 NO HEAT. IGNITION FAILURE NOT PROVEN:

This condition is usually caused by no voltage between pins 1 and 2 on the board during the trial for ignition. Also, there may be bad wiring connections or weak continuity in the harness and the ignitor.

173–1A

When the is *no voltage between pins 1 and 2 on the control board during the trial for ignition,* the board is faulty. To replace the board, turn off the power supply to the unit. Make certain that you have the correct replacement board. Disconnect the wiring, remove the board. Replace the board and reconnect the wiring in on correct terminals. Turn the power to the unit on. Set the thermostat to demand heat. The unit should go through the complete lighting sequence.

173–1B

Check for *bad connections or continuity in the harness and ignitor circuit.* All the connections must have good electrical contact and they must not be burned or corroded. Repair all connections and remove any corrosion from them. Also check the continuity of the harness and the ignitor circuit. If either the harness or the ignitor does not have the correct continuity as indicated by the manufacturer's instructions, replace it with the correct replacement. Set the thermostat to demand heating. The unit should go through the complete lighting sequence.

174–1 NO HEAT. IGNITION FAILURE PROVEN:

There are several possibilities that could cause this problem. Be sure to check each of them to make certain that only one is causing the problem.

174–1A

When *no voltage is found to the gas valve,* check the transformer to make certain that secondary voltage is being produced. If voltage is found at the transformer secondary, the problem can be in the control board. Turn off the power to the unit and replace the control board. Be sure to use an exact replacement. Make the wiring connections exactly as they are shown on the unit wiring diagram. Turn the power to the unit back on. Set the thermostat to demand heating. The unit should go through the complete lighting sequence.

174–1B

Check the *wiring to the gas valve to make certain that the terminals are all tight and are not bad or corroded.* Repair any loose or burned connections and remove all corrosion from them. The corrosion will only reduce the voltage passing through them. It will also eventually ruin the connection that would require replacement of the wiring and perhaps the control.

174–1C

A *gas valve not opening* will cause a "no heat" problem. Make certain that all the gas valves are turned to the open position. If they are found in the correct operating position, check the voltage to each of the valve terminals. If voltage is not found, trace the wiring back to the point where the voltage stops and repair or replace the wiring. If correct voltage is found at all the gas valve terminals, replace the gas valves. They are either sticking closed or the valve coil is bad.

174–1D

Main burners not igniting could be caused by two problems. First, inadequate flame carryover on rough ignition. Second, low inlet gas pressure.

Inadequate flame carryover or rough ignition could be caused by either too high or too low inlet gas pressure or by dirty carry over wings on the burners. To solve this problem, check the inlet gas pressure at the gas pressure connection on the gas valve. Perform this test with the burners operating to detect any unnecessary obstructions in the line. If it is not within the designated limits for the type of gas being used, make the adjustment to bring it within the limits. The limits are 3.5 to 4 water column on natural gas, and on LP gas it is 11″ water column. Or what ever the equipment manufacturer recommends. Also make certain that the burners are clean and present no obstruction to the gas flowing through the ports and burner orifices.

Most of the time when the burners do not light, the problem is caused by a low inlet gas pressure. Check the gas pressure at the outlet of the gas valve. See Figure 2–39.

The pressure must be within the limits recommended by the equipment manufacturer. If it is not, make the necessary adjustments to provide the recommended gas pressure. This test must be performed with the burners in operation. If the line is too small or has restrictions in it, they would not be found if the burners are not operating.

174–1E

After Ignition the Main Burners go off: To check this problem, set the thermostat below room temperature. Wait a few minutes and set the thermostat to demand

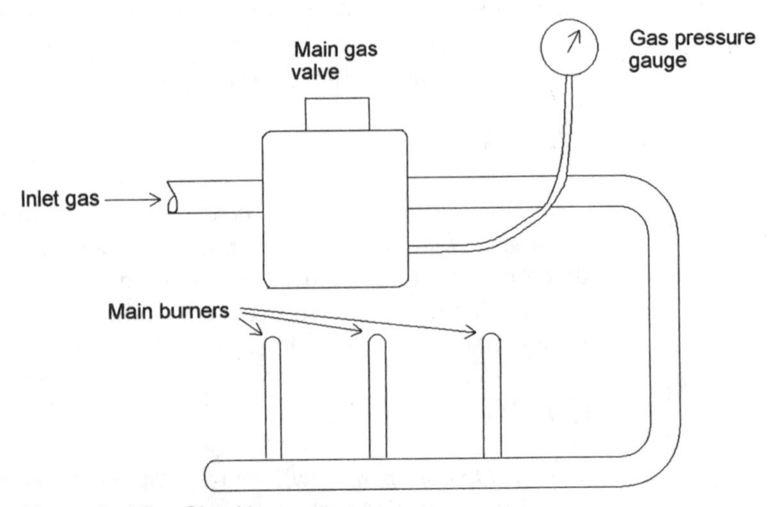

Figure 2–39. Checking gas pressure.

heating. While the burners are igniting check the flame sensor current. If it is not within the manufacturer's recommendations, either place the sensor in another position or replace the sensor. It may be possible to clean the sensor if it is very dirty. This dirt will sometimes reduce the amount of current flowing through the circuit.

174–1F

A *low micro amp flow on the sensor* can be caused by two conditions: loose connections and a dirty flame sensor. See Figure 2–40.

First check for bad connections. Tighten and repair all the connections. If the unit still does not operate, replace the control board. Turn off all the power to the unit. Be sure that you have the correct control board for the application. Remove all the wiring from the board. Remove the board and replace it with the correct one. Rewire the new control board according to the instructions that came with it. Turn the power to the unit back on. The unit should complete the ignition sequence and the burners remain lit. Check the micro amperes through the electrode. If they are still too low, replace the electrode.

A dirty flame sensor can cause this same problem. When the flame sensor is dirty, remove it from the furnace and clean it with either small grit sand paper or steel wool. Replace the sensor. Set the thermostat to demand heat. The unit should go through the complete ignition sequence. Be sure to check the microamp flow through the electrode. If it remains low, replace the the ignition electrode. See Figure 2–41.

174–1G

When the *main burners do not stay on after ignition,* about the only action to take is to replace the control board. To replace the control board, get one that is an exact replacement for the one being replaced. Turn off all power to the unit. Disconnect the wiring from the control board. Remove the control board and install the new one. Reconnect the wiring according to the directions with the board. Turn the

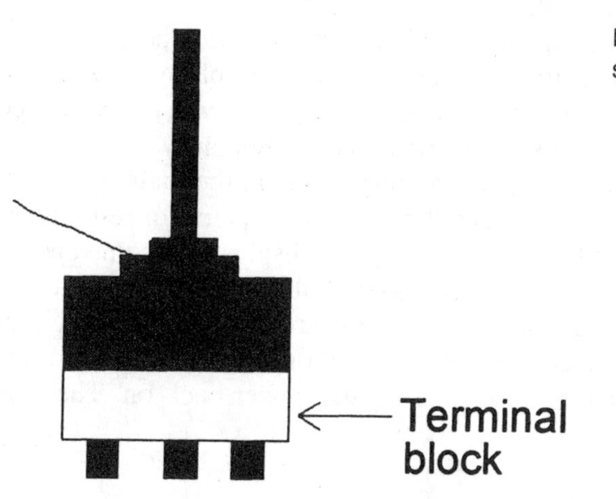

Figure 2–40. Typical flame sensor.

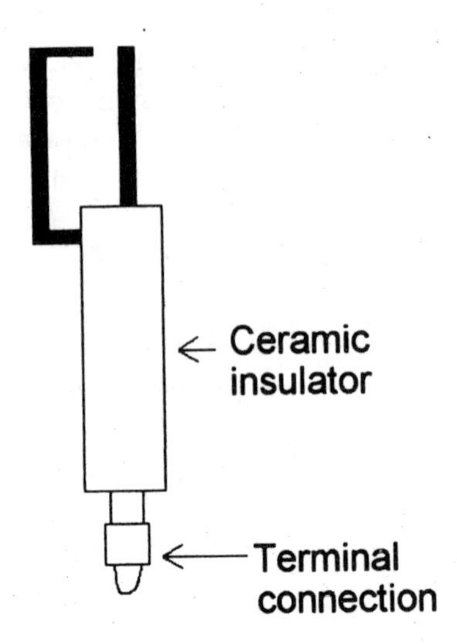

Figure 2–41. Typical ignition electrode.

power to the unit back on. Set the thermostat to demand heat. The furnace should come on and go through the complete ignition sequence.

175–1 TROUBLESHOOTING ELECTRONIC THERMOSTATS:

Electronic thermostats are becoming more popular all the time. Their accuracy, trouble free operation, and ability to operate without being reset for each timing cycle are the main reasons for their popularity. However, there are several things that can go wrong with them that will require the services of a trained technician. The major problems will be discussed as follows.

175–1A

Several things could cause a *blank display;* such as, no power from the thermostat or power supply, the thermostat is not mounted properly, bad electrical connections, or a bad thermostat.

When the thermostat is not receiving power from the transformer or power supply, first check the batteries in the thermostat. Replace them and make certain that the holders are tight and are making good contact with the battery contacts. Be sure that the batteries are installed with the correct polarity.

A blown fuse or a tripped circuit breaker to the main power source will cause the display to be blank. Check the fuses and replace, or reset the breaker. If either of these were the problem, the thermostat display will come back on.

Check for power at the heating or cooling unit. If no power is found at either of these places, check the power wire toward the fuse panel or circuit breaker panel until the loose connection, broken wire, or other problem is found. Turn off the power and repair the problem. Turn the power back on. The thermostat display should come on.

Check to make certain that the thermostat is mounted on the subbase properly. If not, make the necessary changes to properly mount the thermostat on the subbase. When the correct mounting is made, the thermostat display should light.

Bad electrical connections or wiring between the thermostat and the equipment will also cause the display to sometimes be blank. Check these connections and correct any problems found with the wiring in this circuit. When properly connected, the thermostat display will come on.

Another component that may cause a blank display is a bad thermostat. Check the power at the transformer secondary terminals to be sure that power is being supplied to the thermostat. If no power is found here replace the transformer. Be sure to turn off the power to the unit before replacing the transformer to prevent electrical shock and possible equipment damage. When the transformer is replaced and the wiring completed, turn the power to the unit back on. The thermostat display should light and operate. If the display does not light, check the thermostat to see if it is bad. If it is, replace it. The system should now come on and operate properly.

175–1B

The *thermostat microprocessor has been locked-up by a large static discharge.* A static discharge may be caused by the user walking across a wool rug then touching the thermostat. Low humidity in the building will also cause static electricity.

Reset the thermostat microprocessor by turning off the 24 VAC power for at least 20 minutes. When the power is turned back on the thermostat should operate properly. If this problem continues to happen it may be necessary to install additional static electricity protection. Be sure to check with the equipment manufacturer or its representative for the correct procedure for this modification.

175–1C

There are a couple of things that could cause *all parts of the display to be lit, but the time not being updated.* The thermostat may be either stuck in a test or setup mode or it could be faulty.

To check to see if the thermostat is stuck in the test mode, press the Run Program button or run program to exit the test mode.

After these two tests are made and the thermostat still does not operate properly, the thermostat is bad. Replace the thermostat with the correct make and model for the application. When a wrong type thermostat is installed, there will be at least as many, if not more, problems with the unit as before the replacement.

175–1D

The only thing that could be occurring when *only parts of the display flash during operation* is that there is a bad electrical connection between the thermostat display and the microprocessor. The only correction for this problem is to replace both the thermostat and the subbase. Be sure to use the correct model and make for the installation. Otherwise the system will not operate as designed. Be sure to turn off the electrical power to the unit before starting to replace the thermostat. When the installation is completed, turn the power to the unit back on. Reprogram the thermostat according to the users needs. Test the thermostat for proper operation before leaving the job to make certain that the problem has been solved.

175–1E

Generally *when the display flashes during operation,* a large discharge of static electricity has caused the thermostat to lose its program.

Remove the batteries from the thermostat and leave them out for 20 minutes. This will allow the thermostat to reset. If this does not solve the problem, turn off the 24 VAC power to the thermostat so the microprocessor can reset. If this problem continues to recur, additional static protection may be needed. Be sure to use the protection method recommended by the thermostat manufacturer so the system will operate properly.

Another thing that could cause this problem is the capacitor that stores and updates the display has been discharged. This only requires resetting the time and day on the thermostat.

175–1F

When the *display is flashing BAT, REPL BAT, BAT LO, LO BAT, or BATTERY,* the thermostat battery is low and must be replaced. If it is not soon replaced, the thermostat will stop operating. Always use new alkaline batteries in electronic thermostats.

175–1G

When the thermostat *display reads CHECK MALF, EM. HEAT, LEDS OR TROUBLE INDICATOR,* this indicates that there is trouble with the heat pump compressor. It has either stopped or something has happened to keep it from operating at its peak output. Check the heat pump unit to find and repair the problem. Otherwise the system will heat on emergency heat causing a high electric bill.

175–1H

When the thermostat *temperature settings will not change,* either the upper or lower temperature settings have been reached. When this problem occurs, check the thermostat manufacturer's specifications for the temperature range of the thermostat.

If the thermostat is an automatic changeover type, the heating and cooling settings are too close together. Check the manufacturer's specifications to verify how close these settings can be set. Be sure to keep the settings within the limits recommended by the manufacturer. Use the method recommended by the thermostat manufacturer to make the needed changes.

175–1I

When the thermostat *clock does not keep accurate time,* this usually indicates a discharged capacitor. The discharge is usually caused by a power failure.

Try setting the time and day on the thermostat clock. If it cannot be reset, the capacitor is probably ruined requiring replacement of the thermostat. If a new subbase is needed, be sure to wire it according to the manufacturer's recommendations or it may not operate properly. Reprogram the thermostat to meet the users needs. Place the system back in operation and check the system to make sure that it is operating properly.

175–1J

Usually when the *thermostat program was lost during a power failure* this indicates that the thermostat batteries are bad. The batteries keep the thermostat settings until power is returned to the circuit.

Replace the batteries with new alkaline batteries and reprogram the thermostat. To replace the batteries, remove the thermostat from the subbase. Remove the old batteries and install the new ones. Be sure to install them with the correct polarity as indicated on the battery holder. Replace the thermostat on the subbase and reprogram it using the thermostat manufacturer's instructions.

This problem could also be caused by the batteries being installed with the wrong polarity. To change the polarity, remove the thermostat from the subbase, remove the batteries and reinstall them with the polarity of the batteries matching the markings on the battery holder. Replace the thermostat on the subbase and reprogram the thermostat using the thermostat manufacturer's recommendations. If the thermostat will not reprogram, replace the batteries with new alkaline type. If the thermostat cannot now be reprogrammed, it is bad and must be replaced.

175–1K

An *incorrect temperature displayed* can be caused by several different problems: the thermostat is configured for °F or °C display, bad thermostat location, the display shows actual room temperature and not the temperature setting, and many thermostats have a feature that will bring the room temperature up in small increments called the ramp temperature.

Check the thermostat to see if it is configured for °F or °C. When the wrong range is used, set it to the correct range. When the thermostat is installed in a place where it will sense the outdoor temperature more than the indoor temperature, there may be hot water pipes inside the wall behind the thermostat. Also, the sun could be shining through a window and hitting the wall close to the thermostat or the thermostat itself. Sometimes the wall will vibrate when the door is opened or closed, or the outdoor air may blow on the thermostat when the door is opened. There are many other conditions that might upset the proper operation of the thermostat. When any of these conditions occur, the thermostat must be installed in a better location.

Mount the thermostat on an inside wall close to the return air grill or in the return air flow. Mount it about 5′ from the floor and make sure it is level. Be sure that there are no hot or cold water pipes inside the wall behind the thermostat. When the thermostat wires are run down the wall from the attic, or up the wall from the crawl space, be sure to plug the hole behind the thermostat to prevent cold air from entering and affecting its operation.

175–1L

When *the display shows actual room temperature and not the temperature setting,* this usually indicates normal operation. Pressing the Current Setting, Present Setting, or the Temperature Setting key will show the temperature setting of the thermostat. Actually there is nothing wrong with the thermostat when this condition occurs.

175–1M

When the *thermostat does not follow the program,* this is sometimes caused by the thermostat being set in the hold or manual mode. To correct this problem, press the Run Program button and the problem will usually be corrected. Some thermostats may have a button marked differently. Be sure to use the thermostat manufacturer's instructions.

This problem could also be caused by the thermostat not being properly programed. Reprogram the thermostat using the manufacturer's instructions. Be sure to check the thermostat operation before leaving the job to make certain that the unit is operating as it should.

175–1N

When *the temperature changes at the wrong time,* it is usually caused by the thermostat not being properly programed. This problem is simply corrected by reprogramming the thermostat. Be sure to use the thermostat manufacturer's instructions.

Some thermostats have a feature that controls the heating equipment to maximize comfort. This feature causes the thermostat to determine how long it will take to bring the space temperature up to the normal occupied temperature. This is also known as the ramp temperature. It is calculated by the thermostat microprocessor. In most thermostats it will bring the temperature up three degrees at a time until the

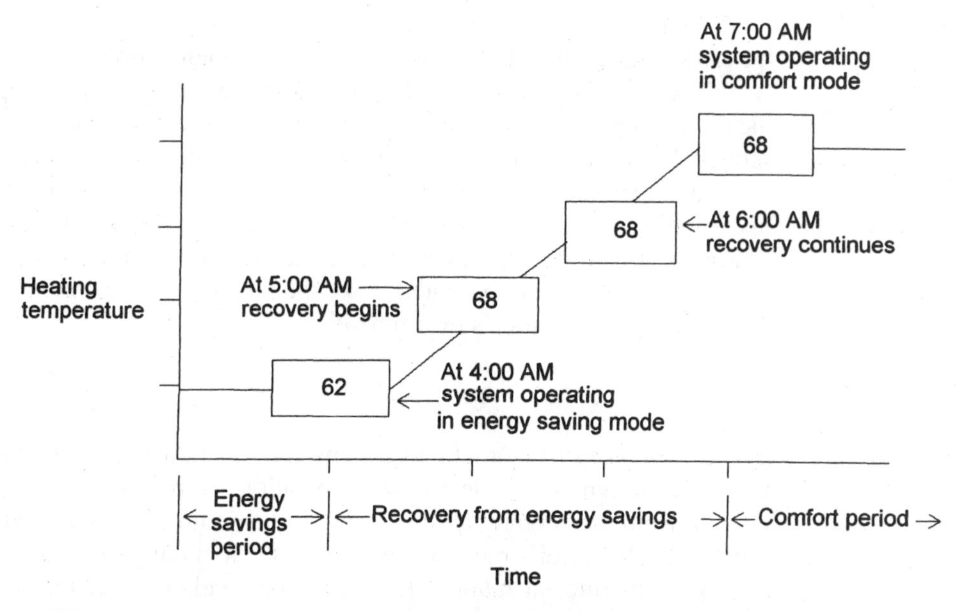

The thermostat uses the reverse steps to gradually lower the temperature for cooling.
Figure 2–42. Typical control temperature setting (ramp temperature setting) during the energy savings period and recovery from energy savings for heating.

occupied space temperature setting is reached. If the user does not like or want this feature, remove the thermostat from the subbase and make the adjustment on the back of the thermostat. See Figure 2–42.

Place the unit in operation and check to make certain that it is operating properly before leaving the job.

175–1P

When *the heating and cooling units are on the same time setting*, it is sometimes cause by the thermostat being wired incorrectly. To correct the problem, wire the thermostat and the unit according to the equipment and thermostat manufacturer's instructions. Be sure to start the unit and make certain that it is operating as it should before leaving the job.

This problem could also be the result of the wrong thermostat being used. The only solution to this problem is to install a correct thermostat. Be sure to use the correct wiring diagram when rewiring the thermostat so that the unit will operate as it should. Reprogram the thermostat using the manufacturer's instructions.

175–1Q

Generally when *the fan will not stop,* the thermostat fan switch has been placed in the ON position, thermostat is calling for heat or cool, the thermostat is programed to run the fan continuously, or the thermostat has an internal fan delay during heating or cooling.

When the problem is caused by the fan switch being in the ON position, move the fan switch to the AUTO position. The fan should stop running.

If the thermostat is calling for heating or cooling, there is no problem unless the room temperature is either too warm or too cool for the thermostat temperature setting. Then the equipment must be checked to find the problem.

Some thermostats are equipped with an internal fan control that will delay fan operation of either the heating or cooling unit. To solve this problem, the thermostat must be reconfigured to overrun the fan delay feature. To do this, remove the thermostat from the subbase and make the necessary adjustment according to the thermostat manufacturer's recommendations.

175–1R

The fan will not start sometimes because fan operation from the thermostat is not properly configured for electric heat operation. To reconfigure the thermostat, remove it from the subbase and make the proper adjustments according to the thermostat manufacturer's recommendations. When making this configuration be sure to use the thermostat manufacturer's recommendations. If the thermostat is not equipped with this feature, a new thermostat with the proper fan configuration must be installed. Be sure to wire the thermostat according to the equipment and the thermostat manufacturer's recommendations for proper operation.

Another method that may be used to correct this problem is to install a switching relay. This is normally the fan switch that is used on other heating units to start the fan when the furnace has warmed sufficiently for fan operation. Be sure to wire the thermostat so that it will start the fan at the correct time. If it is set too low, cool air will be blown into the space during the heating cycle. If the temperature is set too high, the heating elements may get too hot and burn out. Be sure to check the unit to make certain that it is operating properly before leaving the job.

Another thing that could cause this problem on cooling is the thermostat has a built in fan time delay to keep it from starting until the indoor coil is cool enough to prevent blowing warm air into the space. This will also allow the suction pressure to drop to near normal operating pressure before the fan is started to help lower the operating cost of the unit.

To check this problem, set the thermostat 5°F above room temperature. After one minute the fan should start. If it does not, the thermostat is bad and must be replaced with the proper type to provide the desired operation.

Also check the fan switch operation or the fan timer control. If either is bad, replace it with the correct replacement. If the fan switch is bad, the thermostat will need replacing. If the fan timer control is bad, replace it with a time delay control with the correct amount of delay before the fan is started.

175–1S

The *equipment cycling too fast or too slow* is usually caused by an incorrect thermostat cycle rate setting. To solve this problem, set the cycle rate timer properly. Be sure to use the thermostat manufacturer's recommendations.

175–1T

If *changing the dip switch or option screw causes no change in thermostat operation,* then it is usually caused by not turning off the thermostat power to reset the changed option in the microprocessor. To solve this problem, remove the thermostat from the subbase. Remove the batteries from their holder. Wait 20 minutes and reinstall the batteries with the correct polarity as marked on the holder. Reprogram the thermostat and replace it on the subbase. Start the unit and make certain that it is operating as desired before leaving the job.

176–1 NO HEAT:

No heat can be caused by several things: no power to the thermostat, blown fuse or tripped circuit breaker, no power to the heating or cooling unit, thermostat not properly mounted, bad wiring or connections, bad transformer, microprocessor locked up, minimum off time on the thermostat is on, system switch in the wrong position, thermostat improperly programmed, or equipment failure.

176–1A

No power to the thermostat can be caused by low or bad thermostat batteries. Check the batteries, if they are bad or weak replace them with alkaline batteries. When replacing the batteries be sure that they are installed with the correct polarity.

If the batteries check out good, remove them for 20 minutes. Then reinstall them with the correct polarity as marked on the battery holder. Reinstall the thermostat on the subbase and reprogram it to suit the customer's needs. Check the thermostat operation before leaving the job.

176–1B

A blown fuse or tripped circuit breaker can also cause a no heat condition. Check the fuses or circuit breakers. If the fuses are blown replace them. If the circuit breaker is tripped reset it. Then check for proper operation of the unit before leaving the job.

176–1C

When there is no *power to the heating or cooling unit* the equipment will not work. Check for power at both places and if none is found, trace the circuit back to the electrical panel. When the problem is found repair it. Remember, when working on electrical wiring be sure that the power is turned off before the place where the work is being done.

176–1D

When the *thermostat is not properly mounted on the subbase,* the contact points will not mate to the ones that they were designed to mate with. Current will flow through the wrong circuits and will usually damage the thermostat. Remove the thermostat and properly mount it on the subbase. Then check the equipment operation to make certain that it is operating properly before leaving the job. This will usually prevent a call back.

176–1E

Bad wiring or connections between the equipment and the thermostat could cause the equipment to not operate all the time or not operate at all. To solve this problem, check the wiring and connections back to the power panel. Repair any loose or corroded connections. Replace any wiring that has been overheated. Then check the wiring between the equipment and the thermostat. Repair any loose or corroded connections that may be found.

176–1F

A bad transformer will cause the equipment to not operate. To test the transformer, check the voltage at the secondary terminals. There should be 24 VAC indicated. See Figure 2–43.

If 24 VAC are not indicated, check the primary terminals. If line voltage is indicated, the transformer is bad and must be replaced. Be sure to use one that has enough VA to handle the load. If voltage is not on the primary terminals the problem is elsewhere and it must be found and repaired.

176–1G

A locked-up microprocessor can prevent the heater from operating. To check for a locked-up microprocessor, remove the 24 VAC power from the thermostat for 20 minutes to allow the microprocessor to reset itself. Then restore the 24 VAC and

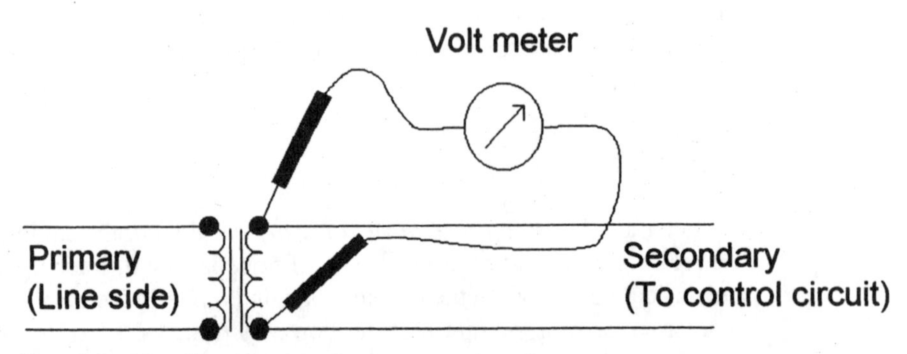

Figure 2–43. Checking transformer secondary for power.

check the thermostat operation. The thermostat may need to be reprogrammed before normal operation can be realized.

176–1H

The minimum off time on the thermostat is what causes the heating equipment to not operate until the off time has lapsed. Wait about 5 minutes to make certain that the minimum off time is not causing the problem. If the unit comes on and starts operating, the minimum off time had the unit off. If the unit does not come on, the problem is due to something else. Find the other condition and repair it.

176–1I

The system switch placed in the wrong position will cause the heating equipment to not operate. Check the system switch and if it is in the wrong position, move it to the HEAT position. The heating unit should come on and start operating. If not, the problem is elsewhere and must be located and repaired before the system will operate properly.

176–1J

When a thermostat is improperly programmed, the heating unit will not operate as desired. To see if something has caused the program to be overridden, press the Run Program button. If the unit comes on and starts operating, the problem is solved. If it does not come on, reprogram the thermostat according to the thermostat manufacturer's instructions and the user's wishes. Be sure to run the unit through a complete cycle before leaving the job.

176–1K

Equipment failure could be the cause of no heat. Check the system to see if there is a problem with it. Some of the safety devices may have the equipment off. The pilot or ignitor may not be working. Be sure to check all the equipment components for problems and make any needed repairs.

177–1 ELECTRONIC OIL BURNER CONTROLS:

Electronic oil burner controls are used to monitor the operation of oil burners to keep them operating safely. When they shut-down the system, there is a problem that must be identified, located, and repaired before the system will perform as it was designed.

177–1A

When the oil burner motor does not start, the trouble may be in the primary control circuit. This circuit includes the cad cell and the stack relay. Check both of these controls and if they are not functioning properly replace the defective one. Be sure to set the control to operate with the same conditions that the one being replaced was set to meet.

CHECKING OUT CAD CELL PRIMARIES: Use the following procedure for checking the operation of these controls. These steps should be done after the installation and at the completion of all service and troubleshooting procedures.

Starting Procedure:

1. Push the red reset button and release it.
2. Open the hand valve in the oil supply line.
3. Set the thermostat to call for heat.
4. Close the line switch; the burner will start.
5. Under normal conditions, the burner operates until the thermostat is satisfied or the line switch is opened.

Troubleshooting Cad Cell Primaries:

For complete troubleshooting of an oil burner installation both the burner and ignition systems, as well as the primary control, must be checked for proper operation and/or condition.

First check out the following parts of the burner and ignition systems:

1. The main power supply and the burner motor fuse.
2. Ignition transformer.
3. The oil filter.
4. The electrode gap and position.
5. The contacts between the ignition transformer and the electrode.
6. The oil pump pressure.
7. The oil piping to the tank.
8. The oil nozzle.
9. The oil supply.

If the trouble does not seem to be in the burner and/or ignition systems, check the cad cell and the primary control.

CHECKING OUT THE PROTECTORELAY (STACK RELAY) CIRCUIT:

1. To check for flame failure operation, shut off the oil supply hand valve while the burner is running. The safety switch should lock out in the switch timing (varies from 15 to 70 seconds depending on the make and model), the ignition stops, the motor stops, and the magnetic oil valve closes. Some intermittent ignition models shut down immediately and attempt one restart before the safety switch locks out. The safety switch must be manually reset before operation can continue.

2. To test for ignition and/or fuel failure, shut off the oil supply while the burner is off. Run the system through the starting procedure, omitting step 2. The safety switch should lock out and it must be manually reset.

3. To check for a power failure, turn off the power supply while the burner is operating. When the flame goes out, restore the power and the burner will restart.

177–1B

Trouble in the thermostat could also cause the burner motor to not start. Check the thermostat to make certain that current is passing through the thermostat heating circuit. If it is not, replace the thermostat with a proper replacement. If there is a circuit through the thermostat, the problem is elsewhere and must be found and repaired.

177–1C

A broken wire or loose thermostat connection can keep the oil burner from starting. Check the thermostat for any loose or corroded connections. Disconnect and clean any corroded connections or the unit still may not operate. Several corroded or loose connections can cause enough resistance to prevent enough current from flowing through the control circuit to operate the controls. If any broken wires are found, check to find the reason. Then repair the broken wire or connection. Turn the unit on and it should operate. Run the unit through a complete operating cycle before leaving the job.

177–1D

Dirty thermostat contacts can cause enough resistance to keep the circuit from operating the controls. To solve this problem, clean the contacts with a business card placed between them. With a small amount of pressure on the movable contact, pull the business card back and forth through the contacts to remove any dirt or corrosion. Do not use a metal file on these contacts. It will only ruin the them. If the cleaning cannot be done or the contacts cannot be reached to be cleaned, replace the thermostat. Be sure to use an exact replacement. Make any required adjustments to allow the thermostat to control the equipment as it was designed.

177–1E

A bad thermostat will keep the burner motor from starting. To check the thermostat, make all the settings required to cause the burner motor to run. If the motor does not start and run, check the thermostat contacts and the circuits through the thermostat. If there is any problem found that cannot be repaired, replace the thermostat. Be sure to use a correct replacement. Make any adjustments to the thermostat so that it will operate the same as the one being replaced. Be sure to allow the thermostat to reach room temperature before making any adjustments. Otherwise the thermostat will be incorrectly set when the thermostat and room temperature are the same. When adjusting a thermostat, determine what adjustments are needed, make them and remove the hands from the thermostat to prevent the hand temperature from affecting the adjustment. Run the unit through a complete operating cycle before leaving the job.

177–1F

The burner may not start because of broken or loose wires inside the oil burner. Check for these conditions and repair any loose or broken connections. Be sure to turn off the power to the oil burner during this procedure to prevent an electrical shock or damage to the equipment. When the repairs are completed turn the electrical power back on. Run the unit through a complete cycle to make certain that it is operating properly before leaving the job.

177–1G

An open thermal overload or a bad motor starting switch may keep the motor from starting. To check the overload, check the voltage to the burner motor terminals. If power is found at this point, then the trouble may be in the internal overload. Let the burner motor cool to room temperature and then try to restart the motor. If it does not restart, check the motor windings for continuity. See Figure 2–44. If the continuity is incorrect replace the motor.

If the motor just hums and tries to start but does not, the problem could be the starting switch. If there is any safe way to turn the motor, turn off the motor and allow it to cool until the overload resets. Then, at the same time, turn on the power and turn the shaft. If the motor starts and runs with normal running amperage, the starting switch is bad and must be replaced. It will usually be necessary to take the motor to a motor repair shop to have this done. When the motor is replaced, run the system through a complete cycle to make certain that it is operating properly before leaving the job.

Sometimes a weak or bad starting capacitor, if used, will cause this same problem. Be sure to check the capacitor before having the starting switch replaced. It is a good idea to replace the starting capacitor when replacing the starting switch because the capacitor could have been overheated and weakened.

177–1H

A defective pump motor can also keep the pump from starting. To check the motor, turn off the electrical power to the motor, then remove the terminal plate. Remove

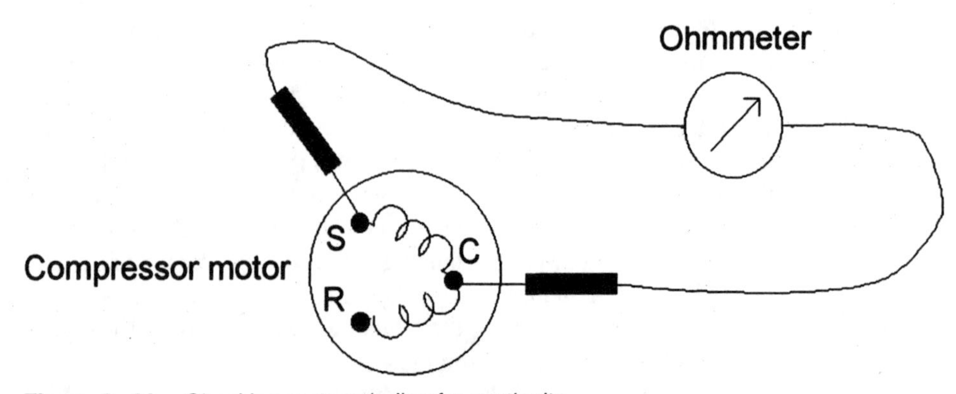

Figure 2–44. Checking motor winding for continuity.

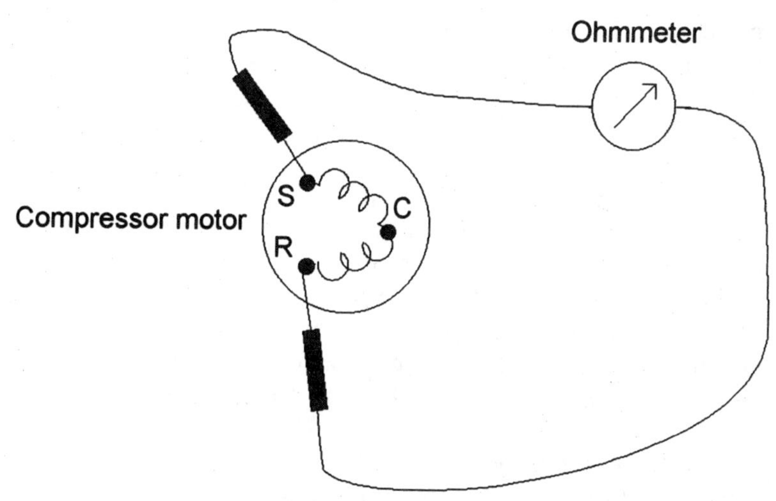

Figure 2–45. Checking motor windings for continuity.

the wiring from the terminals. Be sure to mark which wire goes to each terminal. Then check the continuity between each of the motor terminals. See Figure 2–45.

If the correct continuity is not found, replace the motor. It has a bad winding. Be sure to use an exact replacement or the pump may not fit the motor shaft and frame.

177–1I

A defective oil pump can also prevent the motor from starting and running. To determine if the oil pump is the problem, disconnect the oil pump from the motor shaft and turn on the motor. Check the running amperage. If the amperage is within the normal running amperage the pump is the problem. Either repair or replace the oil pump. When a replacement is used, be sure to use an exact replacement so that it will fit the motor shaft and frame, and the burner tube. Run the unit through a complete operating cycle to see that it is operating properly.

177–2 BURNER MOTOR STARTS, BUT NO FLAME IS ESTABLISHED:

There are several things that could cause this problem: trouble in the primary circuit, loose connections or broken wires, bad ignition transformer, faulty ignition electrodes, or a problem with the ceramic insulators.

177–2A

When it is suspected that there is some type of *trouble in the primary,* check the cad cell, and/or the stack relay as outlined in section 177-1. If either of these are causing the problem either make the necessary adjustments or replace the control with an exact replacement. Do not try to repair these controls in the field. Be sure to run the unit through a complete operating cycle before leaving the job.

177–2B

Loose connections or broken wires between the ignition transformer and the primary circuit can cause a lot of trouble. Check the wiring system to the oil burner and repair or tighten any loose or corroded connections. Replace any wiring that shows to have been overheated. Broken wires are sometimes very hard to find. The insulation will look perfect, but the wire inside can be broken. If an offset or an unusual looking piece of wire is found, pull the insulation on each side of the offset. If the wire is broken at this point the insulation will usually stretch. See Figure 2–46.

Sometimes it may be necessary to remove the insulation to find out if the wire is broken at that place. Be sure to repair any wiring that has been cut or removed to prevent future shorting.

177–2C

A *bad ignition transformer* can also cause the flame not to start. To check the ignition transformer, check the voltage on the primary terminals. Line voltage should be found. If not, turn the thermostat to demand heat and check the voltage at both the primary terminals and the secondary terminals. If voltage is found at the primary terminals and not the secondary terminals, the transformer is bad and must be replaced. Be sure to use an exact replacement for the proper ignition spark to be given.

177–2D

The *ignition electrodes* could be causing the problem. There are several areas to check for faulty ignition electrodes: the position, improperly spaced electrodes, loose electrodes, or dirty electrodes.

Figure 2–46. Checking for a broken wire.

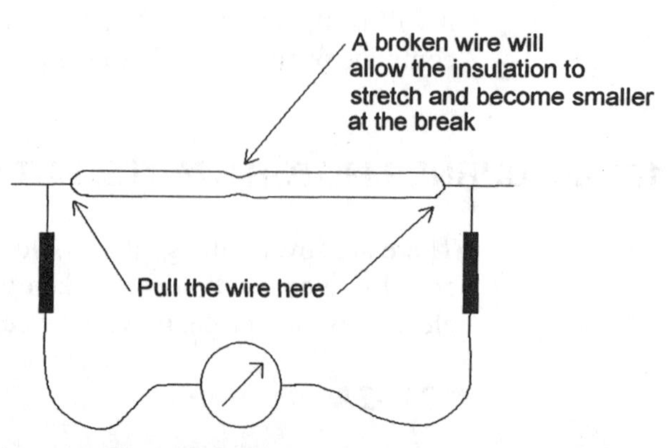
A broken wire will allow the insulation to stretch and become smaller at the break
Pull the wire here
Also check the continuity of the wire with an ohmmeter connected to the wire on each side of the suspected break. When broken, an infinite resistance will be indicated

Usually it is best to use the burner manufacturer's specifications to set the gap and position of ignition electrodes. When the position of the electrodes is found wrong, reinstall them according to the burner manufacturer's specifications. They must be positioned so that they will spark across the fuel-air mixture as it comes from the burner tube.

When the ignition electrodes have too much space between them, the spark will probably not jump the gap to provide ignition for the fuel-air mixture. Set the electrodes according to the burner manufacturer's specifications.

Loose electrodes will not give the correct amount of heat across the fuel-air mixture to ignite the fuel. Usually this condition is hard to find because it will normally cause intermittent problems. Sometimes it will light the burner and sometimes it will not. It is a good practice to check the tightness of the electrodes on every start-up of the equipment.

Dirty electrodes can keep the fuel-air mixture from igniting because the dirt and soot will insulate the electrodes and the spark will not jump from one to the other. They must be cleaned before the system will operate correctly.

177–2E

Dirty, shorted, or damaged ceramic insulators will cause the fuel-air mixture not to light because the electric current is not getting to the electrodes. Check the insulators to make certain that they are not broken. If they are not, it may be possible for them to be cleaned and that should allow the current to pass through them. If there is still no spark between the electrodes and everything else checks good, replace the electrode assembly. Be sure to use an exact replacement so the burner will operate as designed.

177–3 BURNER MOTOR STARTS, BUT THE FLAME GOES ON AND OFF AFTER STARTUP:

There are several things that could cause this problem including dirty ignition or motor relay contacts, burned out contacts in the primary control, bad limit control, bad thermostat, or bad burner components.

177–3A

Dirty ignition or motor relay contacts will reduce or completely prevent the current from passing through them. This reduction in current to the contacts will not leave enough to operate the device it is controlling. The best remedy for this problem is to clean the contacts. Do not file these contacts. Place a business card, or other stiff paper, between the contacts and with a slight pressure on the movable contact pull the card or paper through the contacts to remove any corrosion or pitting that may be present. See Figure 2–47.

Do not use slick paper because it will not clean the contacts. Start the unit and check it though a complete operating cycle before leaving the job.

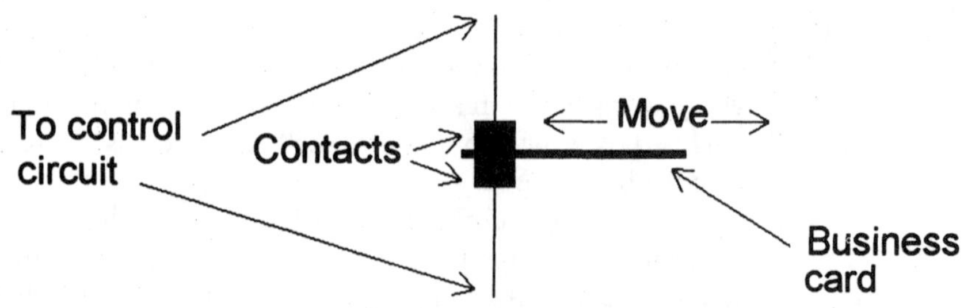

Figure 2–47. Cleaning contacts with a business card.

177–3B

Burned out contacts in the primary control will reduce the amount of current passing through them. This condition could quite possibly cause the motor winding or some control being overheated and burned out. The only way to solve this problem is to replace the primary control.

177–3C

Dirty limit control contacts may also cause the flame to go on and off after ignition. Check the control contacts and clean them if possible. Use a business card, or other rough type paper, and place it between the contacts. Apply a slight pressure on the movable contact and move the card back and forth between the contacts. Then inspect them for pitting or overheating. If either is present, replace the control. Be sure to adjust the settings to match those of the one being replaced or the unit may nor operate as designed. Run the unit through a complete operating cycle before leaving the job.

177–3D

Limit control high and low settings are too close or the differential is set too wide requiring adjustment to the proper settings. If the control cannot be adjusted properly, replace it. Be sure to adjust the new control to the proper settings or the unit will not operate as designed. Be sure to run the unit through a complete operating cycle before leaving the job.

Check out Procedures for the High Limit Control:

1. Close the line switch and set the thermostat to demand heat.
2. Set the high limit at its lowest setting.
3. When the water temperature reaches this setting, the burner will stop.

NOTE: In a hydronic heating system, the water circulator will continue to run as long as the thermostat demands heating and the water temperature is above the low limit setting.

Check Out Procedures for the Low Limit and the Water Circulator Switch:

1. Set the low limit switch at the highest setting.
2. The burner, but not the water circulator, will start if the water temperature is at least -10°F below the set point of the control.
3. Return the low limit switch to the lowest setting. The burner will stop.
4. Set the room thermostat to demand heat. The burner and water circulator will come on.
5. With the burner and water circulator running, raise the low limit setting; the circulator will stop. Lower the setting; the water circulator will come back on.

Return all switches to their normal setting before leaving the job.

177–3E

Broken wires or loose connections in or to the thermostat can cause a lot of intermittent problems. Solving this problem requires that the wiring from the transformer through the control circuit through the thermostat and back must be checked for loose connections, corrosion, or broken wires. Test the loose connections by applying a slight pressure to the terminal to see if it will move. Loosen the terminal a couple of turns and check for corrosion while there. If corrosion is present, remove the wire from the terminal and clean both the wire and the terminal on the control. An overheated terminal can be detected by simply looking at the connection. If the wire insulation is brown or shows signs of overheating, remove the wire and install a new terminal. See Figure 2–48.

A broken wire is sometimes more difficult to locate. Usually the wire is broken inside the insulation and cannot be easily seen. This condition is usually indicated by a slight turn or twist in the insulation. If either of these conditions are found, simply pull on the insulation on each side of the distorted place. If the wire is broken, a slight movement of the insulation can be detected.. Either replace the complete wire or make the required repair properly.

When a bad connection, broken wire, or dirty or burnt contacts are found in the thermostat, try cleaning the contacts first. If this does not correct the problem, replace the thermostat. Be sure to check the terminals behind the thermostat to make certain that they are tight, not corroded, or broken.

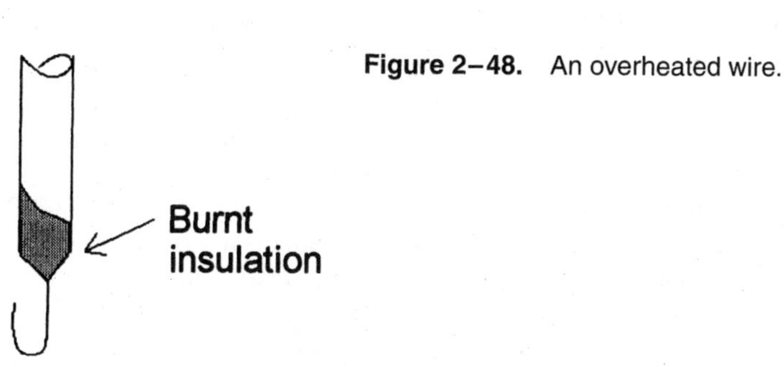

Figure 2–48. An overheated wire.

177–3F

When the *thermostat differential is set too close,* it causes a condition that would cause the flame to go on and off after starting. Check the thermostat; the differential must be at least 2°F. Check to see if the thermostat has an adjustable differential. If so, adjust it to about 3°F. If it is adjusted wider than this, over-and-undershooting of the space temperature could result. When there is no differential adjustment, the thermostat must be replaced. Allow the new thermostat to reach room temperature after installation. Then make any adjustments required. Run the unit through a complete operating cycle before leaving the job.

177–3G

A bad heat anticipator in the thermostat will sometimes allow the burner to go on and off too rapidly. It will usually be accompanied by wide swings in the space temperature. To check the heat anticipator, remove the thermostat cover and look at the heat anticipator.

If it is burnt or has been overheated, it will be a brown color. When this condition is present, replace the thermostat. Set the heat anticipator according to the current flow through the control circuit during the heating cycle at full heat. If the heat anticipator does not have a setting high enough, another thermostat must be used that will handle the current draw in the control circuit during the cycle during the highest current draw.

177–3H

A *bad thermostat* will cause several problems, such as the flame going on and off during the running cycle. This is usually caused by the heat anticipator being shorted and causing the thermostat to think the space temperature is higher than it is. Check the heat anticipator and if it is shorted, replace the thermostat. Be careful when wiring the new thermostat to make certain that the new anticipator is not accidentally burnt or shorted. Turn off the electricity to the control circuit before wiring the new thermostat. Be sure to stop the hole behind the thermostat so that cool or warm air cannot enter the thermostat from behind.

177–4 SYSTEM OVERHEATS THE HOUSE:

This problem could be caused by any number of things. Each of the most often found problems will be discussed in the following text.

177–4A

A *defective primary* may cause the unit to overheat the house. To check out the primary control, with the unit running, disconnect one of the thermostat wires from the primary control terminal. If the relay does not drop out, replace the primary control with a proper replacement. Make any adjustments needed so that the new primary control will operate the system as designed.

177–4B

Shorted thermostat wiring can cause a lot of system problems. When the unit is not operating properly and no reason can be found for the problem, check the control circuit wiring for shorts. When a low voltage control circuit is used, a shorted condition is sometimes difficult to locate. Be sure to check where the wire passes through partitions and openings in the unit panels. If the wire has been stapled to the building framing, it may be necessary to loosen or remove all these staples to find the short. Locating shorts is usually a hard and time consuming procedure. When the short is found, either replace the wiring completely or properly repair the place where the short is located. If a repair is made, be sure that the wire will not short there again.

177–4D

A defective, incorrect, or out of adjustment heat anticipator can cause the unit to overheat the building. The purpose of the heat anticipator is to provide false heat to the thermostat sensing element. When the heat anticipator is bad, the wrong type, or out of adjustment, it will cause the thermostat to maintain an undesirable temperature inside the space. To solve this problem, check the current draw through the control circuit. Then check the heat anticipator to see if it will handle the current draw and/or if it is bad. If the anticipator is bad, or the wrong one is used, replace the thermostat. The heat anticipator cannot usually be replaced in a thermostat. Be sure to set the heat anticipator to the correct current draw through the control circuit. Check the unit through a complete operating cycle to make certain that it is operating properly before leaving the job.

177–4E

Thermostat hung in the ON position will cause the unit to overheat the building because the unit never stops. It continues to blow warm air into the space after the thermostat has been satisfied. To check the thermostat to see if it is stuck in the ON position, simply move the temperature selector to a temperature higher than the space temperature. The thermostat contacts should open and stop the heating unit. If the system does not stop, there is something holding the contacts closed. Replace the thermostat with the correct replacement. Adjust the heat anticipator. Allow the thermostat to reach room temperature, then run the system through a complete operating cycle.

177–4F

The thermostat may not be located to properly control the area, there may be a draft affecting the thermostat. The solution to this problem is to relocate the thermostat to a place where it is not affected by drafts or radiant heat from lights, televisions, warm water pipes, cool air entering the thermostat from behind through the hole used to pull the wiring, or the wall may vibrate each time a door is closed. Be sure to eliminate all of these things. If the problem still persists, relocate the thermostat where it will sense the average return air temperature and where none of the above conditions exist. This may involve running more wire and making new holes in the walls and other building structures, but it must be done before the system will operate as it was designed.

177–4G

The hot water valve may be stuck in the open position allowing hot water to flow through the coil when the thermostat is satisfied. In this case, the water valve must either be repaired or replaced. The choice will usually depend on the age and condition of the valve. Be sure to use one that is compatible with the control circuit so that the system will operate correctly. To check the hot water valve, remove the sensing bulb from the water pipe and place it in a container of cool water. If the valve does not close, or start closing, the valve operator is bad or the valve is stuck and must be replaced. Also, the valve may be stuck in the open position. This can be tested by removing the valve operator and trying to move the stem by hand. It should move fairly easily. If the operator starts closing when it is removed from the valve stem, the operator is working properly. This is a good indication that the valve is causing the problem. In this case the valve must be taken apart and either cleaned or replaced. Be sure to valve off the water flow to the valve before attempting to check the internal parts.

177–4H

When the water circulator runs continuously, it is an indication that the water circulator circuits or the circulator switching device is not operating properly. To solve this problem, check all the connections in the circulator circuit to make certain that they are not loose or corroded. Corrosion is a problem with water systems. Tighten all connections and remove any corrosion found on the terminals. If this does not solve the problem, check the circulator switching device. When the control power is removed from its coil terminals, the water circulator motor should stop running. If it does not, check the voltage across the contacts of the circulator switching device. If there is none indicated, replace the switching device. The contacts are stuck closed. Be sure to turn off all power to the unit before working on electrical circuits.

177–4I

The limit control may be set too high or not operating properly. Reset the limit control to see if this will solve the problem. If the unit starts operating, check the setting of the limit control. On gas-fired heating units, it should never be set above 200°F. When the limit control is used with a boiler or hot water system, check to make certain that the off temperature is set at the temperature recommended by the boiler manufacturer or the design engineer. When this does not solve the problem, replace the limit control. Be sure to use an exact replacement so that normal operation can be maintained.

177–4J

The aquastat limit control in the wrong location or set too high can also cause the unit to overheat the house. Check the temperature off setting of the aquastat and make any adjustments required. If this does not repair the problem, try relocating the aquastat to a position that will sense the water temperature better. The design engineer may need to be consulted for the correct location or temperature setting. It is probably allowing the water temperature to get too high and will cause the temperature to override the thermostat setting.

177–5 SYSTEM DOES NOT HEAT THE HOUSE PROPERLY:

This condition could be caused by any number of problems with the unit.The following are the ones that are most common to this condition.

177–5A

An open wire or loose connection in the thermostat circuit will cause the system to not heat correctly. Check and tighten all connections. Also, check and remove any corrosion that may have formed on any of the connections. The corrosion will reduce the amount of voltage to the control and prevent it from operating properly. A broken wire can also cause the unit to not operate properly. Check for broken wires by looking for any unusual turns in the wiring. Check for a broken wire by pulling on each side of the unusual turns. When a broken wire is inside the insulation, the wire will move apart and cause the insulation to stretch at this point. See Figure 2–49.

If the wire is not long enough to remove the broken end and properly make the connection, replace the wire and leave some extra for any future problems that may occur in this wire.

177–5B

Dirty thermostat contacts will also cause the unit to not heat the space. The dirt between the contacts will reduce the amount of voltage flowing through them and prevent the unit from operating. If the thermostat is equipped with open type contacts, they may be cleaned by placing a rough business card between them, putting a slight pressure on the movable contact, and pulling the card back and forth through the contacts to remove any dust or dirt on them. Do not file these contacts. This will

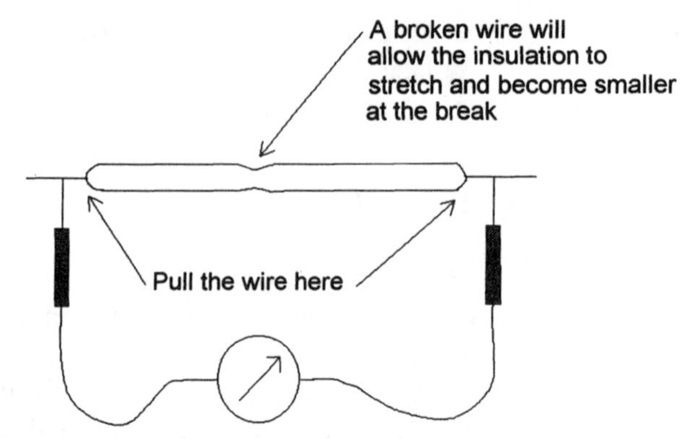
A broken wire will allow the insulation to stretch and become smaller at the break
Pull the wire here
Also check the continuity of the wire with an ohmmeter connected to the wire on each side of the suspected break. When broken, an infinite resistance will be indicated

Figure 2–49. Checking for a broken wire.

ruin the contacts and the thermostat will need to be replaced. If the contacts are so badly pitted that they cannot be cleaned with a business card or other rough paper, then the thermostat must be replaced. Be sure to use an exact replacement or the system will not work as it was designed. When the thermostat is replaced, allow it to reach room temperature before making any type of adjustments.

177-5C

A *thermostat that is out of calibration* will keep the unit from heating the space properly. Check the temperature where the thermostat contacts open and close with an accurate thermometer. If the thermostat is out of calibration, recalibrate it if possible. Most thermostats cannot be calibrated in the field. In this case replace the thermostat with a correct replacement. Do not make any calibrations on the new thermostat until the thermostat has reached room temperature.

177–5D

A wrong heat anticipator or a heat anticipator that is not set properly could cause the unit to not operate properly. Check the heat anticipator setting against the current draw through the heating control circuit. To do this either use a factory made current multiplier on the ammeter, or take 10 turns of the heating wire or the power (R) wire and connect each end of this coil into the heating control circuit. When the current multiplier is used, read the amperage and divide it by the number of turns made into the coil.

Example:

The current flow through the circuit with a ten turn multiplier is 4.5 amperes. The actual current flow will be $4.5 \div 10 = 0.45$ amps.

This is the correct setting for the heat anticipator. If the setting of the anticipator is outside of this current range the wrong heat anticipator is used. This requires that the thermostat be replaced to correct the problem. If the current flow is within the rating of the heat anticipator, set the anticipator to match the current flow. When a new thermostat is installed, set the heat anticipator to the current flow indicated by the measurement made with the ammeter. Any number of turns can be used on the current multiplier. However, the more turns the more accurate the measurement. Always divide the current draw by the number of turns taken on the ammeter tong.

177–5E

A thermostat stuck in the off or open position causes the space to be too cool. Remove the cover of the thermostat and check to see if the contacts are open or closed while the temperature lever is set higher than the space temperature. If the contacts are open, the thermostat contacts are stuck this position. Check for the reason. If it is simply binding, replace the thermostat. If there are wires or other parts of the thermostat keeping the contacts from closing, move the components to allow the contacts to move easily. When replacing the thermostat, be sure to use an exact replacement or the system will not operate as designed. Do not make any adjustments on the replacement thermostat until it has reached the space temperature.

177–5F

An improperly located thermostat will also cause the thermostat to not heat the space properly. If the thermostat is located in a place where it is sensing heat from other sources such as lights, televisions, warm pipes inside the wall, or where rays of the sun will heat the thermostat it must be relocated. Relocate the thermostat to a place where there's no interference with its operation. Mount it where it will sense the average space temperature, such as close to the return air grill. Mount it about 5′ from the floor, on an inside wall away from water heaters or any warm pipes inside the wall and where the closing of doors will not cause a vibration making the thermostat contacts to open and close rapidly.

177–5G

A chronotherm thermostat 12 hours out-of-time is an indication that the thermostat has either stopped at some time or it has been programmed wrong. To solve this problem, simply advance the time setting by 12 hours. Use the thermostat manufacturer's recommendations when programming an electronic thermostat.

177–5H

A low or high limit controller that is set too low will cause the system to underheat the space. The controller will not allow the unit to run long enough to raise the temperature. This problem can be solved by raising the control set point to the proper temperature settings. When the controls are in good condition, the unit will start heating the space and bring it to the desired temperature.

177–5I

When *either the low or the high limit is slow to return to the ON position,* it causes the unit to stay off longer than necessary. The space temperature will drop below the desired point. Check to see which controller is causing the problem. Adjust the controller differential and check to make certain that the unit will operate properly. If not, the controller will probably need to be replaced. Be sure to use an exact replacement so the system will operate as designed.

177–5J

A bad high or low limit controller will cause the unit to not heat the space as wanted. Check to see which controller is defective and replace it with the correct replacement so that the system will operate properly. Be sure to check the unit for any damage that may have been caused by the defective controller.

177–6 SYSTEM FREQUENTLY CYCLES:

A system that frequently cycles is not operating efficiently, nor is it producing the amount of heating or cooling of which it is capable. The cause for the cycling must be found and corrected so that it will operate as designed.

177–6A

A poorly located thermostat could cause the system to cycle on and off frequently. Check to see if the air from one of the air grills is blowing on the thermostat. This will cause it to cycle the equipment too rapidly. Also check to see if something else could be causing the problem, such as the opening of an outside door that will allow a draft to hit the thermostat. If any of these conditions are found, the thermostat must be relocated before proper operation will be realized. Be sure to place the thermostat about 5′ from the floor, where it will sense the average return air temperature, and where there are no warm or cool water pipes inside the wall. There should be no source of heating or cooling on the other side of the wall. Be sure to use an exact replacement or the system will not operate as designed.

177–6B

A loose thermostat heat anticipator that is shorting or not properly adjusted will cause the unit to cycle frequently. This frequent cycling is due to the thermostat constantly getting false readings from the anticipator. If the heat anticipator connections can be tightened, tighten them. If not, replace the thermostat. Be sure to use an exact replacement or the system will not operate as designed. Set the heat anticipator to the actual current draw in the heating part of the control circuit. Do not make any other adjustments until the thermostat has reached space temperature. Use a current amplifier to determine the correct current flow.

177–6C

Dirty thermostat contacts will allow the system to cycle frequently because of the drop in voltage passing through them. Clean open dirty contacts with a rough business card or some other rough piece of cardboard. Place the business card between the contacts and apply a slight pressure on the movable contact. Move the card back-and-forth between them until they are cleaned. If the contacts cannot be cleaned in this manner, replace the thermostat. Do not file these contacts. Be sure to use an exact replacement or the system will not operate as designed. Adjust the heat anticipator to match the current draw through the heating part of the control circuit. Do not make any other adjustments until the thermostat has reached the space temperature.

177–6D

Loose connections or bad wiring in the thermostat circuit will cause the system to cycle frequently because the circuit is opened and closed at random, regardless of what the space temperature is. Check all connections for tightness and corrosion. When corrosion is found, be sure to remove it or the resistance in the connection will increase. Tighten all connections because a loose connection will reduce the amount of voltage reaching the controls. A loose connection will also cause overheating of the connection that could ruin the control or at least cause the connection to loosen further and burn the wire. Also check for any broken wiring by looking for any strange looking turns in the wiring. When one is found, pull the insulation on each side of the turn. If the insulation stretches, the wire inside is broken. If the broken end is too short to make a proper connection, replace the complete wire. Always leave a little extra wire for use later on.

177–6E

When the high limit is set too low, the heating unit will usually go off earlier than it would if it were controlled by the thermostat. Check the limit control to make certain that it is in good working order. This is usually done by placing an accurate thermometer is the discharge air plenum, reducing the air flow, and allowing the high limit control to cycle the burners. Most furnace manufacturers do not recommend that the air flow through the furnace be completely stopped. This will cause the heat exchanger to overheat and possibly be ruined. If the off temperature is below the normal operating temperature of the air passing through the discharge plenum, adjust the limit control to a higher temperature. Never set the high limit off temperature higher than 200°F to keep the unit from overheating. If this adjustment is not possible, replace the high limit control with an exact replacement so that the system will operate as designed. If the sensing element is not the same length as the one being replaced, the control will probably not work properly. This is because the element is not sensing the heat exchanger at the hottest point as determined by the manufacturer. Set the cut-off temperature at 200°F or below. The cut-on temperature is not usually adjustable; it follows the cut-off temperature adjustment at about 25°F differential.

177–6F

When the high limit differential is set too close there is usually something wrong with the control. Limit controls are safety devices and should not be bypassed or repaired in the field. It is better to replace it with an exact replacement. When an exact replacement is not used, the unit may not operate as it was designed. Any type of adjustment on this control should only be temporary. Usually someone, or something has caused the differential to change. They are typically set at about 25°F and cannot be easily changed.

177–6G

When the contacts in the high limit control are dirty, enough contact is not made to allow the needed amount of current to flow through the circuit. The controls in the circuit will not receive enough current to operate properly. If the contacts are dirty, clean them with a business card or other rough piece of cardboard. Do not file them. Be sure to turn off the electricity to the unit before attempting this procedure to prevent personal shock or damage to the unit or property. If the contacts are pitted so that the cardboard cannot clean them, replace the high limit control. Be sure to use an exact replacement so that the unit will operate as designed. Adjust the control to go off at a temperature no higher than 200°F. This will produce a cut-on setting of about 175°F. Operate the unit through a complete operating cycle to make certain that the system is operating properly.

177–7 RELAY CHATTERS AFTER PULLING IN:

When a relay chatters, its contacts will be ruined as well as those in the other circuit controls. The problem must be found and corrected or the contacts will only be ruined again in a short period of time.

177–7A

Relay chattering can be caused by bad contacts in the thermostat because the thermostat contacts will not allow the proper voltage through to operate the other controls. A thermostat that has bad contacts must be replaced. Otherwise the temperature will never be quite satisfactory and the other controls could be ruined. When replacing a thermostat, be sure to use an exact replacement so that the system will operate as it was designed. Program the thermostat to meet the needs of the user.

177–7B

Low voltage to the primary control will usually cause the relay contacts to chatter. When low voltage is found, contact the local power company. When the voltage has been brought back to normal, check the unit to make certain that everything is working properly.

177–7C

Dirty contacts in the primary control will cause relay contacts to chatter because there is not enough voltage reaching the relay coil to continuously hold it in. Clean both the detector and the timing switch contacts so that the correct amount of current will reach the control circuit. To clean the contacts, first turn off the electric power to the unit. Insert a business card or other rough cardboard between the contacts. Place a small amount of pressure on the movable contact and move the cardboard back and forth between the contacts. If the contacts are so badly pitted that they cannot be cleaned in this manner, replace the defective part. Do not file these contacts or they will certainly be ruined and must be replaced. Place the unit back in operation and run it through a complete operating cycle before leaving the job.

177–7D

A *bad primary control load relay* will cause relays to chatter. Check the contacts in the primary control to make certain that there is no resistance between them. To do this, turn off all electric power to the unit. Remove the wiring from one of the relay contact terminals. Check the continuity between the contacts with an ohmmeter. Be sure to zero the ohmmeter. If there is any resistance indicated between the contacts, replace the relay. If there is not infinite resistance between the contacts, either replace them or the control. Repair any bad connections. Run the unit through a complete operating cycle to make certain that it is operating properly.

178–1 TROUBLESHOOTING ELECTRONIC DEFROST CONTROLS:

Electronic defrost controls are unique to heat pump systems. Their operation is critical to both the performance and reliability of the equipment. The following is a discussion of the operation and troubleshooting of the most common electronic defrost controls used on heat pump systems.

Figure 2–50. HSC 621-550 time/temperature defrost control. (Courtesy of Rheem Airconditioning Division)

There are basically two types of defrost controls used, one is a time and temperature (T/T) and the other is the demand (Dem.). See Figure 2–50. The time and temperature control checks for any needed defrost conditions after a set time interval. If the conditions are ready for a defrost cycle at that time the control powers the defrost relay to start the unit through a defrost cycle. The demand defrost control constantly monitors the outdoor air temperature and outdoor coil temperature. When the outdoor coil temperature drops to a set temperature differential below the outdoor air temperature, the control operates the unit through a defrost cycle.

178–1A

Some conditions, other than normal operation, that can cause a build-up of ice or frost on the outdoor coil are: low refrigerant charge, low air flow over the outdoor coil, improper condensate drainage, roof run-off, blowing snow or freezing rain, a reversing valve stuck in the heating position, or a thermostat cycling the unit off during a defrost cycle.

178–1B

The basic tools recommended for troubleshooting defrost controls are:
digital volt-ohm meter, digital thermometer (quick reacting type), and several low voltage jumper wires with alligator clips.

IMPORTANT: When connecting jumper wires to the various defrost control board terminals, **DO NOT** let the jumper wires touch any terminals except those intended. For example, jumpering terminals such as "OUT" to "24VAC" will damage the board or the transformer. Turn all power off before connecting or removing test jumpers.

178–1C

The sequence of operation of the time-temperature defrost controls follows:

1. When the indoor thermostat is switched to the heating position, 24 volts of electricity is directed to the defrost control board terminals "24VAC" and "Com."
2. When the indoor thermostat calls for compressor operation, the power side of the control transformer applies 24 VAC to the "HLD" terminal to time the compressor operation.
3. When the accumulated run time of the compressor reaches the board setting (50 min., 70 min., 90 min.), the defrost control board checks the "SEN" terminal.
4. If the defrost sensor temperature is below 28°F anytime during the accumulated compressor run time, it closes sending a 24 VAC signal to the "OUT" terminal. The "OUT" terminal powers the defrost relay. If the sensor does not close, the board resets and begins its accumulated compressor run time again.
5. The defrost relay de-energizes the 24VAC reversing valve solenoid, stops the outdoor fan, and sends power to the defrost heat control in the indoor air handler through the purple wire.
6. The unit is now in the cooling mode and hot refrigerant gas is directed to the outdoor coil to melt the ice.
7. When the unit has run in the defrost mode for ten minutes, or the outdoor coil temperature reaches approximately 50°F, the outdoor coil sensor opens its contacts interrupting the power to the "SEN" terminal.
8. The opening of these contacts de-energizes the defrost relay and the systems returns to the heat mode.

178–1D

Before starting any troubleshooting procedures, disconnect all electric power to the outdoor unit.

WARNING: The following procedures involve testing "LIVE" electrical circuits! These procedures must be performed by service personnel taking all safety precautions to prevent personal injury or property damage!

1. Remove the compressor and control panel covers.
2. Locate the defrost sensor:
   a. The sensor should be attached to the top coil in the refrigerant circuit on the same side as the metering device distribution tubes. See Figure 2–51.
   b. The sensor should have a snug fit to the refrigerant line (if not, reposition it and use wire ties to "snug" the sensor to the line).

Figure 2–51. Defrost sensor mounting on outdoor coil. (Courtesy of Rheem Airconditioning Division)

3. Fasten a temperature probe at or near the sensor location.
4. Locate the defrost relay in the electrical control box.
5. Locate the defrost control board in the electrical control box. See Figure 2–52.
6. Turn the electrical power on to the unit.
7. Initiate a first stage call for heat at the indoor thermostat.

178–2 TESTING THE DEFROST CONTROLS:

178–2A

To test the defrost relay, the unit must be operating to perform the following steps.

1. Disconnect the defrost relay coil from the defrost control board.
2. Use jumper wires to power the coil of the defrost relay. The unit should go into the defrost cycle. If not, replace the defrost relay.
3. Remove the jumper wires and place the unit back in operation.

178–2B

To test the defrost control board and sensor use the following steps:

1. Set the indoor thermostat to demand heat.
2. Check for 24VAC at the defrost control board terminals "24VAC" and "COM," and terminals "HLD" and "COM." If no voltage is present, make

Figure 2–52. Defrost relay and defrost control.(Courtesy of Rheem Airconditioning Division)

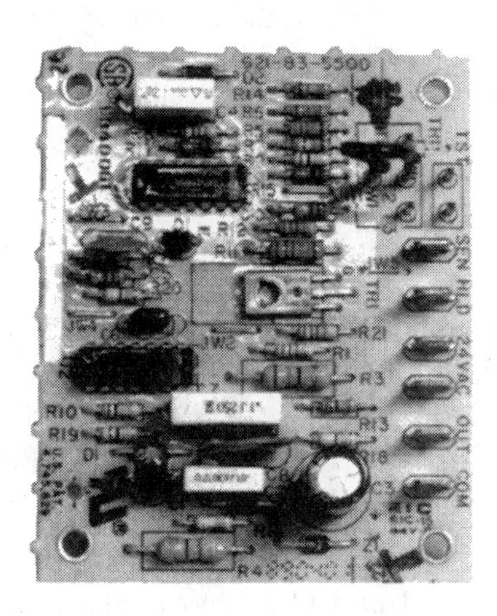

Figure 2–53. HSC 621-550 control board. (Courtesy of Rheem Airconditioning Division)

sure that the indoor thermostat is calling for first stage heat and that all control circuit wiring is correct with good connections and wiring. See Figure 2–53.

3. Block the air flow through the outdoor coil with cardboard or de-energize the outdoor fan. To de-energize the outdoor fan, disconnect the black wire from the #1 terminal on the defrost relay.
4. Monitor the outdoor coil temperature at the defrost sensor location with a digital thermometer.
5. When the outdoor coil temperature drops below 28°F (preferably 20°-25°F), initiate a defrost cycle by jumpering the two “TST” pins on the defrost control board until a defrost cycle is initiated. All timing functions are sped up from 50, 70, or 90 minutes to 11.7, 16.4, or 21.1 seconds.

6. After a defrost cycle has been initiated, immediately remove the jumper from the "TST" pins or the defrost cycle will stop after 2.3 seconds. The factory setting may be adjusted in the field if conditions warrant the change.
7. If the unit still does not go into a defrost cycle, verify that the defrost sensor is closed with an ohmmeter. Check the outdoor coil temperature at the sensor location. The sensor contacts close at 28°F +/− 3°. Replace the sensor if its temperature is below 25°F and its contacts are not closed.
8. If the sensor contacts do not close, connect a jumper wire between the "COM" and "SEN" terminals on the defrost control board. This bypasses the defrost sensor.
9. Jumper the "TST" terminals again to put the unit into a defrost cycle. If the unit goes into defrost now, the defrost control board is good.
10. If the unit still does not go into a defrost cycle, use a voltmeter to check for 24VAC between terminals "COM" and "24VAC," and between terminals "COM" and "HLD." If 24VAC is not present in both points, check for a failed transformer, thermostat, and improper control wiring.
11. If 24VAC is present at both points in the above step, check for 24VAC between terminals "HLD" and "OUT." If there is voltage present and still no defrost, recheck the defrost relay and wiring.
12. If 24VAC is not present at terminals "HLD" and "OUT," replace the defrost control board.
13. If the unit goes into a defrost, allow it to terminate the defrost cycle. It should terminate the defrost when the defrost sensor reaches 50°F.

NOTE: The sensor sometimes lags the coil temperature and may not terminate the defrost cycle until the coil actually reaches 60° to 65°F. This is normal. If the defrost sensor does not open to terminate the defrost cycle at this temperature, replace the sensor.

14. If the sensor contacts open and the defrost is not terminated, replace the defrost control board.
15. After completing this test, **disconnect the power** and replace the black wire to the defrost relay terminal 1.

A chart that may be used for troubleshooting time-temperature defrost controls is shown in Figure 2–54.

178–3 EARLY RANCO DEMAND DEFROST CONTROLS:

The demand defrost control does not use a timer with a selectable setting. A temperature difference between two sensors is used to control the defrost cycle. One

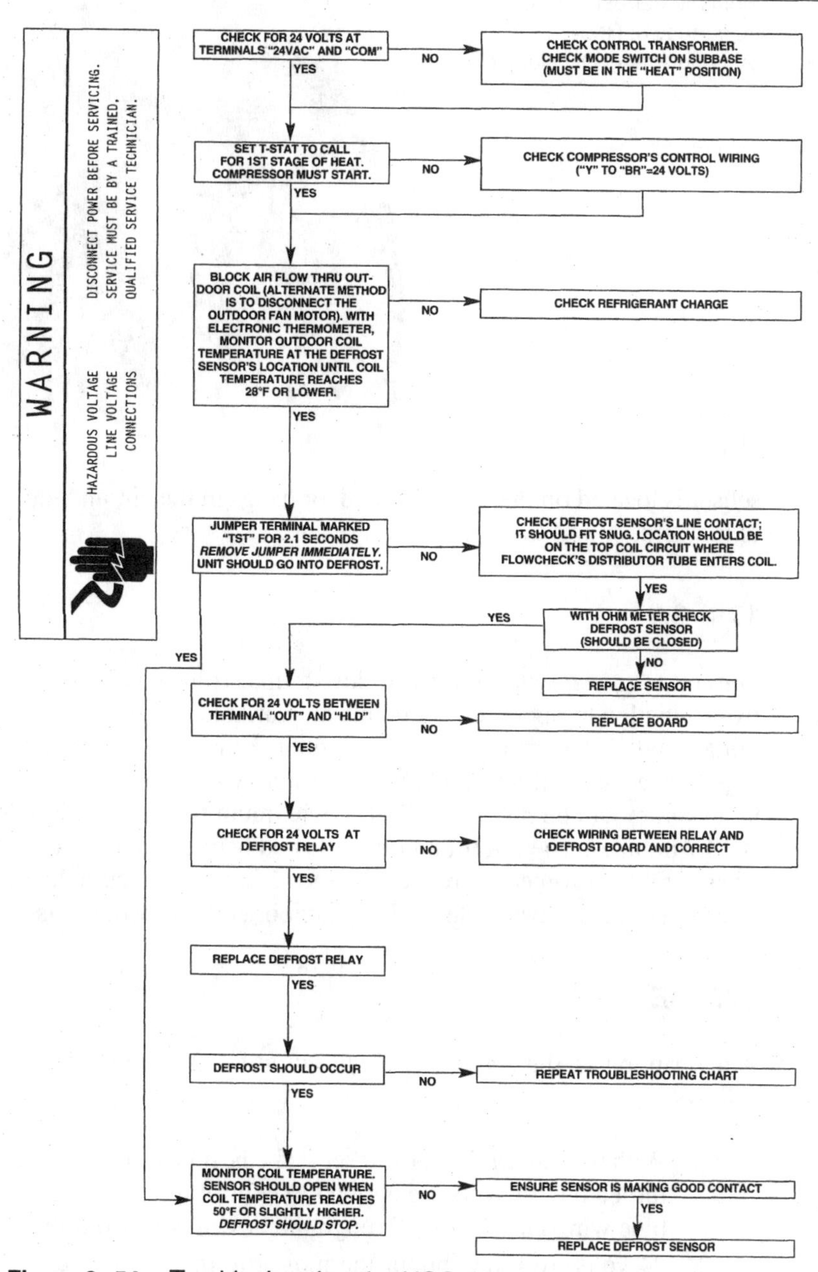

Figure 2–54. Troubleshooting the HSC 621-550 time/temperature defrost control. (Courtesy of Rheem Airconditioning Division)

Figure 2–55. HSC 1030-1 demand defrost control. (Courtesy of Rheem Airconditioning Division)

sensor is located on the control board, or hangs in the air, and the other one clamps to the outdoor coil tubing. See Figure 2–55.

178–3A

The control specifications are as follows: Input voltage is 24 VAC +/−25%. The defrost initiation temperature set points are variable, as follows, air sensor is set at 0 initiate and 40°F terminate; coil temperature sensor is −12 initiate and 23°F terminate. The actual coil termination temperature will be 50° +/− 5°F. The coil inhibit temperature will be 38° +/− 5°F. The maximum time allowed in defrost (override) is 13.7 minutes. The minimum time allowed between defrost cycles is 40 minutes. There will be a forced defrost cycle after 6 hours of accumulated compressor run time without a defrost cycle, and the outdoor coil temperature is below 38°F.

178–3B

A description of the *sequence of operation* for the demand defrost control is as follows:

1. With the indoor thermostat set to the heat position and the electrical power turned on, 24 volts are applied to control terminals "24VAC" through the blue wire and "COM" through the brown wire from the transformer.
2. 24 volts from the indoor thermostat to the "HLD" terminal measures the run time of the compressor.
3. The difference in temperature between the two defrost sensors initiates the defrost cycle. The coil sensor attached to the outdoor coil near the top of the coil. The ambient air sensor is part of the defrost control board.
4. When ice forms on the outdoor coil, it blocks the air flow, causing the coil temperature to drop.

5. When the outdoor coil temperature reaches a set differential below the ambient temperature and is maintained for a minimum of 4.5 minutes, the defrost control board powers the defrost relay through the "OUT" terminal.
6. The defrost relay de-energizes the 24-volt reversing valve solenoid, stopping the outdoor fan, and sends power to the defrost heat control in the indoor air handler by the purple wire.
7. Hot refrigerant gas is then directed into the outdoor coil to melt the accumulated ice.
8. When the outdoor coil reaches a temperature of approximately 50°F, the coil defrost sensor breaks the power to the "OUT" terminal. After 13.7 minutes with the system in the defrost mode, the defrost will automatically be terminated even if the outdoor sensor has not reached 50°F.
9. This action de-energizes the defrost relay, returning the system to the heat mode.
10. After the defrost cycle has been terminated, an internal mechanism in the defrost control board prevents another defrost cycle for 40 minutes of accumulated compressor run time regardless of the temperatures of the sensors.
11. After six hours of compressor run time without a defrost cycle and with the outdoor coil sensor below 38°F, the control puts the unit through a forced defrost cycle.

CAUTION: The following procedures involve the testing of "live" electrical circuits! These procedures must be performed by service personnel taking all safety precautions to prevent personal injury or equipment damage.

178–3C

Prior to initiating the following troubleshooting procedures, disconnect the electrical power from the outdoor unit.

1. Remove the compressor and control box cover panels.
2. Locate the defrost sensor.
   a. The sensor should be attached to the top of the outdoor coil circuit on the same side as the metering device distribution tubes.
   b. The sensor should have a "snug fit" to the refrigerant line (if not, secure it and use wire ties to "snug" the sensor to the line).
3. Place a temperature probe on the coil at/near the sensor location.
4. Locate the defrost relay in the electrical control panel.
5. Locate the defrost control board in the electrical control panel.
6. Turn the electrical power on to the unit.
7. Initiate a first stage call for heat at the indoor thermostat.

178–3D

Use the following steps when *testing the defrost relay.* (The unit must be operating to perform the following tests.)

1. Disconnect the defrost relay coil from the defrost control board.
2. Use a jumper wire to route the 24VAC power to the coil of the defrost relay. The unit should go into a defrost cycle. If not, replace the defrost relay.
3. Remove all jumper wires.

178–3E

Use the following steps when testing the *defrost control board.*

1. Check for 24VAC at the defrost control board terminals "24VAC" and "COM" and at terminals "HLD" and "COM." If voltage is not present, make sure that the indoor thermostat is calling for first stage heat and that all wiring is correct.
2. Block the air flow through the outdoor coil with cardboard or de-energize the outdoor fan. To de-energize the outdoor fan, disconnect the black wire from the #1 terminal on the defrost relay.
3. Jumper the two "DEFROST" pins on the defrost control board until a defrost cycle is initiated.
4. If the outdoor coil sensor is below 38°F, the system will remain in the defrost cycle until it reaches 50°F or 13.7 minutes have passed. If the coil sensor is above 38°F, the defrost cycle will be terminated as soon as the jumper is removed from the "DEFROST" terminals.
5. If the outdoor coil does not defrost, check for 24VAC at the "24VAC" and the "OUT" terminals. If no voltage is present, replace the control board and its attached sensor.
6. If 24VAC are present at the "24VAC" and the "OUT" terminals, check for voltage at the defrost relay coil. If 24VAC are present here but there is still no defrost cycle, replace the defrost relay.
7. If 24VAC is not indicated at both points, check for a failed transformer, indoor thermostat, and improper control wiring.
8. Allow the unit to terminate the defrost cycle. It should terminate defrost when the outdoor coil defrost sensor reaches 50 to 55°F. If it does not, replace the complete defrost control board and its attached sensor.

Figure 2–56 is a chart for troubleshooting the demand defrost control board.

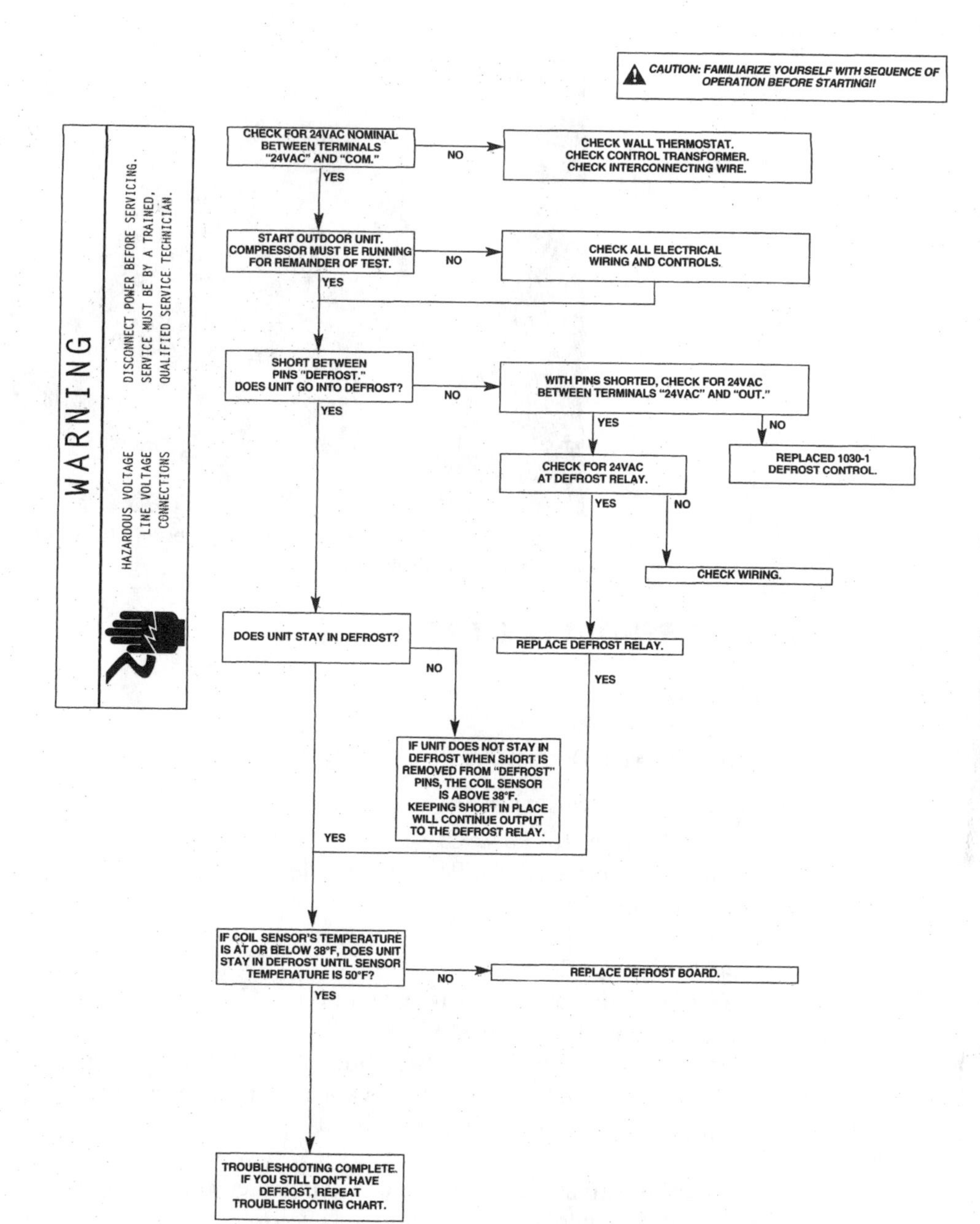

Figure 2–56. Troubleshooting the HSC 1030-I demand defrost control. (Courtesy of Rheem Airconditioning Division)

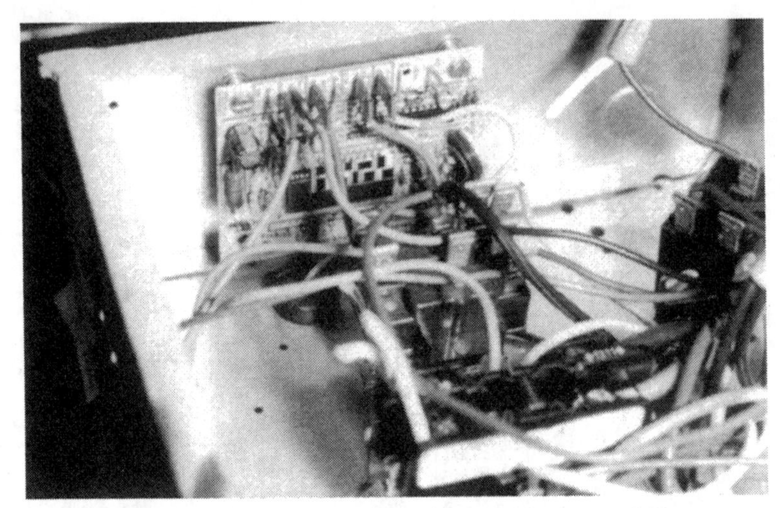

Figure 2–57. Demand defrost control board. (Courtesy of Rheem Airconditioning Division)

178-4 DEMAND DEFROST CONTROL:

178–4A

Ranco Demand Defrost Control. See Figure 2–57.

Control Specifications:

Input Voltage: 18–32 VAC.
Initiate Temperature Set Points: Variable.

Air Sensor	Coil sensor
0°F	−12°F
40°F	24°F

Defrost Termination Coil Temperature: 55°F
Coil Enable Temperature: 35°F
Maximum Time in Defrost (override): 15 minutes
Minimum Time between defrost varies: 20 minutes above 34°F ambient to 120 minutes below 19°F ambient.

Forced defrost after six hours of accumulated compressor run time with no defrost and the outdoor coil temperature is below 36°F.

178–4B

The following steps are the *sequence of operation of the Ranco demand defrost control board*:

1. With the indoor thermostat system switch set to demand heat and the electrical power is turned on, 24 volts are directed to the defrost control board from the "B" terminal of the indoor thermostat subbase and from the com-

Figure 2–58. Sensors mounted on Ranco DD1 control. (Courtesy of Rheem Airconditioning Division)

mon side of the transformer by a wire nut assembly in the outdoor unit control panel.

2. A call for heat by the thermostat sends 24VAC to the "HLD" terminal on the control board which measures the accumulated run time of the compressor.
3. The required difference in temperature between the two defrost sensors initiates a defrost cycle. The blue coil sensor attaches to the outdoor coil near the top. The red ambient air sensor hangs free in the outdoor air flow. See Figure 2–58.
4. As ice forms on the outdoor coil it blocks the air flow through the coil, causing the coil temperature to drop.
5. When the outdoor coil temperature drops to the set differential below the ambient air temperature and is maintained for a minimum of 8 minutes, the defrost control board powers the defrost relay from the "DEF RLY" terminal.
6. The defrost relay de-energizes the 24 volt reversing valve solenoid, stops the outdoor fan, and sends power to the defrost heat control in the indoor air handler by the purple wire.
7. The unit is now operating in the cooling mode and hot refrigerant gas goes directly to the outdoor coil to melt the accumulated ice.
8. When the outdoor coil temperature rises to approximately 55°F, the defrost sensor breaks the power to the "DEF RLY" terminal, or defrost termination will automatically occur after 15 minutes if the sensor never reaches 55°F.
9. This de-energizes the defrost relay, returning the system to the heat mode.
10. After defrost termination, an internal timing mechanism in the defrost control board prevents another defrost cycle for 20 minutes to 120 minutes of accumulated compressor run time regardless of the sensor temperatures.
11. After six hours of accumulated compressor run time with no defrost and with the outdoor coil sensor temperature below 36°F, the defrost control operates the unit through a forced defrost cycle.

> **WARNING:** The following procedures involve testing "live" electrical circuits! These procedures must be performed by service personnel taking all safety precautions to prevent personal injury and equipment damage.

Prior to initiating the troubleshooting procedures, disconnect the electrical power to the outdoor unit.

1. Remove the compressor and control panel covers.
2. Locate the defrost sensor.
   a. The outdoor coil sensor should be attached to the top coil circuit on the same side as the metering device distribution tubes.
   b. The sensor should have a "snug fit" to the refrigerant line (if not, secure it and use a hose clamp to "snug" the sensor to the line).
   c. The air sensor should not touch anything, but should hang freely in the outdoor air flow.
3. Connect some thermometer probes to the refrigerant line at/near the sensor location.
4. Locate the defrost relay in the electrical control panel.
5. Locate the defrost control board in the electrical control panel.
6. Turn the electrical power on to the unit.
7. Initiate a first stage call for heat at the indoor thermostat.

178–4C

The following steps are a description for *testing the defrost relay.* The unit must be in operation to perform the following tests.

1. Disconnect the defrost relay from the defrost control board.
2. Use jumper wires to route 24VAC power to the coil of the defrost relay. The unit should go into a defrost cycle. If not, replace the defrost relay.
3. Remove the jumper wires.

178–4D

The following is a description of the steps for testing the *Ranco demand defrost control board.* The unit must be operating to perform the following tests.

1. Set the indoor thermostat to call for heat.
2. Check for 24VAC at the defrost control board terminals "24VAC" and "COM" and at the "HLD" and "COM" terminals. If no voltage is present, make sure that the indoor thermostat is calling for first stage heat and that all wiring is correct.

3. Momentarily "jumper" the two "test" pins on the control board until a defrost cycle is initiated. A red LED glows.
4. If the outdoor coil sensor is below 35°F, the system will remain in defrost until the coil sensor temperature reaches 55°F, or until 15 minutes have passed. If the coil sensor is above 35°F the defrost will terminate after 30 seconds.
5. If the outdoor coil does not defrost, check for 24VAC at terminals "OUT" and "COM." If there is no voltage present, replace the defrost control board.
6. If there is 24VAC present at terminals "OUT" and "COM," check for 24VAC at the defrost relay coil. If 24VAC is present but still no defrost cycle, replace the defrost relay.

IMPORTANT: Do not jumper terminals "OUT" and "COM," as the defrost control board will be damaged.

7. If 24VAC is not indicated at both points in step 6, check for a failed transformer, indoor thermostat, or improper control wiring.
8. Allow the unit to terminate the defrost cycle. It should terminate defrost when the outdoor coil sensor temperature reaches 55°F, or 15 minutes has passed. If not, replace the control board. See demand defrost troubleshooting chart in Figure 2–59.

178–5 RANCO DEMAND DEFROST CONTROLS:

The following is a description of the *Ranco demand defrost control.* See Figure 2–60.

The power input to the defrost control board is 24VAC (=/−25%) and 240VAC. See Figure 2–61.

1. The 24VAC is applied between the "R" and "C" terminals on the board.
2. When 24VAC is applied to terminal "Y" the compressor run time is calculated.
3. 24VAC is applied to the "B" terminal on the control board and the reversing valve terminal on the control board.
4. 240VAC is applied to the "FAN" terminals on the board. See Figure 2–62.

The power output is 24VAC/240VAC. See Figure 2–63.

1. 24VAC is supplied to the compressor contactor from the "Yout" terminal.
2. 24VAC is supplied to the reversing valve from the "RV" terminal.
3. 24VAC is supplied to the auxiliary heat from terminal "D."
4. 240VAC is supplied to the outdoor fan from the "FAN" terminals.

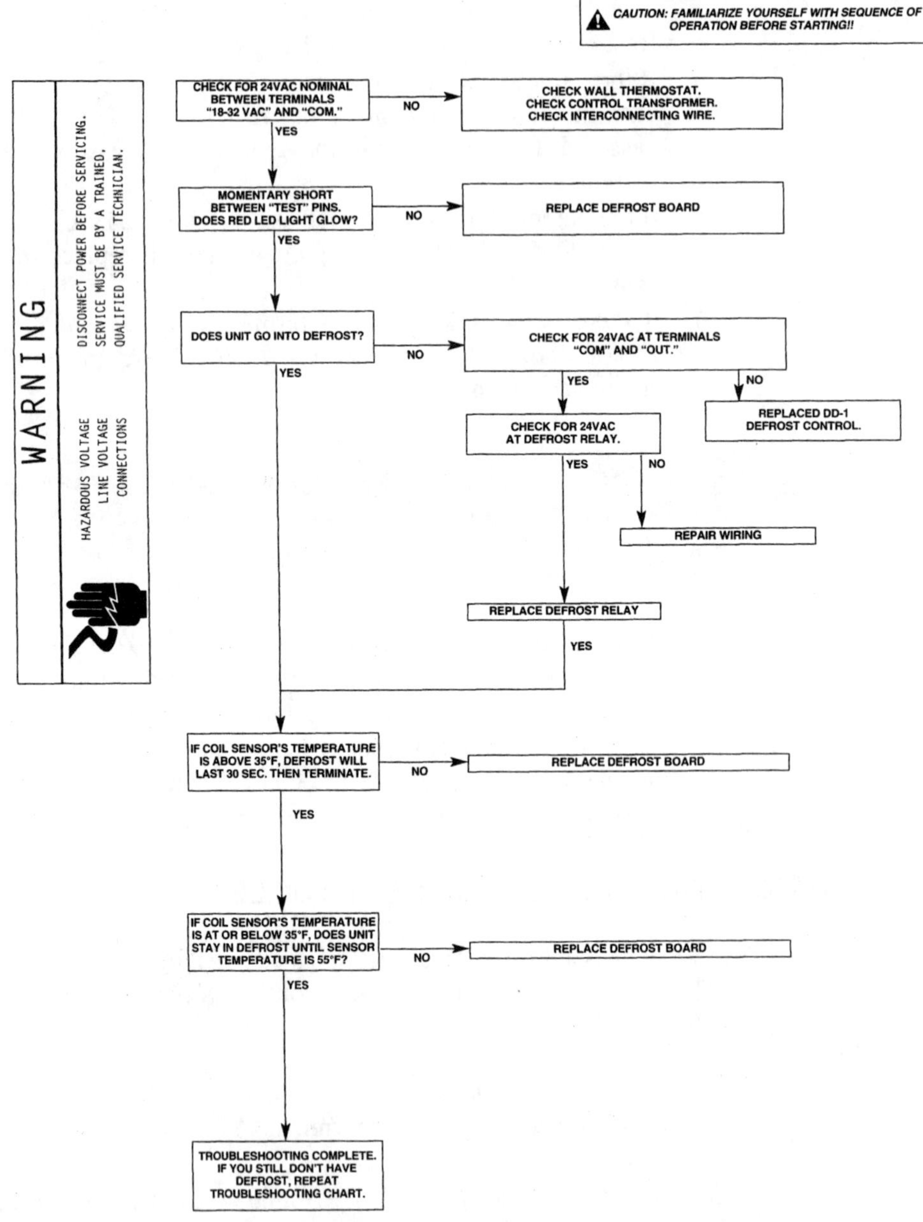

Figure 2–59. Troubleshooting the Ranco DD1-032001-000 demand defrost control. (Courtesy of Rheem Airconditioning Division)

178–5A

The *temperature measurement* is made with two sensors. See Figure 2–64.

1. The two temperature sensors are:
   a. The air sensor which is mounted on the control board (Y2A);
   b. The coil sensor (Y1) which is mounted on the outdoor coil to a copper tube (the coil sensor wire is 24 inches in length).

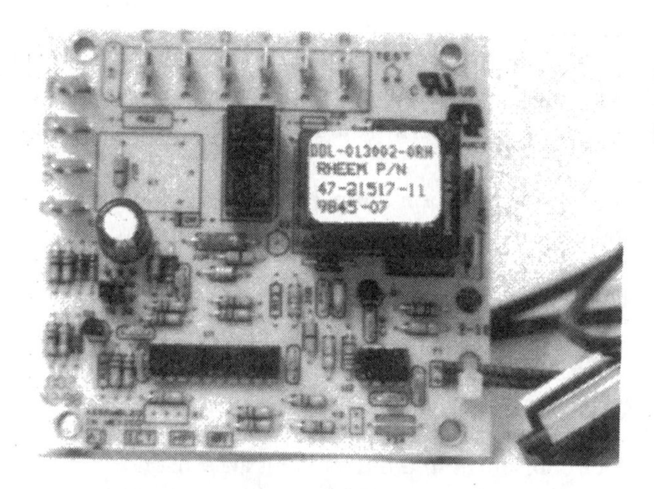

Figure 2–60. Ranco DDL defrost control. (Courtesy of Rheem Airconditioning Division)

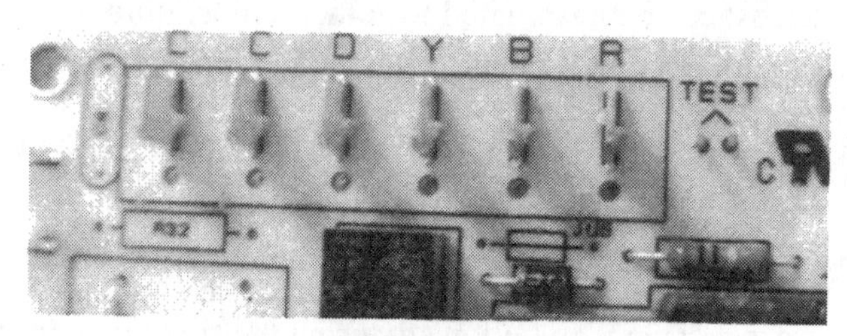

Figure 2–61. Low voltage connections. (Courtesy of Rheem Airconditioning Division)

Figure 2–62. Fan terminals. (Courtesy of Rheem Airconditioning Division)

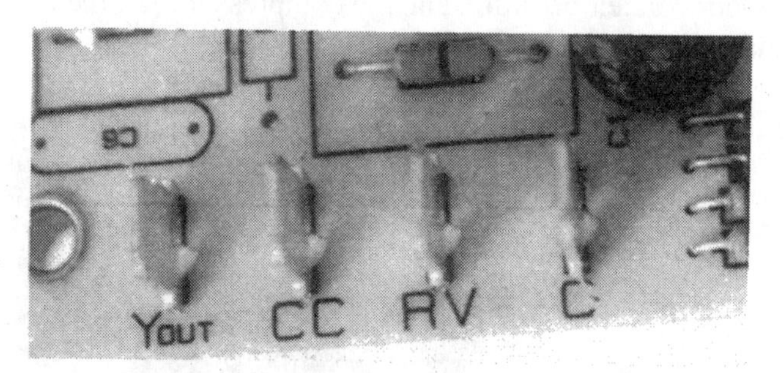

Figure 2–63. Output terminals. (Courtesy of Rheem Airconditioning Division)

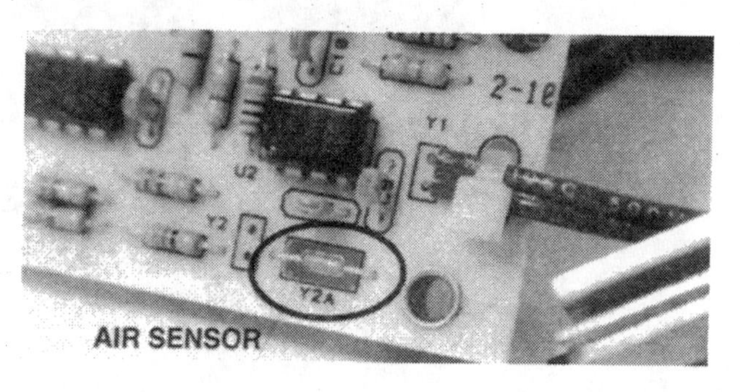

Figure 2–64. Air sensor. (Courtesy of Rheem Airconditioning Division)

2. The defrost is initiated by a difference between the air and coil sensors (if the coil is below 35°F and the 34 minute lockout time has been exceeded).
3. The defrost cycle is terminated by the coil sensor temperature or time (14 minutes).
4. The defrost termination temperature is field adjustable (50°F, 60°F, 70°F, and 80°F); the factory setting is 70°F. See Figure 2–65.

The system parameters are as follows:

1. The outdoor coil temperature must be below 35°F for a defrost cycle to be initiated.
2. The defrost cycle is terminated by the temperature difference between the outdoor coil sensor and the ambient air sensor.
3. The defrost can be terminated by either temperature or time.
   a. The outdoor coil termination temperature may be field selected (the factory setting is 70°F).
   b. The defrost cycle is terminated by time if it exceeds 14 minutes.
4. The defrost control has a lockout time period.
   a. There must be a minimum of 34 minutes of accumulated compressor run time before another defrost cycle can be initiated.
   b. A defrost cycle is not allowed during the first two minutes of compressor operation to allow the system pressures and temperatures to stabilize.

178–5B

The *defrost function of the defrost control* is to monitor the outdoor ambient temperature, outdoor coil temperature, and the compressor run time to determine when to initiate and to terminate a defrost cycle.

Figure 2–65. Termination settings. (Courtesy of Rheem Airconditioning Division)

1. When there is an initial power up (or after a power loss), a sacrificial defrost will occur in order to calibrate the defrost control board. It checks the time needed to warm the outdoor coil to the termination temperature with what should be a clean, frost-free coil.
2. A defrost cycle cannot occur unless the outdoor coil sensor is below 35°F.
3. The defrost control constantly monitors both the outdoor coil temperature and the outdoor air temperature. When the correct temperature difference between them occurs, the control initiates a defrost cycle. The correct temperature difference is not fixed as with some other demand defrost controls. The initiation temperature difference changes as the outdoor ambient temperature becomes colder.
4. Upon defrost initiation, the defrost control stops the outdoor fan, de-energizes the reversing valve, and sends power to the indoor air handler to cycle on the auxilliary heat.
5. During the defrost period the defrost control continues to monitor the outdoor coil temperature and the amount of time in the defrost cycle.
6. The defrost cycle is terminated when the outdoor coil sensor reaches the predetermined termination temperature (factory set for 70°F). The defrost control then powers the outdoor fan, the reversing valve, and cycles off the auxilliary heat.
7. If the defrost cycle continues for more than 14 minutes, it will be terminated mechanically.
8. If the defrost cycle terminates on the outdoor coil sensor temperature, or after 14 minutes with the outdoor coil temperature above 35°F for more than 4 minutes, the defrost control assumes that the outdoor coil is free of frost.
9. The defrost control does not allow another defrost cycle for 34 minutes. This is to avoid any unnecessary defrost periods caused by transient system conditions.
10. Should the defrost cycle be terminated after 14 minutes, and the outdoor coil remains below 35°F for less than 4 minutes, the control assumes that the defrost was not completed and, after 34 minutes of accumulated compressor run time in the heat mode, the system initiates another sacrificial defrost cycle.
11. After 6 hours of accumulated compressor run time with no defrost period and the outdoor coil sensor is below 35°F, the defrost control initiates a forced defrost cycle. This makes sure that there is adequate oil return to the compressor.
12. Should the indoor thermostat become satisfied during the defrost cycle because of the auxilliary heat, the defrost control immediately (no 34 minute wait) goes into a defrost cycle on the next call for unit operation.
13. Should the outdoor coil temperature become greater than the termination temperature or if the coil temperature exceeds the outdoor ambient air temperature (cooling operation) all defrost operations are suspended.

178–5C

To *test the demand defrost control* use the following steps:

1. There must be a "Y" call for compressor operation to perform the "Test." To test, momentarily jumper the "TEST" pins with a metal strip. See Figure 2–66.

   During this test the defrost initiation temperature is ignored. The unit responds to normal time and temperature conditions while in the "TEST" mode. If a sensor failure is detected, no defrost cycle will occur.

IMPORTANT: As long as the "TEST" pins remain shorted, the termination temperature is ignored and the unit will remain in the defrost cycle for only 3.5 seconds.

2. Remove the "TEST" pin jumper within the 3.5 seconds, and monitor the outdoor coil temperature at the defrost sensor location. A termination should occur when the outdoor coil reaches the termination temperature. Be aware that there is a temperature and time lag between the coil reaching termination temperature and when the sensor responds to terminate the defrost cycle.
3. Defrost termination should occur when the defrost sensor reaches the termination temperature (70°F factory setting).
4. If the defrost control does not perform as previously described, replace it.

NOTE: If the ambient sensor fails, the defrost control will initiate a defrost cycle every 34 minutes with the outdoor coil temperature below 35°F.

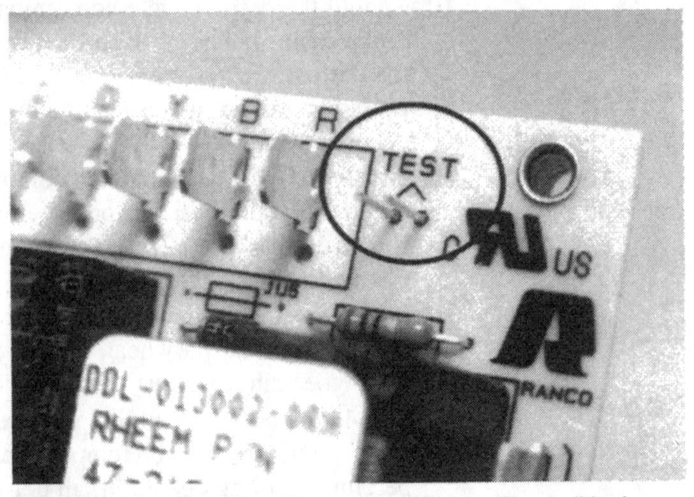

Figure 2–66. Test pins. (Courtesy of Rheem Airconditioning Division)

If the outdoor coil sensor fails, the defrost control cannot initiate a defrost cycle.

178–5D

Use the following procedure for *defrost control replacement.* This Ranco defrost control must be powered continuously, a minimum of 5 conductor low voltage wires must be run to the outdoor unit. Previous Ranco defrost control wiring used 4 conductor low voltage wire.

If replacing a defrost control on a unit with only 4 conductor low voltage wiring, and installing a 5 conductor wire is impossible, connect the "red" and "blue" control wires at the outdoor unit to the control wire attached to the "B" terminal from the thermostat. Use the following wiring diagram in such cases. See Figure 2–67.

A troubleshooting chart for the Ranco defrost control is shown in Figure 2–68.

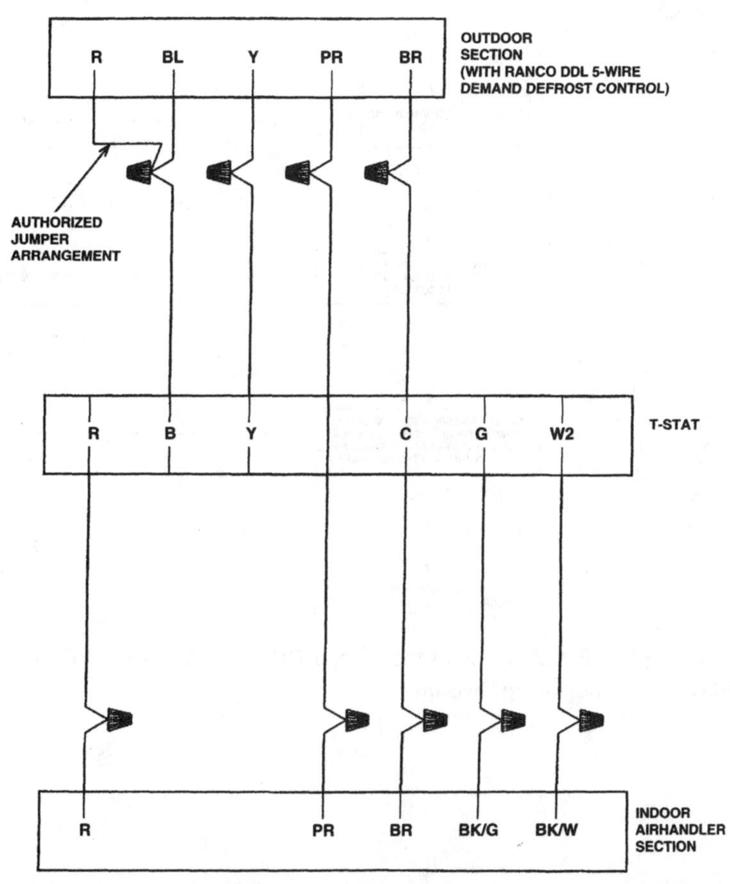

Figure 2–67. Ranco DDL low voltage wiring diagram. (Courtesy of Rheem Airconditioning Division)

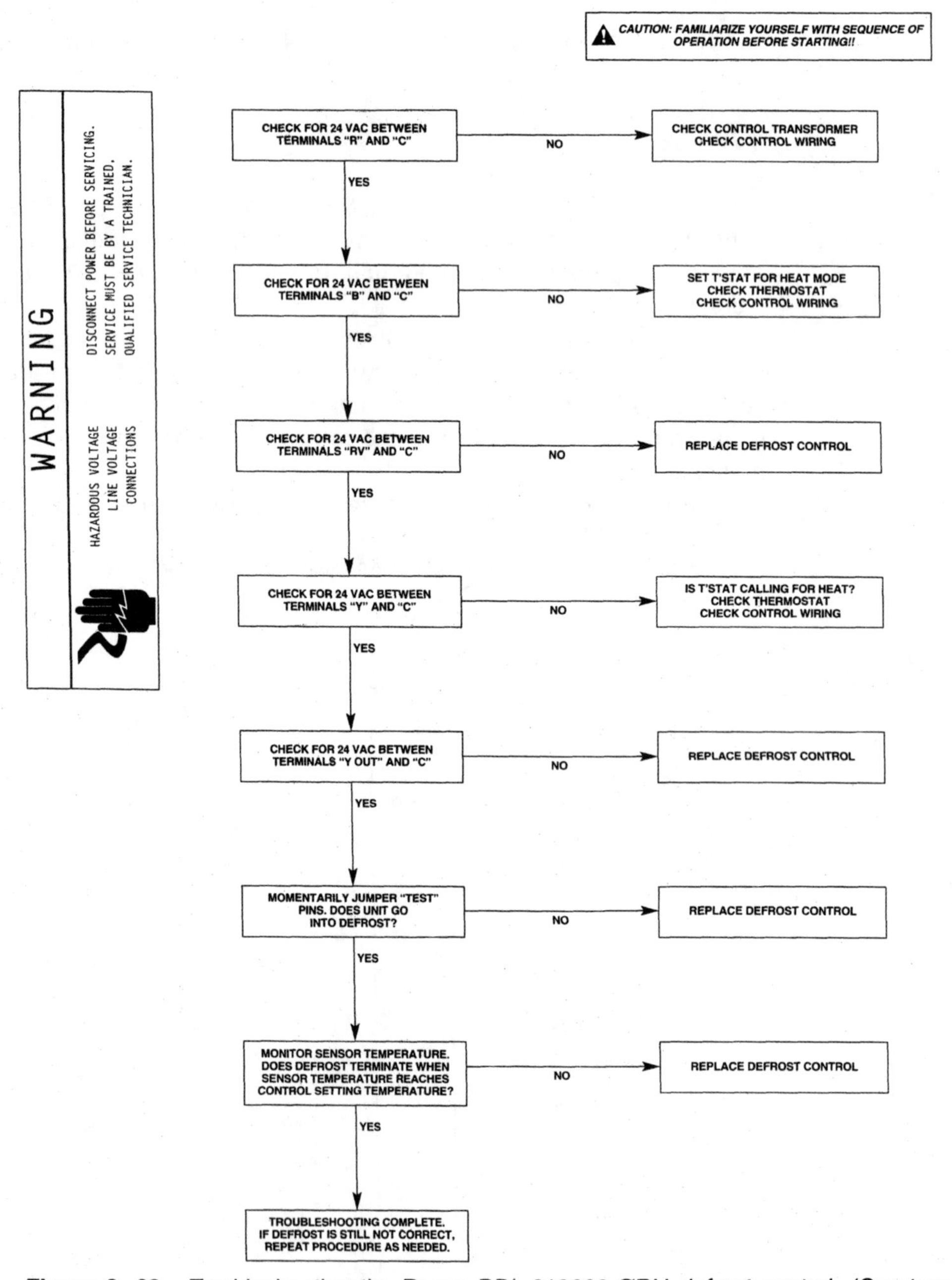

Figure 2–68. Troubleshooting the Ranco DDL-013002-ØRH defrost control. (Courtesy of Rheem Airconditioning Division)

3

Start-Up Procedures

GAS HEATING

The following steps are recommended for start-up procedures for starting a GAS HEATING unit. The technician will most likely rearrange, delete some, or add some to suit his/her technique after becoming experienced.

1. Close off all gas valves.
2. Wait approximately five minutes for any unburned gas to escape from combustion area.
3. Open the gas cock in the gas line.
4. Turn off-on-pilot knob to "pilot" position.
5. Strike match and hold flame by the pilot burner.
6. Depress pilot knob. See Figure 3–1.
7. Hold pilot knob in for approximately one minute after lit.
8. Release pilot knob.
9. Turn thermostat down below room temperature or turn off electrical power to furnace.
10. Turn gas valve knob to "on" position.
11. Turn on electrical power or raise temperature setting of thermostat above room temperature.
12. Check flame on main burners. Adjust primary air shutter if there is any yellow in the flame. If burner cannot be properly adjusted, close off the gas valves, remove and clean the burners. Reinstall burners, complete the above steps, and adjust the primary air.
13. Check for carbon monoxide in the flue gases inside the draft diverter. Make any adjustments indicated. See Figure 3–2.
14. Check the fan "on" and "off" temperatures, and the limit "off" temperature with a thermometer inserted into the circulating air stream as close as possible to the sensing element. See Figure 3–3.

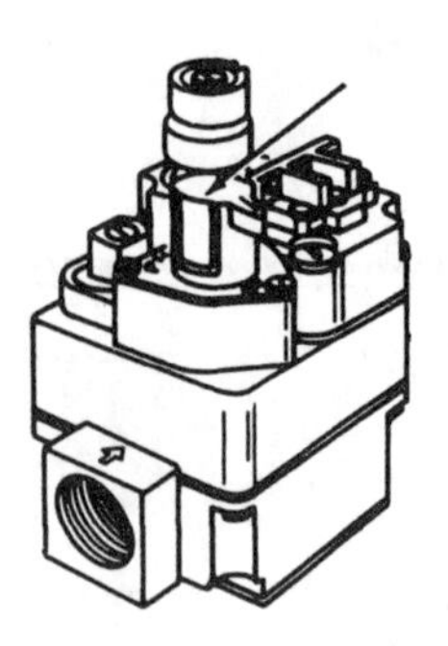

Figure 3–1. Pressing knob on gas valve to light pilot.

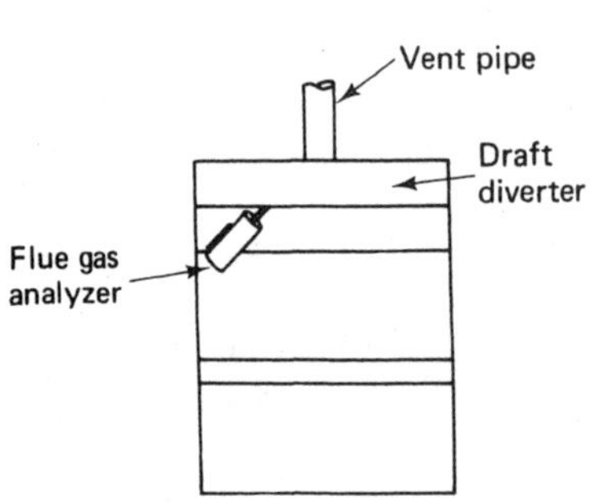

Figure 3–2. Checking flue gases.

Flue pipe
Discharge air plenum
Thermometer
Fan-limit control

Figure 3–3. Checking fan-limit control operation.

15. Check the heat anticipator.
16. Turn off the electrical power to furnace after fan has stopped running.
17. Clean or replace the air filter.
18. Check condition of all bearings. Replace or repair as required.
19. Lubricate all bearings requiring service.
20. Check belt condition and tension. Replace or adjust as required. See Figure 3–4.
21. Check calibration of thermostat. Calibrate as required.
22. Clean and vacuum furnace and area around furnace.
23. Replace all covers.

REFRIGERATION

The following steps are recommended for start-up procedures for starting a REFRIGERATION SYSTEM. The technician will most likely rearrange, delete some, or add some to suit his/her technique after becoming experienced.

1. Check condenser and clean if necessary.
2. Check all bearings; make necessary repairs.
3. Lubricate all bearings requiring service.
4. Check condition and tension of all belts. Replace or adjust as required. See Figure 3–4.

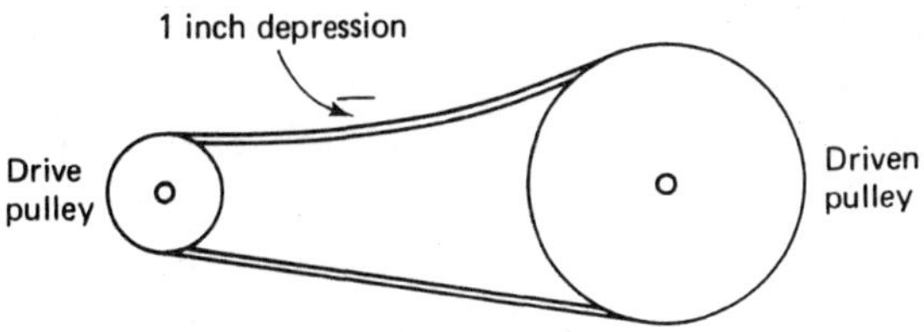

Figure 3–4. Checking belt tension.

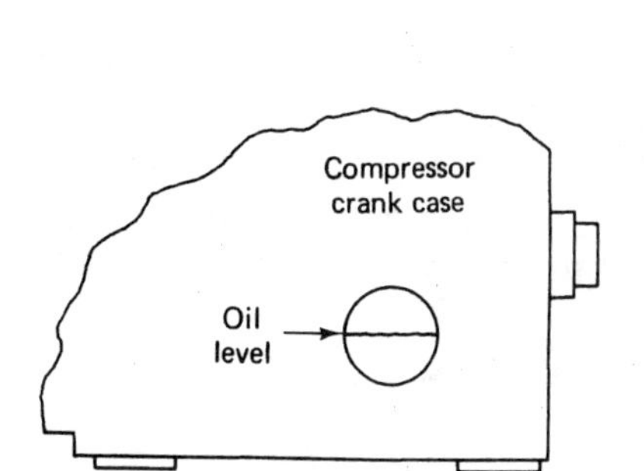

Figure 3–5. Checking compressor oil level.

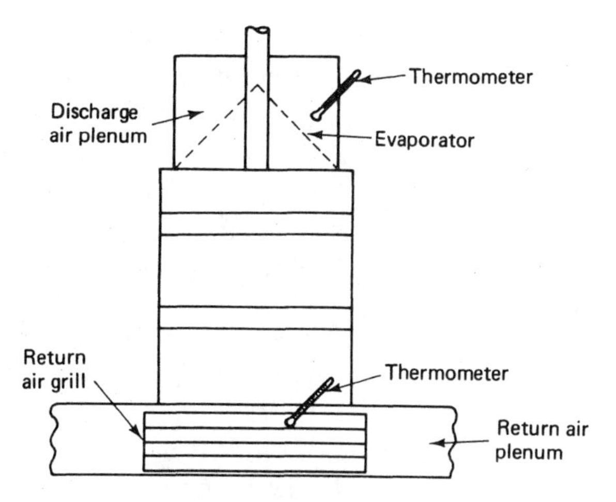

Figure 3–6. Checking temperature drop on evaporator.

5. Check condition of compressor contactor contacts. Replace if necessary.
6. Check tightness of all electrical connections. Tighten loose connections.
7. Check and repair burnt or frayed wiring.
8. Start unit and check suction and discharge pressures.
9. Check compressor oil level. See Figure 3–5.
10. Check refrigerant charge. If short, repair leak and add refrigerant.
11. Check amperage draw to all motors.
12. Check temperature drop on evaporator. See Figure 3–6.
13. Clean and vacuum indoor unit and area around unit.
14. Replace all covers.
15. Allow unit to operate for 24 hours and check the fixture temperature to see if it corresponds to the thermostat setting.

AIR CONDITIONING—HEAT PUMP

The following steps are recommended for start-up procedures for starting an AIR CONDITIONING-HEAT PUMP unit. The technician will most likely rearrange, delete some, or add some to suit his/her technique after becoming experienced.

1. Check to be sure that electricity has been on to the outdoor unit for 24 hours.
2. Check the outdoor coil and clean if necessary.

3. Check all bearings; make necessary repairs.
4. Lubricate all bearings requiring service.
5. Check condition and tension of all belts. Replace or adjust as required. See Figure 3–7.
6. Check condition of compressor contactor contacts. Replace if necessary.
7. Check tightness of all electrical connections. Tighten loose connections.
8. Check and repair burnt or frayed wiring.
9. Set thermostat for cooling, and lower temperature setting below room temperature.
10. Check temperature rise on outdoor coil. See Figure 3–8.
11. Check suction and discharge pressures.
12. Check refrigerant charge in system. If short, repair leak and add refrigerant.
13. Check the amperage draw to all motors.
14. Check temperature drop on indoor coil. If more than 24°F (13.33°C), check for dirty indoor coil. If less than 18°F (10°C), check for inefficient compressor or refrigerant shortage on cooling. On heating, the temperature rise should be approximately 30°F (16.67°C). See Figure 3–9.
15. Check thermostat calibration, Calibrate if necessary.
16. Clean or replace air filter.
17. Clean and vacuum indoor unit and area around unit.
18. Replace all covers.

ELECTRIC HEATING

The following steps are recommended for start-up procedures for starting an ELECTRIC HEATING unit. The technician will most likely rearrange, delete some, or add some to suit his/her technique after becoming experienced.

1. Turn off all electric power to unit.
2. Visually check for any burnt or bad wiring and replace as required.

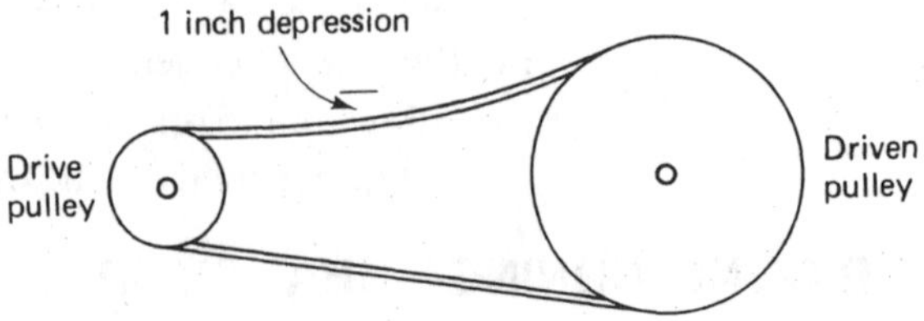

Figure 3–7. Checking belt tension.

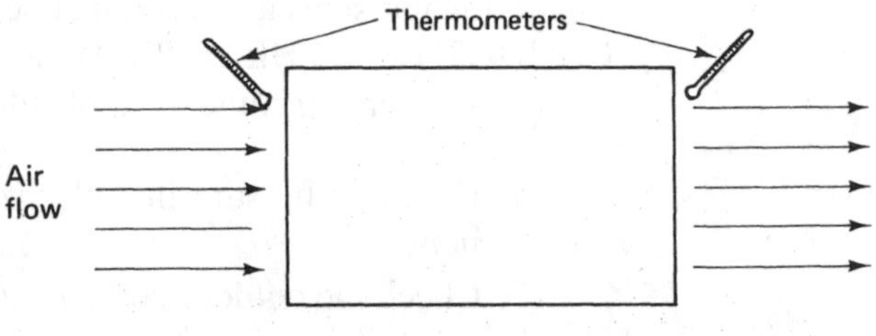

Figure 3–8. Checking temperature rise on outdoor coil.

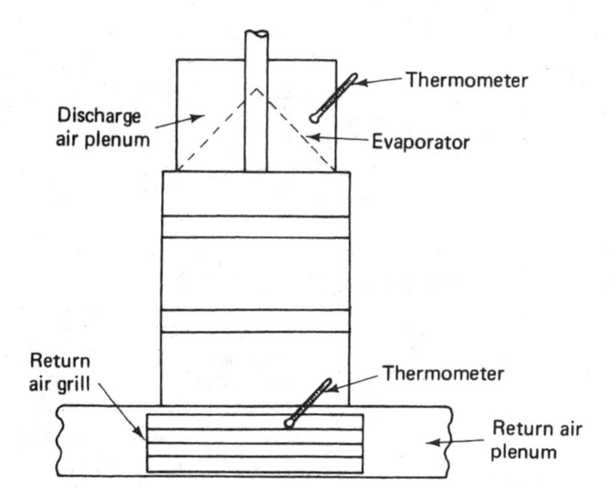

Figure 3–9. Checking temperature drop on evaporator.

3. Tighten all electrical connections.
4. Turn on electric power.
5. Set thermostat well above room temperature to insure that all stages are demanding.
6. Check voltage drop on all elements after all relays are closed.
7. Check amperage to all elements. Replace any bad elements.
8. Check setting of heat anticipator. Adjust as required.
9. Set thermostat temperature selector below room temperature.
10. After fan has stopped, turn off all electricity.
11. Check the condition of all bearings.
12. Lubricate all bearings requiring service.
13. Check condition and tension of belt. Replace or adjust as required. See Figure 3–10.
14. Replace or clean air filter.
15. Clean and vacuum furnace and area around furnace.
16. Replace all covers.
17. Turn on electricity.

OIL BURNER

The following steps are recommended for start-up procedures for starting an OIL BURNER system. The technician will most likely rearrange, delete some, or add some to suit his/her technique after becoming experienced.

1. Check oil level in storage tank.
2. Open valve in oil supply line to burner.
3. Bleed air from fuel pump. One-pipe systems must be bled. Two-pipe systems will usually bleed themselves.
4. Turn on electric switch.

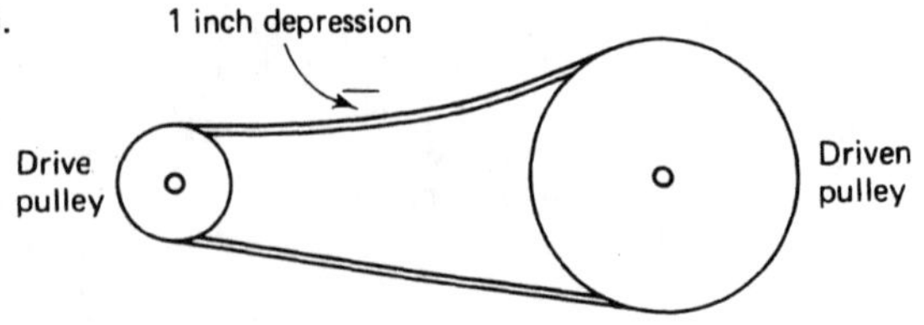

Figure 3–10. Checking belt tension.

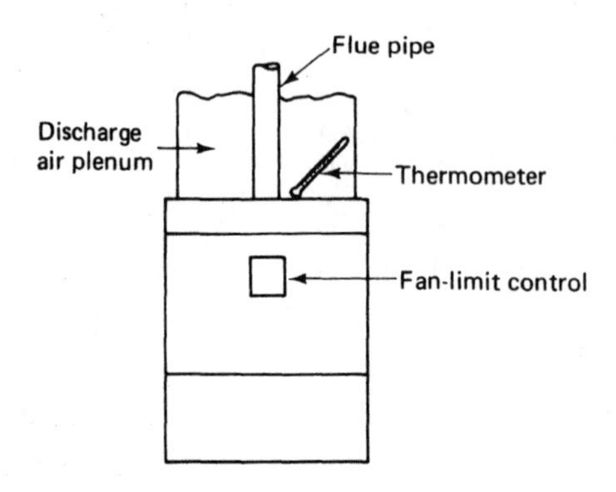

Figure 3–11. Checking fan-limit control operation.

5. Place a can under the bleed port to catch purged oil. Loosen the bleed plug and start the burner. Allow burner to run until a solid stream of oil is purged from the port. Turn off burner and screw in bleed port.
6. Set thermostat to call for heat.
7. If burner does not start, reset primary safety control and reset oil burner motor overload.
8. After the oil burner is started, make the adjustments outlined on page 140.
9. Check operation of the draft control.
10. Check operation of the primary safety control.
11. Clean or replace the air filters.
12. Adjust or replace the fan belt. See Figure 3–10.
13. Lubricate all bearings.
14. Check the fan control settings for accuracy. See Figure 3–11.
15. Check the limit control setting for accuracy.

4

Standard Service Procedures

REFRIGERANT RECOVERY/RECYCLING: Refrigerant recovery means that the refrigerant is removed from the system and stored in an EPA approved cylinder until the needed repairs are completed. This may be done in two ways: 1) liquid recovery, and 2) vapor recovery. The units used for this purpose must be EPA certified. Also, note that a hermetic or semihermetic compressor should never be used to recover refrigerant. A hermetic or semihermetic compressor can be damaged in several ways when used for this purpose. 1) The lubricating oil can be pumped from the compressor crankcase causing possible damage to the compressor surfaces needing lubrication; 2) the compressor valves could possibly be damaged due to the oil passing through them; and 3) the motor winding insulation is much less effective when electric power is applied when the windings are in a vacuum, especially when they are not surrounded by refrigerant. Thus, when the pressure inside the system is lowered sufficiently to reach that required by EPA; the compressor-motor winding would probably short out, ruining the motor and requiring the compressor-motor to be replaced.

Recovery Definition: To recover refrigerant means to remove it from the system being repaired regardless of its condition and store it in an EPA approved recovery cylinder. This method does not necessarily include testing or processing it in any way.

Recycle Definition: To recycle refrigerant means to clean it for reuse by oil separation and single or multiple passes through cleaning devices, such as replaceable core filter-driers, to remove moisture, any acid present, and particulate matter. Generally, these procedures are done at the job site or in the company service shop.

Reclaim Definition: To reclaim refrigerant means to reprocess it to new refrigerant specifications by means of a distillation process. Reclaiming refrigerant requires that a chemical analysis be made to determine if it meets the appropriate refrigerant specifications. Generally, refrigerant can only be reclaimed at a facility that is designed specifically for this purpose.

RECOVERY EQUIPMENT: There are two different types of refrigerant recovery units. The type of work most often done will determine the one that should be purchased. In any case the one used must be approved by EPA.

Recovery units are designed in several different ways. They may have an internal system that can have several different uses. There may be an oil separator included in these units. Some will have an oil separator with a multiple bypass filter-drying arrangement. They may also include an oil separator with a single pass filter-drying arrangement. See Figures 4–1 through 4–3.

RECOVERY METHODS: The two methods of recovering refrigerant from a system are liquid and vapor. Usually the amount of refrigerant contained in the system will determine which method is used. It is usually best to use the liquid method when a large amount of refrigerant is to be recovered.

Liquid Recovery Method: This is the faster method of refrigerant removal. When the quantity of refrigerant warrants, this is the method that should be used. See Figure 4–4 for a suggested method of connecting the recovery unit to the system for liquid refrigerant recovery.

Always use caution to prevent liquid refrigerant from entering the recovery unit compressor. It is the pressure difference between the recovery cylinder and the refrigeration system that causes the refrigerant to flow into the cylinder. The suction of the recovery unit pumps the vapor from the top of the recovery cylinder and the recovery unit discharge pressure is directed back into the system causing the liquid

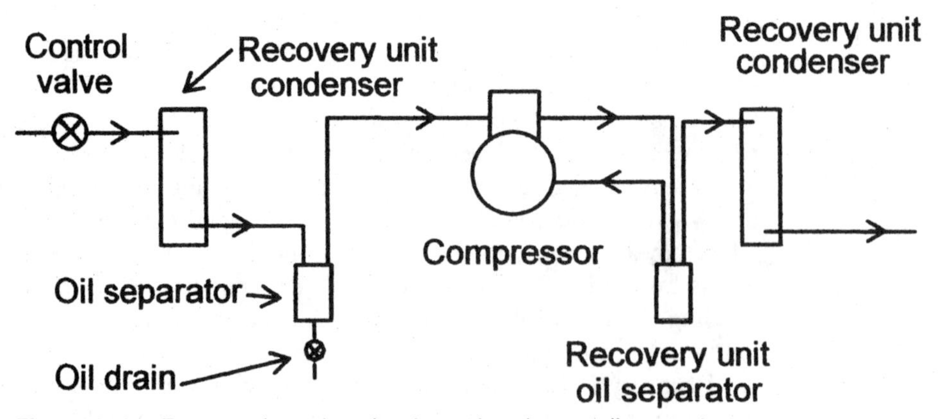

Figure 4–1. Recovery/recycle unit schematic using an oil separator.

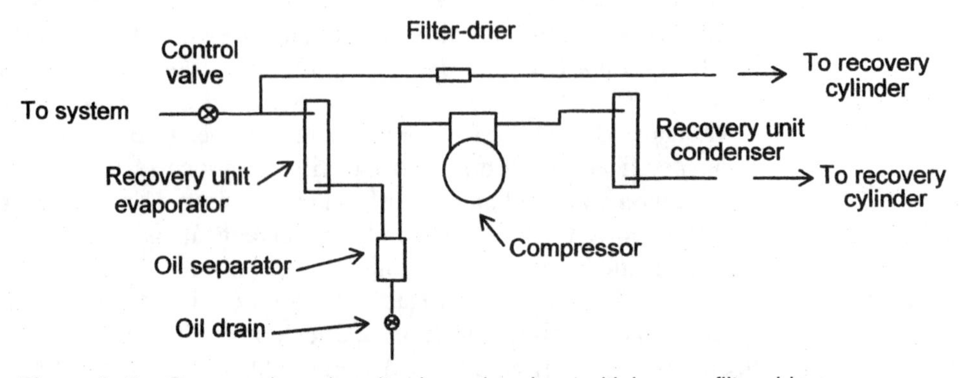

Figure 4–2. Recovery/recycle unit schematic using multiple pass filter-driers.

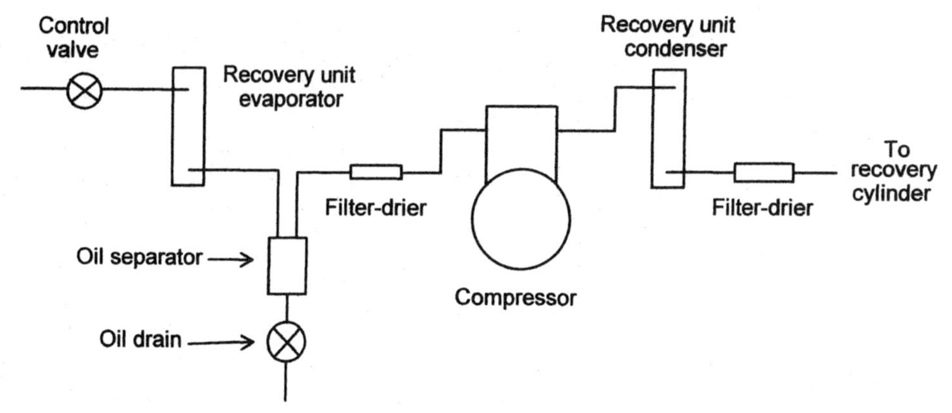

Figure 4–3. Recovery/recycle unit schematic using single pass filter-drier.

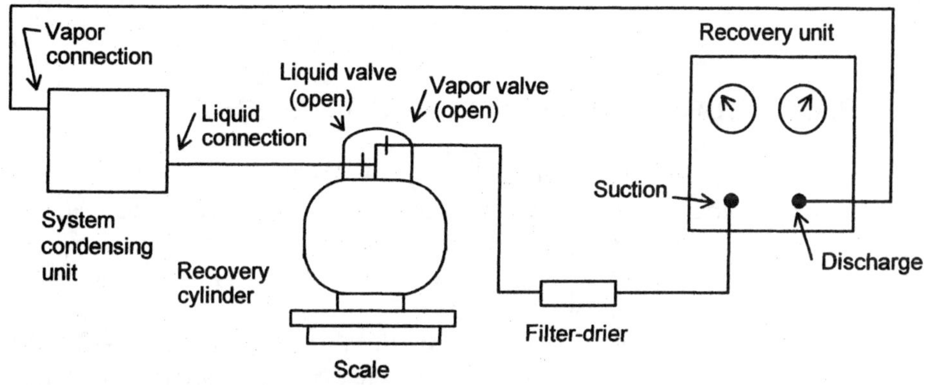

Figure 4–4. Liquid recovery connections.

to flow from the system and into the recovery cylinder. To remove any contaminants from the refrigerant during the recovery process, place a filter-drier in the vapor line to the recovery unit.

Vapor Recovery Method: When using the vapor recovery method, use the connections shown in Figure 4–5.

When this method is used, the refrigerant vapor is pumped from the system by the recovery unit compressor. The vapor is then condensed in the recovery unit, then forced into the refrigerant recovery cylinder. This method is usually used when only a small amount of refrigerant is to be recovered or when there is no liquid service connection provided on the system. To clean the refrigerant during the recovery process, place a filter-drier in the vapor line between the system and the recovery unit. The vapor recovery method is almost always more time consuming than the liquid recovery method, but it is the only method that can be used on all systems.

When recovering refrigerant from a system, the evacuation standards for the type of system being serviced must meet EPA standards. Always check the appropriate evacuation standards.

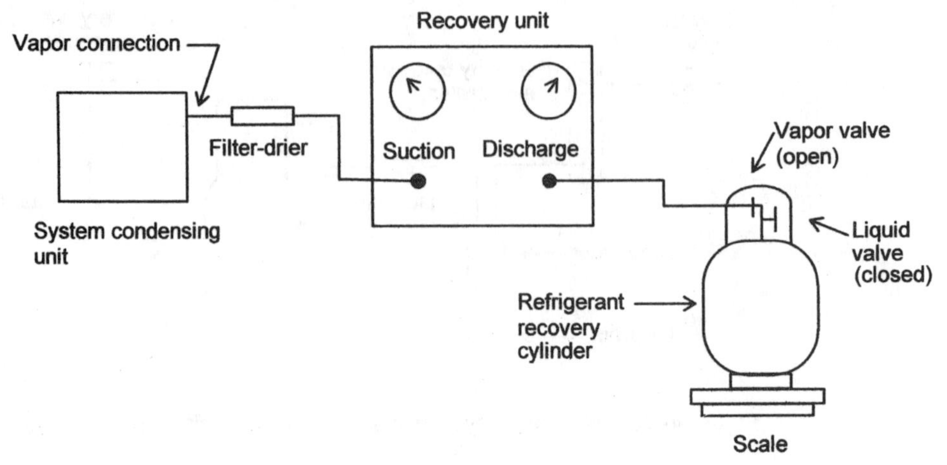

Figure 4–5. Vapor recovery connections.

1-1 PROCEDURE FOR REPLACING COMPRESSORS:

Past experiences have demonstrated that after a hermetic motor burnout has occurred, the refrigeration system must be cleaned to remove all contaminants. Without removal of these contaminants, a repeat burnout will occur. Failure to follow minimum cleaning recommendations as quickly as possible will result in an excessive risk of a repeat burnout.

Cleaning of the system by flushing with refrigerant—R11, or similar refrigerants, has been used in the past under certain conditions and with mixed degrees of success. This flushing method is seldom used today and is not recommended here.

The first step is to be certain that the compressor is burned. The following list is a general procedure used:

1. When a compressor fails to start, it may appear to be burnt out. However, the fault can be in many other areas.
2. Eliminate all other possibilities. Check for electrical and mechanical misapplication or malfunction.
3. Check the compressor motor for shorted, grounded, and open windings. If none are found, check the winding resistance with a precision ohmmeter to determine if turn-to-turn shorts exist.
4. Next, remove a small amount of refrigerant gas from the compressor discharge. Smell the gas in small whiffs. A burned motor is usually indicated when a characteristic acrid odor is detected.

If the motor is burned, refrigerant should be discharged from the system. The service technician should adhere to certain safety practices described following. In addition to the electrical hazards, the service technician should be aware of the dangers of receiving acid burns.

1. When recovering refrigerant from a system in which a burnout has occurred, avoid injury by not allowing the refrigerant to touch the skin. Do not allow the burnt refrigerant vapors to escape inside the building. They will cause damage to the metal surfaces as well as contaminate the air causing it to be unfit for breathing.

2. When it is necessary for the service technician to come in contact with the oil or sludge in a burned-out compressor, rubber gloves should be worn to prevent possible acid burns.

When replacing the compressor in a system that has had a burnout, thoroughly clean the system to remove as many of the contaminants as possible. The following is a list of recommended procedures:

1. When it is impossible to valve the system off, the total refrigerant charge must be recovered using the proper EPA procedures. When the compressor can be valved off, (both the suction and discharge service valves front-seated) the refrigerant must then be properly recovered from the compressor before it is replaced.

2. When the system has been valved off (front-seating the compressor service valves) and the refrigerant saved in the system, and the compressor has been replaced, the system can be pumped down and an approved clean-up suction line filter-drier installed in the line at the compressor suction. See Figure 4–6.

3. Install an approved oversized liquid line filter-drier in the liquid line. See Figure 4–7. On units that have been valved off, this cannot be done until the refrigerant has been pumped into the high side of the system.

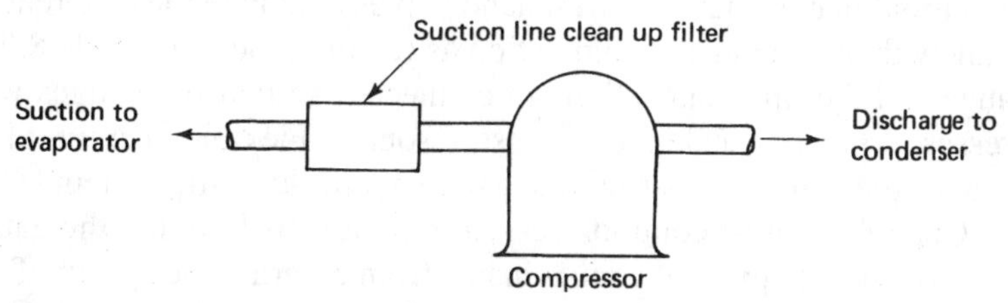

Figure 4–6. Suction line clean up filter-drier.

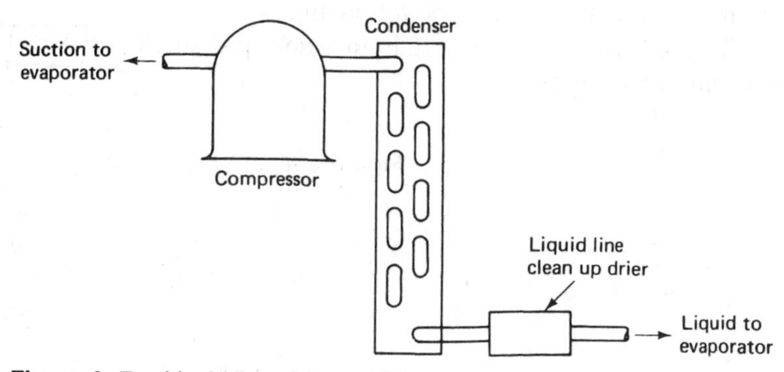

Figure 4–7. Liquid line clean up drier.

4. Check the refrigerant flow control device and clean it thoroughly or replace it.
5. Remove the inoperative compressor and install the proper replacement. Make all the connections leak-tight.
6. Purge the system by blowing refrigerant through the system, forcing the refrigerant completely through all piping and coils.
7. Triple-evacuate the system. The last evacuation should lower the system pressure to 1,000 microns or lower, if time permits.
8. Recharge the system and put it back in operation. On systems larger than 5 hp, add refrigerant to complete the charge.
9. After the system has been in operation for approximately 24 hours, check the acid content of the oil and the condition of the clean-up driers and filters. If acid is found in the oil, noted by discoloration, or the clean-up filters and driers show signs of stoppage, replace them, drain the oil, and if possible put a new charge in the system. Repeat this procedure until a clean system is indicated.

2-1 PROCEDURE FOR USING THE GAUGE MANIFOLD:

The gauge manifold is probably the most important tool in the serviceman's toolbox. The various "gauges" can be used to check system pressures, to charge refrigerant into the system, to evacuate the system, to add oil to the system, to purge noncondensables from the system, and for many other uses.

The gauge manifold set consists of a compound gauge, a pressure gauge, and a manifold that is equipped with hand valves to isolate the different connections or to allow their use in any combination as required. See Figure 4–8. The ports to the gauge and the line connection are connected so that the gauges will indicate the pressure when connected to a pressure source. Flexible, leak-proof hoses are used to make the connections from the gauge manifold to the system.

One of the most common service functions is hooking the gauges to the system. Take care to prevent contaminants from entering the system. The hoses should be purged by allowing a small amount of refrigerant to escape from the fittings before tightening. The specific procedures for connecting the gauges to a system containing refrigerant are discussed following.

On systems where it is certain that both pressures are above 0 psig, use the following procedures:

Figure 4–8. Gauge manifold.

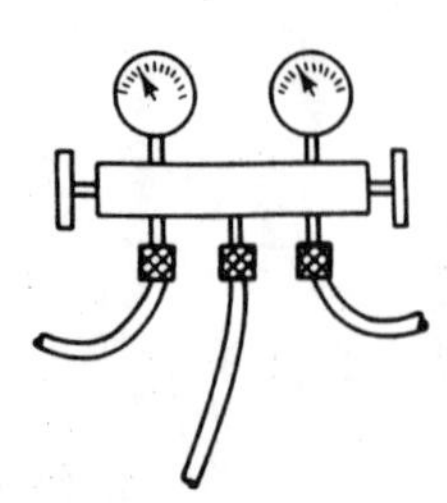

1. Front-seat the valves on the gauge manifold.
2. Back-seat the system service valves to isolate the gauge ports from the rest of the system.
3. Make the hose connections to the system. Loosen one end of the center hose.
4. Crack the system service valves off the back seat. Do this slowly to prevent a sudden inrush of high pressure gas to the gauge.
5. Crack open one valve on the gauge manifold and allow a small amount of refrigerant vapor to escape out the center hose for a few seconds. Close the gauge manifold valve and repeat this process with the other valve.
6. The gauge manifold is now connected to the system and is ready for use. See Figure 4–9.

On systems where the low side pressure is below 0 psig, use these procedures:

1. Front-seat the valves on the gauge manifold.
2. Back-seat the system service valves to isolate the gauge ports from the rest of the system.
3. Make the hose connections to the system. Tighten the hose connection to the discharge service valve, loosen the hose connection to the suction service valve, and plug the center hose connection to prevent the escape of refrigerant at this point.
4. Crack the system discharge service valve off the back seat. Do this slowly to prevent a sudden inrush of high pressure gas to the gauge.
5. Crack open the high side gauge manifold valve and allow a small amount of refrigerant vapor to escape out the low side hose connection at the system service valve. Tighten the hose connection after a few seconds. Close the gauge manifold valve.
6. Crack the system suction service valve off the back seat.
7. The gauge manifold is now connected to the system and is ready for use. See Figure 4–9.

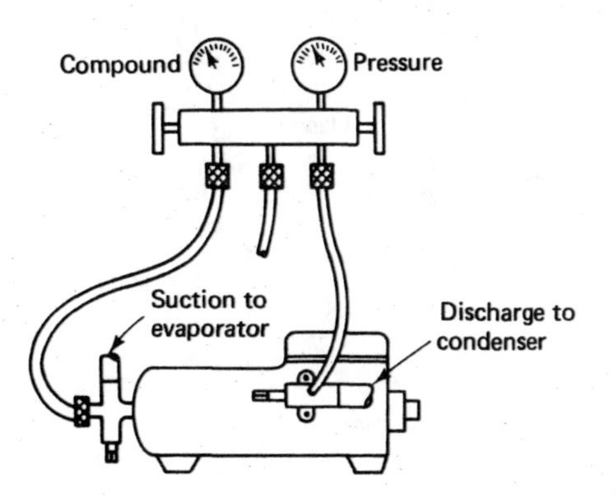

Figure 4–9. Gauge manifold connected to system.

Treat gauges with care. Do not drop the gauges or subject them to pressures higher than the maximum pressure shown on the scale. Gauges should be kept in adjustment so that the proper pressures are indicated.

3-1 PROCEDURE FOR CHECKING A THERMOCOUPLE:

The following steps are a recommended procedure for checking A THERMOCOUPLE. The technician will most likely rearrange, delete some, or add some to suit his/her technique after becoming experienced.

1. Install an adapter between the thermocouple and power unit.
2. Light pilot. Follow lighting instructions on furnace.
3. Set meter on millivoltage scale.
4. To check open circuit voltage, remove the thermocouple from adapter and touch one meter lead to the outer sheath, and the other meter lead to the inner wire of the thermocouple. This is DC voltage. If a negative reading is indicated, change meter leads. Be sure to keep the pilot burning during this test. The meter should indicate approximately 30 mV. If less than 25 mV is indicated, replace the thermocouple. See Figure 4–10.
5. To check closed circuit voltage, connect the thermocouple to the adapter with the pilot burning. Touch one meter lead to the adapter terminal and the other lead to the outer sheath of the thermocouple. If the reading indicated is less than 17 mV, replace the thermocouple or clean the pilot burner, or both. See Figure 4–10.

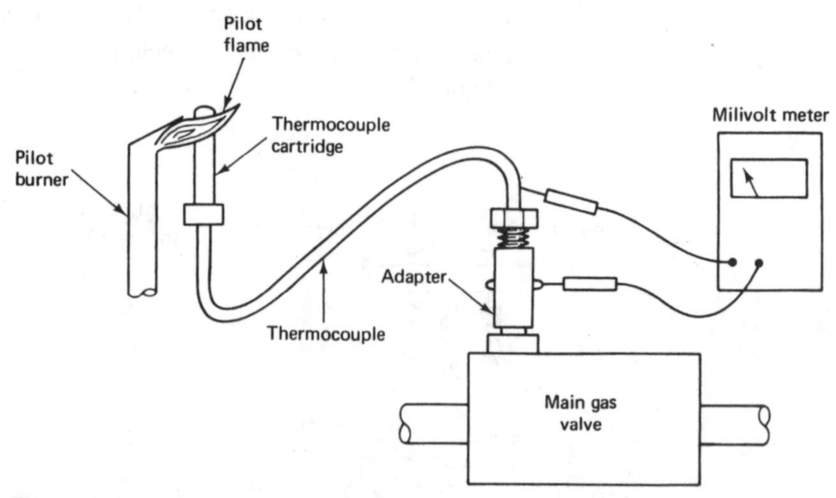

Figure 4–10. Closed circuit thermocouple test.

6. To check the thermocouple drop-out voltage, connect the meter as outlined in step 5. Turn off the pilot and observe the meter. When a dull thud in the gas valve is heard, the drop-out voltage is indicated on the meter. The drop-out voltage should be about 8 mV. If the drop-out is above 8 mV, the pilot stat power unit is defective and needs replacement. There is no correction for a low drop-out voltage below this.
7. Remove the adapter, reconnect the thermocouple to the power unit, and replace covers and panels.

4–1 PROCEDURE FOR ADJUSTING FAN AND LIMIT CONTROL:

The following steps are a recommended procedure for ADJUSTING A FAN AND LIMIT CONTROL. The technician will most likely rearrange, delete some, or add some to suit his/her technique after becoming experienced.

1. Light pilot. Follow lighting instructions on furnace.
2. Insert a thermometer as close as possible to the control-sensing element. This will generally be in the discharge air plenum directly above the fan and limit control. See Figure 4–11.
3. Turn thermostat up above room temperature.
4. Observe the thermometer. The fan should start at 150°F (66°C). If not, adjust the control to compensate for the difference. *Example*: If the fan starts at 140°F (60°C), adjust the fan "on" setting 10°F (5.5°C) higher.
5. To check the limit control, restrict the flow of air through the furnace by blocking the return air opening.
6. Observe the thermometer. The main burner flame should go out at a maximum of 200°F (93°C). If the flame goes out between 180°F (82°C) and 200°F (93°C), no adjustment is required. If adjustment is required, follow the example in step 4.

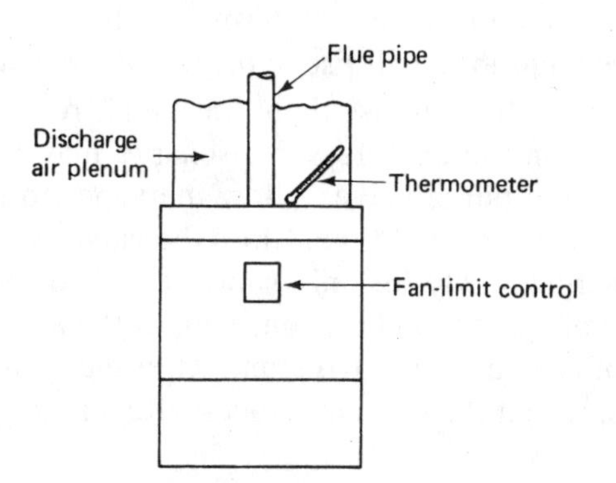

Figure 4–11. Checking fan-limit control operation.

7. To check the fan "off" setting, set the thermostat below room temperature and remove the return air restriction. Allow the fan to run and cool the furnace down.
8. Observe the thermometer. The fan should stop running at 100°F (37.8°C). If not, adjust the control to compensate for the difference, using the example in step 4.
9. Repeat all the steps outlined above and check the settings. Make any adjustments required.
10. Remove the thermometer and replace all panels and covers.

5-1 PROCEDURE FOR REMOVING NONCONDENSABLES FROM THE SYSTEM:

Air is considered a noncondensable under the pressures and temperatures normally encountered in an air conditioning or refrigeration system. Air can enter the refrigerant circuit in several ways. The most common ways are a leak in the low side of the system or improper connection and use of the service gauges. In some cases it may not be practical or desirable to purge the complete refrigerant charge and evacuate the system. However, the air must be removed to prevent damage to the system due to chemical reactions and to help keep the system operating efficiently.

The air will normally be trapped in the top of the receiver and the condenser because of the liquid seal at the receiver or condenser outlet. Air in the system may be detected by a higher than normal condensing pressure caused by the trapped air. The amount of increase in pressure will be determined by the amount of trapped air.

To remove noncondensables from the system, use the following procedure:

1. Locate and remove the source of noncondensables.
2. Connect the gauges to the system. See Figure 4–9.
3. If possible, pump the system down.
4. Stop the unit. Leave the condenser fan running on air-cooled units or block the water valve open on water-cooled units and leave the water pump running. Allow the unit to cool down for approximately ten minutes. During this time the noncondensables will rise to the top of the condenser.
5. If purge valves are on the unit, use them for the air removal process. If not, the gauge port on the compressor discharge may be used. To remove, slowly open the purge valve. Allow the vapor to bleed off very slowly for only a short period of time. Close the valve and let the unit sit idle. Be sure to purge slowly for only short periods of time to prevent the boiling off of refrigerant and the remixing of the air and refrigerant and the removal of an excess of refrigerant. After the system has sat idle for a few minutes, repeat the removal process. Repeat this process three or four times.

6. Start the system and check the discharge pressure after a few minutes of operation. If the discharge pressure is still abnormally high, repeat the purging process starting with step 4. Repeat steps 4, 5, and 6 until satisfactory operation is obtained. It may be necessary to remove the complete refrigerant charge, using the proper procedure, and recharge the system with new refrigerant.
7. Put the system back in normal operating condition.

6-1 PROCEDURE FOR PUMPING A SYSTEM DOWN:

Pumping a system down is the process generally used to put all the refrigerant into the condenser or receiver. Pumping down is used to save the refrigerant when work is required on components in the low side of the refrigerant system. Pumping a system down can be accomplished on systems equipped with service valves and a liquid line shut-off valve.

To pump a system down, use the following procedure:

1. Install the gauge manifold on the system. See Figure 4–9.
2. Start the unit.
3. Front-seat the liquid line (king) valve.
4. Observe both the suction and discharge pressures. If there is a sharp increase in the discharge pressure, stop the compressor. Check to determine the reason for the increase in discharge pressure. If the receiver and condenser are full of refrigerant, the remaining charge must be removed from the system using EPA mandated procedure.

   When the suction pressure is reduced to about 1 or 2 psig (6.895 or 13.790 KN/m^2), stop the compressor and observe the gauges. If the suction pressure increases to 10 or 15 psig (68.948 or 103.421 KN/m^2), start the compressor and pump the system down to 1 or 2 psig again and stop the compressor. The suction pressure should remain at around this pressure. If not, repeat the pump-down process. If the pump-down process is repeated more than three times, the compressor discharge valves may be leaking. In this case the discharge service valve must be closed to prevent refrigerant bleeding into the low side of the system.

CAUTION: Do not start the compressor when the discharge valve is closed.

5. On units equipped with a low-pressure control set at a pressure higher than 1 or 2 psig, it will be necessary to electrically bypass the control to keep the compressor running while pumping the system down.

6. Relieve any remaining pressure on the low side of the system by opening the low side hand valve on the gauge manifold. Do not attempt to weld or solder on a system having refrigerant pressure inside.
7. The necessary repairs can now be made. It is desirable to install a new liquid line drier before recharging the system.
8. To put the system back in operation, open the compressor discharge service valve and open the liquid line (king) valve. Allow a small amount of refrigerant to escape through the gauge manifold; then close the gauge manifold low side hand valve.
9. Start the unit and check the refrigerant charge. Add any required refrigerant to bring the system to full charge.

7-1 PROCEDURE FOR PUMPING A SYSTEM OUT:

Pumping a system out is the process used to save the refrigerant in a system that does not have service valves, when repairs are to be made. This procedure is also used when repairs are to be made to the high side of the system and a system pump-down cannot be done. To accomplish this process, an EPA certified refrigerant recovery unit is required. The recovery unit is used to pump the refrigerant from the recovery system and discharge it into the refrigerant recovery cylinder where it is stored while the repairs are being made. The refrigerant may then be charged back into the system.

To pump a system out, use the following procedure:

1. Stop the unit.
2. Connect the gauge manifold to the system. See Figure 4–9.
3. Connect the center hose on the gauge manifold to the suction service valve on the portable condensing unit.
4. Connect the liquid line connection on the portable condensing unit to the valve on the refrigerant cylinder. Leave this connection loose for purging air from the lines and from the portable condensing unit. See Figure 4–12.
5. Slowly crack the system service valves off the back seat until pressure is indicated on the gauges. Open the service valves on the refrigerant recovery unit.
6. Open the hand valves on the gauge manifold and allow the refrigerant to blow out the connection on the refrigerant cylinder for a few seconds. Tighten the hose connection on the refrigerant cylinder.
7. Open the refrigerant cylinder valve.
8. Start the portable condensing unit and pump the refrigerant from the system and into the cylinder. To prevent overloading of the recovery unit, regulate the suction pressure by partially closing the hand valves on the gauge manifold.

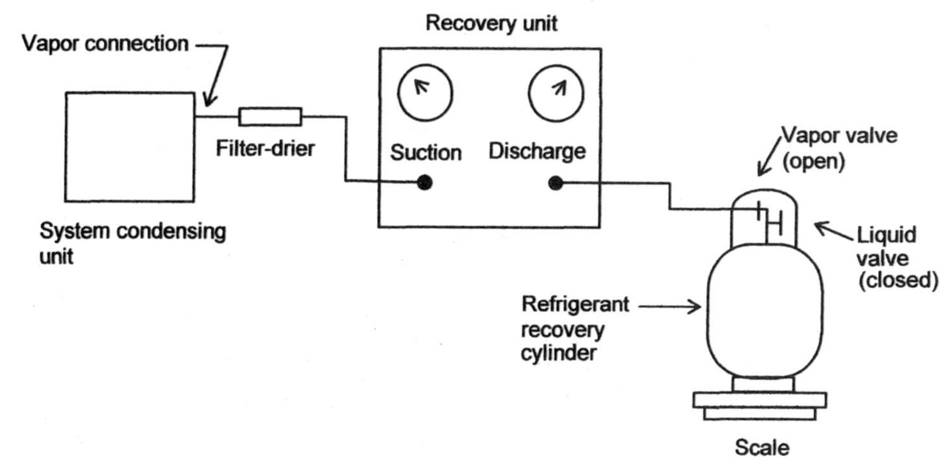

Figure 4–12. Connections for removing refrigerant with a recovery unit.

CAUTION: Be sure not to overcharge a cylinder with refrigerant. Use as many cylinders as required by weight.

9. Remove the refrigerant until only 1 or 2 psig pressure (6.895 or 13.790 KN/m^2) is left inside the system. Close off all valves to prevent the refrigerant escaping from the cylinder and the portable condensing unit.
10. Release any remaining pressure from the refrigeration system and make the required repairs. It is desirable to install a new liquid line drier before charging the system with refrigerant.
11. To put the system back in operation, a complete evacuation should be completed. Then charge the refrigerant from the cylinder and the recovery unit back into the system. It may be desirable to install a drier in the charging line to remove any contaminants from the system. See Figure 4–13.

8-1 PROCEDURE FOR LEAK TESTING:

When the refrigerant has escaped from a system, the leak must be found and be repaired or the refrigerant will escape again. Refrigeration systems must be gas-tight for two reasons: First, any leakage will result in a loss of refrigerant charge. Second, leaks will allow air and moisture to enter the system when the pressure is reduced below 0 psig. Refrigerant leaks can and do occur at any time and at any point due to low-quality workmanship, age of equipment, and vibration.

Because leak detection is a common service procedure, the service technician must be familiar with the various methods used. Each method has its advantages

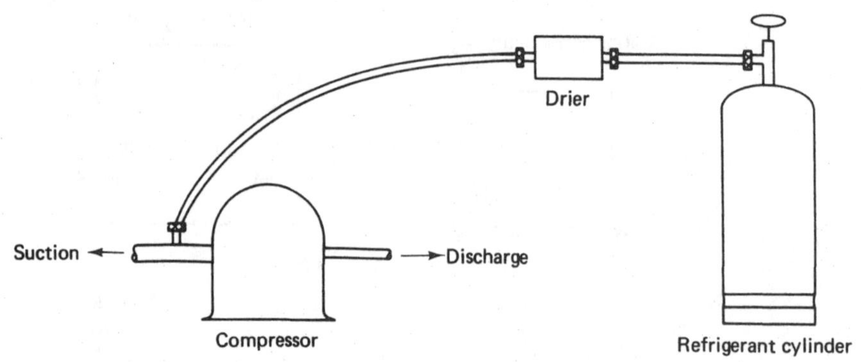

Figure 4–13. Liquid drier in charging line.

and disadvantages, depending on the circumstances surrounding the system. All the preferred methods require that pressure be introduced into the system. The three most popular methods are: electronic leak detection, halide-torch leak detection, and soap-bubble leak detection. However, the electronic method is not satisfactory in a heavy concentration of refrigerant because it is extremely sensitive, and the halide torch is not generally satisfactory when small leaks or blowing winds are encountered. Under most other conditions these two tests are found most satisfactory. Because electronic detectors are very sensitive, the soap-bubble test is most satisfactory for finding small leaks in a suspected area. Sometimes more than one of these methods may be used to locate one leak.

Evacuation is not a recommended leak detection method for many reasons. First, evacuation will not pinpoint a leak. Second, a flake of paint or a grain of sand may cover the leak during evacuation, preventing the loss of vacuum in the system. Third, more air and moisture are drawn into the system requiring more evacuation for their removal. Therefore, one of the above three methods is preferred.

To leak-test a system, use the following procedure:

1. Stop the unit to reduce air movement as much as possible.
2. Choose the leak detection method most suitable for the situation encountered.
3. Connect the gauge manifold to the refrigeration system. See Figure 4–9.
4. Be sure that the system has at least 35 psig pressure (241.316 KN/m^2) inside it. If not, increase the pressure by adding refrigerant vapor to the system. Be sure to use the same refrigerant as that in the system.
5. Test each joint, fitting, and gasket in the entire system. Mark each leak as it is found. The repairs can be made after all leaks have been located. Any joints or areas that have signs of oil residue on them should be given special attention. If no leak is found with the halide torch or the electronic leak detector, use a soap-bubble solution to be sure no leaks exist.
6. Either pump the system down, pump the system out, or recover the refrigerant from the system, depending on the amount of refrigerant in the unit.

Usually a refrigerant charge of less than 10 lb (4.5359 Kg) is removed. A charge of more than 10 lb (4.5359 Kg) is generally saved.

7. Repair the leaks and test to be sure that all leaks have been stopped.
8. Evacuate the system.
9. Install new liquid line driers on the system.
10. Recharge the system with the proper amount of clean, dry refrigerant.

9-1 PROCEDURE FOR EVACUATING A SYSTEM:

Evacuation is the process used to remove air and moisture from a refrigeration system. Evacuation is accomplished by use of pumps specially designed for this purpose. A discarded refrigeration compressor is not suitable. Never run a motor compressor while the system is evacuated. To do so may result in serious damage to the motor winding. Evacuation is required any time a system has become contaminated or the compressor or system has been exposed to the atmosphere for long periods of time.

Recovering the refrigerant from a system will remove a good portion of the air, and driers will remove any moisture from the system—up to and only to the capacity of the drier. Therefore, there are still contaminants left in the system, and evacuation is the best means of being reasonably sure that the system is free of these contaminants.

There are basically two evacuation procedures: (1) simple evacuation, and (2) triple evacuation. Simple evacuation is used on systems containing only a minimum of contaminants. Triple evacuation is used on systems containing a greater amount of contaminants.

To use the simple evacuation method, use the following procedure:

1. Connect the gauge manifold to the system. See Figure 4–9.
2. Recover the refrigerant from the system.
3. Connect the center hose on the gauge manifold to the vacuum pump. (Figure 4–14)
4. Start the vacuum pump and pump a vacuum of at least 1,000 microns. A vacuum of 1,500 microns is preferable.

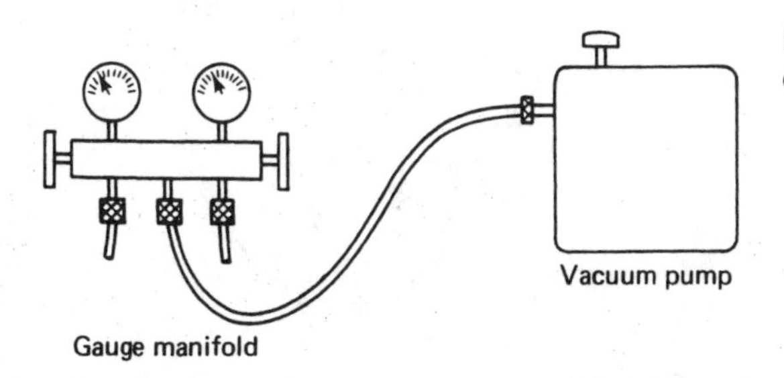

Figure 4–14. Gauge manifold connected to vacuum pump.

5. Close off the gauge manifold hand valves.
6. Stop the vacuum pump. Do not stop the vacuum pump before closing the gauge manifold hand valves to prevent air from entering the system.
7. Disconnect the center hose of the gauge manifold from the vacuum pump and connect it to a cylinder containing the proper refrigerant. See Figure 4–15.
8. Open the cylinder valve.
9. Loosen the center hose connection at the gauge manifold. Bleed the hose for a few seconds; then tighten the connection.
10. Open the gauge manifold hand valves and admit refrigerant into the system.
11. Close the high side hand valve on the gauge manifold.
12. Start the unit and add the proper charge of refrigerant.

To use the triple-evacuation method, use the following procedure:

1. Connect the gauge manifold to the system. See Figure 4–9.
2. Recover all refrigerant from the system.
3. Connect the center hose on the gauge manifold to the vacuum pump. See Figure 4–14.
4. Start the vacuum pump and pump a vacuum of approximately 500 microns.
5. Close off the gauge manifold hand valves.
6. Stop the vacuum pump. Do not stop the vacuum pump before closing the gauge manifold hand valves. This is to prevent air entering the system.
7. Disconnect the center hose of the gauge manifold from the vacuum pump and connect it to a cylinder containing the proper refrigerant. See Figure 4–15.
8. Open the cylinder valve.
9. Loosen the center hose connection at the gauge manifold. Bleed the hose for a few seconds; then tighten the connection.
10. Open the gauge manifold hand valves and admit refrigerant into the system until a pressure of about 5 psig is indicated on the gauges.

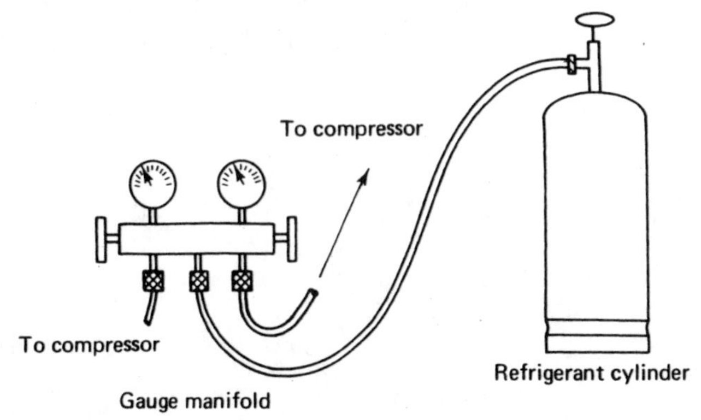

Figure 4–15. Gauge manifold connected to refrigerant cylinder.

11. Close the refrigerant cylinder valve and the gauge manifold hand valves.
12. Disconnect the hose from the cylinder.
13. Open the gauge manifold hand valves and bleed the pressure from the system.
14. Repeat steps 3 through 13.
15. Repeat steps 3 through 9 only. Pump a vacuum of 500 microns.
16. Open the gauge manifold hand valves and admit refrigerant into the system until cylinder pressure is indicated on the gauges.
17. Close the high side gauge manifold hand valve.
18. Start the unit and add the proper charge of refrigerant.

10-1 PROCEDURE FOR CHARGING REFRIGERANT INTO A SYSTEM:

The performance of refrigeration and air conditioning systems is highly dependent on the proper charge of refrigerant in the system. A system that is undercharged will operate with a starved evaporator. Low suction pressures, loss of system capacity, and possible compressor overheating are the results of an undercharged system. On the other hand, an overcharge of refrigerant will back up in the condenser. High discharge pressures and liquid refrigerant flooding of the compressor, with possible compressor damage, are the results of an over-charged system. Larger systems can tolerate a reasonable amount of overcharging or under-charging without severe effects, but some of the smaller systems have a critical charge. The system must be properly charged to obtain proper operation.

The amount of charge will depend on the size of the system in Btu, the length of the lines, the type of refrigerant, and the operating temperature. Therefore, each system must be considered separately. The unit nameplate will usually indicate what type of refrigerant is required and the approximate weight of the required refrigerant.

There are two methods of charging refrigerant into a system: (1) liquid charging, and (2) vapor charging. Liquid charging is much faster than vapor charging and is, therefore, used extensively on large field build-up systems. Never charge liquid refrigerant into the compressor suction or discharge service valves. Liquid entering the compressor can damage the compressor valves. Vapor charging is normally done only when small amounts of refrigerant are to be added to a system. Vapor charging also allows the refrigerant to be charged into the compressor suction service valve.

To use the liquid-charging method, use the following procedure:

1. Connect the gauge manifold to the system. See Figure 4–9.
2. Open the system service valves and purge air from the lines.
3. Connect a charging line to the liquid line (king) valve. This line should include a liquid drier to prevent contaminants entering the system.
4. Connect the charging line to the liquid valve on a cylinder of the proper type of refrigerant. See Figure 4–16.

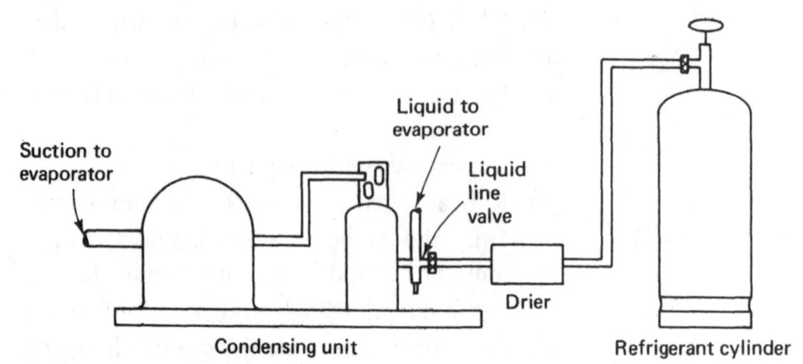

Figure 4–16. Connections for liquid charging a system.

5. Open the refrigerant cylinder liquid valve.
6. Loosen the charging line connection at the liquid line (king) valve and allow the refrigerant to escape for a few seconds.
7. If the correct charge weight is known, the refrigerant cylinder can be weighed to see when sufficient refrigerant has been charged into the system.
8. Close the liquid line (king) valve and start the compressor. Allow the liquid refrigerant to enter the liquid line until the correct weight has been charged into the system. See step 10.
9. If the correct charge weight is not known, the liquid line valve must be opened periodically and the system operation observed. If more refrigerant is needed, close the liquid line valve again and charge more refrigerant into the system. Repeat this process until the proper charge is indicated. See step 10.
10. Closely watch the discharge pressure gauge. A sudden increase in discharge pressure indicates that the capacity of the condenser and receiver has been reached. Stop charging the unit immediately and open the liquid line valve. Any additional refrigerant must be charged into the system by the vapor method.

To use the vapor-charging method, use the following procedure:

1. Connect the gauge manifold to the system. See Figure 4–9.
2. Connect the center line on the gauge manifold to a cylinder of the proper refrigerant. See Figure 4–15.
3. Open the refrigerant cylinder valve and the hand valves on the gauge manifold.
4. Loosen the hose connection at the system service valves and allow the refrigerant to escape for a few seconds. Retighten the connections.
5. Close the gauge manifold hand valves.
6. Crack the system service valves.
7. Start the unit.

8. Open the low side gauge manifold hand valve and charge refrigerant into the system until the proper amount has been charged into the system.
9. Closely watch the discharge pressure gauge during the charging process to be certain that the system is not overcharged.

11-1 PROCEDURE FOR DETERMINING THE PROPER REFRIGERANT CHARGE:

Determining the proper refrigerant charge is an important procedure. A system that does not contain the proper charge of refrigerant will not operate to maximum efficiency. There are several ways to determine if a system is properly charged, such as weighing the charge, using a sight glass, using a liquid level indicator, using the liquid subcooling method, using the superheat method, and using the manufacturers charging charts.

To use the charge weight method, use the following procedure:

1. Connect the gauge manifold to the system service valves. See Figure 4–9.
2. Purge all pressure from the system. Open both hand valves on the gauge manifold.
3. Connect the center line from the gauge manifold to a vacuum pump. See Figure 4–14.
4. Start the vacuum pump and pump a deep vacuum on the system.
5. Close the gauge manifold hand valves.
6. Stop the vacuum pump. Do not stop the vacuum pump before closing the gauge manifold hand valves. This is to prevent air entering the system.
7. Disconnect the center line from the vacuum pump and connect it to a cylinder of the proper type of refrigerant. See Figure 4–15. This cylinder may be a charging cylinder containing the proper amount of refrigerant or a large cylinder placed on an accurate scale. Small systems require a charging cylinder, whereas large systems require a large cylinder.
8. Open the cylinder valve.
9. Loosen the center hose connection at the gauge manifold. Bleed the hose for a few seconds, then tighten the connection.
10. Open both hand valves on the gauge manifold and charge the refrigerant into the system. This must be done suddenly so that all the refrigerant will enter the system before the vacuum vanishes.
11. Start the unit and check the operation.

To use the sight glass method, use the following procedure:

1. Start the unit and allow it to operate for several minutes.
2. Check the flow of refrigerant through the sight glass. A flashlight may be needed to adequately see the flow. See Figure 4–17.

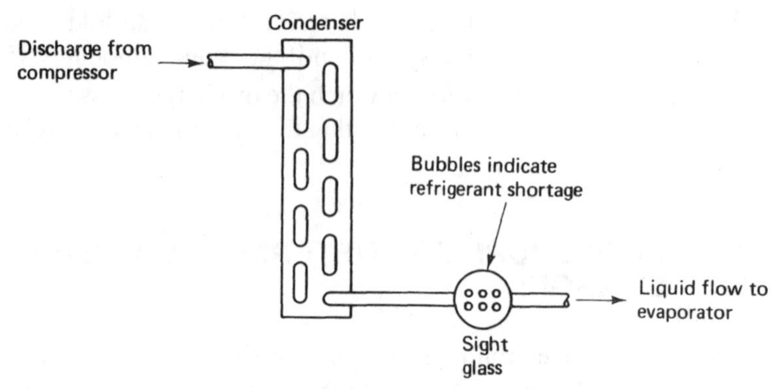

Figure 4–17. Checking refrigerant charge using a sight glass.

3. A steady stream of bubbles indicates the system is low on refrigerant. If these bubbles are intermittent, allow the system to operate awhile longer to see if they will disappear. If the bubbles remain, the system is low on refrigerant.
4. Connect the gauge manifold to the system service valves. See Figure 4–9.
5. Connect the center line to a cylinder of the proper type of refrigerant. See Figure 4–15.
6. Crack the system service valves and open the gauge manifold hand valves and loosen the connection on the refrigerant cylinder. Bleed the lines for a few seconds. Tighten the connection.
7. Close the hand valves on the gauge manifold.
8. Open the refrigerant cylinder valve.
9. Open the low side hand valve on the gauge manifold and admit refrigerant into the system while observing the sight glass and the discharge pressure gauge. When the bubbles disappear, close the low side hand valve. Observe the sight glass. If the bubbles reappear, add more refrigerant into the system. Repeat this process until the bubbles do not reappear. A sudden increase in the discharge pressure indicates a system overcharge. Stop charging the unit and remove some of the refrigerant.

To use the liquid level indicator method, use the following procedure:

1. Start the unit and allow it to operate 10 to 15 minutes.
2. After the system has been in operation long enough for the pressures to stabilize, crack the liquid level test port. A continuous flow of liquid refrigerant from the port indicates sufficient charge. A continuous flow of vapor from the port indicates a shortage of refrigerant. The test port is usually located in the lower section of the condenser. Some units may have a liquid level indicator in the receiver tank. See Figure 4–18.

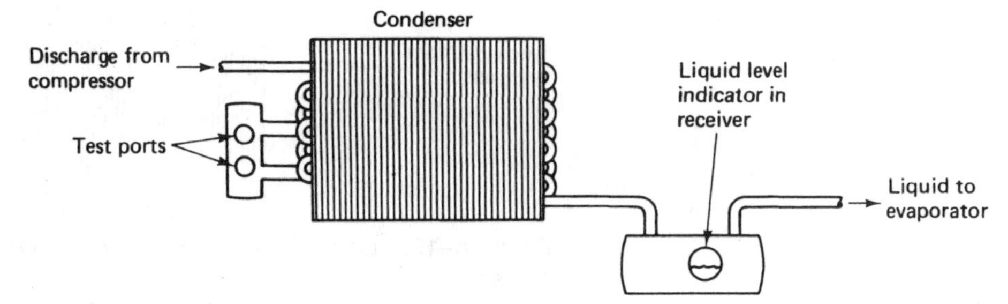

Figure 4–18. Liquid level indicator parts.

3. To add refrigerant to the system, connect the gauge manifold to the system service valves. Close the gauge manifold hand valves. Connect the center charging line to a cylinder containing refrigerant of the proper type. Leave this connection loose. See Figure 4–15.
4. Crack the system service valves open until pressure is indicated on the gauges.
5. Open the gauge manifold hand valves. Allow refrigerant to escape from the hose connection on the cylinder for a few seconds, then tighten the connection.
6. Close the high side gauge manifold hand valve.
7. Open the refrigerant cylinder valve and charge refrigerant into the system.
8. Periodically open the liquid level test port and check for liquid refrigerant.
9. Continue to add refrigerant until a continuous stream of liquid refrigerant is indicated.
10. When liquid is indicated, close the low side gauge manifold hand valve. Allow the unit to operate a few minutes and check for liquid again. If a continuous stream of liquid is not indicated, add more refrigerant and check again. Continue this process until the proper charge is indicated.

To use the liquid subcooling method, use the following procedure:

1. Start the unit and allow it to operate for 10 to 15 minutes.
2. Strap a thermometer to the liquid line at the condenser outlet. See Figure 4–19.
3. After the system has operated long enough for the pressures to stabilize, check the liquid temperature leaving the condenser with a thermometer.
4. Connect the gauge manifold to the system service valves. Close the gauge manifold hand valves. See Figure 4–9.
5. Open the system service valves until pressure is indicated on the gauges.
6. Loosen the hose connections on the gauge manifold and allow the refrigerant vapor to escape for a few seconds, then retighten the connections.
7. Compare the liquid line temperature to the condensing temperature. The condensing temperature is determined by relating the discharge pressure to

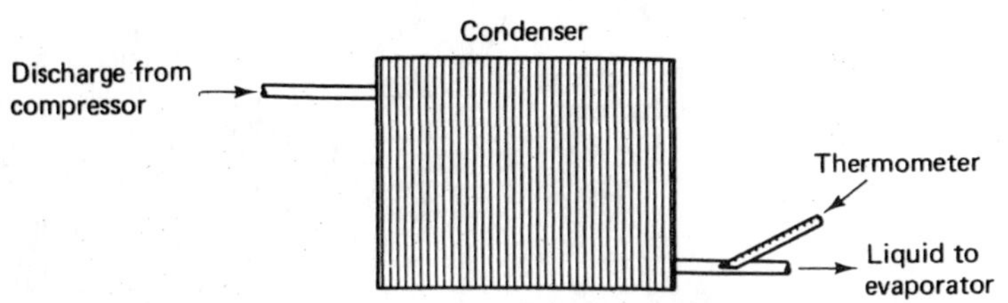

Figure 4–19. Checking liquid and subcooling during charging process.

the corresponding temperature on a pressure-temperature chart. The liquid line temperature is indicated on the thermometer installed in step 2 on the previous page. The liquid line temperature should be approximately 5°F (2.35°C) below the condensing temperature. If less than 5°F (2.35°C), more refrigerant is needed in the system. See Table 4–1.

8. To add refrigerant, connect the center charging hose on the gauge manifold to the valve on a cylinder containing the proper type of refrigerant. Do not tighten this connection. See Figure 4–15.
9. Open the hand valves on the gauge manifold. Allow the refrigerant to escape a few seconds, then tighten the connection on the cylinder valve.
10. Close the high side hand valve on the gauge manifold.
11. Open the cylinder valve and charge refrigerant into the system.
12. Observe the thermometer on the liquid line and the discharge temperature. When the desired subcooling is obtained, close the low side hand valve on the gauge manifold to stop adding refrigerant to the system.
13. Allow the system to operate a few minutes to allow the pressures to stabilize, and check the amount of subcooling. If additional subcooling is required, add more refrigerant to the system. Continue this process until the 5°F (2.35°C) subcooling remains stable.

To use the superheat method, use the following procedure:

1. Start the unit and allow it to operate for 10 to 15 minutes.
2. Strap a thermometer on the suction line about 6 in. from the compressor. Insulate the thermometer bulb so that accurate readings are indicated. See Figure 4–20.
3. If a low side pressure port is available, connect the low side gauge to the system by connecting the low side hose on the gauge manifold to the pressure port. If a pressure port is not available, strap a thermometer to a return bend about midway on the evaporator. Do not put the thermometer on a fin. Insulate the thermometer bulb so that accurate readings may be indicated. See Figure 4–21.
4. If a pressure port is available, determine the difference between the suction line temperature and the saturation temperature equivalent to the suction pressure with the unit running. The saturation temperature is determined by

Table 4–1. Refrigerant Pressure Temperature Relationship

°F	22	410A	12	134a	401a	409A	502	404a	507	408A	402A
–40	***4.8***	6.9	***14.6***	***17.7***	***17.7***	***16.4***	0.7	0.8	1.7	*1.0*	2.1
–44	***2.0***	9.1	***12.9***	***16.2***	***16.0***	***14.8***	2.3	2.5	3.5	*1.1*	3.9
–40	0.5	11.6	***11.0***	***14.5***	***14.5***	***13.1***	4.1	5.5	5.5	2.8	5.9
–36	2.2	14.2	***8.9***	***12.8***	***12.5***	***11.2***	6.0	7.5	7.6	4.6	8.0
–32	4.0	17.1	***6.7***	***10.8***	***10.6***	***9.2***	8.1	9.7	9.9	6.6	10.3
–28	5.9	20.1	***4.3***	***8.6***	***8.3***	***6.9***	10.3	12.0	12.4	8.7	12.8
–24	7.9	23.4	***1.6***	***6.2***	***6.0***	***4.5***	12.7	14.5	15.0	11.0	15.5
–20	10.1	26.9	0.6	***3.6***	***3.5***	***1.9***	15.3	17.1	17.8	13.5	18.4
–16	12.5	30.7	2.1	***0.8***	***0.5***	0.5	18.1	20.0	20.9	16.1	21.5
–12	15.1	34.7	3.7	1.1	1.4	2.0	21.0	23.0	24.1	18.9	24.8
–8	17.9	39.0	5.4	2.8	3.1	3.6	24.2	26.3	27.6	21.9	28.3
–4	20.8	43.7	7.2	4.5	4.8	5.3	27.5	29.8	31.3	25.2	32.1
0	24.0	48.6	9.2	6.5	6.7	7.2	31.1	33.5	35.2	28.7	36.1
2	25.6	51.1	10.2	7.5	8.0	8.2	32.9	34.8	37.3	30.5	38.1
4	27.3	53.8	11.2	8.5	8.8	9.2	34.9	37.4	39.4	32.3	40.4
6	29.1	56.7	12.3	9.6	9.9	10.2	36.9	39.4	41.6	34.3	42.6
8	30.9	59.4	13.5	10.8	11.0	11.3	38.9	41.6	43.8	36.3	44.9
10	32.8	62.3	14.6	12.0	12.2	12.5	41.0	43.7	46.2	38.3	47.3
12	34.7	65.4	15.8	13.1	13.4	13.6	43.2	46.0	48.5	40.4	49.7
14	36.7	68.6	17.1	14.4	14.6	14.8	45.4	48.3	51.0	42.6	52.2
16	38.7	71.9	18.4	15.7	15.9	16.1	47.7	50.7	53.5	44.9	54.8
18	40.9	75.2	19.7	17.0	17.2	17.4	50.0	53.1	56.1	47.2	57.5
20	43.0	78.3	21.0	18.4	18.6	18.7	52.5	55.6	58.8	49.6	60.2
22	45.3	82.3	22.4	19.9	20.0	20.1	54.9	58.2	61.5	52.0	63.0
24	47.6	85.9	23.9	21.4	21.5	21.5	57.5	60.9	64.3	54.5	65.9
26	49.9	89.7	25.4	22.9	23.0	22.9	60.1	63.6	67.2	57.1	68.9
28	52.4	93.5	26.9	24.5	24.6	24.4	62.8	66.5	70.2	59.8	72.0
30	54.9	96.8	28.5	26.1	26.2	26.0	65.6	69.4	73.3	62.5	75.1
32	57.5	101.6	30.1	27.8	27.9	27.6	68.4	72.3	76.4	65.3	78.3
34	60.1	105.7	31.7	29.5	29.6	29.2	71.3	75.4	79.6	68.2	81.6
36	62.8	110.0	33.4	31.3	31.3	30.9	74.3	78.5	82.9	71.2	85.0
38	65.6	114.4	35.2	33.2	33.2	32.7	77.4	81.8	86.3	74.2	88.5
40	68.5	118.0	36.9	35.1	35.0	34.5	80.5	85.1	89.8	77.4	92.1

(Continued)

Table 4–1. Refrigerant Pressure Temperature Relationship—Continued

°F	22	410A	12	134a	401a	409A	502	404a	507	408A	402A
42	71.5	123.6	38.8	37.0	37.0	36.3	83.8	88.5	93.4	80.6	95.7
44	74.5	128.3	40.7	39.1	39.0	38.2	87.0	91.9	97.0	83.9	99.5
46	77.6	133.2	42.7	41.1	41.0	40.2	90.4	95.5	100.8	87.3	103.4
48	80.7	138.2	44.7	43.3	43.1	42.2	93.9	99.2	104.6	90.7	107.3
50	84.0	142.2	46.7	45.5	45.3	44.3	97.4	102.9	108.6	94.3	111.4
52	87.3	148.5	48.8	47.7	**60.0**	**63.6**	101.0	**109.0**	112.6	97.9	**120.0**
56	94.3	159.3	53.2	52.3	**65.0**	**68.9**	108.6	**117.0**	121.0	105.5	**129.0**
60	101.6	169.6	57.7	57.5	**70.0**	**74.5**	116.4	**125.0**	129.7	113.5	**138.0**
64	109.3	182.6	62.5	62.7	**76.0**	**80.3**	124.6	**134.0**	139.0	121.8	**147.0**
68	117.3	195.0	67.6	68.3	**82.0**	**86.3**	133.2	**144.0**	148.6	130.6	**157.0**
72	125.7	208.1	72.9	74.2	**89.0**	**92.8**	142.2	**153.0**	158.7	139.7	**168.0**
76	134.5	221.7	78.4	80.3	**95.0**	**99.4**	151.5	**164.0**	169.3	149.3	**179.0**
80	143.6	235.3	84.2	86.8	**102.0**	**106.4**	161.2	**174.0**	180.3	159.4	**190.0**
84	153.2	250.8	90.2	93.6	**109.0**	**113.7**	171.4	**185.0**	191.9	169.8	**202.0**
88	163.2	266.3	96.5	100.7	**117.0**	**121.2**	181.9	**197.0**	203.9	180.8	**214.0**
92	173.7	282.5	103.1	108.2	**125.0**	**129.1**	192.9	**209.9**	216.6	192.2	**227.0**
96	184.6	299.3	110.0	116.1	**133.0**	**137.4**	204.3	**222.0**	229.8	204.1	**240.0**
100	195.9	317.2	117.2	124.3	**142.0**	**146.0**	216.2	**235.0**	243.5	216.6	**254.0**
104	207.7	335.2	124.7	132.9	**151.0**	**154.9**	228.5	**249.0**	257.9	229.5	**269.0**
108	220.0	354.3	132.4	142.8	**160.0**	**164.2**	241.3	**264.0**	272.9	243.0	**284.0**
112	232.8	374.2	140.5	151.3	**170.0**	**173.9**	254.6	**279.0**	288.6	257.0	**299.0**
116	246.1	394.8	148.9	161.1	**180.0**	**183.9**	268.4	**294.0**	304.9	271.6	**316.0**
120	259.9	417.7	157.7	171.3	**191.0**	**194.4**	282.7	**311.0**	321.9	286.8	**332.0**
124	274.3	438.7	166.7	182.0	**202.0**	**205.2**	297.6	**328.0**	339.7	302.6	**350.0**
128	289.1	462.0	176.2	193.1	**213.0**	**216.5**	312.9	**345.0**	358.2	319.0	**368.0**
132	304.6	486.2	185.9	204.7	**225.0**	**228.2**	328.9	**364.0**	377.6	336.0	**387.0**
136	320.6	511.4	196.1	216.8	**237.0**	**240.3**	345.4	**383.0**	397.7	353.6	**406.0**
140	337.3	539.0	206.6	229.4	**250.0**	**252.9**	362.6	**402.0**	418.7	372.0	**426.0**
144	354.5	564.8	217.5	242.4	**263.0**	**265.9**	380.4	**423.0**	440.6	390.9	**447.0**
148	372.3	593.8	228.8	256.0	**277.0**	**279.5**	398.9	**444.0**	462.0	410.6	**468.0**

BOLD ITALIC = VACUUM

BLACK = VAPOR PSIG (CALCULATING SUPERHEAT)

BLACK BOLD = LIQUID PSIG (CALCULATING SUBCOOLING)

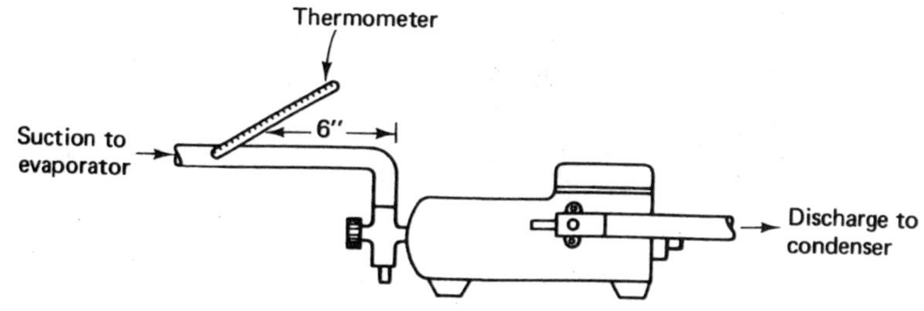

Figure 4–20. Thermometer strapped to suction line.

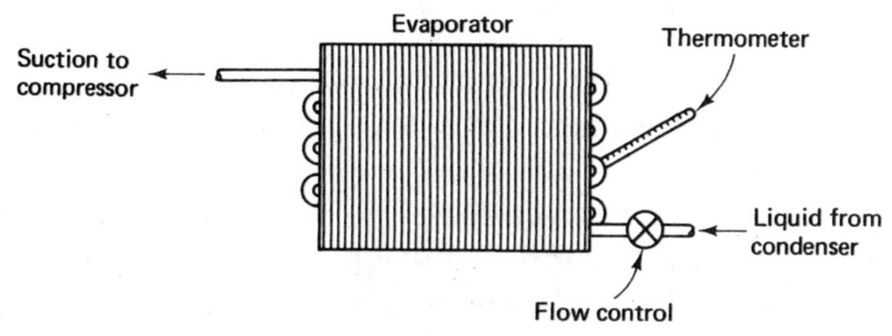

Figure 4–21. Thermometer strapped to evaporator return bend.

relating the suction pressure to the corresponding temperature on a pressure-temperature chart. The temperature difference should be approximately 20° to 30°F (11.11° to 16.67°C). See Table 4–1.

If no pressure port is available, check the difference between the two thermometers. The difference in temperature should be approximately 20°F to 30°F (11.11°C to 16.67°C). When the unit is operating at a normal operating condition, a superheat of 20°F to 30°F (−6.6°C to −1.1°C) is satisfactory. A superheat lower than approximately 20°F (−6.6°C) indicates an overcharge of refrigerant. A superheat higher than approximately 30°F (−1.1°C) indicates an undercharge of refrigerant.

5. To add refrigerant, connect the low side gauge on the gauge manifold to the pressure port on the system. If a low side pressure port is not available, a saddle valve must be installed. See Figure 4–22.
6. Connect the center charging hose on the gauge manifold to the valve on a cylinder of the proper type of refrigerant. Do not tighten this connection. See Figure 4–15.
7. Open the gauge manifold hand valve and allow the refrigerant to escape for a few seconds. Tighten the connection on the cylinder valve.
8. Open the refrigerant cylinder valve and charge refrigerant into the system.
9. Observe the superheat by the method used previously in step 4. When the proper superheat is reached, stop charging refrigerant into the system.
10. Allow the unit to continue running until the pressures and temperatures have stabilized. If more refrigerant is needed, repeat the charging process until the desired superheat is stabilized.

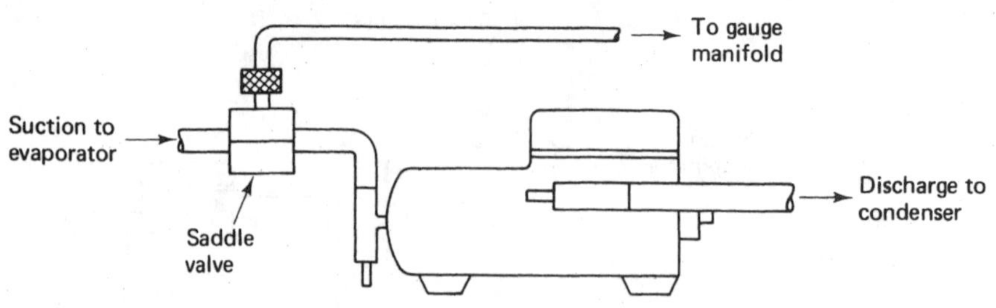

Figure 4–22. Installation of saddle valve.

To use the manufacturer's charging chart method, use the following procedure:

1. Start the unit and allow it to operate for 10 to 15 minutes.
2. Connect the gauge manifold to the system service valves. See Figure 4–19.
3. Connect the center charging hose on the gauge manifold to the valve on a cylinder of the proper type of refrigerant. Do not tighten this connection.
4. Open the hand valves on the gauge manifold and allow refrigerant to escape from the loose connection a few seconds. Tighten the connection on the cylinder valve.
5. Close the high side hand valve on the gauge manifold.
6. Obtain a copy of the manufacturer's charging chart for that model unit.
7. Compare the pressures to those indicated on the chart.
8. If more refrigerant is needed, open the valve on the refrigerant cylinder and add refrigerant to the system.
9. Stop charging after a few minutes by closing the low side hand valve on the gauge manifold and compare the pressure readings indicated on the chart. Repeat this procedure until the system pressures compare with those indicated on the chart.

12-1 PROCEDURE FOR DETERMINING THE COMPRESSOR OIL LEVEL:

All refrigeration compressors require a specified amount of oil. This oil is required for lubrication of the moving parts, and helps to make a refrigerant seal between the components. An abnormally low oil level will probably result in a loss of lubrication and compressor damage. An excess of lubricating oil will result in oil slugging, probable damage to the compressor valves, and lost system efficiency due to oil logging of the evaporator.

To use the sight glass method, use the following procedure:

1. Start the unit and allow it to operate for 10 to 15 minutes.
2. Check the oil level in the sight glass. See Figure 4–23. A flashlight may be needed to accurately determine the level. The oil level should be at or slightly above the center of the sight glass while the unit is operating. If less than this, add oil. If more than this, remove the excess oil.

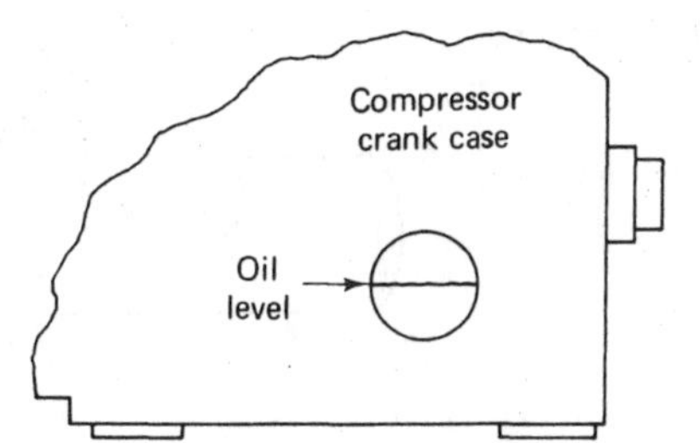

Figure 4–23. Checking compressor oil level.

On sealed systems that do not have an oil sight glass, determining the amount of oil in the compressor is difficult. When a leak occurs and the amount of oil lost is small and can be reasonably calculated, add that amount to the system. If there has been a large amount of oil lost, the compressor must be removed, the oil drained, and the correct amount added to the compressor before placing it back in operation.

13-1 PROCEDURE FOR ADDING OIL TO A COMPRESSOR:

Adding oil to a compressor is often required, and the service technician should be familiar with the different procedures used. There are three procedures used, depending on the type of system encountered and the type of tools at hand. These procedures are: (1) the open system method, (2) the closed system method, and (3) the oil pump method.

To use the open system method, use the following procedure:

1. Connect the gauge manifold to the system service valves. See Figure 4–9.
2. Close the gauge manifold hand valves and open the compressor service valves.
3. Start the unit.
4. Front-seat the compressor suction service valve and run the unit until the low side pressure is reduced to 1 or 2 psig (6.895 KNm^2 or 13.790 KNm^2). The low pressure switch may need to be bypassed.
5. Stop the compressor.
6. Front-seat the compressor discharge service valve.
7. Open the low side hand valve on the gauge manifold and relieve the slight pressure remaining in the system.
8. Remove the oil fill plug and pour the oil into the compressor crankcase until the proper level is reached. Use precaution to prevent contamination of the oil. Figure 4–24.
9. Close the low side hand valve on the gauge manifold.
10. Slightly open the compressor suction service valve and allow a small amount of refrigerant to escape out of the oil fill hole.
11. Close off the compressor suction service valve.

Figure 4–24. Pouring oil into a compressor.

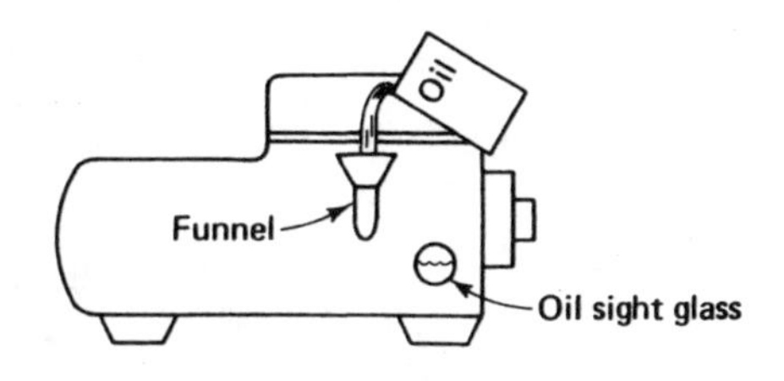

12. Replace the oil fill plug and tighten.
13. Back-seat the compressor service valves.
14. Start the compressor and check the oil level in the compressor.
15. Remove the gauge manifold from the system.

To use the closed system method, use the following procedure:

1. Connect the gauge manifold to the system service valves. See Figure 4–9.
2. Place the center charging line in a container of clean dry oil. See Figure 4–25.
3. Open the system service valves until a pressure is indicated on the gauges.
4. Slightly open the low side hand valve on the gauge manifold and purge a small amount of refrigerant through the lines and the oil.
5. Front-seat the system service valve.
6. Start the unit and draw a vacuum on the compressor crankcase.
7. Open the low side hand valve on the gauge manifold and draw the oil into the compressor through the service valve. Be sure that the center line remains in the oil to prevent air entering the system.
8. When a sufficient amount of oil has been drawn into the compressor, close the hand valve on the gauge manifold.
9. Back-seat the system service valve, and place the unit in normal operation.

To use the oil pump method, use the following procedure:

1. Connect the low side gauge of the gauge manifold to the system suction service valve.
2. Close the low side hand valve on the gauge manifold.
3. Crack the system service valve until a pressure is indicated on the gauge.
4. Connect the center charging line to the oil pump. Do not tighten the connection.
5. Open the low side gauge manifold hand valve and purge refrigerant through the loose connection for a few seconds, then tighten the connection.
6. Place the oil pump in a container of clean dry oil.

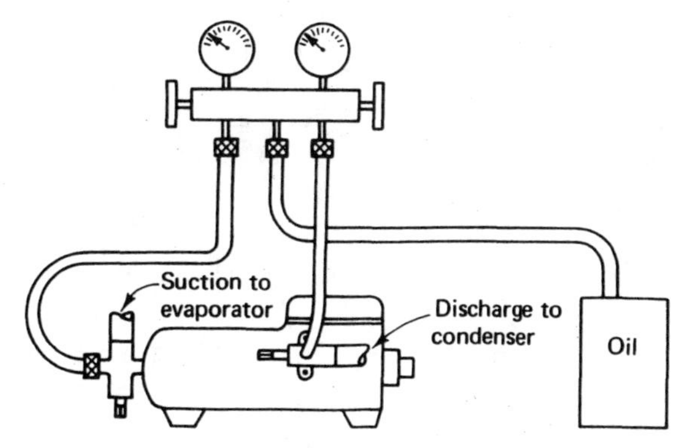

Figure 4–25. Adding to a closed system.

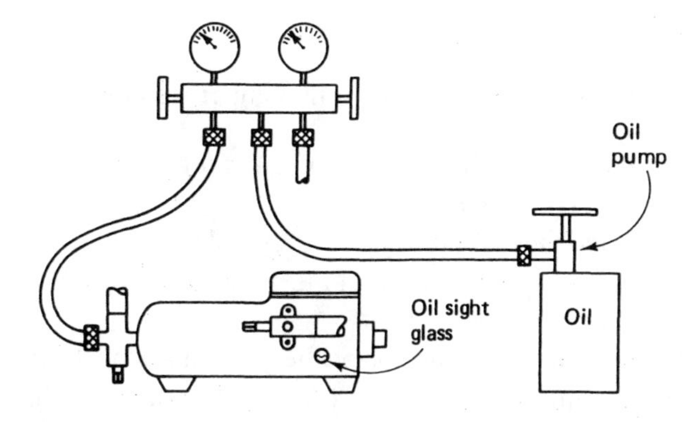

Figure 4–26. Adding oil to a system with an oil pump.

7. Open the gauge manifold hand valve completely.
8. Move the system service valve to the midway position.
9. Pump oil into the system until the proper level is reached. See Figure 4–26.
10. Back-seat the system service valve.
11. Close the low side hand valve on the gauge manifold.
12. Remove the gauge manifold and place the system in normal operation.

14–1 PROCEDURE FOR LOADING A CHARGING CYLINDER:

Charging cylinders are ideal for charging systems that use less than 5 lb (.1417 Kg) of refrigerant. Charging cylinders are calibrated in ounces for each type of refrigerant. Therefore, it is possible to charge the exact amount of refrigerant into the system. The service technician should be familiar with the steps involved in loading a charging cylinder.

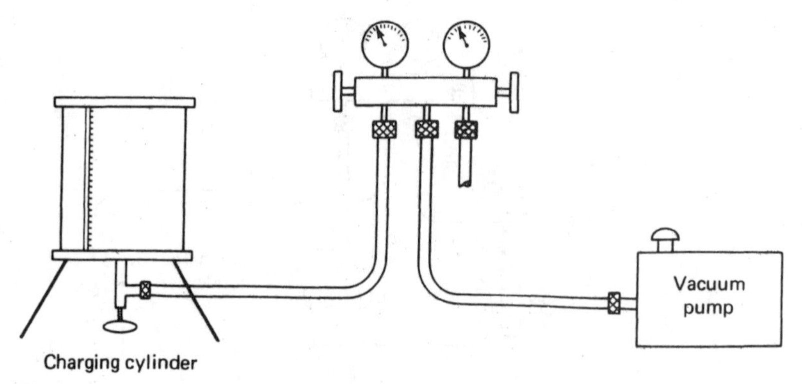

Figure 4–27. Connections for evacuating a charging cylinder.

To load a charging cylinder, use the following procedure:

1. Connect the low pressure gauge of the gauge manifold to the charging cylinder. See Figure 4–27.
2. Open the hand valve on the charging cylinder.
3. Open the low side hand valve on the gauge manifold and relieve all pressure from the charging cylinder.
4. Connect the center charging line on the gauge manifold to the vacuum pump.
5. Start the vacuum pump and draw as deep a vacuum as possible with the pump.
6. Close the low side hand valve on the gauge manifold.
7. Remove the center charging line from the vacuum pump and connect it to the valve on a cylinder of the proper type of refrigerant. If there is a liquid valve, connect the line to it. See Figure 4–28.
8. Open the refrigerant cylinder valve and loosen the center charging line connection at the gauge manifold. Allow the refrigerant to escape for a few seconds, and then retighten the connection.
9. Invert the refrigerant cylinder so that liquid refrigerant will enter the charging line. If the line is connected to a liquid valve, the cylinder does not need to be inverted.
10. Open the low side hand valve on the gauge manifold and allow liquid refrigerant to be drawn into the charging cylinder.
11. When the desired amount of refrigerant is drawn into the charging cylinder, close the refrigerant cylinder valve. If the refrigerant stops flowing before the desired amount is in the charging cylinder, the vent valve on top of the charging cylinder may be opened to permit the escape of vapor, thus allowing more liquid to enter. See Figure 4–28.
12. Close the charging cylinder hand valve and disconnect the lines from the charging cylinder. Be sure that any liquid refrigerant in the lines does not come in contact with the skin or eyes.

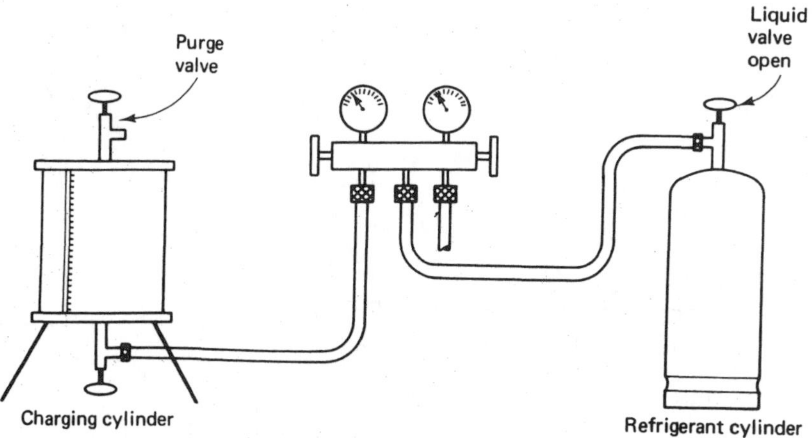

Figure 4–28. Connections for loading a charging cylinder.

15-1 PROCEDURE FOR CHECKING COMPRESSOR ELECTRICAL SYSTEMS:

Potential Starting Relay System with Two-Terminal Overload

Potential (voltage) starting relays are nonpositional. The contacts are normally closed. These relays are generally used on single-phase compressors of ½ hp and larger. Use the following procedure to check this type of system. See Figure 4–29.

1. Check to make certain that the proper voltage is being supplied to the unit. If proper voltage is supplied, continue to make the following checks. If the wrong voltage is being supplied, correct the problem with the power supply.
2. Disconnect the electrical power from the unit.
3. If a fan motor is used, disconnect one of the leads.
4. Make the following tests with an ohmmeter.
5. Check the continuity between points 1 and 2. See Figure 4–29. If no continuity is found, the control contacts are open. Close the contacts by setting the control to demand operation. If continuity is found, continue to make the following checks. If no continuity is found with the control on demand, replace the control.
6. Check the continuity between points 3 and 4. If continuity is found, continue to make the following checks. If no continuity is found, wait about ten minutes and check the continuity again. The overload may be tripped. If no continuity now, replace the overload. Be sure to use an exact replacement.
7. Check the continuity between points 4 and 5. If continuity is found, continue to make the following checks. If no continuity is found, repair broken wire or loose connection.

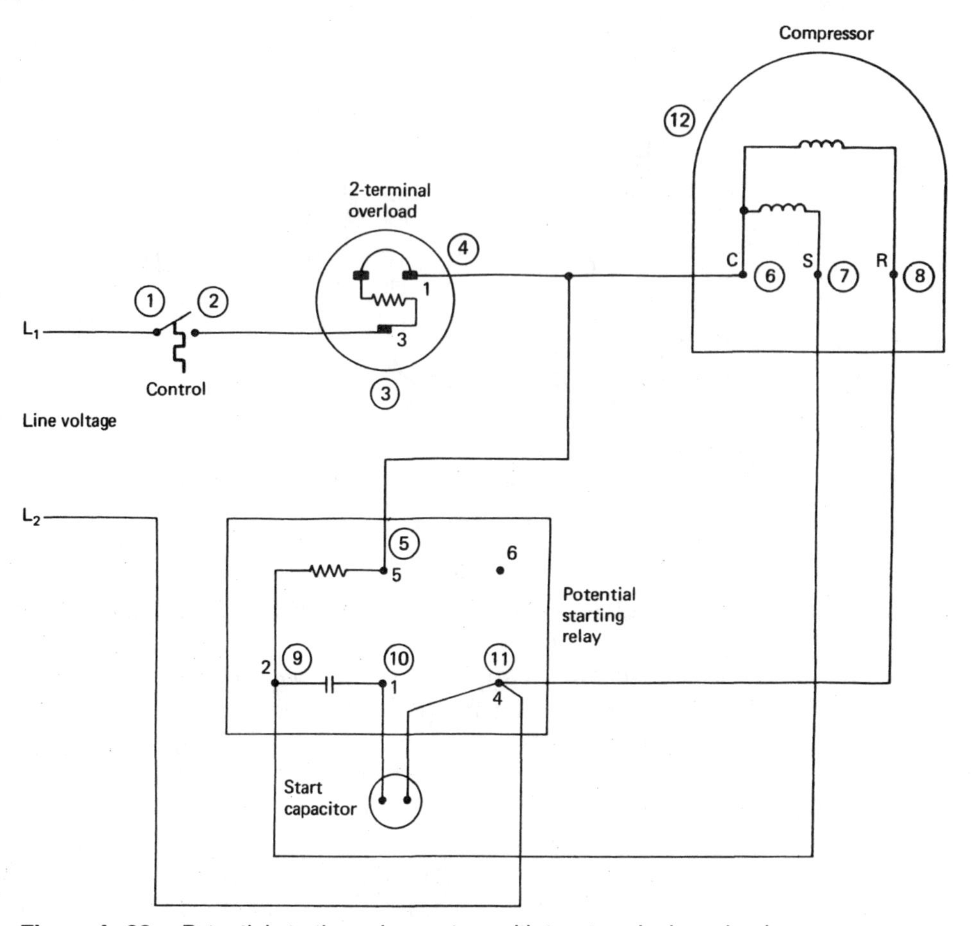

Figure 4–29. Potential starting relay system with two-terminal overload.

8. Check the continuity between points 4 and 6. If continuity is found, continue to make the following checks. If no continuity is found, repair broken wire or loose connection.
9. Remove the wiring from the start terminal (point 7). Check the continuity between points 6 and 7. Compare this reading to that given by the compressor manufacturer for that particular compressor. If proper continuity is found, continue to make the following checks. If no or improper continuity is found, replace the compressor. The start winding is defective.
10. Remove the wiring from the run terminal (point 8). Check the continuity between points 6 and 8. Compare this reading to that given by the compressor manufacturer for that particular compressor. If proper continuity is found, continue to make the following checks. If no or improper continuity is found, replace the compressor. The run winding is defective.
11. Check the continuity between points 5 and 9. If continuity is found, continue to make the following checks. If no continuity is found, the start relay coil is open. Replace the relay. Be sure to use an exact replacement.

12. Check the continuity between points 9 and 10. If continuity is found, continue to make the following checks. If no continuity is found, the start relay contacts are defective. Replace the relay. Be sure to use an exact replacement.
13. Set the ohmmeter on R × 100,000. Check the continuity between points 10 and 11. If needle deflection is found, continue to make the following checks. If no needle deflection is found, the capacitor is open. Replace the capacitor. Be sure to use a proper replacement.
14. Set the ohmmeter on R × 1. Check the continuity between points 10 and 11. If no continuity is found, continue to make the following checks. If continuity is found, the capacitor is shorted. Replace the capacitor. Be sure to use a proper replacement.
15. Check the continuity between points 6 and 12. If no continuity is found, continue to make the following checks. If continuity is found, the compressor motor is shorted. Replace the compressor.
16. Check the continuity between point 9 and the wire removed from point 7 (start terminal). If continuity is found, replace the wiring on the terminal and continue to make the following checks. If no continuity is found, repair the wiring or loose connection.
17. Check the continuity between point 11 and the wire removed from point 8 (run terminal). If continuity is found, replace the wiring on the terminal and continue to make the following checks. If no continuity is found, repair the wiring or loose connection.
18. If the previous steps do not indicate the trouble and the compressor will operate for a short period of time, disconnect the wire from point number 10. Touch the wire to the same terminal and turn on the electricity. When the compressor starts, remove the wire from the terminal. A slight spark may occur. If the compressor continues to operate, replace the start relay. The contacts are not opening. Be sure to use an exact replacement. If the compressor does not continue to run when the wire is removed, reconnect the wire and proceed to the next step.
19. Check the amperage draw through the wire to point 6 (common terminal) while the compressor is trying to run. Compare the amperage draw to the locked rotor (LR) amperage of the motor. If the LR amperage is very close to that listed by the manufacturer, replace the compressor. The compressor has mechanical problems.

Potential Starting Relay System with Three-Terminal Overload

Potential (voltage) starting relays are nonpositional. The contacts are normally closed. These relays are generally used on single-phase compressors of ½ hp and larger. Use the following procedure to check this type of system. See Figure 4–30.

1. Check to make sure that the proper voltage is being supplied to the unit. If the proper voltage is being supplied, continue to make the following checks. If the wrong voltage is being supplied, correct the problem with the power supply.

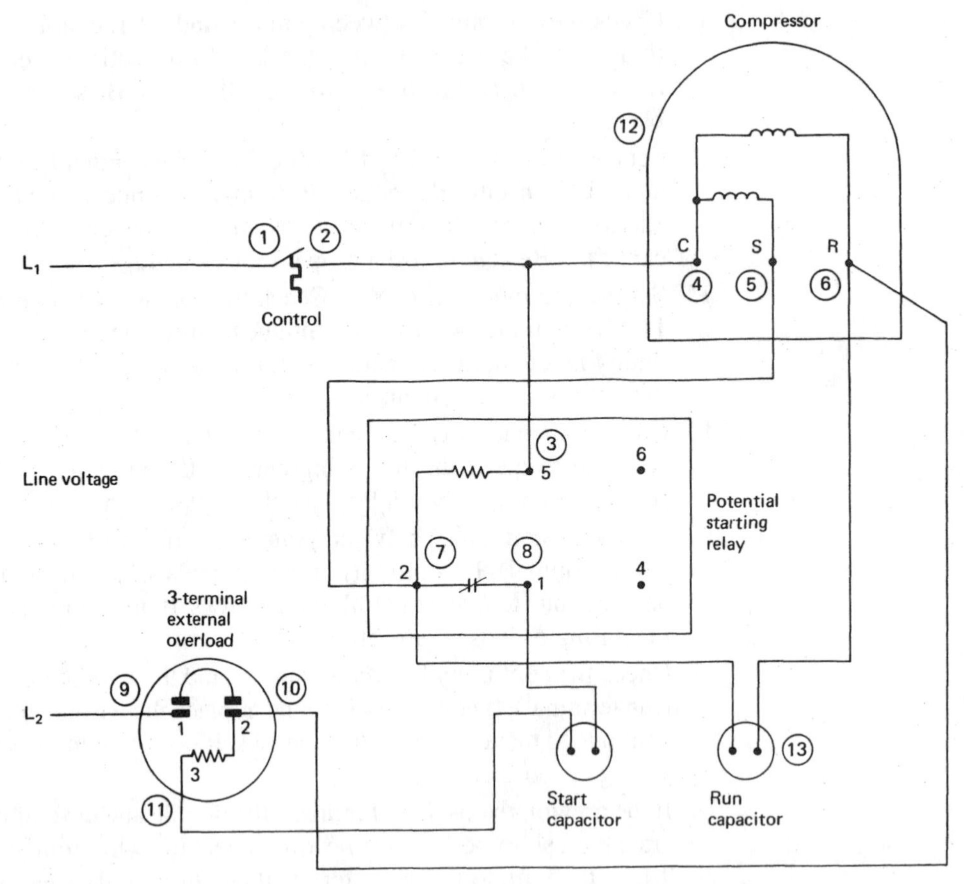

Figure 4–30. Potential starting relay system with three-terminal overload.

2. Disconnect the electrical power from the unit.
3. If a fan is used, disconnect one of the leads.
4. Make the following checks with an ohmmeter.
5. Check the continuity between points 1 and 2. See Figure 4–30. If continuity is found, continue to make the following checks. No continuity indicates the control contacts are open. Close the contacts by setting the control to demand operation. If there is no continuity with the control on demand, replace the control.
6. Check the continuity between points 2 and 3. If continuity is found, continue to make the following checks. If no continuity is found, repair the broken wire or loose connection.
7. Check the continuity between points 2 and 4. If continuity is found, continue to make the following checks. If no continuity is found, repair the broken wire or loose connection.

8. Remove the wiring from point 5 (start terminal). Check the continuity between points 4 and 5. Compare this reading to that given by the compressor manufacturer for that particular compressor. If proper continuity is found, continue to make the following checks. If no continuity or improper continuity is found, replace the compressor. The start winding is defective.
9. Remove the wiring from point 6 (run terminal). Check the continuity between points 4 and 6. Compare this reading to that given by the manufacturer for that particular compressor. If proper continuity is found, continue to make the following checks. If no or improper continuity is found replace the compressor. The run winding is defective.
10. Check the continuity between points 3 and 7. If continuity is found, continue to make the following checks. If no continuity is found, replace the relay. The coil is faulty. Be sure to use an exact replacement.
11. Check the continuity between points 7 and 8. If continuity is found, continue to make the following checks. If no continuity is found, replace the relay. The contacts are faulty. Be sure to use an exact replacement.
12. Set the ohmmeter on R × 100,000. Check the continuity between points 8 and 11. If needle deflection is found, continue to make the following checks. If no needle deflection is found, the start capacitor is open. Replace the capacitor. Be sure to use an exact replacement.
13. Set the ohmmeter on R × 1. Check the continuity between points 9 and 11. If no continuity is found, continue to make the following checks. If continuity is found, the start capacitor is shorted. Replace the capacitor. Be sure to use a proper replacement.
14. Set the ohmmeter on R × 100,000. Check the continuity between point 7 and the wire removed from point 6 (run terminal). If needle deflection is found, continue to make the following checks. If no needle deflection is found, the run capacitor is open. Replace the capacitor. Be sure to use a proper replacement.
15. Set the ohmmeter on R × 1. Check the continuity between point 7 and the wire removed from point 6 (run terminal). If no continuity is found, continue to make the following checks. If continuity is found, the run capacitor is shorted. Replace the capacitor. Be sure to use a proper replacement.
16. Check the continuity between points 9 and 11. If continuity is found, continue to make the following checks. If no continuity is found, replace the overload. Be sure that the unit has been off for about ten minutes to allow the overload to cool. Be sure to use an exact replacement.
17. Check the continuity between points 4 and 12. If no continuity is found, continue to make the following checks. If continuity is found, replace the compressor. The motor is grounded.
18. Check the continuity between point 7 and the wire removed from point 5 (start terminal). If continuity is found, continue to make the following checks. If no continuity is found, repair the wiring or loose connection. Reconnect the wiring to terminal (point 5).

19. Check the continuity between point 13 and the wire removed from point 6 (run terminal). If continuity is found, continue to make the following checks. If no continuity is found, repair the wiring or loose connection. Reconnect this wire to the terminal (point 6).
20. Check the continuity between point 10 and the wire removed from point 6 (run terminal). If continuity is found, continue to make the following checks. If no continuity is found, repair the wiring or loose connection. Reconnect this wire to the terminal (point 6).
21. If the previous steps do not indicate the trouble and the compressor will operate for a short period of time, disconnect the wire from point number 8. Touch the wire to the same terminal and turn on the electricity. When the compressor starts, remove the wire from the terminal. A slight spark may occur. If the compressor continues to operate, replace the start relay. The contacts are not operating. Be sure to replace it with an exact replacement. If the compressor does not continue to run when the wire is removed, reconnect the wire and proceed to the next step.
22. Check the amperage through the wire to point number 4 (common terminal) while the compressor is trying to run. Compare the amperage draw to the locked rotor (LR) amperage of the motor. If the LR amperage is very close to that listed by the manufacturer, replace the compressor. The compressor has mechanical problems.

Current-Type Starting Relay System with a Two-Terminal Overload

Current-type starting relays are positional-type relays. They must be mounted in the position indicated by arrows on the relay. The contacts are normally open, and they are closed by an electromagnetic coil. They are opened by gravity—thus, the purpose of mounting them in the proper position. These relays are generally used on fractional hp motors up to about ½ hp. Use the following procedure to check this type of system. See Figure 4–31.

1. Check to make certain that the proper voltage is being supplied to the unit. If proper voltage is being supplied to the unit, continue to make the following checks. If the wrong voltage is being supplied, correct the problem with the power supply.
2. Disconnect the electrical power from the unit.
3. If a fan motor is being used, disconnect one of the leads.
4. Make the following checks with an ohmmeter.
5. Check the continuity between points 1 and 2. See Figure 4–31. No continuity indicates the control contacts are open. Close the contacts by setting the control to demand operation. If continuity is found, continue to make the following checks. If no continuity is found with the control on demand, replace the control.
6. Check the continuity between points 3 and 4. If continuity is found, continue to make the following checks. If no continuity is found, wait about

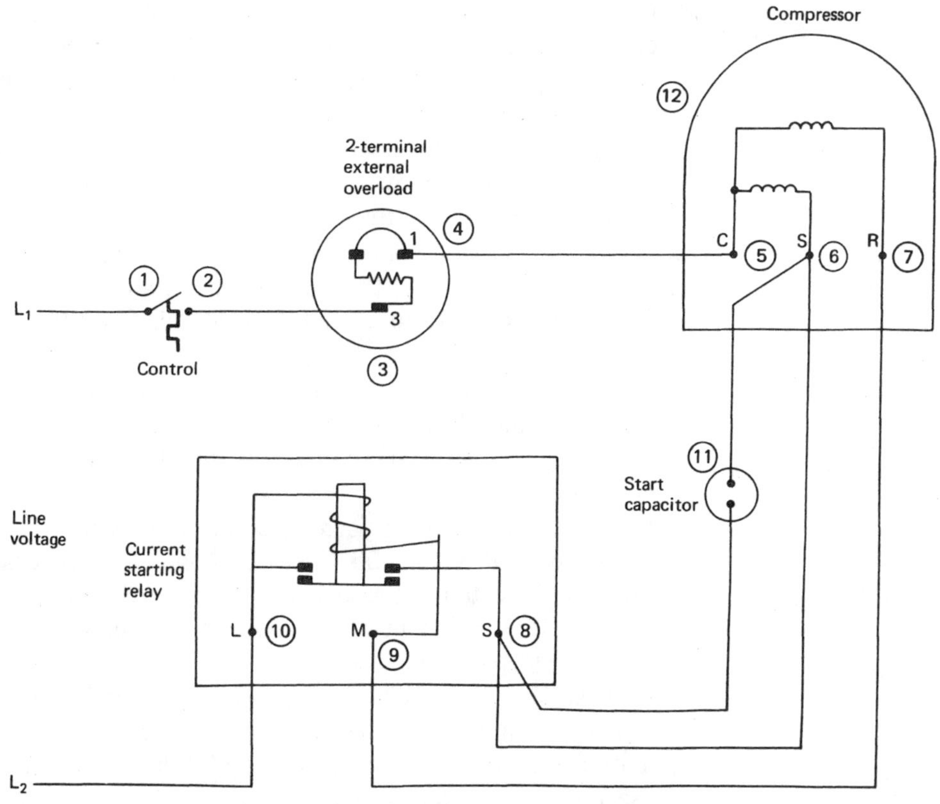

Figure 4–31. Current-type starting relay system with a two-terminal overload.

ten minutes and check the continuity again. The overload may be tripped. If no continuity now, replace the overload. Be sure to use an exact replacement.

7. Check the continuity between points 4 and 5. If continuity is found, continue to make the following checks. If no continuity is found, repair the wire or loose connections.
8. Remove the wire from point 6 (start terminal). Check the continuity between points 5 and 6. Compare this reading to that given by the manufacturer for that particular compressor. If proper continuity is found, continue to make the following checks. If no or improper continuity is found, replace the compressor. The start winding is defective.
9. Remove the wire from point 7 (run terminal). Check the continuity between points 5 and 7. Compare this reading to that given by the compressor manufacturer for that particular compressor. If the proper continuity is found, continue to make the following checks. If no or improper continuity is found, replace the compressor. The run winding is defective.
10. Check the continuity between point 5 (common terminal) and point 12 (compressor housing). If no continuity is found, continue to make the following checks. If continuity is found, replace the compressor. The motor winding is grounded.

11. Check the continuity between points 10 and 8. If no continuity is found, continue to make the following checks. If continuity is found, replace the start relay. The contacts are closed and should be open. Be sure to use an exact replacement.
12. Check the continuity between points 10 and 9. If continuity is found, continue to make the following checks. If no continuity is found, replace the relay. The coil is open. Be sure to use an exact replacement.
13. Check the continuity between point 7 (run terminal) and point 9. If continuity is found, replace the wire on the terminal and continue to make the following checks. If no continuity is found, repair the wire or loose connections.
14. Check the continuity between point 6 (start terminal) and point 8. If continuity is found, replace the wire on the terminal and continue to make the following checks. If no continuity is found, repair the wire or loose connections.
15. If a starting capacitor is used, disconnect the wire from point 11. Set the ohmmeter on R × 100,000. Check the continuity between points 8 and 11. If needle deflection is found, continue to make the following checks. If no needle deflection is found, the capacitor is open. Replace the capacitor. Be sure to use a proper replacement.
16. Set the ohmmeter on R × 1. Check the continuity between points 8 and 11. If no continuity is found, continue to make the following checks. If continuity is found, the capacitor is shorted. Replace the capacitor. Be sure to use a proper replacement.
17. If the above steps do not indicate the trouble and the compressor hums but does not start, disconnect the wire from point 8 and touch it to point 10. Turn on the electricity. If the compressor starts, remove the wire from the terminal. A slight spark may occur. If the compressor continues to operate, replace the relay. Be sure to use an exact replacement. If the compressor does not start, reconnect the wire to point 8 and proceed to the next step.
18. Check the amperage drawn in the wire to point 5 (common terminal) while the compressor is trying to run. Compare the amperage draw to the locked rotor (LR) amperage of the motor. If the LR amperage is very close to that listed by the manufacturer, replace the compressor. The compressor has mechanical problems.

Permanent Split Capacitor (PSC) System with Two-Terminal Overload: PSC motors are low starting torque motors that have no starting relay or starting capacitor. This system is used on builders' models and room-type air conditioning units. Use the following procedure to check this type of system. See Figure 4–32.

1. Check to make certain that the proper voltage is being supplied to the unit. If proper voltage is being supplied, continue to make the following checks. If the wrong voltage is being supplied, correct the problem with the power supply.

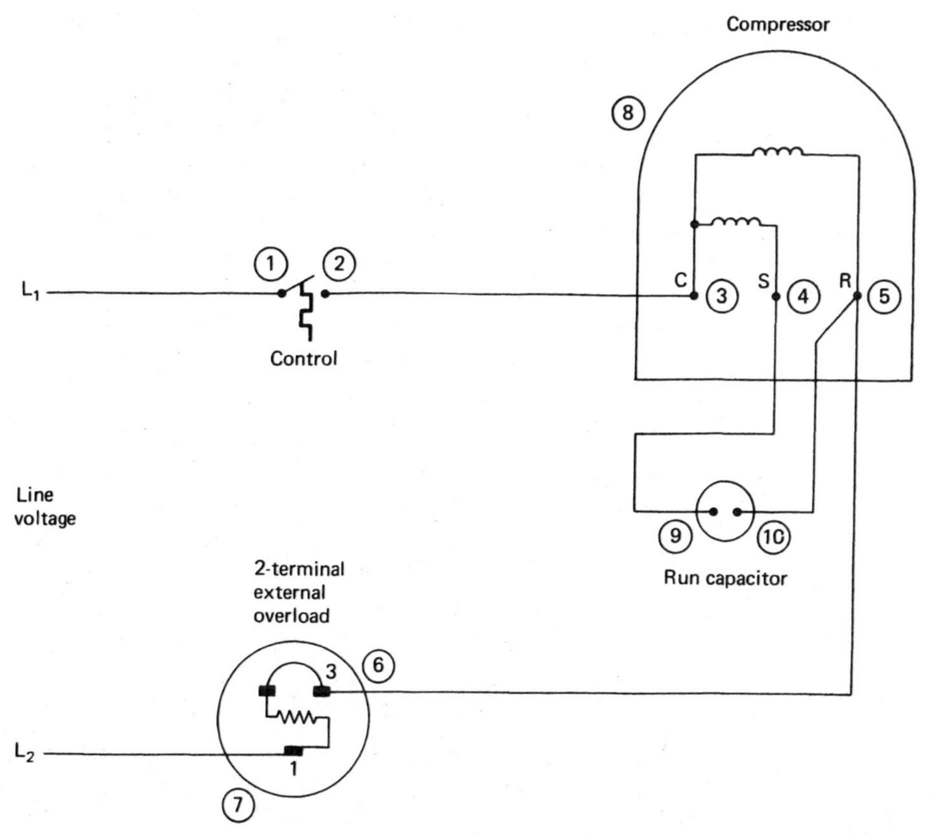

Figure 4–32. Permanent split capacitor (PSC) system with two-terminal overload.

2. Disconnect the electrical power from the unit.
3. If a fan motor is being used, disconnect one of the leads.
4. Make the following checks with an ohmmeter. Be sure to zero the ohmmeter.
5. Check the continuity between points 1 and 2. See Figure 4–32. If continuity is found, continue to make the following checks. No continuity indicates the control contacts are open. Close the contacts by setting the control to demand operation. If no continuity is found with the control on demand, replace the control.
6. Check the continuity between points 2 and 3. If continuity is found continue to make the following checks. If no continuity is found, repair the wiring or loose connections.
7. Remove the wire from point 4 (start terminal). Check the continuity between points 3 and 4. Compare this reading to that given by the compressor manufacturer for that particular compressor. If proper continuity is found, continue to make the following checks. If no or improper continuity is found, replace the compressor. The start winding is defective.
8. Remove the wiring from point 5 (run terminal). Check the continuity between points 3 and 5. Compare this reading to that given by the compressor

manufacturer for that particular compressor. If proper continuity is found, continue to make the following checks. If no or improper continuity is found, replace the compressor. The run winding is defective.

9. Check the continuity between point 3 (common terminal) and point 8 (compressor housing). If no continuity is found, continue to make the following checks. If continuity is found, replace the compressor. The motor winding is grounded.
10. Set the ohmmeter on R × 100,000. Check the continuity between points 9 and 10. If needle deflection is found, continue to make the following checks. If no needle deflection is found, the capacitor is open. Replace the capacitor. Be sure to use a proper replacement.
11. Set the ohmmeter on R × 1. Check the continuity between points 9 and 10. If no continuity is found, continue to make the following checks. If continuity is found, the capacitor is shorted. Replace the capacitor. Be sure to use a proper replacement.
12. Check the continuity between points 5 and 6. If continuity is found, continue to make the following checks. If no continuity is found, repair the wire or loose connections.
13. Check the continuity between points 6 and 7. If continuity is found, continue to make the following checks. If no continuity is found, wait about ten minutes and check the continuity again. The overload may be tripped. If no continuity now, replace the overload. Be sure to use an exact replacement.
14. Reconnect all wiring. Connect a 12-in. piece of electrical wire to each terminal on a 130 mfd × 370 VAC start capacitor. Touch the loose ends of one wire to point 4 (start terminal) and the other wire to point 5 (run terminal). Turn on the electricity. If the compressor starts immediately, remove the two wires from points 4 and 5. If the compressor continues to run, install a hard-start kit. See Figure 4–33. If the compressor does not start or stops running when the wires are removed, proceed to the next step.
15. Check the amperage draw in the wire to point 3 (common terminal) while the compressor is trying to run. Compare the amperage to the locked (LR) amperage of the motor. If the LR amperage is very close to that listed by the manufacturer, replace the compressor. The compressor has mechanical problems.

Permanent Split Capacitor (PSC) System with Internal Thermostat Overload

PSC motors are low starting torque motors that have no starting capacitor or starting relay. However, they do use a run capacitor. These systems have no external overload because a more sensitive internal overload is located inside the compressor housing. This system is used on builders' models and room air conditioning units. Use the following procedure to check this type of system. See Figure 4–34.

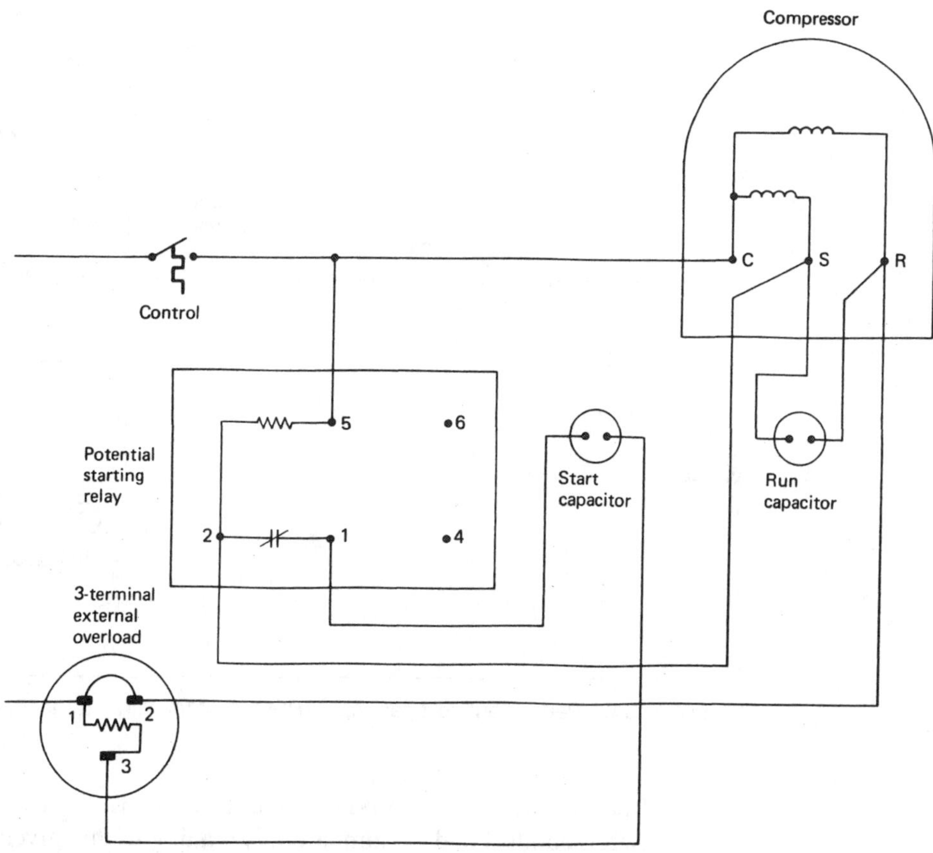

Figure 4–33. Connections used for installing a hard-start kit.

1. Check to make certain that the proper voltage is being supplied to the unit. If proper voltage is being supplied, continue to make the following checks. If the wrong voltage is being supplied, correct the problem with the power supply.
2. Disconnect the electrical power from the unit.
3. If a fan motor is being used, disconnect one of the leads.
4. Make the following checks with an ohmmeter. Be sure to zero the ohmmeter.
5. Check the continuity between points 1 and 2. Refer to Figure 4–29. If continuity is found, continue to make the following checks. No continuity indicates the control contacts are open. Close the contacts by setting the control to demand operation. If no continuity is found with the control on demand, replace the control.
6. Check the continuity between points 2 and 3. If continuity is found, continue to make the following checks. If no continuity is found, repair the wiring or loose connections.

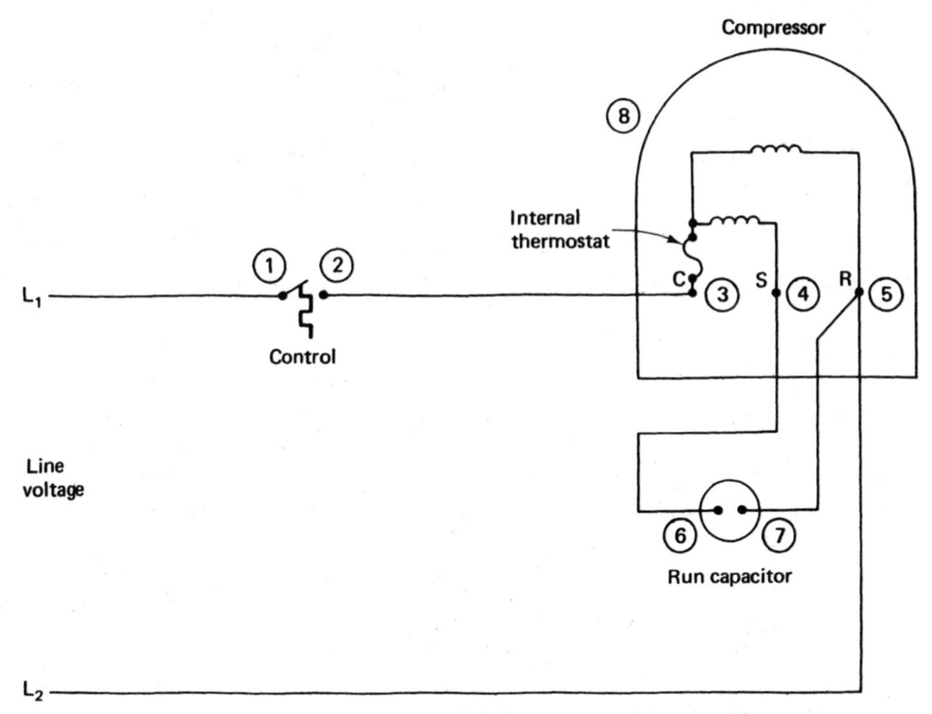

Figure 4–34. Permanent split capacitor (PSC) system with internal thermostat.

7. Remove the wire from point 4 (start terminal). Check the continuity between points 3 and 4. Compare this reading to that given by the compressor manufacturer for that particular compressor. If continuity is found, continue to make the following checks. If no or improper continuity is found, remove the wire from point 5 (run terminal). Check the continuity between points 4 and 5. If no continuity is found, it can be assumed that the start winding is open. Replace the compressor. Then, if no continuity is found, the internal overload may be tripped. Cool the compressor shell below 125°F (51.78°C) or until the hand may be held on the compressor without much discomfort. Again, check the continuity between points 3 and 4. If no continuity is found, the internal overload is defective. Replace the compressor.
8. Remove the wire from point 5 (run terminal). Check the continuity between points 3 and 5. Compare this reading to that given by compressor manufacturer for that particular compressor. If proper continuity is found, continue to make the following checks. If no or improper continuity is found, check the continuity between points 4 and 5. If continuity is found, assume that the internal overload is open. Cool the compressor shell below 125°F (51.78°C) or until the hand may be held on the compressor without much discomfort. Again, check the continuity between points 3 and 5. If no continuity is found, the internal overload is defective. Replace the compressor.

9. Check the continuity between points 3 and 8. If no continuity is found, continue to make the following checks. If continuity is found, replace the compressor. The motor winding is grounded.
10. Set the ohmmeter on R × 100,000. Check the continuity between points 6 and 7. If needle deflection is found, continue to make the following checks. If no needle deflection is found, the capacitor is open. Replace the capacitor. Be sure to use a proper replacement.
11. Set the ohmmeter on R × 1. Check the continuity between points 6 and 7. If no continuity is found, continue to make the following checks. If continuity is found, the capacitor is shorted. Replace the capacitor. Be sure to use a proper replacement. If the capacitor is swollen, it should be replaced.
12. Reconnect all wiring. Connect a 12-in. piece of wire to each terminal on a 130 mfd × 370 VAC start capacitor. Touch the loose end of one wire to point 4 (start terminal) and touch the loose end of the other wire to point 5 (run terminal). Turn on the electricity. If the compressor starts immediately, remove the two wires from points 4 and 5. If the compressor continues to run, install a hard-start kit. See Figure 4–33. If the compressor does not start or stops running when the wires are removed, proceed to the next step.
13. Check the amperage draw in the wire to point 3 (common terminal) while the compressor is trying to run. Compare the amperage draw to the locked rotor (LR) amperage of the motor. If the LR amperage is very close to that indicated by the manufacturer, replace the compressor. The compressor has mechanical problems.

Permanent Split Capacitor (PSC) System with Internal Thermostat Overload and External Overload

PSC motors are low starting torque motors that have no starting relay or starting capacitor. However, they do use a run capacitor. This particular compressor has both an internal thermostat and an external overload for added protection. The internal thermostat is temperature-sensitive and interrupts the line voltage to the compressor motor common terminal. The external overload is both temperature-sensitive and current-sensitive. Its contacts interrupt the control circuit. Use the following procedure to check this type of system. See Figure 4–35.

1. Check to make certain that the proper voltage is being supplied to the unit. If proper voltage is being supplied, continue to make the following checks. If the wrong voltage is being supplied, correct the problem with the power supply.
2. Disconnect the electrical power from the unit.
3. If a fan motor is being used, disconnect one lead.
4. Make the following checks with an ohmmeter. Be sure to zero the ohmmeter.
5. Check the continuity between points 1 and 2. See Figure 4–35. If continuity is found, continue to make the following checks. No continuity indi-

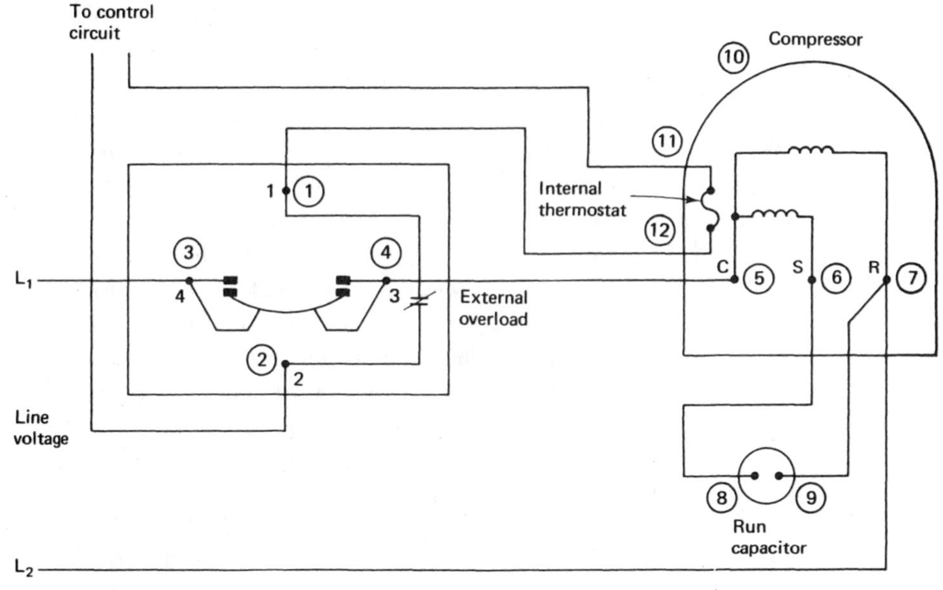

Figure 4–35. Permanent split capacitor (PSC) system with internal thermostat and external overloads.

cates the overload is open. Wait about ten minutes. Check the continuity again. If no continuity now, replace the external overload. Be sure to use an exact replacement.

6. Check the continuity between points 3 and 4. If continuity is found, continue to make the following checks. No continuity indicates the external overload is defective. Replace the overload. Be sure to use an exact replacement.
7. Remove the wire from point 6 (start terminal). Check the continuity between points 5 and 6. Compare this reading to that given by the manufacturer for that particular compressor. If proper continuity is found, continue to make the following checks. If no or improper continuity is found, the start winding is defective. Replace the compressor.
8. Remove the wire from point 7 (run terminal). Check the continuity between points 5 and 7. Compare this reading to that given by the manufacturer for that particular compressor. If proper continuity is found, continue to make the following checks. If no or improper continuity is found, the run winding is defective. Replace the compressor.
9. Remove the wire from point 11. Check the continuity between points 11 and 12. If continuity is found, continue to make the following checks. If no or improper continuity is found, the internal thermostat is open. Cool the compressor below 125°F (51.78°C) or until the hand may be held on the compressor shell without much discomfort. Again, check the continuity between points 11 and 12. If no continuity is found, replace the compressor. The internal thermostat is defective.

10. Check the continuity between points 5 and 10. If no continuity is found, continue to make the following checks. If continuity is found, replace the compressor. The motor winding is grounded.
11. Set the ohmmeter on R × 100,000. Check the continuity between points 8 and 9. If needle deflection is found, continue to make the following checks. If no needle deflection is found, the capacitor is open. Replace the capacitor. Be sure to use a proper replacement.
12. Set the ohmmeter on R × 1. Check the continuity between points 8 and 9. If no continuity is found, continue to make the following checks. If continuity is found, the capacitor is shorted. Replace the capacitor. Be sure to use a proper replacement. If the capacitor is swollen, replace it.
13. Reconnect all wiring. Connect a 12-in. piece of electrical wire to each terminal on a 130 mfd × 370 VAC start capacitor. Touch the loose end of one wire to point 6 (start terminal) and the loose end of the other wire to point 7 (run terminal). Turn on the electricity. If the compressor starts immediately, remove the two wires from points 6 and 7. If the compressor continues to run, install a hard-start kit. See Figure 4–33. If the compressor does not start or stops running when the wires are removed, proceed to the next step.
14. Check the amperage draw in the wire to point 5 (common terminal) while the compressor is trying to run. Compare the amperage draw to the locked rotor (LR) amperage of the motor. If the LR amperage is very close to that indicated by the manufacturer, replace the compressor. The compressor has mechanical problems.

16-1 PROCEDURE FOR CHECKING THERMOSTATIC EXPANSION VALVES:

In checking complaints, if the expansion valve is suspected as the source of trouble, an orderly procedure for locating the exact difficulty and remedying it in the shortest time possible is desirable.

1. Check the suction pressure and discharge pressure. They act as the "pulse" of the refrigerating system and are the best guide to locating trouble.
2. Check the superheat. For this purpose, an accurate pocket thermometer may be taped to the suction line at the remote bulb location.

The following headings group the suggested possibilities of trouble indicated by the gauge and superheat readings.

Low Suction Pressure—High Superheat

A. Expansion valve limiting flow.
    1. Inlet pressure too low from excessive vertical lift, too small liquid line, or excessively low condensing temperature. Resulting pressure difference across valve too small.

2. Gas in liquid line—due to pressure drop in the line or insufficient refrigerant charge. If there is no sight glass in the liquid line, a characteristic whistling noise may be observed at the expansion valve.
3. Valve restricted by pressure drop through coil, requiring change to external equalizer. See Figure 4–36.
4. External equalizer line plugged or external equalizer connection capped without providing a new valve cage or body with internal equalizer.
5. Moisture, wax, oil, or dirt plugging valve orifice. Ice formation or wax at valve seat may be indicated by sudden rise in suction pressure after shutdown and system has warmed up.
6. Valve orifice too small.
7. Superheat adjustment too high.
8. Power assembly failure or partial loss of charge.
9. Remote bulb of gas charged Thermo Expansion Valve has lost control due to the remote bulb tubing or power head being colder than the remote bulb.
10. Filter screen clogged.
11. Wrong oil type.

B. Restriction in system other than thermo valve. (Usually, but not necessarily, indicated by frost or lower than normal temperature at point of restriction.)
   1. Strainers clogged or too small.
   2. Solenoid Valve failure or valve undersized.
   3. King valve at liquid receiver outlet too small or not fully opened.
   4. Plugged lines.
   5. Hand valve stem failure or valve too small or not fully opened.
   6. Liquid line too small.
   7. Suction line too small.

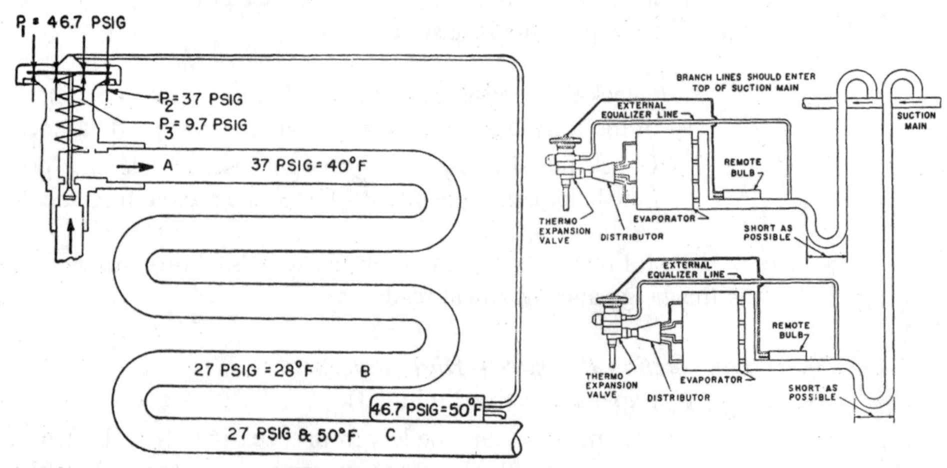

Figure 4–36. External equalizer connection methods: (a) Thermo expansion valve with internal equalizer on evaporator with 10 psi pressure drop. (b) Schematic of recommended piping of rising suction lines to a common suction main.

8. Wrong oil type in system, blocking liquid flow.
9. Discharge or suction service valve on compressor restricted or not fully opened.

Low Suction Pressure—Low Superheat

A. Poor distribution in evaporator, causing liquid to short circuit through favored passes, throttling valve before all passes receive sufficient refrigerant.
B. Compressor oversized or running too fast due to wrong pulley.
C. Uneven or inadequate coil loading, poor air distribution or brine flow.
D. Evaporator too small—indicated sometimes by excessive ice formation.
E. Evaporator oil logged.

High Suction Pressure—High Superheat

A. Compressor undersized.
B. Evaporator too large.
C. Unbalanced system having an oversized evaporator, an undersized compressor and a high load on the evaporator.
D. Compressor discharge valves leaking.

High Suction Pressure—Low Superheat

A. Compressor undersized.
B. Valve superheat setting too low.
C. Gas in liquid line with oversized thermo valve.
D. Compressor discharge valves leaking.
E. Pin and seat of expansion valve wire drawn, eroded, or held open by foreign material resulting in liquid flood-back.
F. Ruptured diaphragm or bellows in a Constant Pressure (Automatic) Expansion Valve resulting in liquid flood-back.
G. External equalizer line plugged or external equalizer connection capped without providing a new valve cage or body with internal equalizer.
H. Moisture freezing valve in open position.

Fluctuating Suction Pressure

A. Poor Control.
    1. Improper superheat adjustment.
    2. Trapped suction line. See Figure 4–37.
    3. Improper remote bulb location or application. See Figure 4–38.
    4. "Flood-back" of liquid refrigerant caused by poorly designed liquid distribution device or uneven coil circuit loading. Also improperly hung evaporator. Evaporator not plumb. See Figure 4–39.
    5. External equalizer tapped at common point on application with more than one valve on same evaporator.
    6. Faulty condensing water regulator, causing change in pressure drop across valve.
    7. Evaporative condenser cycling, causing radical change in pressure difference across expansion valve.
    8. Cycling of blowers or brine pumps.

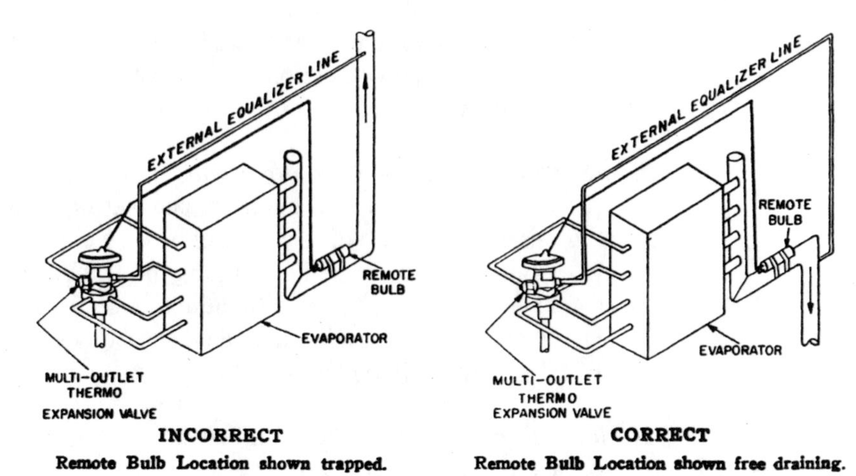

Figure 4–37. Remote bulb installed on a trapped suction line.

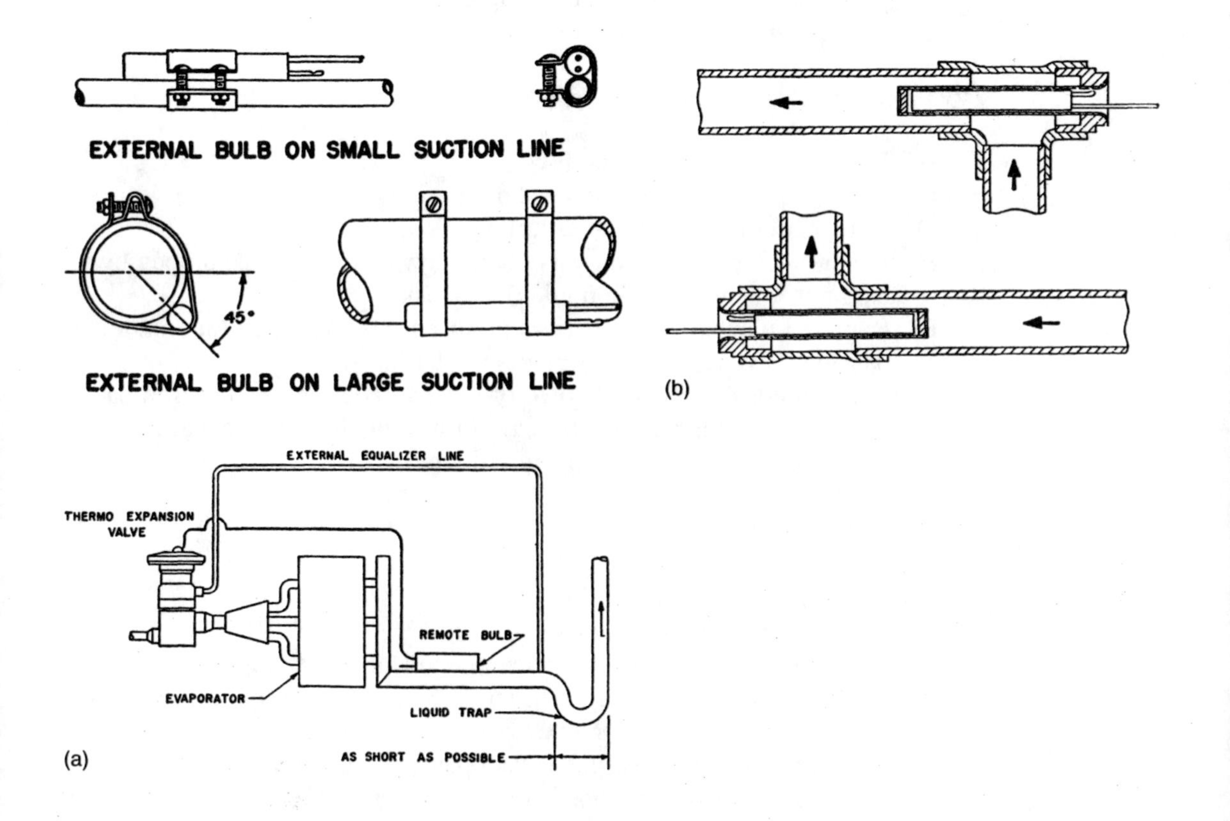

Figure 4–38. Recommended remote bulb location and schematic piping for rising suction line.

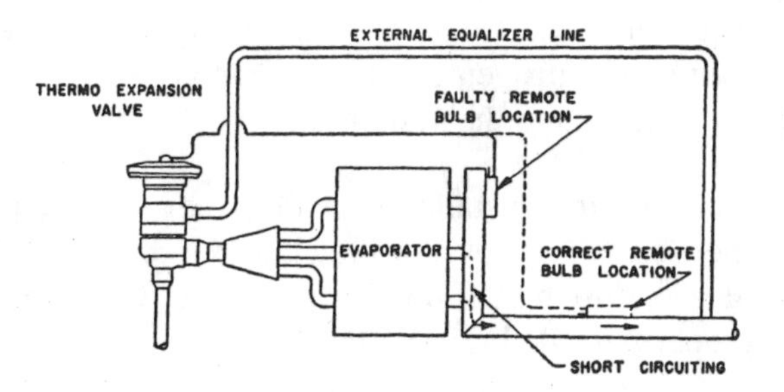

Figure 4–39. Correct remote bulb location on "short circuiting" evaporator to prevent "flood back."

Fluctuating Discharge Pressure

A. Faulty condensing water regulating valve.
B. Insufficient charge—usually accompanied by corresponding fluctuation in suction pressure.
C. Cycling of evaporative condenser.
D. Inadequate and fluctuating supply of cooling water to condenser.

High Discharge Pressure

A. Insufficient cooling water. (Inadequate supply or faulty water valve.)
B. Condenser or liquid receiver too small.
C. Cooling water above design temperature.
D. Air or non-condensable gases in condenser.
E. Overcharge of refrigerant.
F. Condenser dirty.
G. Air type condenser improperly located to dispel hot discharge air.

Expansion Valve "Freeze-Ups"

Expansion valve trouble may occur from the formation of ice crystals or the separation of wax out of the oil at the pin and seat, causing the refrigerant flow to be restricted or stopped entirely. Waxing generally occurs only at extreme low temperatures; however, ice formation due to moisture in the system can occur whenever the evaporator is operating below freezing. Moisture is generally indicated by the starving of the evaporator. However, it may cause a valve to freeze open and cause "flood-back." When the compressor stops and the expansion valve and the system are allowed to warm up, the system will operate satisfactorily until the ice crystals form again. Another indication of moisture in the expansion valve can be noted if tapping the valve body will cause it to feed again for a short time. Under no circumstances should a torch be used on the valve body to melt ice accumulation at the pin and seat. If the ice must be melted to put the system in operation temporarily, then apply hot rags. This is equally effective and will not damage the valve.

To remedy this moisture problem install a dryer of suitable size and operate the system above freezing for a few days to allow the moisture to be trapped in the dryer rather than to freeze at some location in the evaporator. If moisture persists, dump the refrigerant charge and the oil, then dry the system by means of a vacuum and heat or by blowing dry nitrogen gas through the system.

Moisture can be admitted to the system:

A. At the time of original installation from moist air in the piping.
B. At the time of charging by not blowing the air and moisture out of the charging hose before tightening the fitting at the connection.
C. From improperly refilled refrigerant drums.
D. By allowing air to enter the system at the time of adding oil.
E. From leaking or defective shaft seal.
F. By opening the system when in a vacuum.

In a methyl chloride system, regardless of the operating range, the presence of moisture creates an acid condition that attacks the system, creates sludge, and causes copper to plate upon the steel parts of the compressor as well as on the pin and seat of the expansion valve.

17-1 PROCEDURE FOR TORCH BRAZING:

This section describes procedures that are important for making sound brazements of tubing in the air conditioning and refrigeration industry, using the phos-copper and the silver brazing filler metals.

The importance of brazing cannot be overlooked and/or overemphasized since it is a major part of the air conditioning and refrigeration industry.

Brazing is the process used in joining the major components of a refrigeration system into a closed circuit. Since the closed circuit contains a refrigerant, every brazed joint must be leak-free; if not, the refrigerant will escape, creating a severe inconvenience to the customer as well as a costly repair.

The purpose of this section is to provide the basic information about the correct method for torch brazing.

Definitions

Brazing is the application of heat above a temperature of 800°F and below the melting points of the base metals to produce coalescence or bonding by surface adhesion forces between the molten filler metals and the surfaces of the base metals. The filler metal is distributed through the joint by using capillary action.

Brazing is not to be confused with soldering even though the procedures are very similar. Soldering is a term used for metal-joining processes at temperatures below 800°F.

In order for bonding and distribution by capillary attraction to occur, the filler metal must be able to "wet" the base metals. Wetting is the phenomenon in which the forces of attraction between the molecules of the molten filler metal and the molecules of the base metals are greater than the inward forces of attraction existing between the molecules of the filler metal.

The degree of wetting is a function of the compositions of the base metal and the filler metal and of the temperature. Good wetting can only occur on perfectly clean and oxide-free surfaces.

Filler metals

The quality and strength of a brazement are more a function of the physical parameters of the joint and the brazing procedures used than which filler metal is applied to the joint. These parameters determine the selection of the best-suited and the most easily handled filler metal for a particular joint.

The phos-copper brazing alloys were specifically developed for joining copper and copper alloys. They are used for brazing copper, brass, bronze, or combinations of these.

When brazing with either brass or bronze, a flux must be used to prevent the formation of an "oxide coat" over these base metals. This coat would prevent wetting and the flow of the filler metal. However, when brazing copper to copper joints, these alloys are self-fluxing.

Because of the phosphorous embrittlement of the joint, phos-copper filler metals should not be used on ferrous metals or base metals containing more than 10% nickel. These alloys are also not recommended for use on aluminum bronze.

Unlike the phos-copper alloys, the silver brazing alloys contain no phosphorous. These filler metals are used to braze all ferrous, copper, and copper-based metals with the exception of aluminum and magnesium. They require a flux with all applications.

Extreme care should be taken when using the low-temperature, cadmium-bearing silver alloy because of poisonous cadmium fumes being emitted.

Most refrigeration repair work can be accomplished with a couple of filler metals. "Braze-it 15" (15% silver content) is a commonly used phos-copper filler metal, and "Braze-it 45" w/cd (45% silver, cadmium-bearing) is a commonly used silver brazing filler metal.

Torch brazing procedures*

Copper to copper tube joints using the phos-copper brazing alloys

I. Correct oxygen and fuel gas mixture.

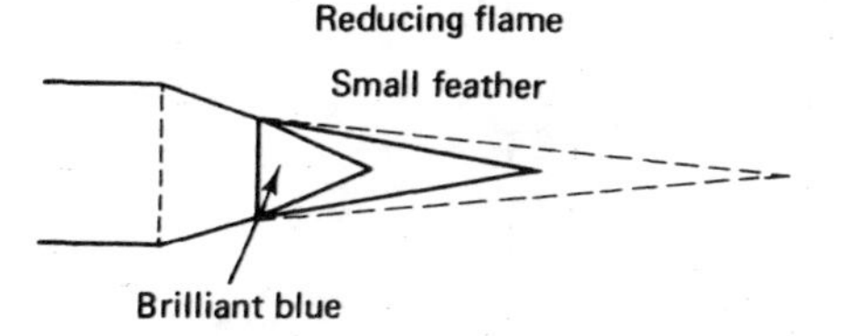

Figure 4–40. Best type of flame for brazing.

Excessive gas mixture—a reducing type of flame denotes excessive fuel gas; a greater amount of fuel gas than oxygen. A slightly reducing flame heats and cleans the metal surface for quicker and better brazing.

---

*Article by Anthony Donofrio, Product Coordinator, Engelhard Industries, Warwick, RI. Used with permission.

Figure 4–41. Flame adjustment.

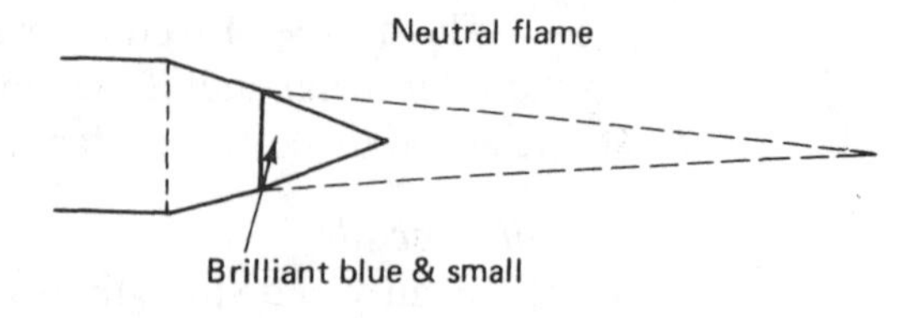

Balanced gas mixture—the gas mixture contains an equal amount of oxygen and fuel gas. It produces a flame that heats the metal, but has no other effect.

Figure 4–42. Neutral flame.

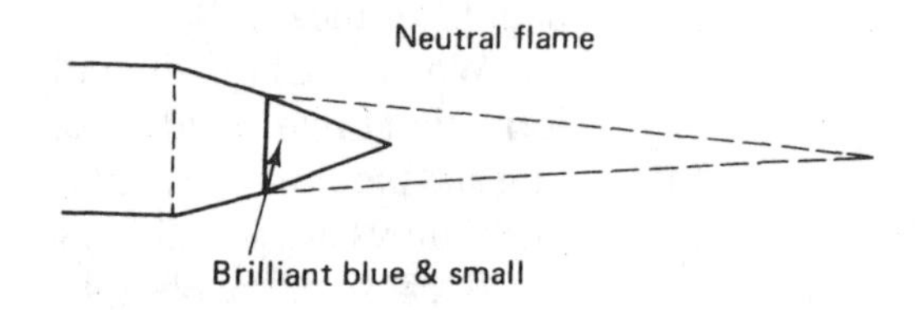

Excessive oxygen mixture—the gas mixture contains an excessive amount of oxygen. It produces a flame that oxidizes the metal surface. A black oxide scale will form on the metal surface.

II. Cleanliness.

General cleanliness is of prime importance to reliable brazing. All metal surfaces to be brazed must be cleaned of all dirt and foreign matter. When repair work is to be done, the metal surfaces to be joined must be either wire brushed and/or cleaned with sand paper. A concerted effort must be made to keep oil, paint, dirt, grease, and aluminum off the surface of metals to be joined for these contaminants will:

1. Keep brazing filler metal from flowing into the joint.
2. Prevent brazing filler metal from wetting or bonding to the metal surfaces.

III. Correct insertion and clearance of parts.

Figure 4–43. Proper clearance for soldered fittings.

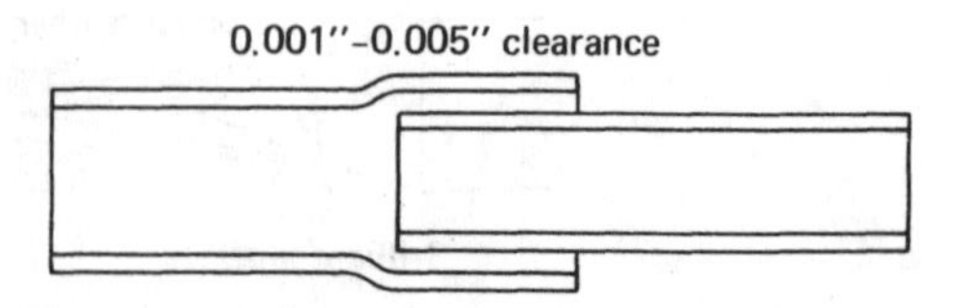

Minimum insertion distance is equal to diameter of inner tube. There should be .001–.005 in. clearance between the walls of the inner and outer tubes.

IV. Correct amount and application of heat

Heat evenly over the entire joint circumference, and joint length. Use enough heat to get entire joint hot enough to melt the brazing filler metal without heating the filler metal directly with the flame.

Heat correctly

Heat zone

Female tube

Male tube

Figure 4–44. Heating tubes to be soldered.

Heat both tubes at the joint, and distribute heat evenly.

Overheating

Figure 4–45. Overheated joint.

DO NOT OVERHEAT the joint to the point that the metal to be joined begins to melt. Always try to use the correct size torch tip, and a slightly reducing flame. Overheating enhances the base metal-filler metal interactions (e.g., formation of chemical compounds). In the long run, these interactions are detrimental to the life of the joint.

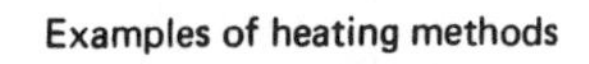

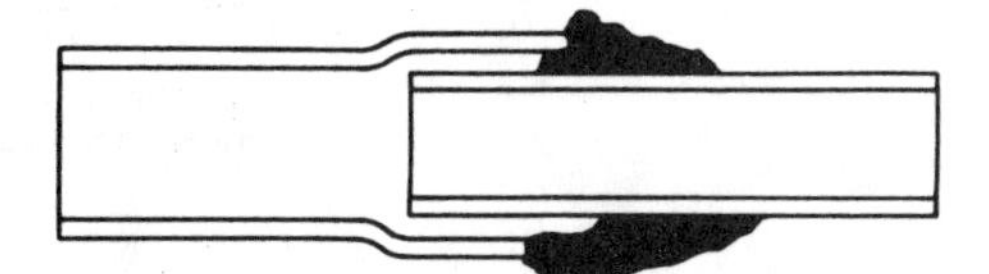

Figure 4–46. Male tube properly heated. Female tube too cool.

Male tube at brazing temperature, and female tube too cool. Result—filler metal ran away from the joint toward the heat source.

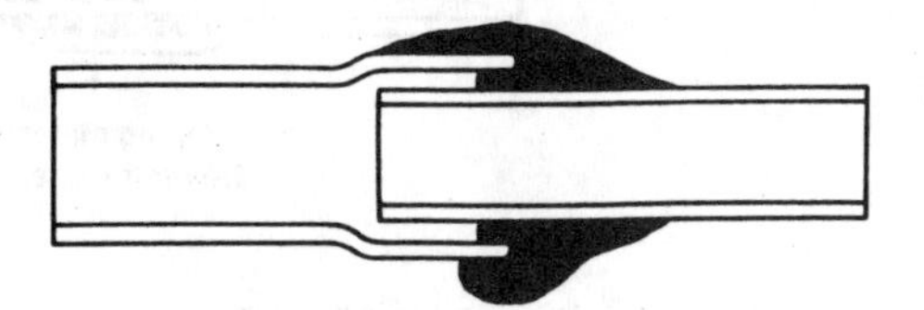

Figure 4–47. Flame and filler metal applied at same time.

Filler metal and torch were applied at the same time. As illustrated, total joint area was not heated properly, and the heat never reached male tube to allow capillary action of the filler metal.

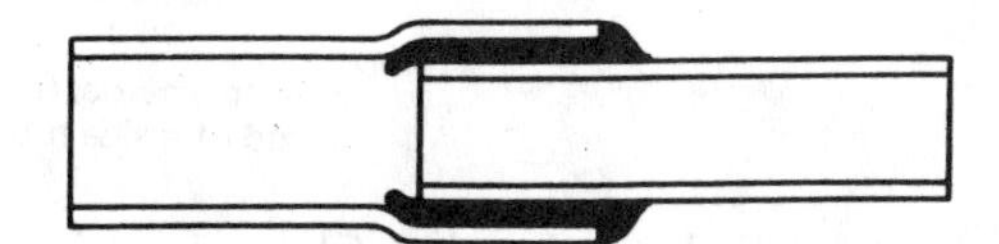

Figure 4–48. Properly heated and fitted joint.

Male tube inserted properly to diameter of female tube. Heat applied evenly to complete joint area. Filler metal was melted by heat of the tube.

V. Correct application of brazing rod.

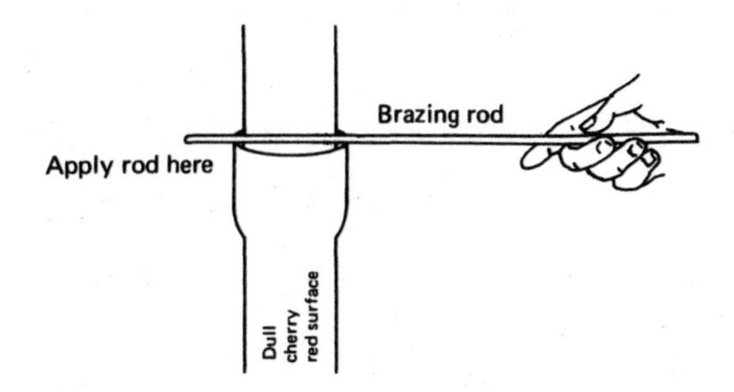

Figure 4–49. Proper application of filler rod.

Use the brazing rod as a temperature indicator. If the brazing rod melts when placed on the hot joint, the tubing is hot enough to begin brazing. For best results, preheat the rod slightly with the outer envelope of the flame, but the hot copper tubing (not the direct flame) should melt the brazing rod.

VI. Capillary attraction.

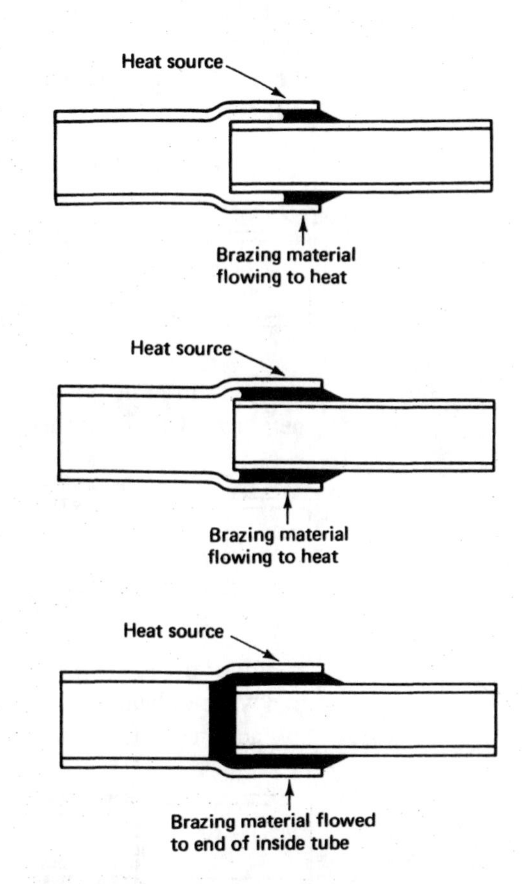

Figure 4–50. Capillary attraction of filler material into a joint.

This is the phenomenon by which the brazing filler metal is drawn into the joint. It is caused by the attraction between the molecules of the brazing filler metal and the molecules of the metal surfaces to be joined. It can only work if: (1) the surface of the metal is clean, (2) the clearance between the metal surfaces is correct, (3) the metal at the joint area is hot enough to melt the brazing filler metal, and (4) the brazing filler metal will flow toward the heat source as illustrated.

Copper to brass tube joints using the phos-copper brazing alloys.

1. The previously mentioned procedures for copper to copper joints are strictly followed.
2. Before application of heat to the joint, a small amount of flux is applied to allow wetting of the brass by the filler metal.
3. Upon completion of the brazement, the flux residue is thoroughly cleaned off. Cleaning can be accomplished with hot water and mechanical brushing. Most fluxes are corrosive and must be completely removed from the joint.

Steel to: steel, copper, brass, or bronze using silver brazing alloys.

1. The above-mentioned procedures for joining copper to copper are strictly followed.
2. Before application of heat to the joint, flux is applied to allow wetting and flow to take place between the filler metal and the base metals.
3. Before application of the filler metal to the joint, a small amount of flux is applied to the rod. Rod is heated up and then dipped into the flux. This coats the filler metal with a thin layer of flux that provides for quick flow by preventing the formation of an oxide coat around the filler metal (zinc oxide).
4. Flux must be completely washed off upon completion of the joint.

Fluxes

A flux is analogous to a sponge absorbing water, only a flux is absorbing oxides. It can only absorb so much before it becomes useless as a flux.

When a flux becomes saturated with oxides, its viscosity increases. The ability of the flux to be displaced by the filler metal is therefore impeded. This leads to flux entrapment in the joint, which eventually causes corrosion and leaks.

One should use the least amount of flux that will get the job done. Then thoroughly clean the flux residue off after completing the brazement. Apply the flux along the surface of the joint and not into the joint itself. Allow the flux to flow into the joint ahead of the filler metal.

Simplified rules for good brazing

1. Use a slightly reducing flame to get the maximum in heating and cleansing action.
2. Make sure metal surfaces are clean.
3. Examine parts for correct insertion and clearances.
4. Flux parts: (1) Use minimum amount and (2) apply to outside of joint. No flux required for joining copper to copper with phos-copper filler metals.
5. Apply heat evenly around the joint to the proper brazing temperature.

6. Apply filler metal to joint. Make sure it is completely distributed throughout the joint by using the torch. Molten filler metals will follow the heat.
7. If flux is used, thoroughly clean off the residue.
8. A major part of good brazing is to complete the brazement as quickly as possible. Keep the heating cycle short. Avoid overheating.
9. Have adequate ventilation. Brazing may result in hazardous fumes: cadmium fumes from cadmium-bearing filler metals and fluoride fumes from fluxes.

5

Representative Wiring Diagrams

REPRESENTATIVE AMANA REFRIGERATION WIRING DIAGRAMS

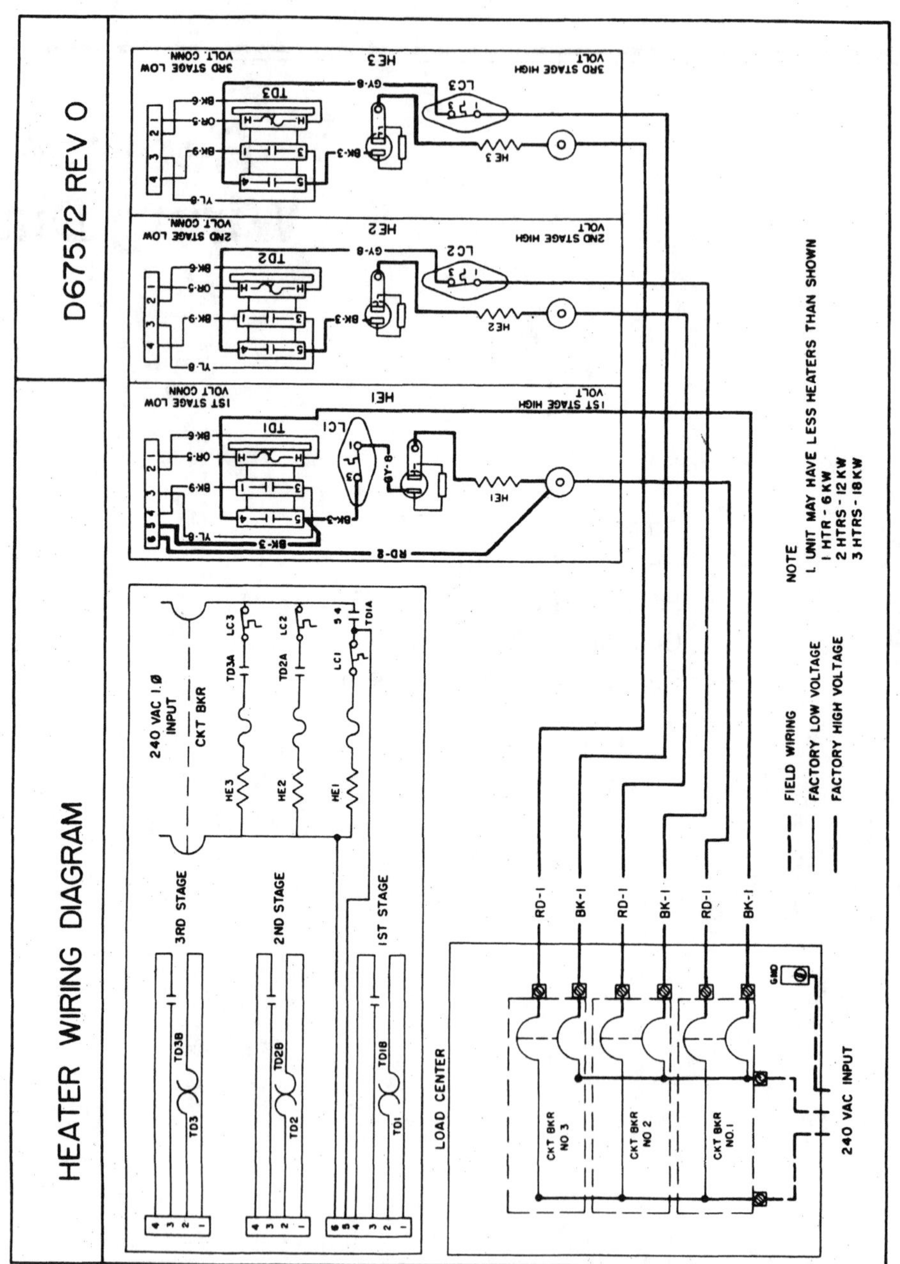

FOR 42, 48, 60M BTUH SINGLE PHASE UNITS

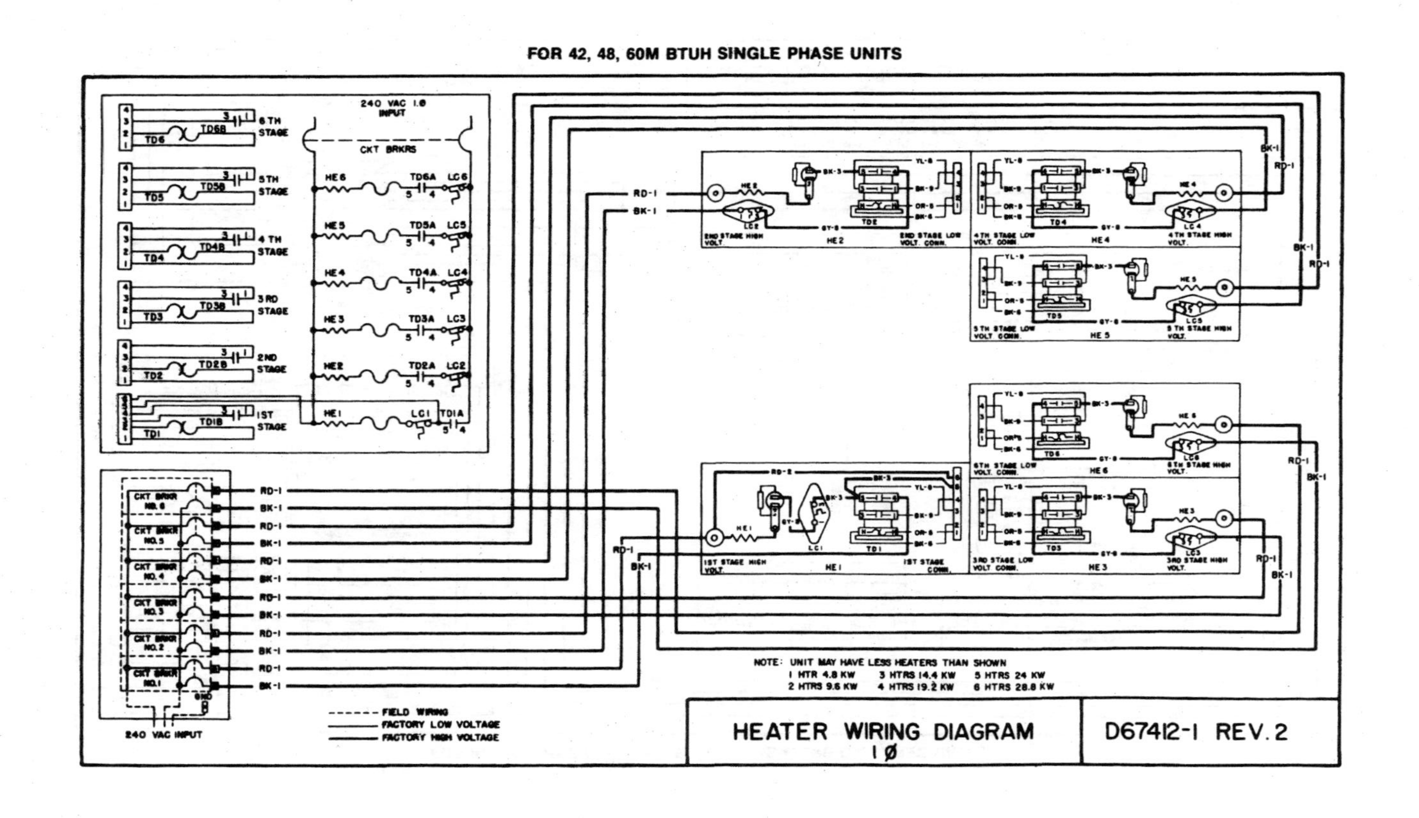

FOR 42, 48, 60M BTUH SINGLE PHASE HEATERS ON THREE UNITS

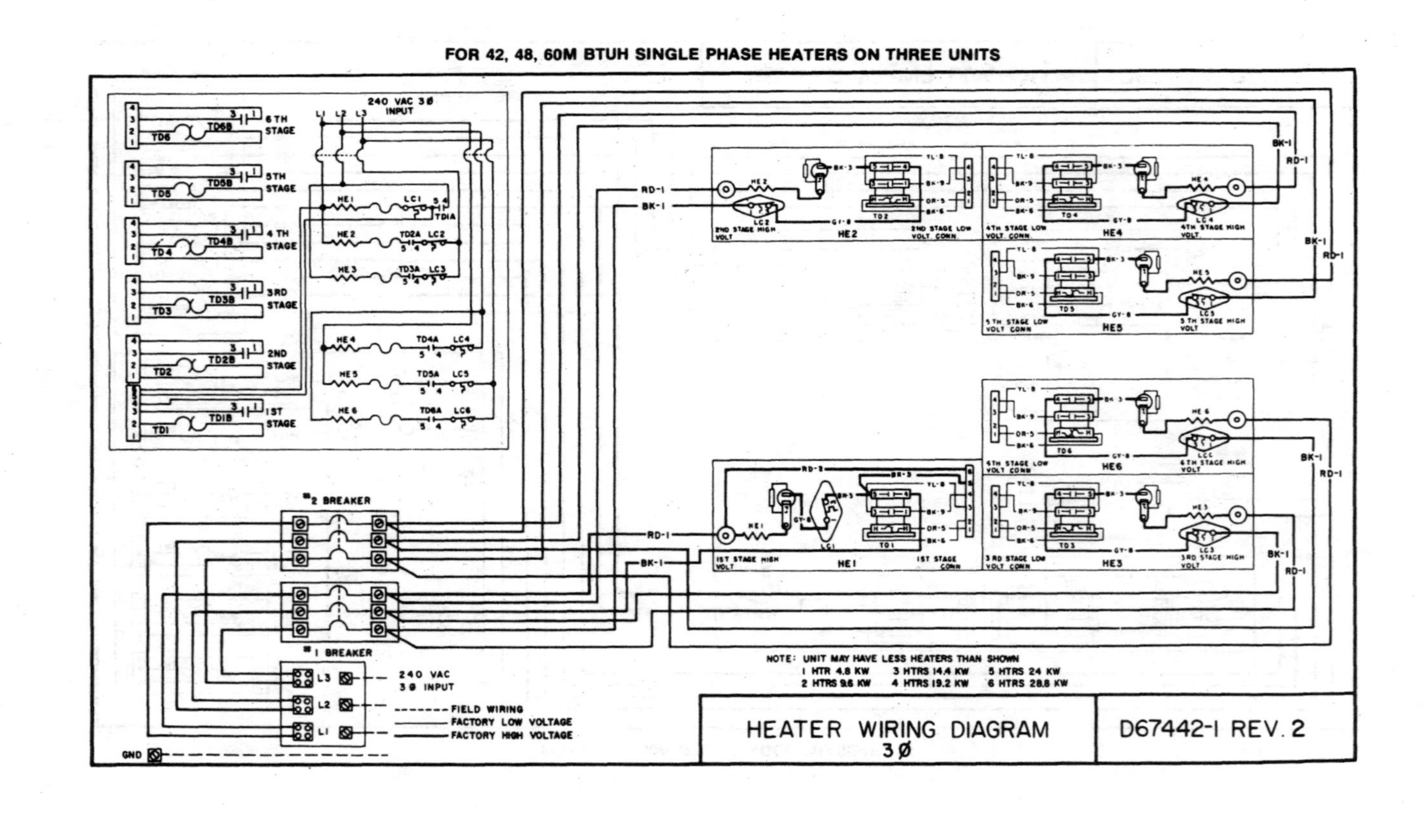

HEATER WIRING DIAGRAM 3Ø

D67442-1 REV. 2

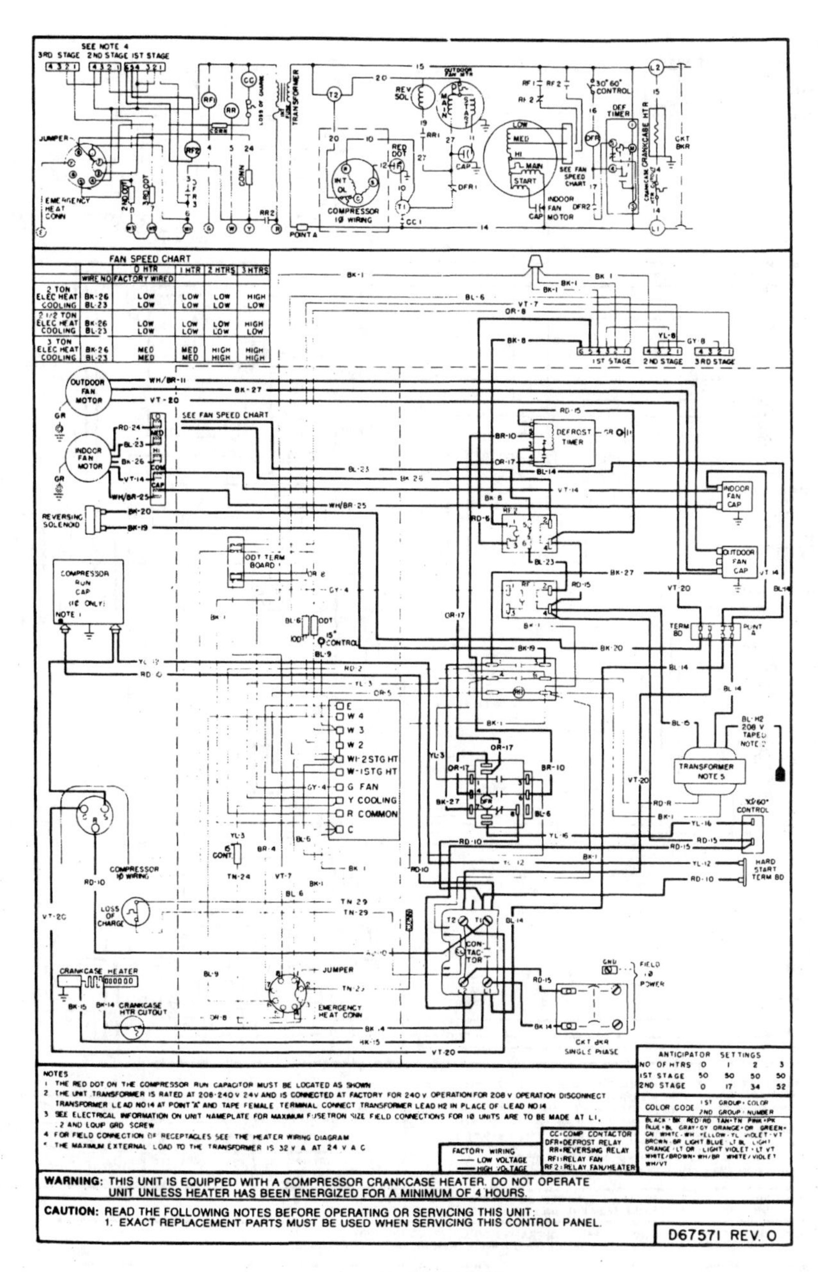

UNIT WIRING DIAGRAM
24M, 30M and 36M BTUH UNITS

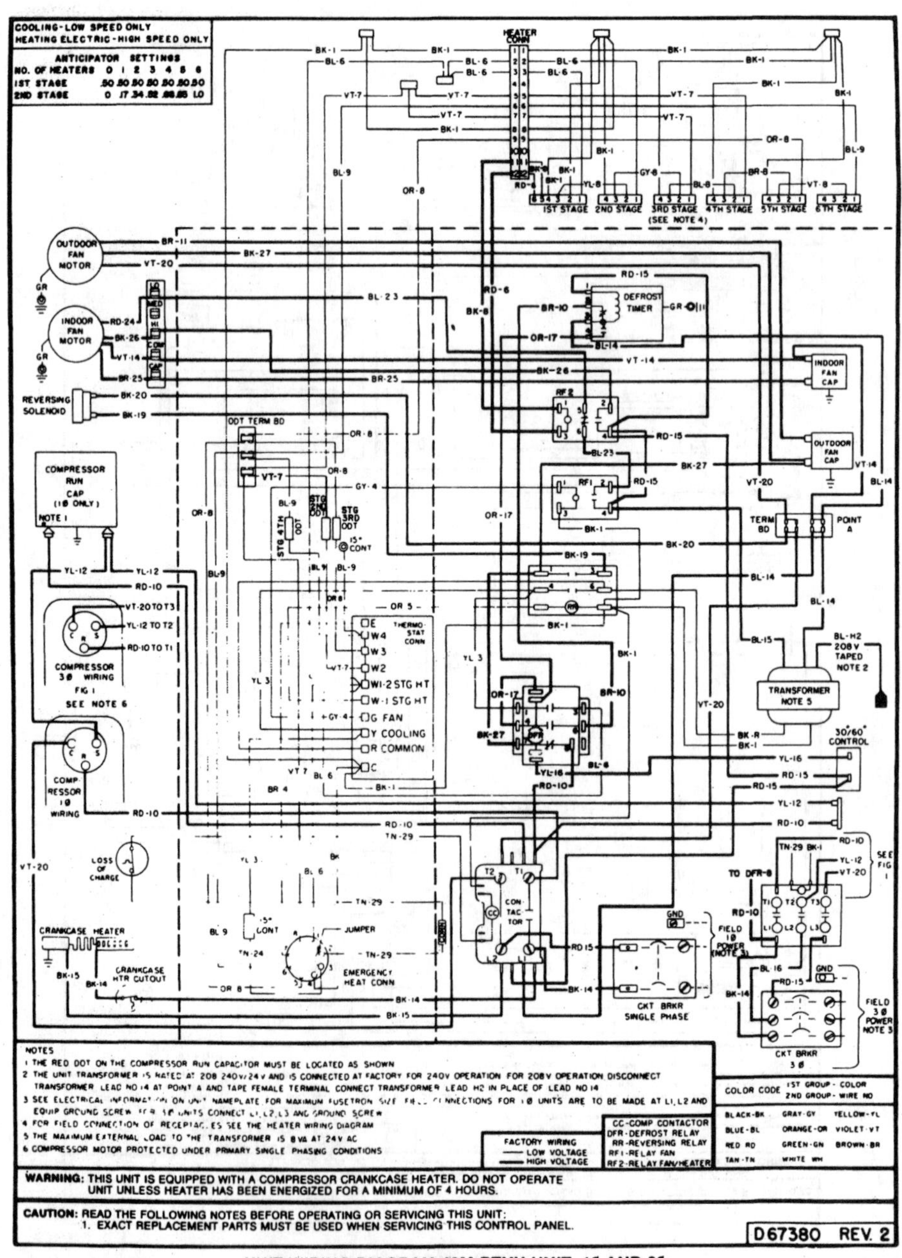

UNIT WIRING DIAGRAM 42M BTUH UNIT, 1Ø AND 3Ø
See page 18 for schematic wiring diagram

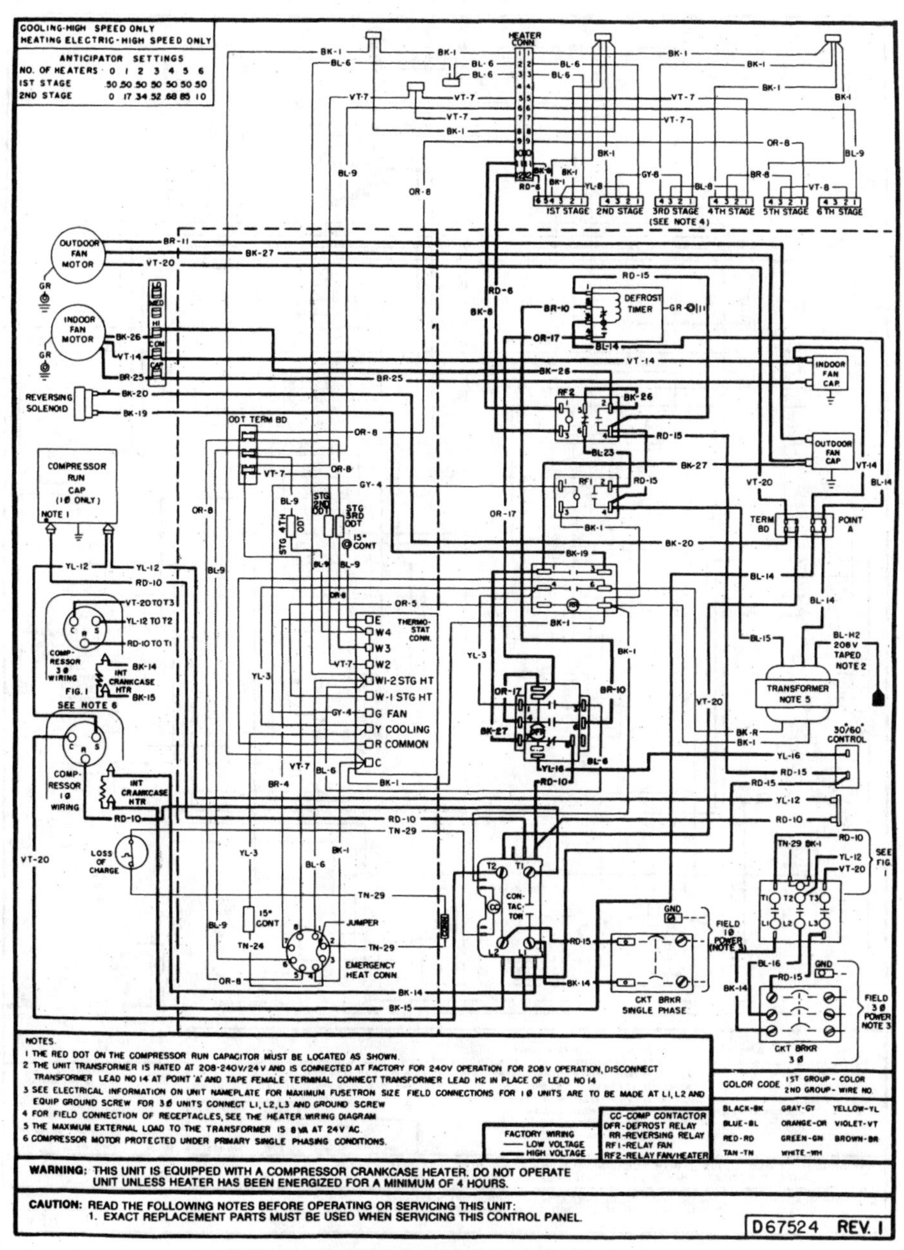

UNIT WIRING DIAGRAM 48M BTUH UNIT, 1Ø AND 3Ø
See page 18 for schematic wiring diagram

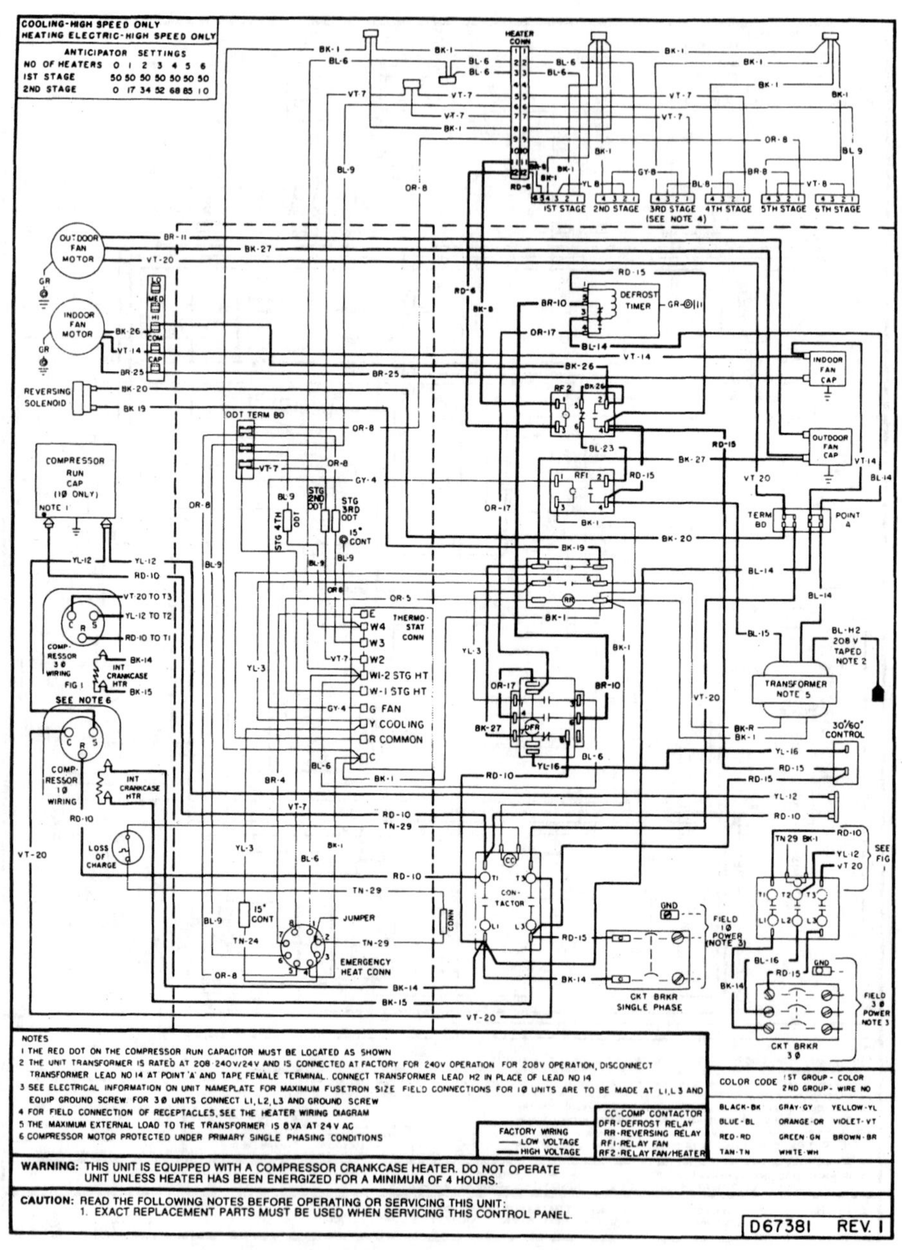

UNIT WIRING DIAGRAM 60M BTUH UNIT, 1Ø AND 3Ø
See page 18 for schematic wiring diagram

SCHEMATIC WIRING DIAGRAMS

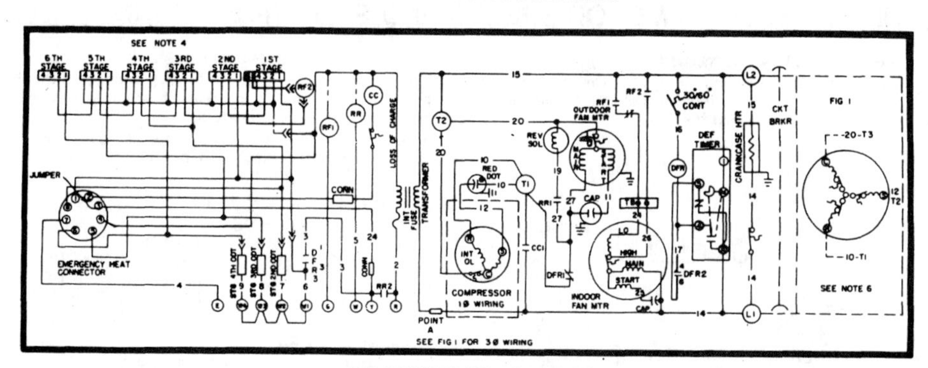

42M BTUH UNIT, 1Ø AND 3Ø

(Reference D67380)

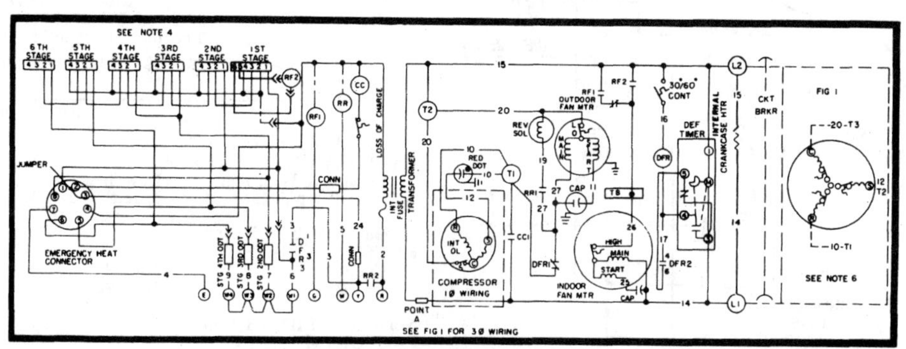

48M BTUH UNIT, 1Ø AND 3Ø

(Reference D67524-1)

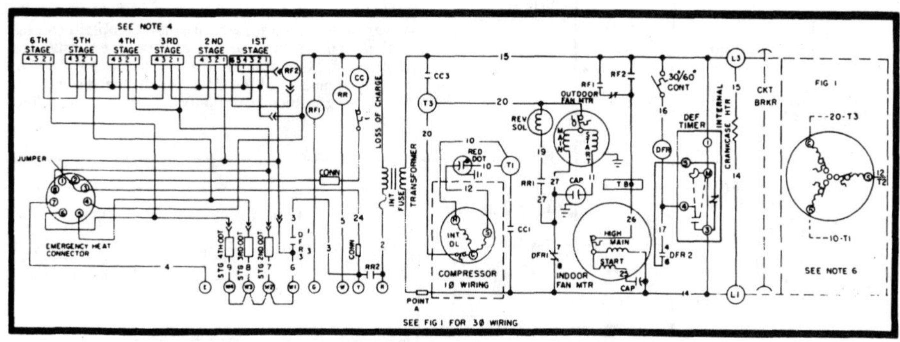

60M BTUH UNIT, 1Ø AND 3Ø

(Reference D67381)

NOTES:

4. For field connection of receptacles, see heater wiring diagram.
6. Compressor motor protected under primary single phasing conditions.

INDOOR BLOWER MOTOR SPEEDS AND ELECTRICAL CONNECTIONS

FACTORY MOTOR CONNECTIONS

The following sketches illustrate how the indoor blower motor circuit wires are factory connected to the unit blower terminal board.

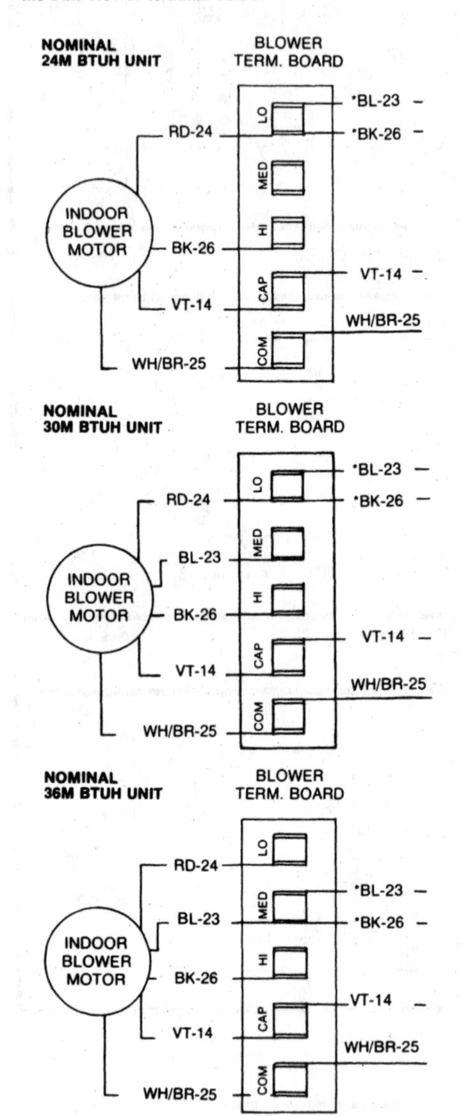

HEATING AND COOLING SPEEDS

The indoor blower motor is factory connected for the recommended cooling speed and heat pump only operation without electric heaters.

After the correct number of heaters have been installed, check the tabulation below and connect the motor leads accordingly.

Nominal 24M and 30M Units

SPEEDS

COOLING	HEATING
BL-23 to LO	0 heater BK-26 to LO
	1 heater BK-26 to LO
	2 heaters BK-26 to LO
	3 heaters BK-26 to HI

Nominal 36M Unit

COOLING	HEATING
BL-23 to MED	0 heater BK-26 to MED
	1 heater BK-26 to MED
	2 heaters BK-26 to HI
	3 heaters BK-26 to HI

Also check the unit wiring diagram for additional information.

Nominal 42M BTUH Unit
FACTORY CONNECTED

Cooling & Heat Pump	Low Speed
Heat Pump & Elec. Heat	High Speed

Nominal 48M BTUH Unit
FACTORY CONNECTED

Cooling	High Speed
Heating	High Speed

Nominal 60M BTUH Unit
FACTORY CONNECTED

Cooling	High Speed
Heating	High Speed

The above information illustrates how the indoor blower motor is factory connected to obtain the rated cooling and heating air quantities at external static pressures to the unit as shown in the air flow data chart. The electrical motor connections should not be changed regardless how many heaters are installed in the unit.

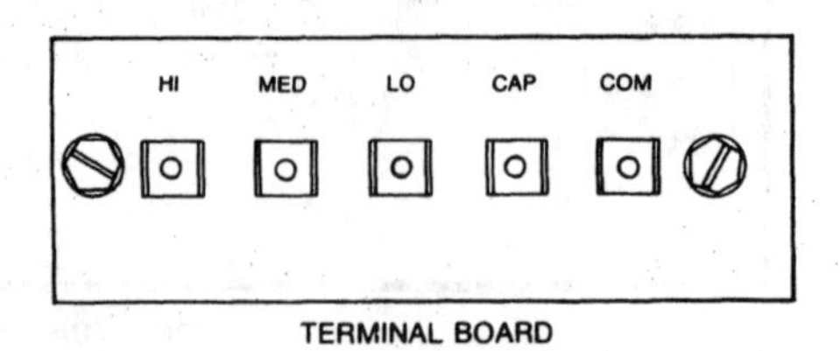

TERMINAL BOARD

*BL-23 and BK-26 have piggy back connectors and are the wires which must be re-connected in relation to the number of heaters installed in the unit.

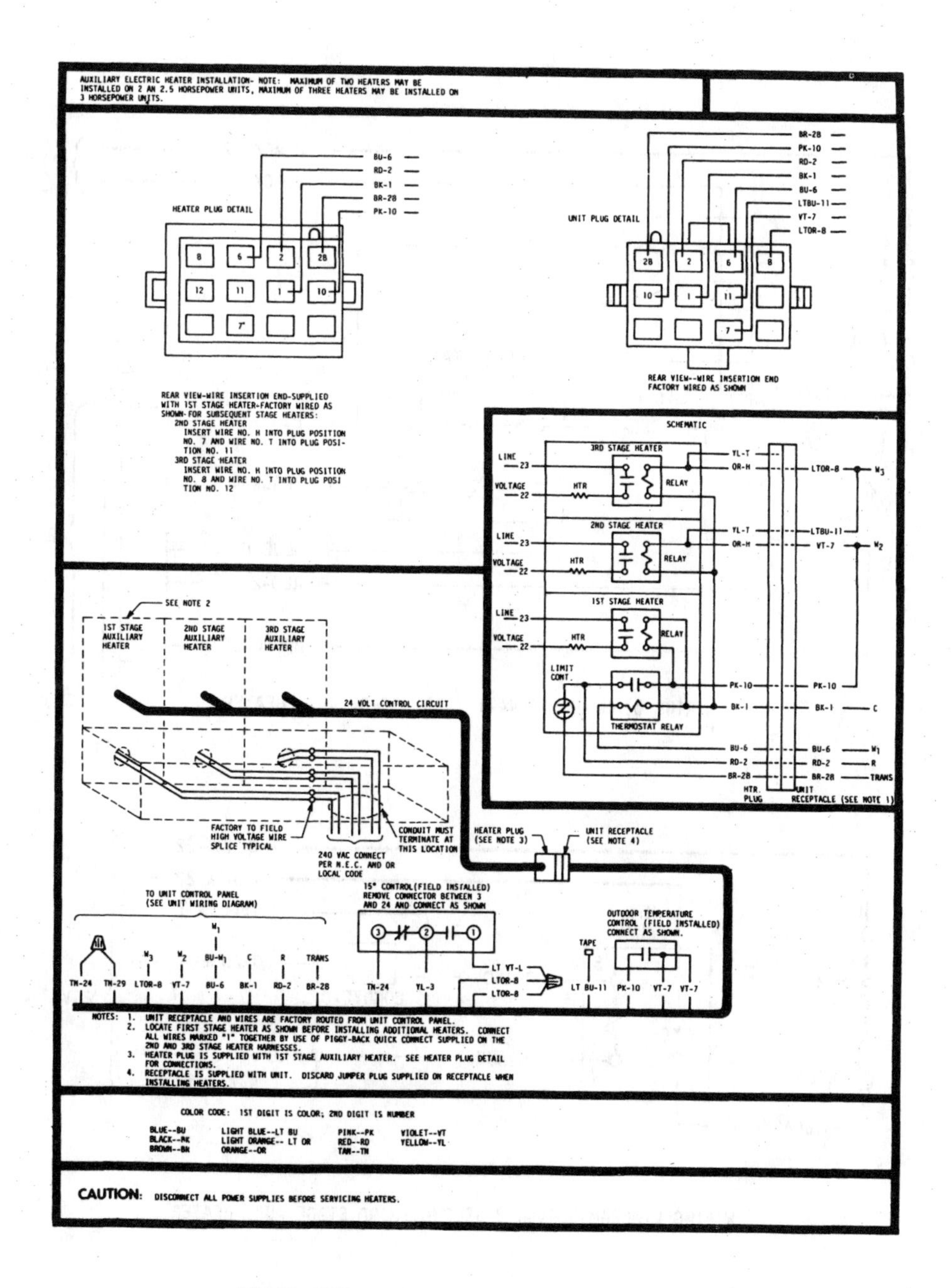

HEATER WIRING DIAGRAM-PACKAGE HEAT PUMP

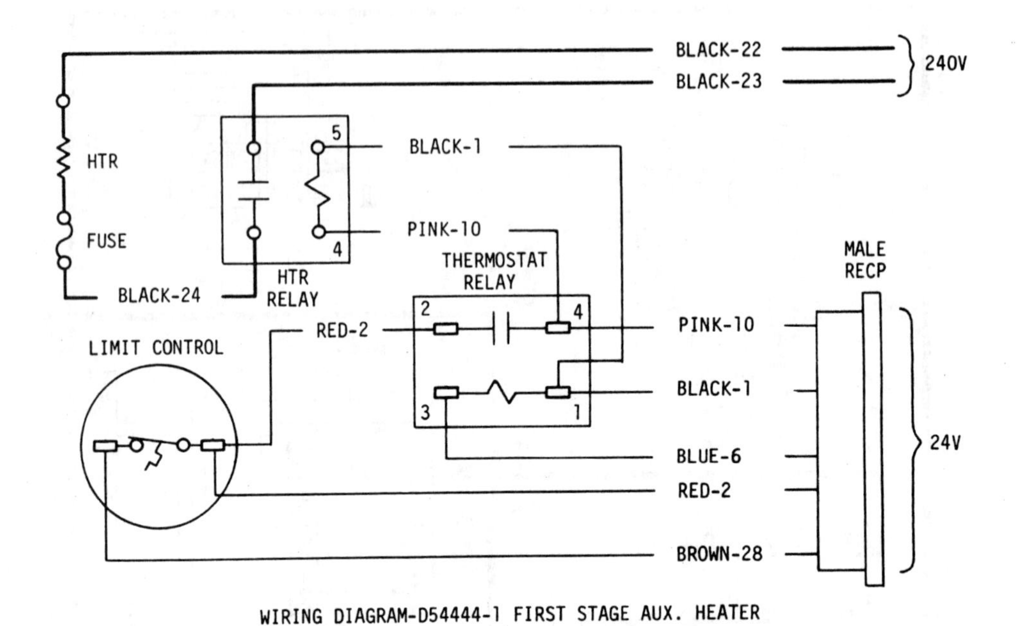

WIRING DIAGRAM-D54444-1 FIRST STAGE AUX. HEATER

BLACK-22
BLACK-23
240V
HTR
RELAY
YELLOW-T
5
HTR
ORANGE-H
ORANGE-H
CONNECTOR
24V
FUSE
BLACK-1
4
BLACK-1
BLACK-24

WIRING DIAGRAM-D54444-2 SECOND,THIRD STAGE AUX. HEATER

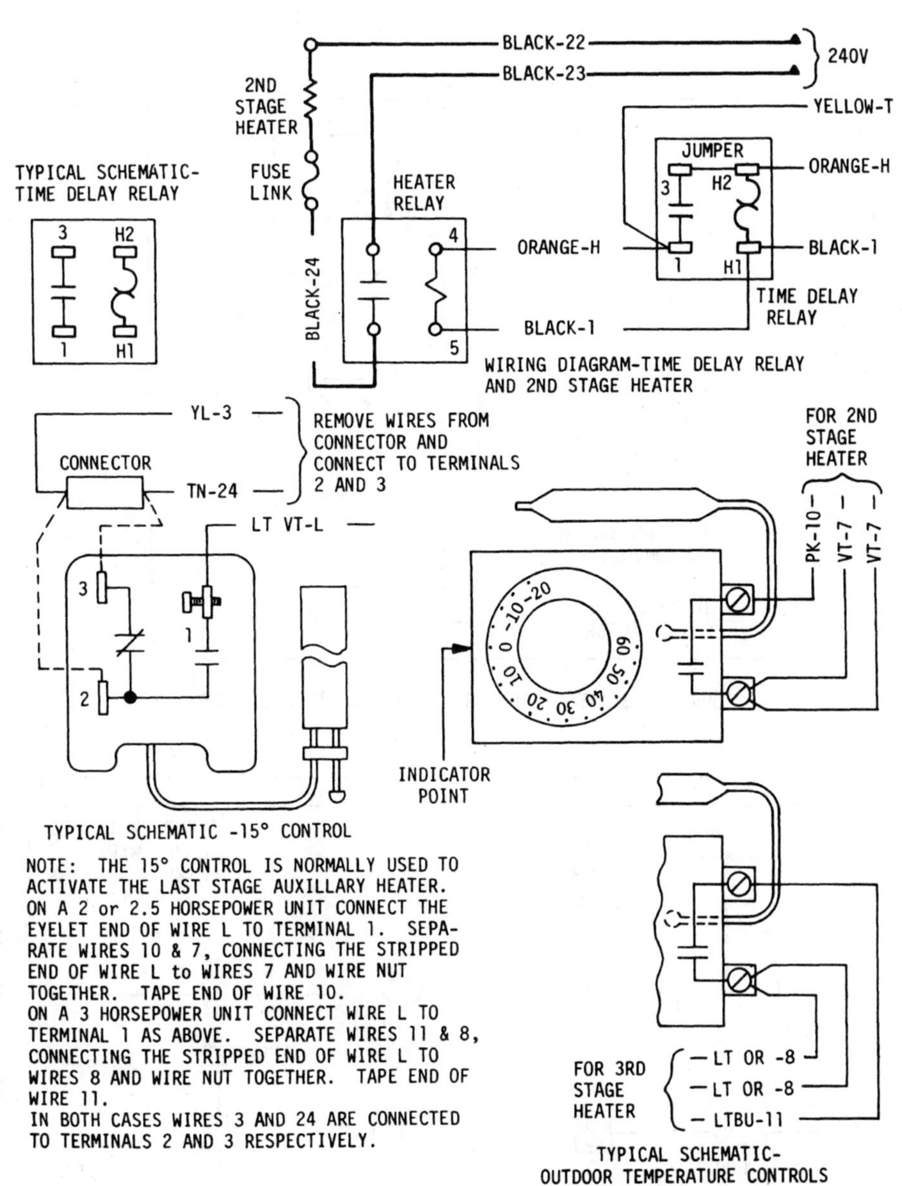

TYPICAL SCHEMATIC -15° CONTROL

NOTE: THE 15° CONTROL IS NORMALLY USED TO ACTIVATE THE LAST STAGE AUXILLARY HEATER. ON A 2 or 2.5 HORSEPOWER UNIT CONNECT THE EYELET END OF WIRE L TO TERMINAL 1. SEPARATE WIRES 10 & 7, CONNECTING THE STRIPPED END OF WIRE L to WIRES 7 AND WIRE NUT TOGETHER. TAPE END OF WIRE 10.
ON A 3 HORSEPOWER UNIT CONNECT WIRE L TO TERMINAL 1 AS ABOVE. SEPARATE WIRES 11 & 8, CONNECTING THE STRIPPED END OF WIRE L TO WIRES 8 AND WIRE NUT TOGETHER. TAPE END OF WIRE 11.
IN BOTH CASES WIRES 3 AND 24 ARE CONNECTED TO TERMINALS 2 AND 3 RESPECTIVELY.

TYPICAL SCHEMATIC-OUTDOOR TEMPERATURE CONTROLS

NOTE: IF ONE OR TWO OUTDOOR TEMPERATURE CONTROLS ARE USED; SEPARATE THE THREE WIRES (PER CONTROL) HELD BY A WIRE NUT, ON THE BULKHEAD WHERE THE CONTROL BOX IS TO BE MOUNTED AND WIRE AS ABOVE.

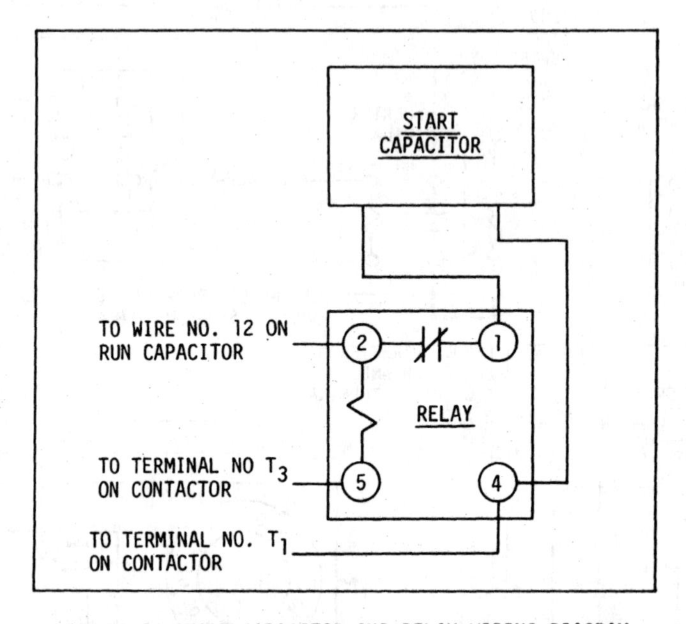

FIGURE 23-START CAPACITOR AND RELAY WIRING DIAGRAM

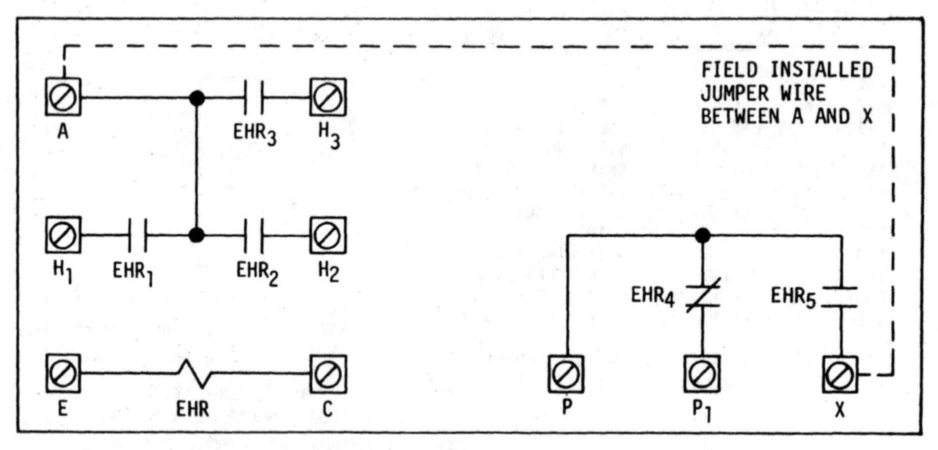

TYPICAL SCHEMATIC-EMERGENCY HEAT RELAY

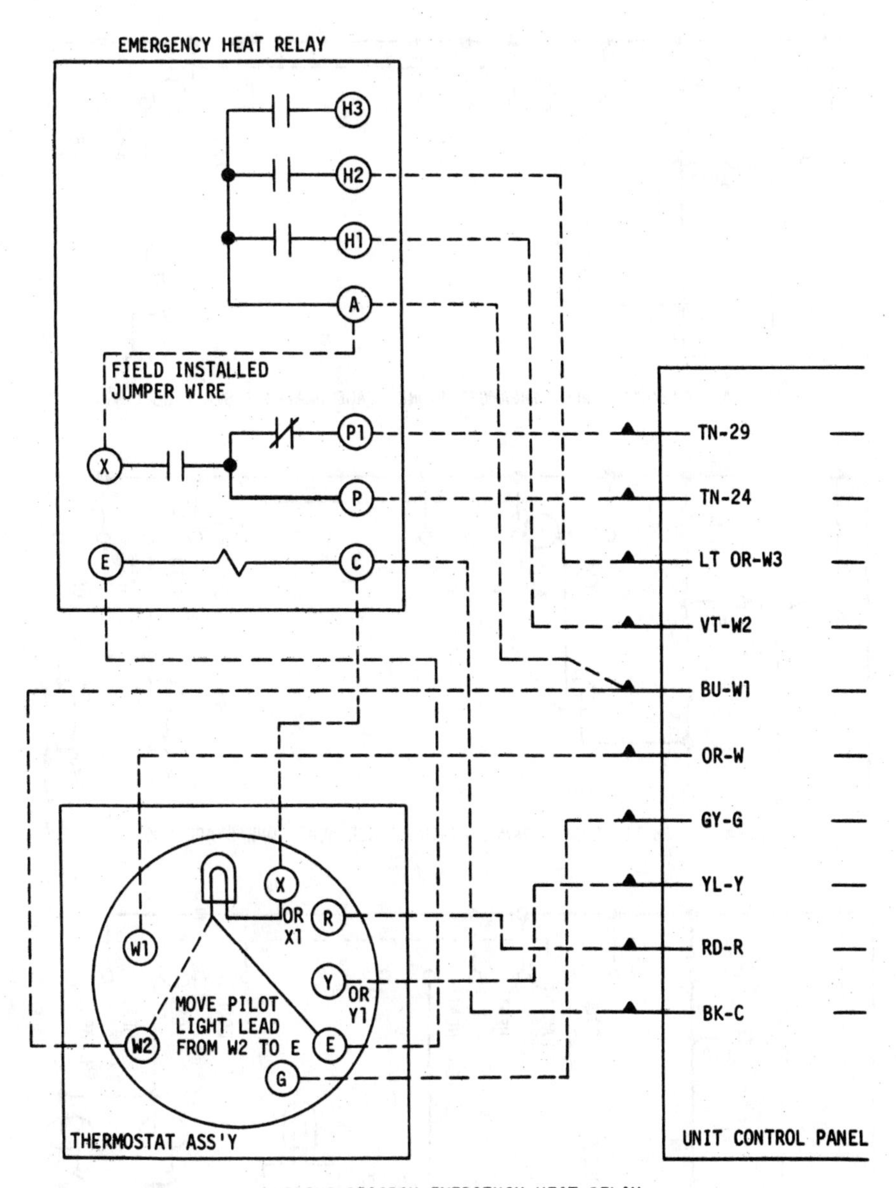

WIRING DIAGRAM-EMERGENCY HEAT RELAY

NOTE: IF OUTDOOR TEMPERATURE OR 15° CONTROLS ARE USED WITH EMERGENCY HEAT RELAY REFER TO TYPICAL SCHEMATIC OF CONTROLS FOR FIELD CONNECTIONS.

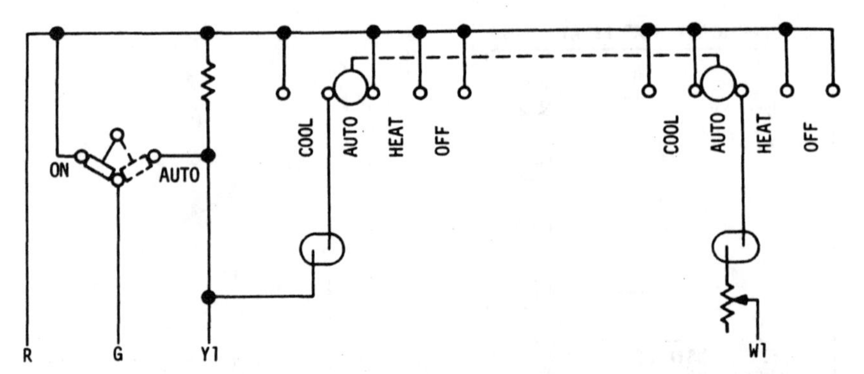

TYPICAL SCHEMATIC- ROOM THERMOSTAT-ONE STAGE HEAT-ONE STAGE COOL

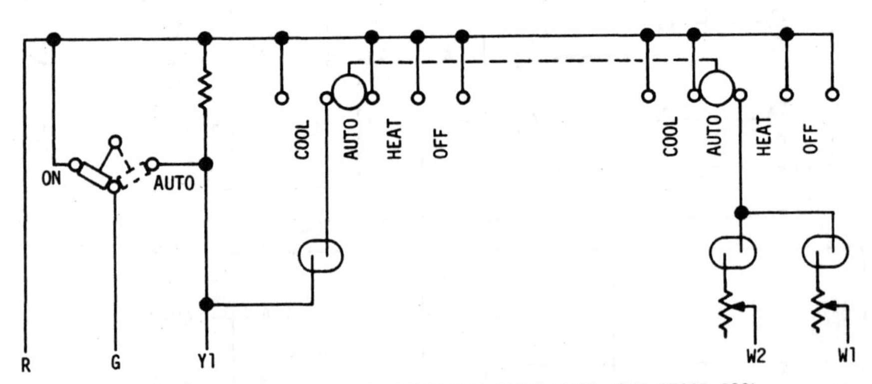

TYPICAL SCHEMATIC-ROOM THERMOSTAT-TWO STAGE HEAT- ONE STAGE COOL

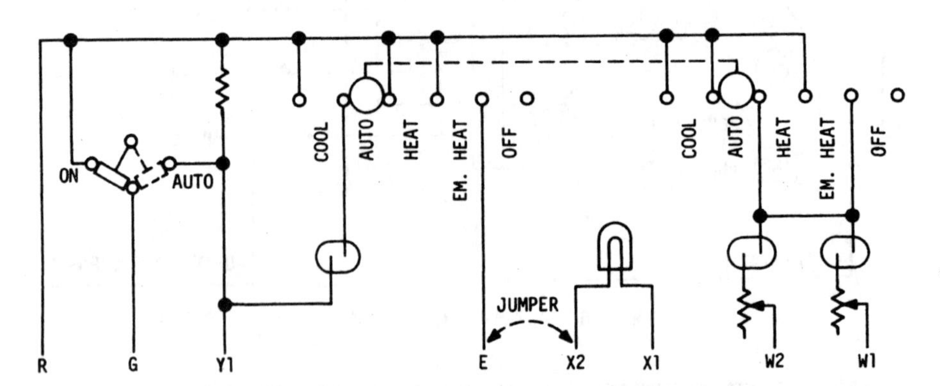

TYPICAL SCHEMATIC-EMERGENCY HEAT ROOM THERMOSTAT-TWO STAGE HEAT- ONE STAGE COOL

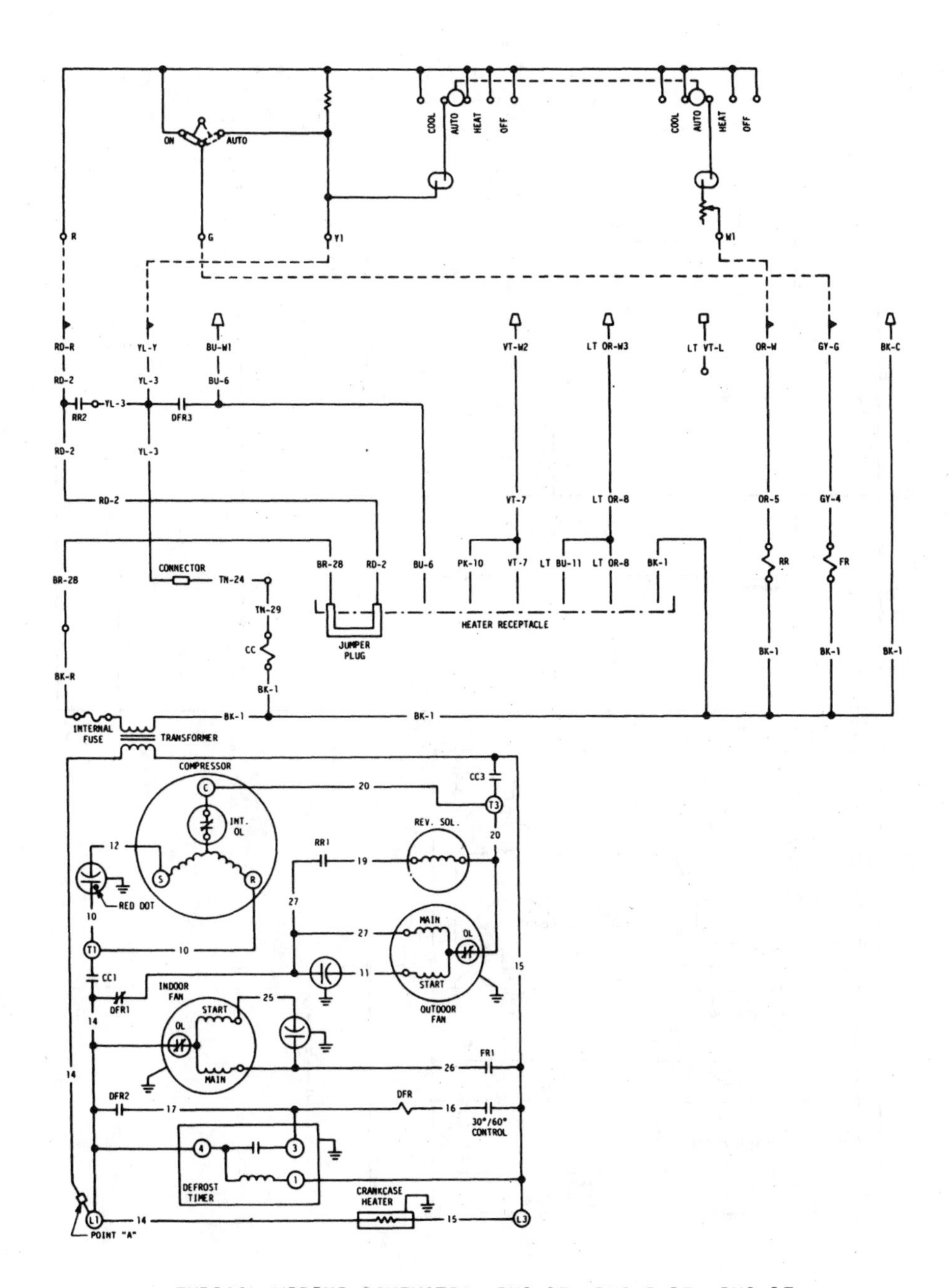

TYPICAL WIRING SCHEMATIC- PH2-1E, PH2.5-1E, PH3-1E

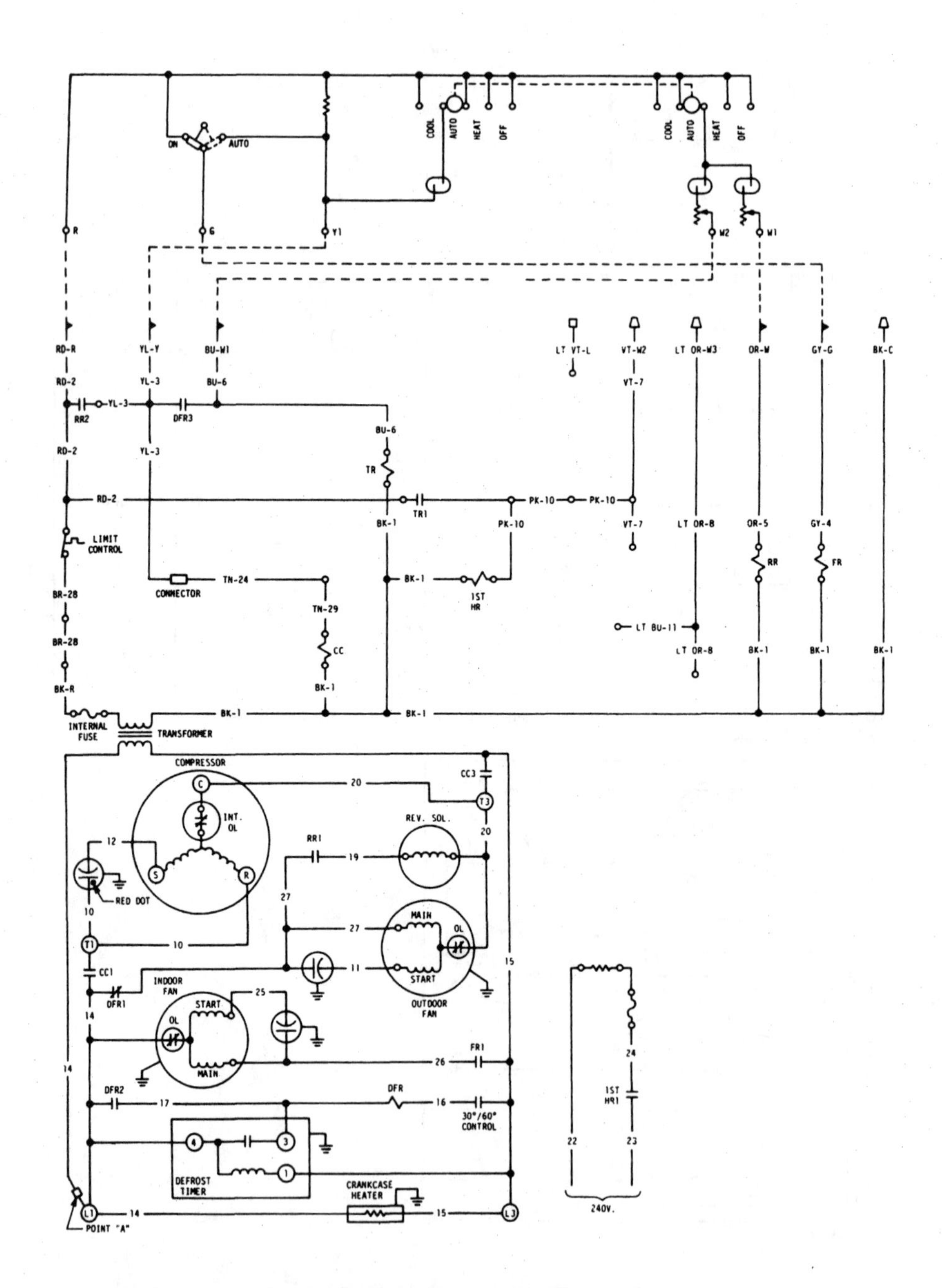

TYPICAL WIRING SCHEMATIC-D54444-1-PH2-1E, PH2.5-1E, PH3-1E

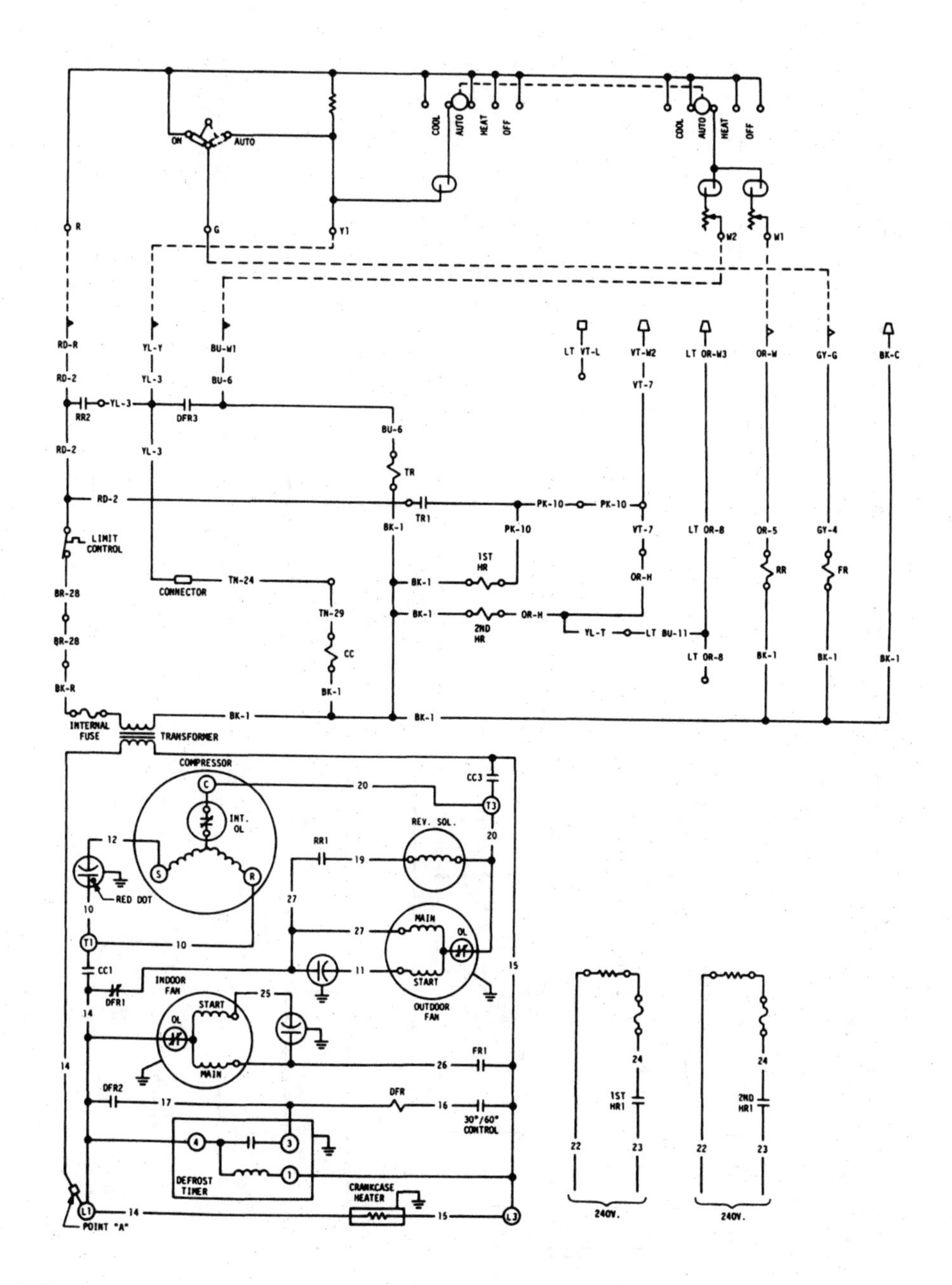

TYPICAL WIRING SCHEMATIC-(1)D54444-1, (1)D54444-2-PH2-1E, PH2.5-1E, PH3-1E

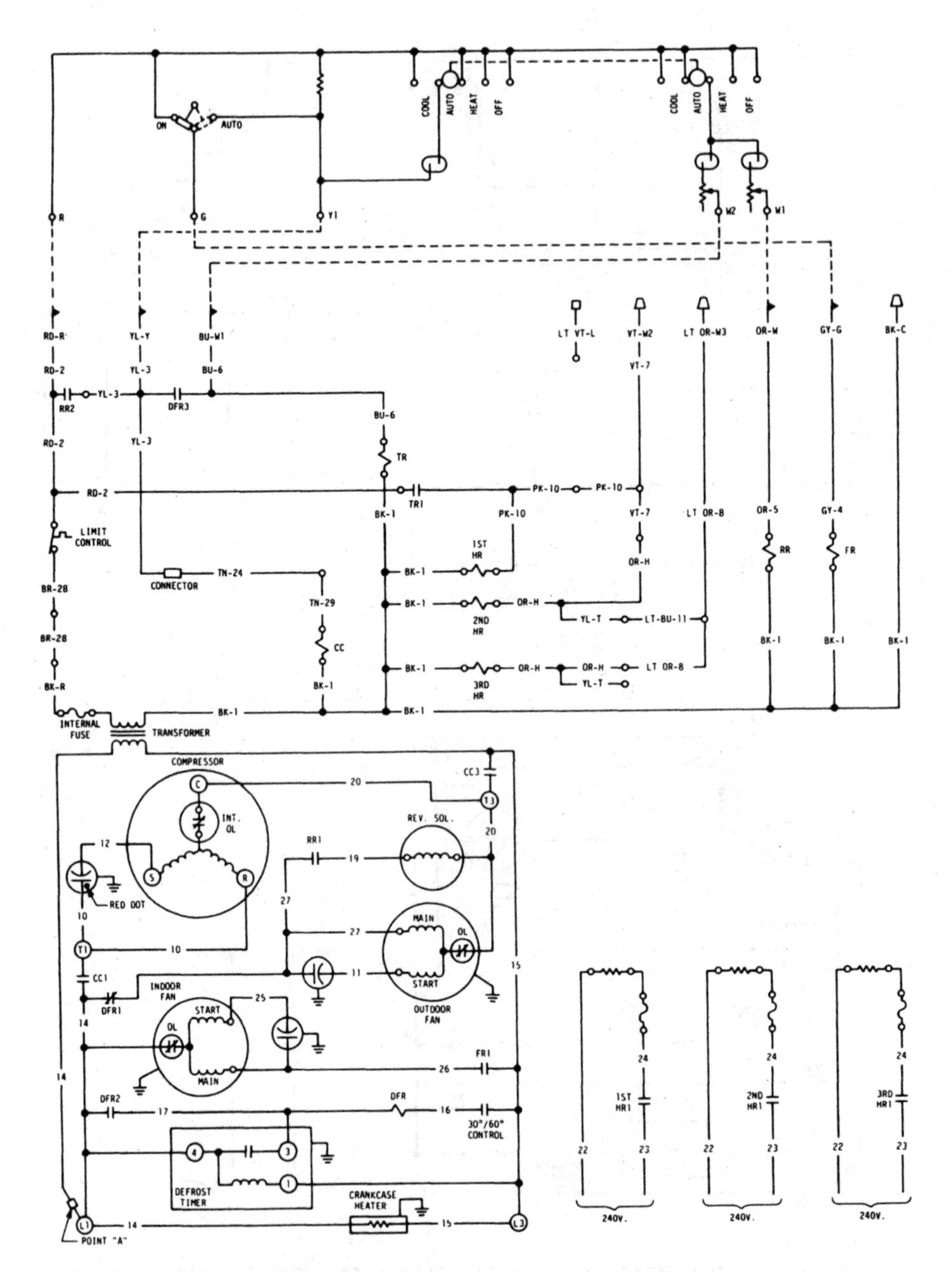

TYPICAL WIRING SCHEMATIC - (1) D54444-1, (2) D54444-2 - PH3-1E

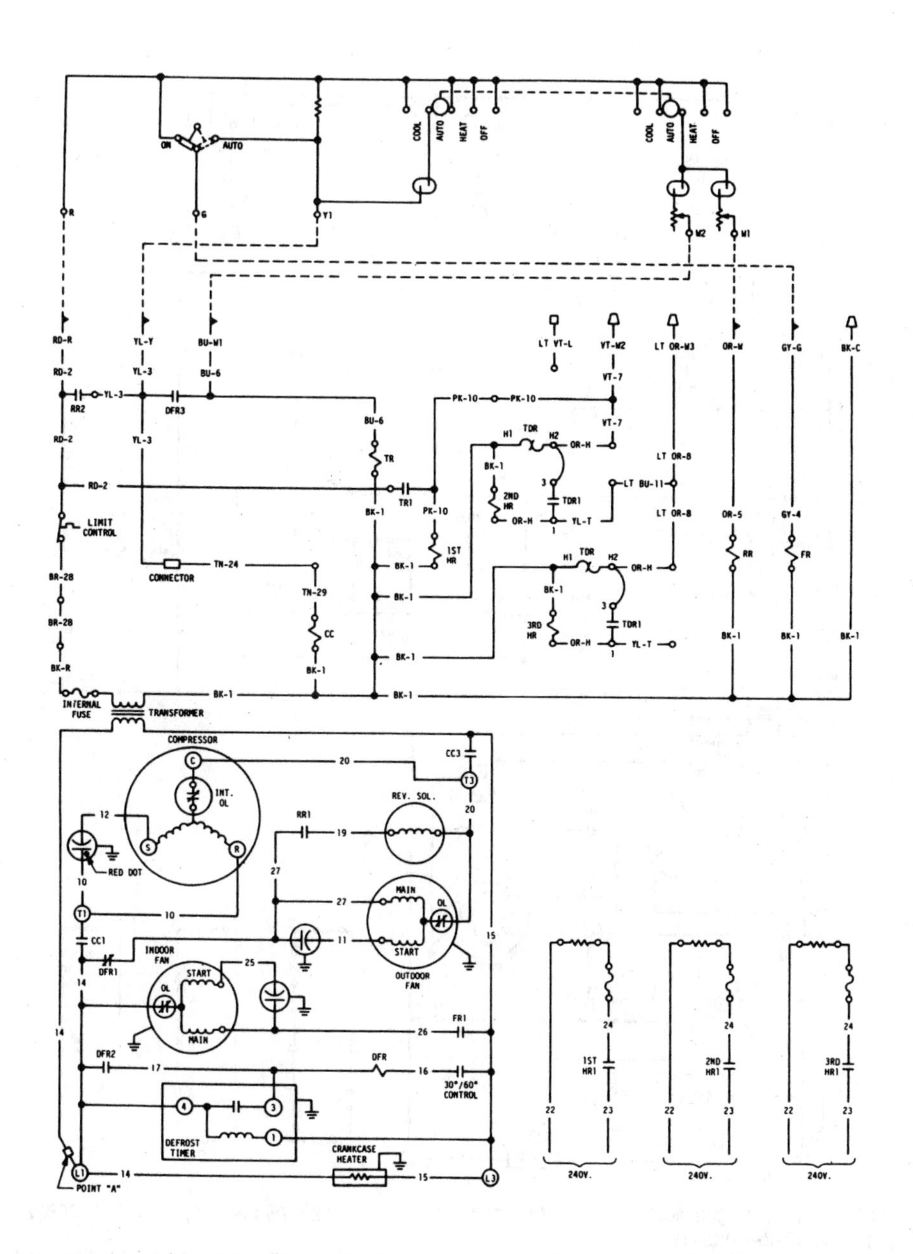

TYPICAL WIRING SCHEMATIC- (1) D54444-1, (2) D54444-2, (2) A46850-1 - PH3-1E
NOTE: WIRING FOR PH2-1E AND PH2.5-1E IS IDENTICAL EXCEPT ONLY 2 HEATERS AND ONE TIME DELAY RELAY MAY BE INSTALLED.

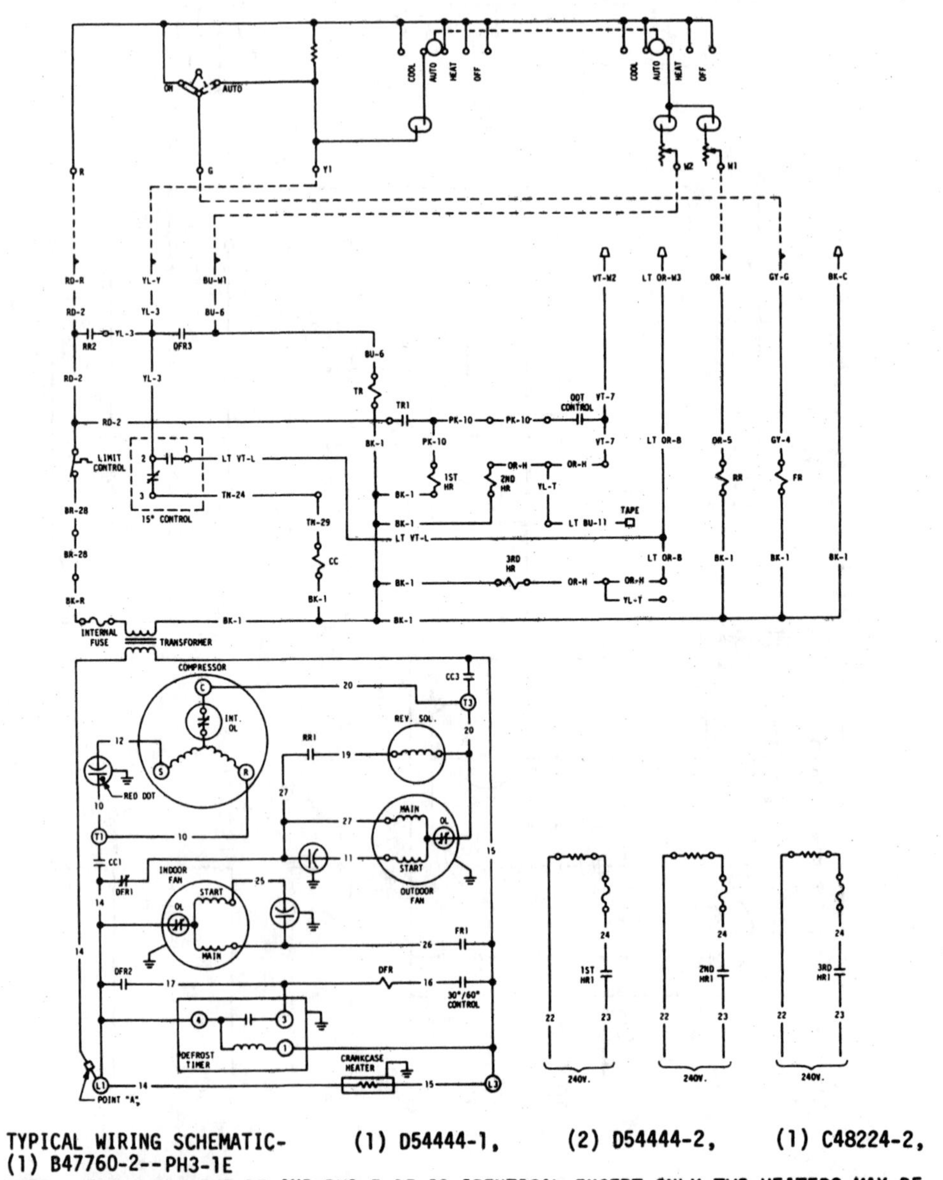

TYPICAL WIRING SCHEMATIC- (1) D54444-1, (2) D54444-2, (1) C48224-2, (1) B47760-2--PH3-1E

NOTE: WIRING FOR PH2-1E AND PH2.5-1E IS IDENTICAL EXCEPT ONLY TWO HEATERS MAY BE USED--THEREFORE EITHER THE 15° CONTROL OR OUTDOOR TEMPERATURE CONTROL MAY BE WIRED WITH THE 2ND STAGE ELECTRIC HEATER.

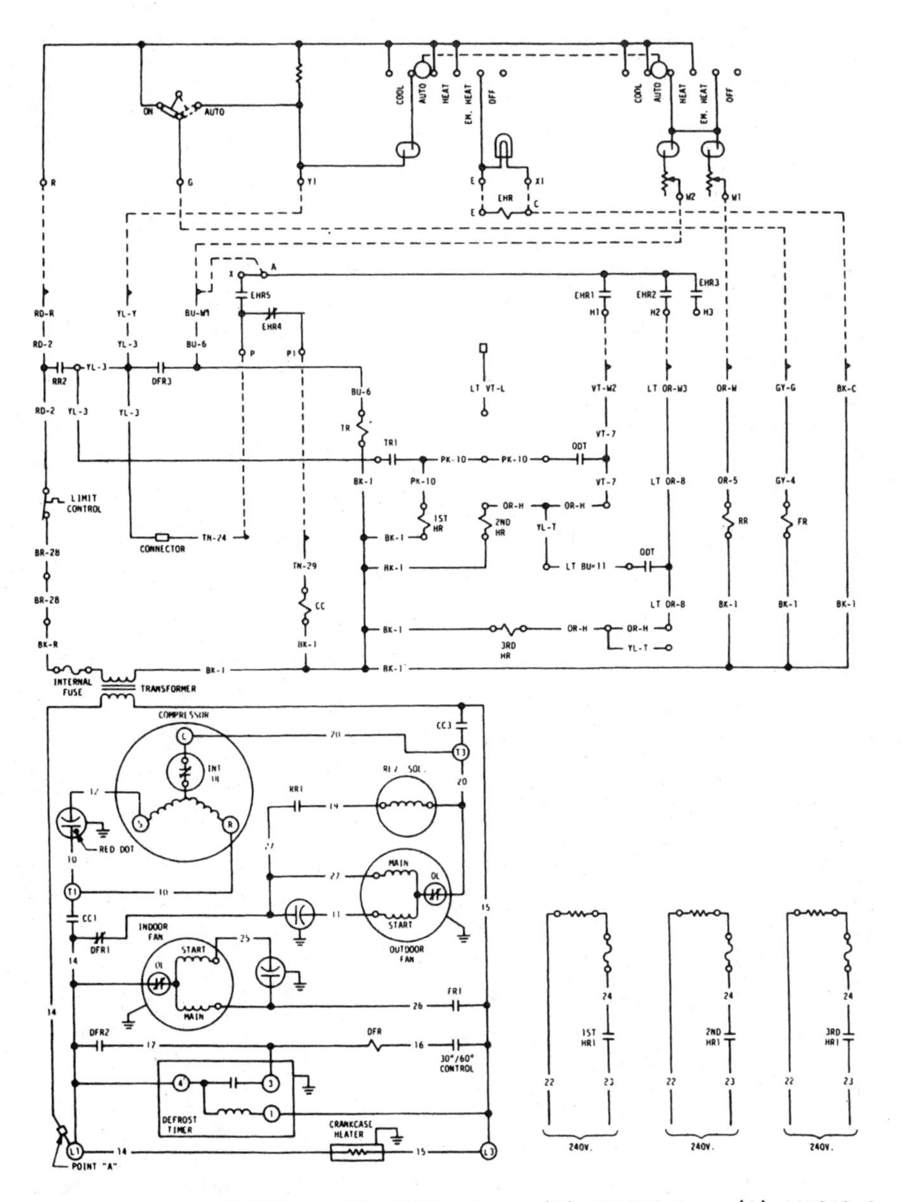

TYPICAL WIRING SCHEMATIC- (1) D54444-1, (2) D54444-2, (1) A46202-2, (2) C48224-2-- PH3-1E

NOTE: WIRING FOR PH2-1E AND PH2.5-1E IS IDENTICAL EXCEPT ONLY TWO HEATERS MAY BE USED.

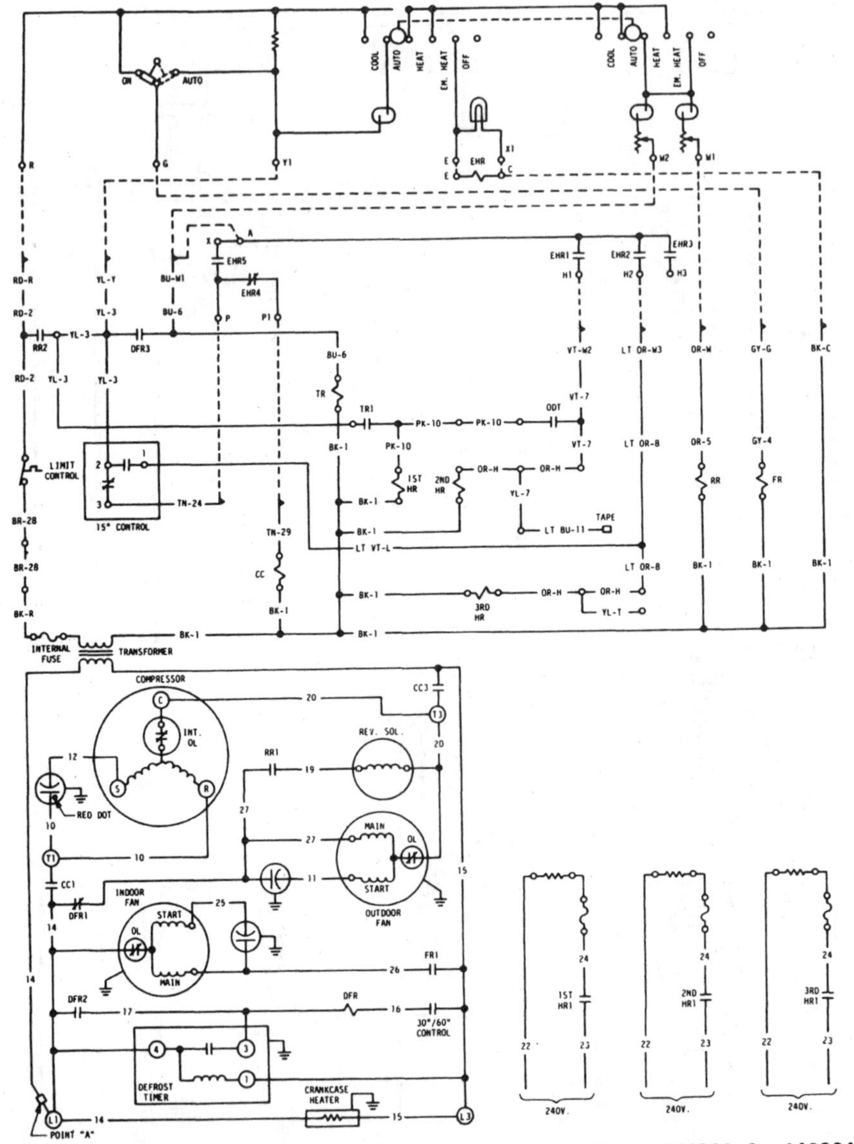

TYPICAL WIRING SCHEMATIC- (1) D54444-1, (2) D54444-2, A46202-2, C48224-2, B47760-2-- PH3-1E

NOTE: WIRING FOR PH2-1E AND PH2.5-1E IS IDENTICAL EXCEPT ONLY TWO HEATERS MAY BE USED--THEREFORE EITHER THE 15° CONTROL OR OUTDOOR TEMPERATURE CONTROL MAY BE WIRED WITH THE 2ND STAGE ELECTRIC HEATER

REPRESENTATIVE HEATWAVE WIRING DIAGRAMS (COURTESY OF SOUTHWEST MANUFACTURING CO.)

General

The following information has been prepared to help the installer save time and obtain the best possible installation to insure continuous trouble-free operation of your forced air electric furnace. Read these instructions and all other instructions on this unit carefully before beginning the installation.

This furnace has been completely assembled and pre-wired at the factory except for thermostat and high voltage connections. It features an easily serviceable slide out heater insert assembly (see Figure 2) which contains the resistance heater strips and controls. The complete blower and motor assembly (see Figure 3) also slides out for easy servicing. Air filter retaining tabs are built into the unit to allow the filter to be mounted in the bottom, back or either right or left sides. Both heating element and blower sections are fully insulated for quiet, efficient operation.

1. Damage in Transit
   a. Upon receipt of shipment of the material, inspect all cartons for external damage. If external damage is noted, open the carton and inspect for damage to equipment. Mark the number of cartons received in this condition on the delivering carrier's bill, and request the services of their inspector.
   b. If upon opening a carton concealed damage is discovered, open the entire shipment and note all equipment so damaged. Contact the delivering carrier and request inspection of the damaged equipment. Do not destroy the carton as the inspector from the freight company will need this to determine reason for damages.

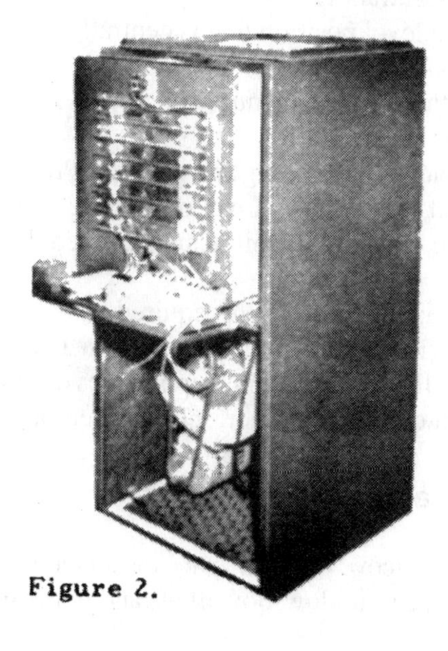

Figure 2.

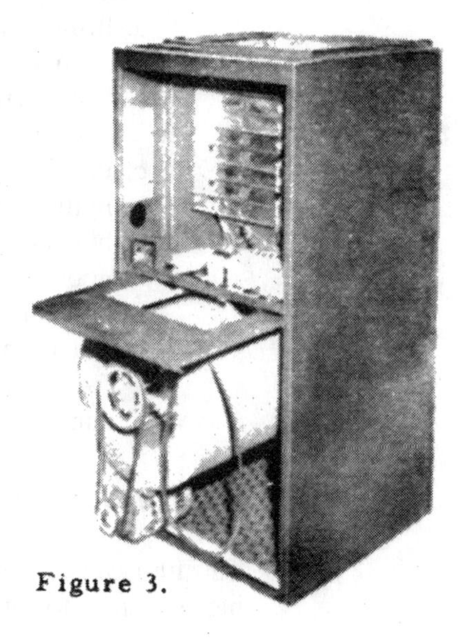

Figure 3.

c. Normally, claims for any and all damages should be filed with the freight company within five working days after receipt of shipment.
d. Since all materials are sold FOB factory, it is the responsibility of the consignee to file claim with the delivering carrier for materials received in a damaged condition.

2. Notice to Installer and Home Owner
   a. Keep This Book. It is intended as a permanent part of your furnace installation and should be affixed on/or adjacent to the furnace.
   b. Install duct work in accordance with Standards of the National Fire Protection Association for the Installation of Air-Conditioning and Ventilating Systems (Pamphlet No. 90A or 90B).
   c. Consult all authorities having jurisdiction before installation is made. Supply wiring must conform to the latest revision of National Electric Code or local code having jurisdiction.
3. Electrical Connections
   The wiring diagram contained in this book and attached to the back of the control access door illustrates complete internal wiring as well as field connections for the power supply and thermostat. Install a separate fuse disconnect switch or multi-breaker near the furnace so the power supply can be turned off for adjustment and servicing. For supply connections, use wire suitable for at least 75° C. (167° F.) and of the size shown on the wiring diagram.

NOTE: The unit must be grounded. If rigid conduit is not used, a separate ground wire must be provided.

4. Selecting Location and Clearances
   a. Install the furnace in a level position that is centrally located for ease in connecting air distribution ducts.
   b. See clearances from the furnace, plenum and ducts to combustible construction, indicated in Chart No. 1.
   c. Provide sufficient space for servicing the furnace. Provide at least 30" in the front of the unit to permit removal of the heater insert assembly, blower section and air filter if the unit is not installed directly behind a door.
   d. A separate non-combustible base *is not* required.
   e. The room thermostat should be centrally located on an inside wall away from any registers, grills, heat producing appliances or direct rays of the sun in order to accurately sense the correct room temperature.

Chart No. 1 Minimum Installation Clearances

This furnace is designed for closet, alcove or crawl space installation with the following clearances to combustible surfaces, upflow, downflow, and horizontal positions.

Furnace front ....0 in.
Furnace back ....0 in.
Furnace sides ....0 in.
Inlet & outlet plenum
top, bottom or sides ....0 in.
Inlet & outlet duct
top, bottom or sides ....0 in.

NOTE: Accessibility clearances for ease of service and installation should take precedence over the minimum clearances to combustible construction shown in the above chart. Refer to paragraph C under section 4, Selecting Location and Clearances, on previous page.

Installation

After checking the furnace for shipping damage and selecting a good central location, the unit is ready to be installed. See the correct wiring diagram, for furnace model number, in this book and on the inside of the control access door. Also see Chart No. 4 on page 351.

IMPORTANT: This furnace is provided with an air filter. When the filter is replaced, the new filter should be of the same type and dimensional size as originally supplied.

A removable metal plate is installed in the bottom of the furnace. If return air is brought in the bottom, remove the metal plate and place filter in the bottom using metal clips furnished on each side to hold filter in place. Discard metal plate. If return air is not brought in the bottom of the unit, the metal plate must not be removed.

When cutting an opening for the return air in either side or back of the furnace casing, emboss marks are provided which locate the corners of the opening.

Summer Cooling Units

When used in connection with a cooling unit, the furnace must be installed parallel with or on the upstream side of the cooling unit to avoid condensation in the furnace. If used with a parallel flow arrangement, the dampers or other means used to control the flow of air must be adequate to prevent chilled air from entering the furnace, and if manually operated, must be equipped with means to prevent operation of either unit unless the damper is in the full heat or cool positions. The Underwriters'

Laboratories, Inc. label pertains to the furnace only. It does not cover any air cooler or evaporator coil that may be integrated with the furnace.

1. Installation Up-Flow Position
   a. Set the unit in the desired location making sure it is reasonably level and resting on a smooth surface.
   b. Secure the return and supply air plenums to the unit with sheet metal screws. Attach an asbestos section to the discharge end to reduce any vibration.
   c. Connect the power supply and low voltage wiring.
2. Installation Down-Flow Position
   a. Turn the unit upside down so the blower is at the top.
   b. Set the unit in the desired location making sure it is reasonably level and resting on a smooth surface.
   c. Secure the return and supply air plenums to the unit with sheet metal screws. Attach an asbestos section to the return air end to reduce any vibration.
   d. Loosen the screws on blower motor rubber end rings and rotate motor until oil cups are in an upright position. Tighten screws.
   e. Connect the power supply and low voltage wiring.
3. Installation Horizontal Position
   a. Rotate the unit 90 degrees and lay it on either side.
   b. Set the unit in the desired location making sure it is reasonably level.
   c. Secure the return air and supply air plenums to the unit with sheet metal screws. Attach an asbestos section to both return and discharge ends to reduce any vibration.
   d. Loosen the screws on blower motor rubber end rings and rotate motor until oil cups are in an upright position. Tighten screws.
   e. Connect the power supply and low voltage wiring.

Operation

After all supply and return air ductwork and wiring has been completed, make sure all supply air registers are open. The unit is now ready to operate.

1. Starting the Furnace
   a. Re-check the wiring diagram supplied with the furnace.
   b. Open disconnect switch to power OFF.
   c. Check all factory and field installed wiring for accuracy and tight connections.
   d. Turn thermostat to lowest setting so the unit will not start on thermostat.
   e. Close disconnect switch to power ON.
   f. Remove thermostat cover and turn thermostat up until first stage contacts close. The blower will start and all heaters in the first stage are energized.

NOTE: If furnace is factory wired for single stage only, the unit will now be operating at full rated capacity. By-pass Steps g & h below and continue with Step i.

g. Continue to turn thermostat up until second stage contacts close. All heaters in the second stage are energized. The unit is now operating at full rated capacity.
h. Reverse Step g. Turn thermostat down until only second stage contacts open. All heaters in the second stage will shut off.
i. Reverse Step f. Turn thermostat down until first stage contacts open. All heaters in the first stage will shut off and the blower will automatically stop.

2. Adjusting Temperature Rise Through Furnace
This furnace has been approved for a temperature rise of 70° F. to 105° F. Adjust the blower speed for the air temperature rise (within the approved range) desired for the particular installation. Make sure furnace doors are in place before proceeding.

To measure this temperature rise through the furnace, place a plenum thermometer in the supply air plenum. Locate the thermometer as close to the outlet air end as possible, making sure it is out of "straight line" sight of the heating elements so it will not register any radiant heat. Place a second plenum thermometer in the return air plenum close to the air filter.

Turn thermostat up to the highest setting to start the unit. After the plenum thermometer has reached its highest reading and become stabilized, subtract the reading on the thermometer in the return air plenum from the reading on the thermometer in the supply air plenum.

If this temperature is low, open the power disconnect switch to the OFF position. Remove the blower access door and open the adjustable motor pulley. If this temperature is high, close the adjustable motor pulley.

NOTE: Adjust motor pulley approximately one-half to one turn at a time until amount of temperature change is determined. Adjust blower belt tension after each pulley change by adjusting 3/8" bolt on motor mount. Belt should be as loose as possible without slipping, approximately 3/4" to 1" deflection.

Install the blower access door and close the power disconnect switch to the ON position. Repeat this procedure until the desired setting within the approved temperature rise range is obtained.

Set the thermostat to the desired room temperature. The furnace is now ready for automatic operation.

With furnace operating, adjust the supply air registers to balance out the amount of air delivery required for the desired temperature in each room.

3. Sequence of Operation of Controls
   a. Sequence Controls
   The control circuit is operated by one sequence control on units with single stage heating and two sequence controls on two stage heating units. The sequence controls are energized by the low voltage circuit. When the thermostat calls for heat, it closes the low voltage circuit and applies 24 volts to the first stage sequence control energizing the 240 volt circuit to the blower motor and heating elements in the first stage. If the unit is equipped with two stage heating and the thermostat is not satisfied by the first stage heat, 24 volts is applied to the second stage sequence control energizing the 240 volt circuit to the second stage heating elements.

   When the room temperature reaches the thermostat setting, the thermostat will open the 24 volt circuit to the second stage sequence control which will open the 240 volt circuit to the heating elements in the second stage. If the room temperature remains sufficiently high enough to keep the thermostat setting satisfied, the thermostat will open the 24 volt circuit to the first stage sequence control which will open the 240 volt circuit to the first stage heating elements and blower motor. The blower will automatically stop.
   b. Limit Control
   This control has been set to maintain a maximum outlet air temperature of less than 200° F. If for any reason the supply air temperature reaches 200° F., the limit control located at the top of the heating elements will open the 240 volt circuit to the transformer. This will in turn open the 24 volt circuit to the thermostat and sequence control(s), causing the 240 volt circuit to the heating elements to open. The heating elements and blower motor will automatically go off. Even if the thermostat is calling for heat.
   c. Auxiliary Limit Control
   A separate auxiliary limit control wired in series with the outlet air limit control is located at the bottom of the heating elements. If for any reason the inside of the heating element or blower compartments should become over-heated, this control will shut the heating elements and blower motor off in the same sequence as described for the outlet air limit control.

   After the limit controls cool sufficiently, they will automatically close the 240 volt and 24 volt circuits. The furnace will again be operational.

   This arrangement provides 100% safe operation of the furnace.

IMPORTANT: If an electrical or mechanical failure should occur, immediately open the disconnect switch or multibreaker to power OFF and contact a qualified serviceman.

Maintenance and Servicing

This furnace is like any piece of machinery in that regular inspection and service will lead to a more satisfactory overall operation and longer life for all parts. Inspect your furnace at least every 60 days during the heating season and check these points.

1. If your furnace should cycle on either of the limit controls, it could be overheating for any of the following reasons:
   a. Dirty filter
   b. Blower belt slipping
   c. Motor pulley not set fast enough
   d. Too much resistance through the unit, such as undersized ductwork, dampers or registers closed or obstructed
2. Filters can and do cause more trouble than anything else in a forced air heating system. Clogged or dirty filters cut down efficiency and lead to poor overall operation.

   When inspecting the furnace, remove the filter from the furnace and clean off surface dirt and lint. Discard the filter immediately if any discoloration shows through from the dirty side to the clean side, as this means the filter has absorbed all of the dirt it possibly can.

   A filter may clog faster in a new home than in an old home. Dust from construction work, sanding floors and lint from new rugs, carpets and blankets can clog a filter within one week. If you notice any decrease in efficiency of your furnace within a few days after it is put into operation, check your filter.

   A forced air furnace does not develop dirt, but rather collects existing dirt in its filters. Dirt on filters is dirt and dust which would otherwise have remained in your home. Filters are truly valuable. Replace them often; the cost is small.

NOTE: When installing air filters, make sure the filter arrows are pointing in the proper direction for air-flow, as indicated on the filter.

3. Motor and Blower
   Lubricate the motor bearings at least every six months with a few drops of good S.A.E. 20 motor oil.

   Occasionally wipe clean the blower housing and wheel of any dust accumulation. Check pulleys, belt and bearings.

NOTE: The blower bearings are permanently self-lubricating. No lubrication is necessary.

The Importance of Correct Belt Tension & Alignment

Literally thousands of field investigations prove that incorrect drive belt tension and alignment is the major cause of bearing and belt failure.

An easy means of checking belt tension is illustrated by Figure 5. When the belt is grasped as shown you should obtain approximately 1" deflection (total) when "finger" pressure is applied. Excessive deflection is an indication of too little tension, leading to belt failure through wear and loss of wheel speed. Too little deflection indicates excessive tension and will surely lead to noisy blower operation, premature bearing failure, and decreased belt life. Bearings and shafting have failed simply because excessive belt tension destroyed the oil film between the bearing and shaft.

Figure 6 shows the recommended method of checking belt and drive alignment. Alignment is equal in importance to tension from both wear and quiet operation considerations. The straight-edge *must* touch at the four points shown to assure good alignment.

Check Belt Tension During Installation

Many furnace and air conditioning units are shipped with belt tension purposely high to help hold the motor in position during shipment. We *strongly recommend* that belt tension be checked carefully during installation of the equipment to prevent expensive call-backs and customer complaints.

Figure 5.

Figure 6.

ELB-35 1Ø ELECTRIC FURNACE W/MH SEQUENCERS

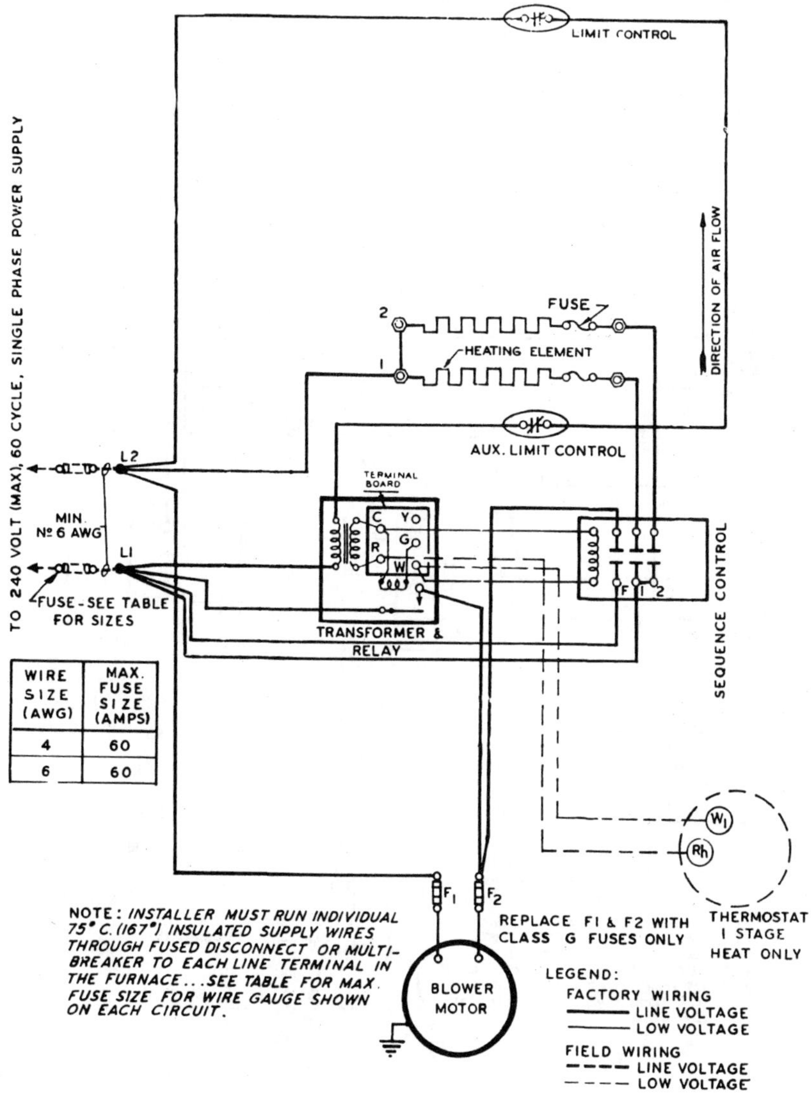

LIMIT CONTROL

DIRECTION OF AIR FLOW

FUSE

HEATING ELEMENT

AUX. LIMIT CONTROL

TERMINAL BOARD

TO 240 VOLT (MAX), 60 CYCLE, SINGLE PHASE POWER SUPPLY

L2

MIN. № 6 AWG

L1

FUSE - SEE TABLE FOR SIZES

TRANSFORMER & RELAY

SEQUENCE CONTROL

WIRE SIZE (AWG)	MAX. FUSE SIZE (AMPS)
4	60
6	60

THERMOSTAT 1 STAGE HEAT ONLY

REPLACE F1 & F2 WITH CLASS G FUSES ONLY

BLOWER MOTOR

NOTE: *INSTALLER MUST RUN INDIVIDUAL 75° C. (167°) INSULATED SUPPLY WIRES THROUGH FUSED DISCONNECT OR MULTI-BREAKER TO EACH LINE TERMINAL IN THE FURNACE... SEE TABLE FOR MAX. FUSE SIZE FOR WIRE GAUGE SHOWN ON EACH CIRCUIT.*

LEGEND:

FACTORY WIRING
- ——— LINE VOLTAGE
- ——— LOW VOLTAGE

FIELD WIRING
- - - - - LINE VOLTAGE
- - - - - LOW VOLTAGE

FORM № 0497-163

ELB-50 1Ø ELECTRIC FURNACE W/MH SEQUENCERS

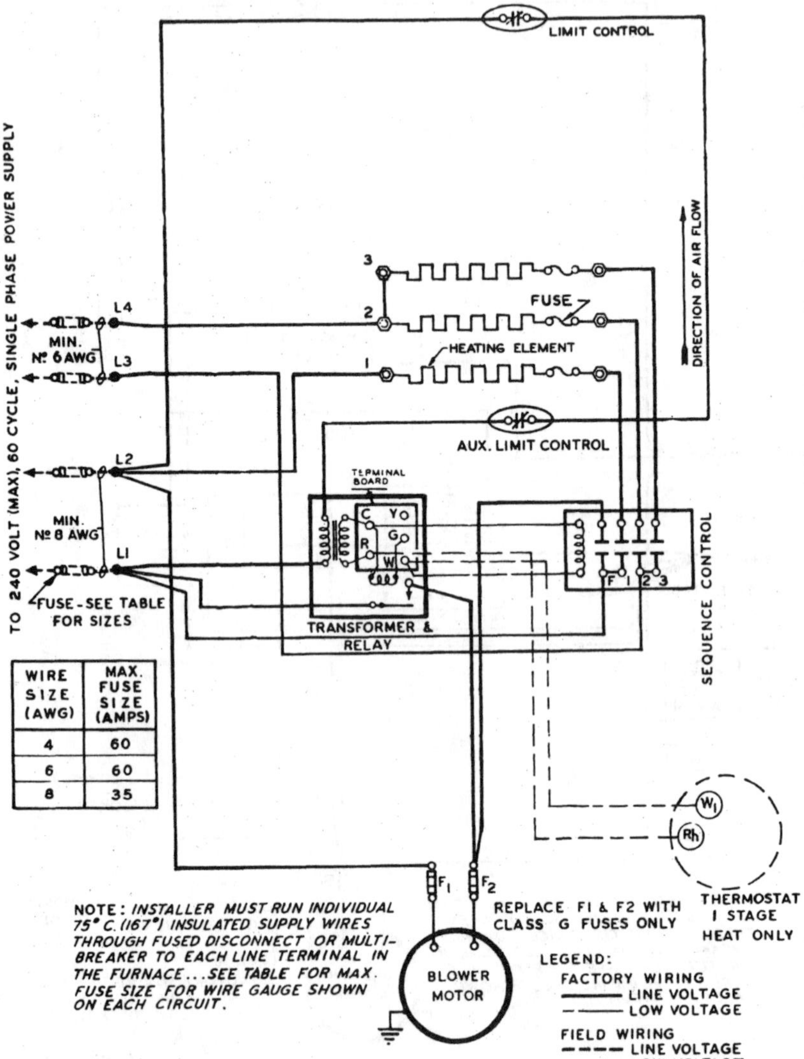

LIMIT CONTROL

TO 240 VOLT (MAX), 60 CYCLE, SINGLE PHASE POWER SUPPLY

L4

MIN. Nº 6 AWG

L3

L2

MIN. Nº 8 AWG

L1

FUSE - SEE TABLE FOR SIZES

3

2

1

FUSE

HEATING ELEMENT

DIRECTION OF AIR FLOW

AUX. LIMIT CONTROL

TERMINAL BOARD

C Y G R W

TRANSFORMER & RELAY

F 1 2 3

SEQUENCE CONTROL

WIRE SIZE (AWG)	MAX. FUSE SIZE (AMPS)
4	60
6	60
8	35

W_1

Rh

THERMOSTAT
1 STAGE
HEAT ONLY

F_1 F_2

NOTE: *INSTALLER MUST RUN INDIVIDUAL 75° C. (167°) INSULATED SUPPLY WIRES THROUGH FUSED DISCONNECT OR MULTI-BREAKER TO EACH LINE TERMINAL IN THE FURNACE... SEE TABLE FOR MAX. FUSE SIZE FOR WIRE GAUGE SHOWN ON EACH CIRCUIT.*

REPLACE F1 & F2 WITH CLASS G FUSES ONLY

BLOWER MOTOR

LEGEND:
FACTORY WIRING
——— LINE VOLTAGE
——— LOW VOLTAGE
FIELD WIRING
- - - - LINE VOLTAGE
- - - - LOW VOLTAGE

FORM Nº 0497-164

ELB-70 1Ø ELECTRIC FURNACE W/MH SEQUENCERS

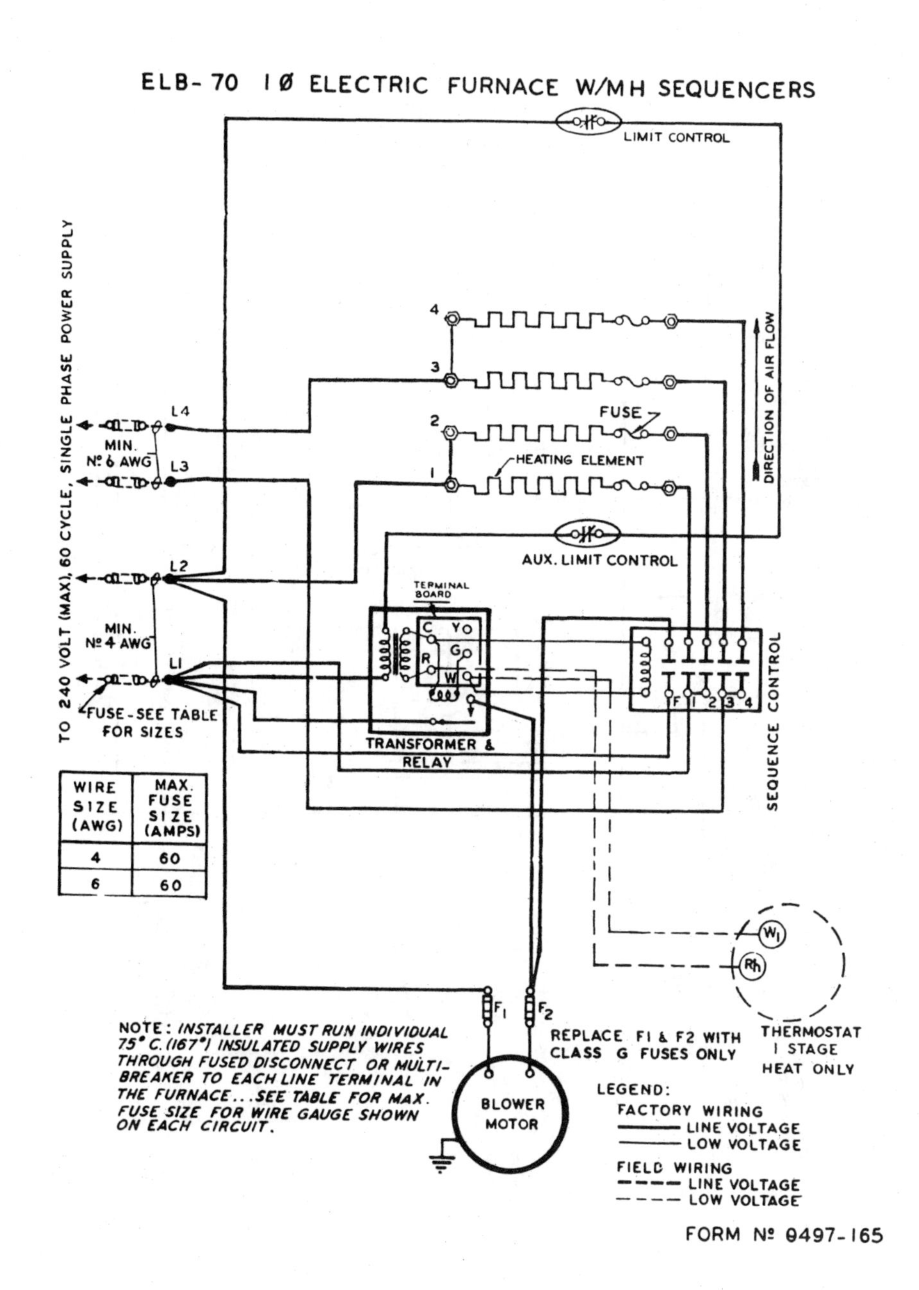

WIRE SIZE (AWG)	MAX. FUSE SIZE (AMPS)
4	60
6	60

FORM Nº 0497-165

ELB-85 1 Ø ELECTRIC FURNACE W/MH SEQUENCERS

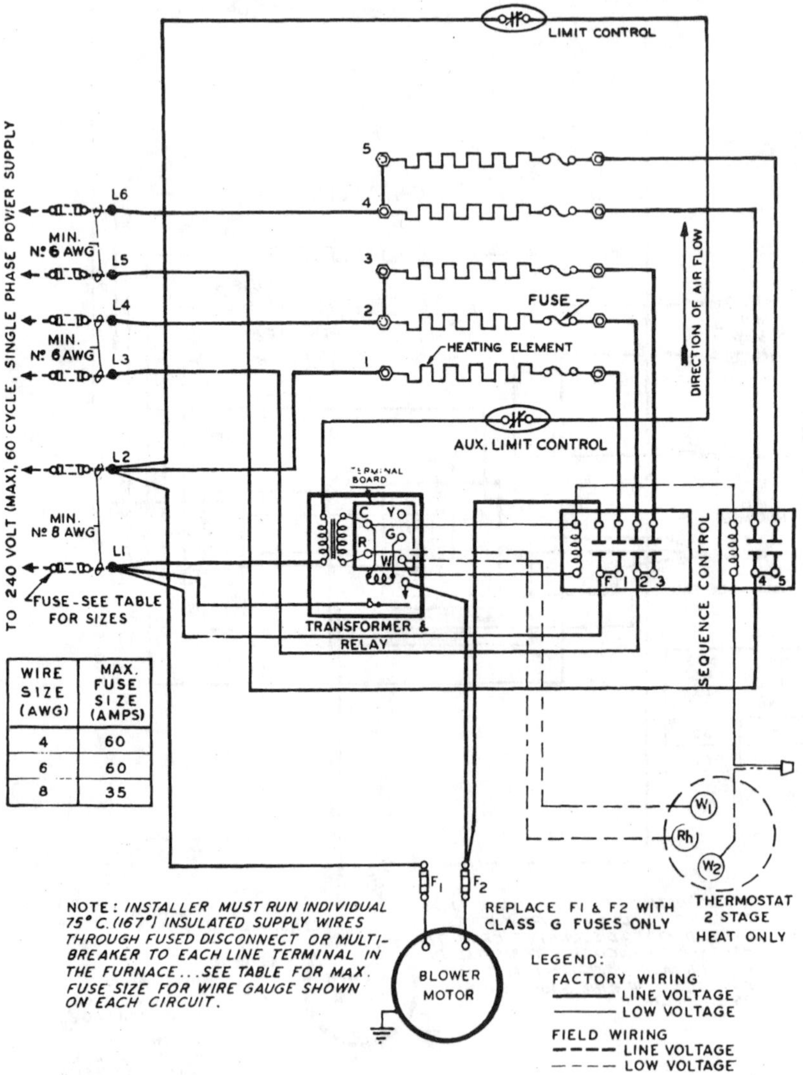

LIMIT CONTROL

TO 240 VOLT (MAX), 60 CYCLE, SINGLE PHASE POWER SUPPLY

L6
MIN. Nº 6 AWG
L5
L4
MIN. Nº 6 AWG
L3
L2
MIN. Nº 8 AWG
L1
FUSE-SEE TABLE FOR SIZES

5
4
3
2
1
FUSE
HEATING ELEMENT
DIRECTION OF AIR FLOW
AUX. LIMIT CONTROL

TERMINAL BOARD
C YO
R G
W
TRANSFORMER & RELAY

F 1 2 3
4 5
SEQUENCE CONTROL

WIRE SIZE (AWG)	MAX. FUSE SIZE (AMPS)
4	60
6	60
8	35

W1
Rh
W2
THERMOSTAT 2 STAGE HEAT ONLY

F1
F2
REPLACE F1 & F2 WITH CLASS G FUSES ONLY

BLOWER MOTOR

NOTE: *INSTALLER MUST RUN INDIVIDUAL 75° C. (167°) INSULATED SUPPLY WIRES THROUGH FUSED DISCONNECT OR MULTI-BREAKER TO EACH LINE TERMINAL IN THE FURNACE... SEE TABLE FOR MAX. FUSE SIZE FOR WIRE GAUGE SHOWN ON EACH CIRCUIT.*

LEGEND:
FACTORY WIRING
LINE VOLTAGE
LOW VOLTAGE
FIELD WIRING
LINE VOLTAGE
LOW VOLTAGE

FORM Nº 0497-166

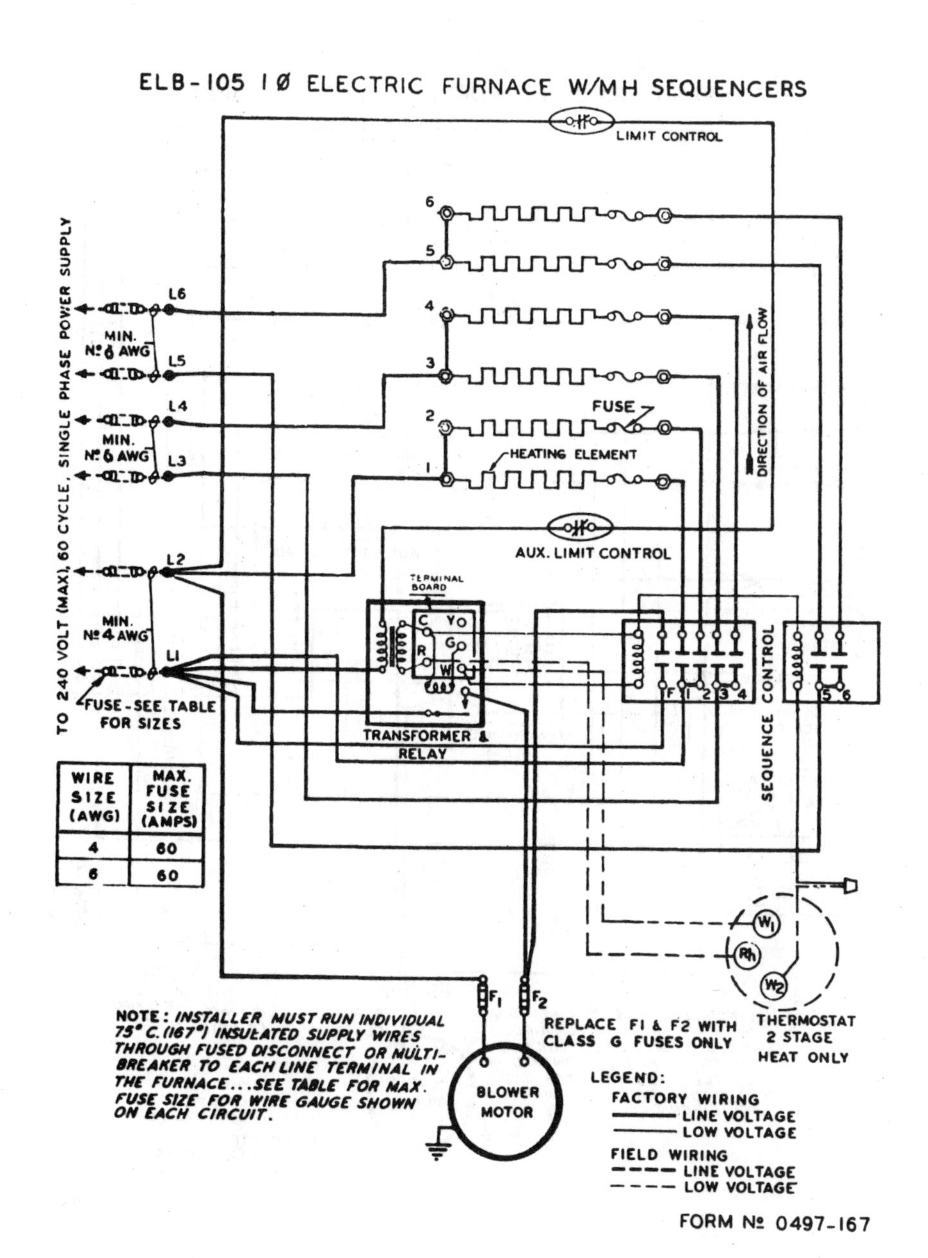

WIRE SIZE (AWG)	MAX. FUSE SIZE (AMPS)
4	60
6	60

ELB-120 1 Ø ELECTRIC FURNACE W/MH SEQUENCERS

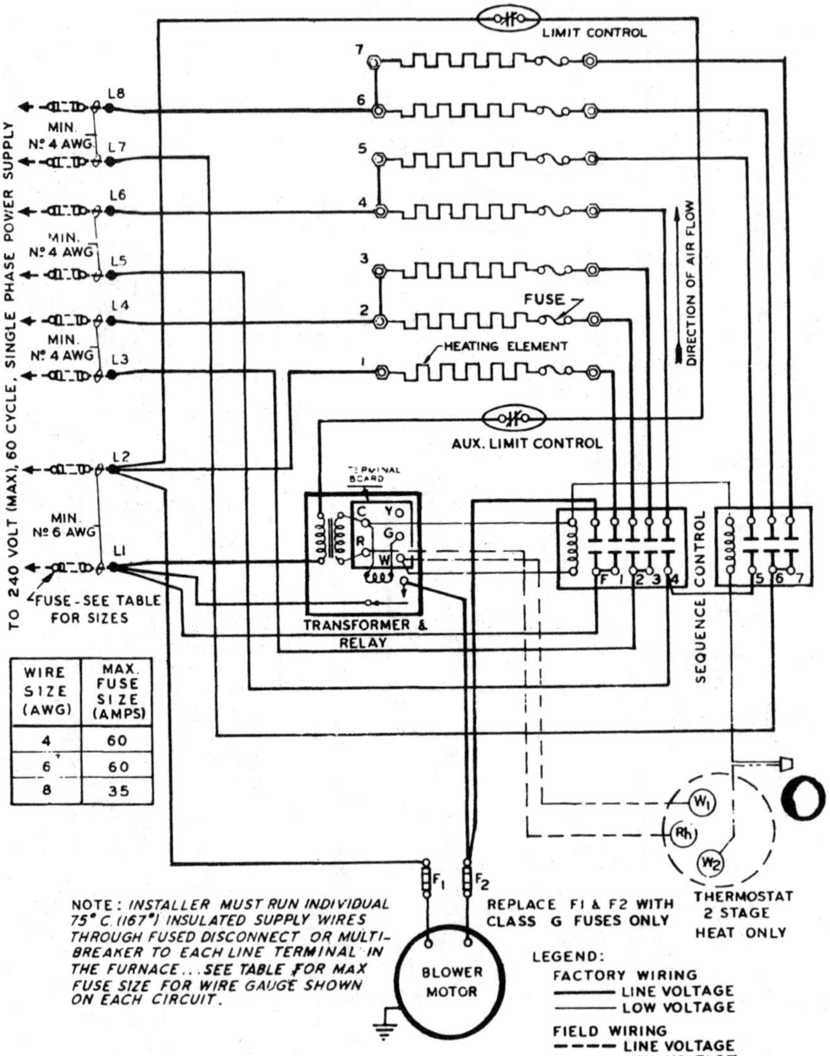

LIMIT CONTROL

7
6
5
4
3
2
1

FUSE

HEATING ELEMENT

DIRECTION OF AIR FLOW

AUX. LIMIT CONTROL

L8
MIN.
Nº 4 AWG
L7
L6
MIN.
Nº 4 AWG
L5
L4
MIN.
Nº 4 AWG
L3
L2
MIN.
Nº 6 AWG
L1

TO 240 VOLT (MAX), 60 CYCLE, SINGLE PHASE POWER SUPPLY

FUSE-SEE TABLE FOR SIZES

TERMINAL BOARD

C Y G R W

TRANSFORMER & RELAY

F 1 2 3 4

5 6 7

SEQUENCE CONTROL

WIRE SIZE (AWG)	MAX. FUSE SIZE (AMPS)
4	60
6	60
8	35

W1
Rh
W2

THERMOSTAT
2 STAGE
HEAT ONLY

F1
F2

REPLACE F1 & F2 WITH CLASS G FUSES ONLY

NOTE: *INSTALLER MUST RUN INDIVIDUAL 75° C (167°) INSULATED SUPPLY WIRES THROUGH FUSED DISCONNECT OR MULTI-BREAKER TO EACH LINE TERMINAL IN THE FURNACE...SEE TABLE FOR MAX FUSE SIZE FOR WIRE GAUGE SHOWN ON EACH CIRCUIT.*

BLOWER MOTOR

LEGEND:
FACTORY WIRING
LINE VOLTAGE
LOW VOLTAGE
FIELD WIRING
LINE VOLTAGE
LOW VOLTAGE

FORM Nº 0497-168

Repair Parts

Parts may be purchased from your local dealer or ordered from:

Southwest Manufacturing
P.O. Box 151 · Aurora, Missouri 65605

To have orders filled promptly and correctly, please furnish the following information with your order: model number and serial number of your furnace as shown on the name plate as well as the name and number shown on the part being ordered.

Chart No. 4

Install one-stage heating thermostat (with models ELB-35, 50 & *70) and set adjustable heat anticipator at .65 amps. (*Available in two-stage on special order.)

Install two-stage heating thermostat (with models ELB-85, 105 & 120) and set adjustable heat anticipator on stage one at .65 amps and on stage two at .65 amps.

Furnace Model No.	Branch Circuit		AWG Wire Size Power	Max. Fuse or Multi Size (Amp)	Staging
	No.	Description			
ELB-35-23	1	Motor and Heater 1 & 2	(2) 6	60	Single Stage
ELB-50-23, ELB-50-4 & ELB-50-5	1	Motor and Heater 1	(2) 8	35	
	2	Heater 2 & 3	(2) 6	60	Single Stage
ELB-70-23, ELB-70-4 & ELB-70-5	1	Motor and Heater 1 & 2	(2) 4	60	
	2	Heater 3 & 4	(2) 6	60	Single Stage
ELB-85-23, ELB-85-4 & ELB-85-5	1	Motor and Heater 1	(2) 8	35	Stage one Blower & Htrs., 1-2-3
	2	Heater 2 & 3	(2) 6	60	
	3	Heater 4 & 5	(2) 6	60	Stage two Htrs., 4-5
ELB-105-23, ELB-105-4 & ELB-105-5	1	Motor and Heater 1 & 2	(2) 4	60	Stage one Blower & Htrs., 1-2-3-4
	2	Heater 3 & 4	(2) 6	60	
	3	Heater 5 & 6	(2) 6	60	Stage two Htrs., 5-6
ELB-120-23 ELB-120-4 & ELB-120-5	1	Motor and Heater 1	(2) 6	35	Stage one Blowers & Htrs., 1-2-3-4
	2	Heater 2 & 3	(2) 4	60	
	3	Heater 4 & 5	(2) 4	60	Stage two Htrs., 5-6-7
	4	Heater 6 & 7	(2) 4	60	

WIRING ARRANGEMENT
FOR HEATING & AIR CONDITIONING
WITH SERIES "ELB" FURNACES AND
CONDENSING UNITS WITHOUT TRANSFORMERS

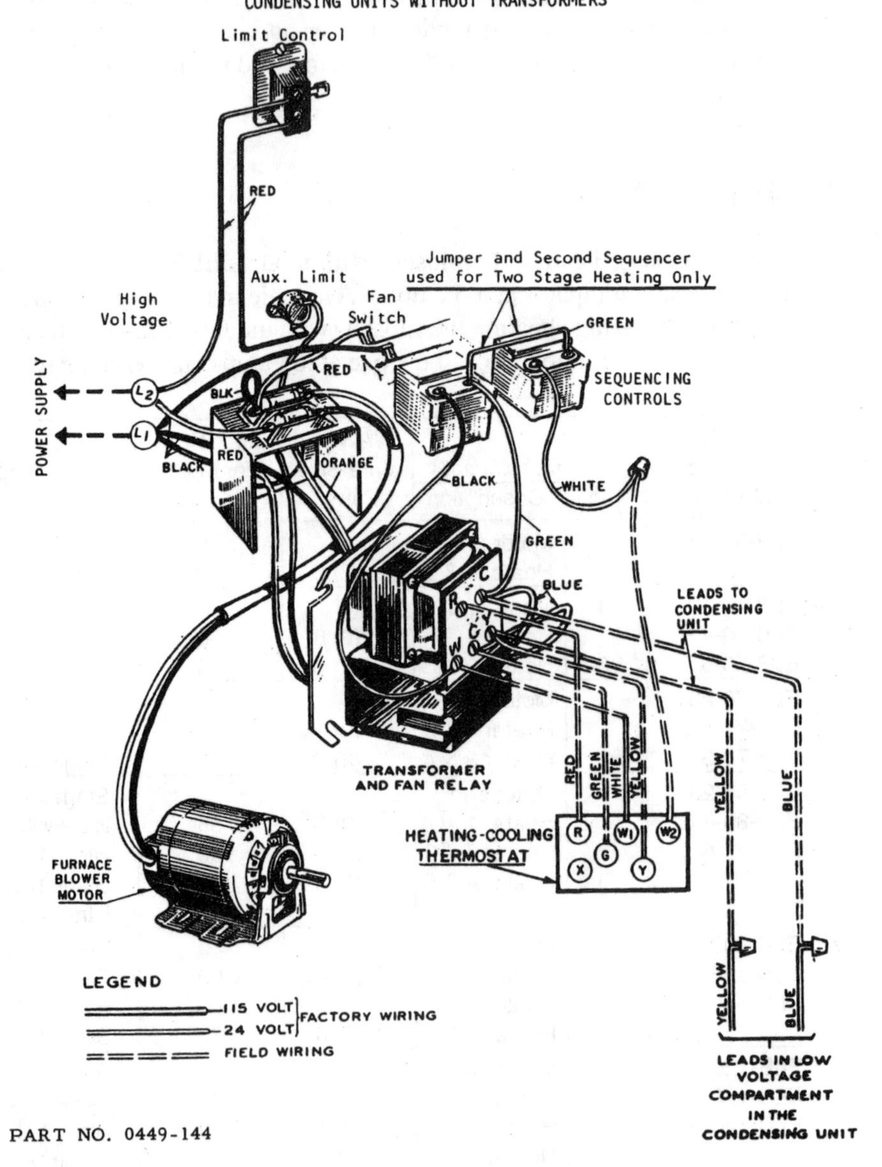

PART NO. 0449-144

115/208 - 230 V. 60 Hz/1 PH., 20 AMP. TIME DELAY FUSES

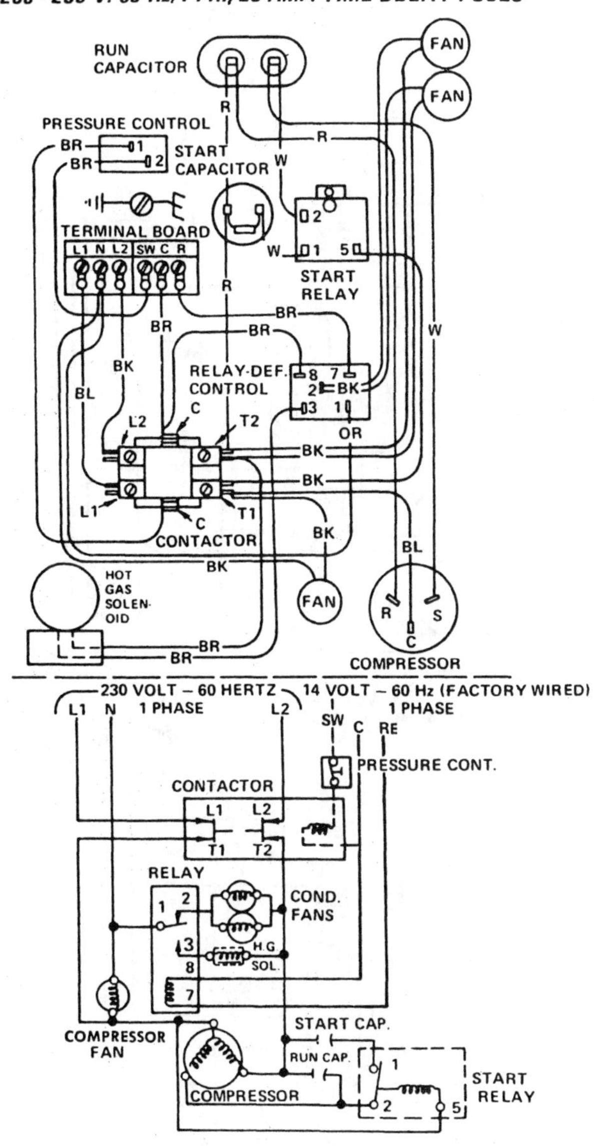

115/208 - 230 V. 60 Hz/1 PH., 20 AMP. TIME DELAY FUSES

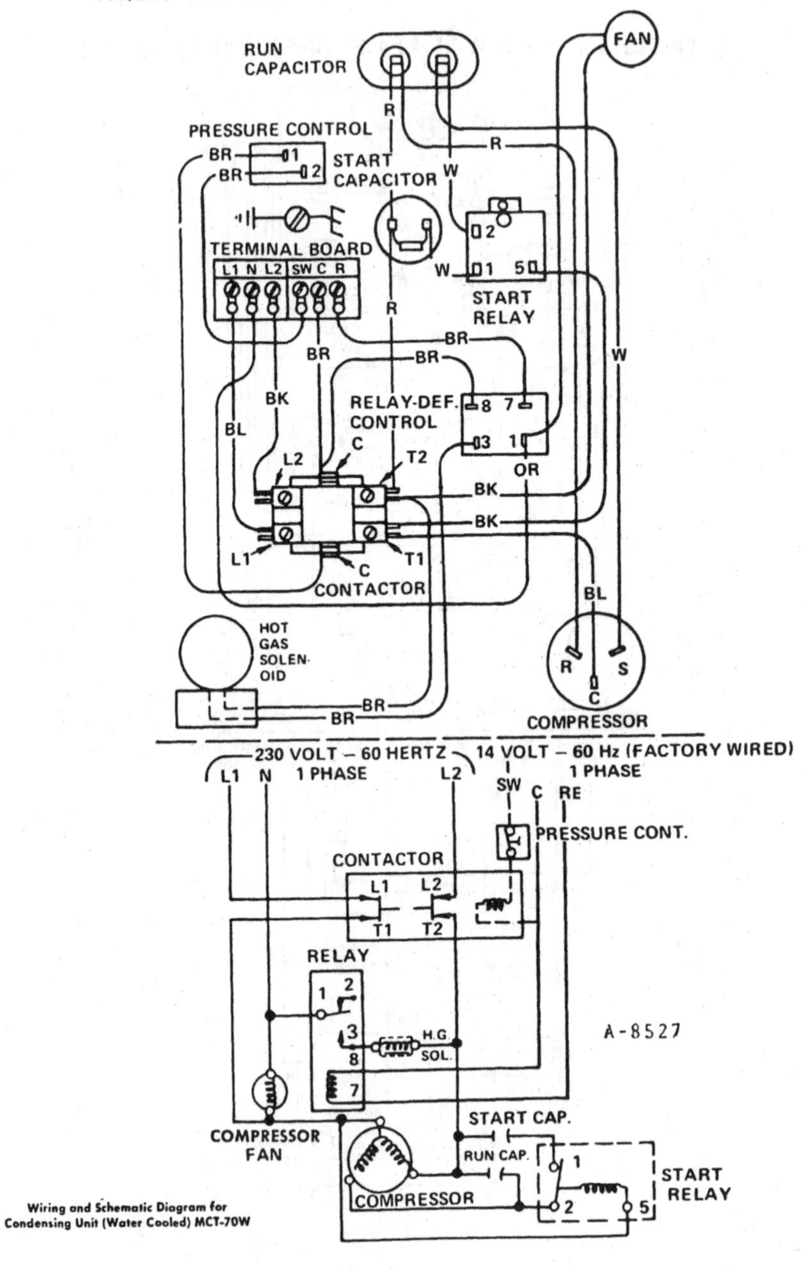

Wiring and Schematic Diagram for Condensing Unit (Water Cooled) MCT-70W

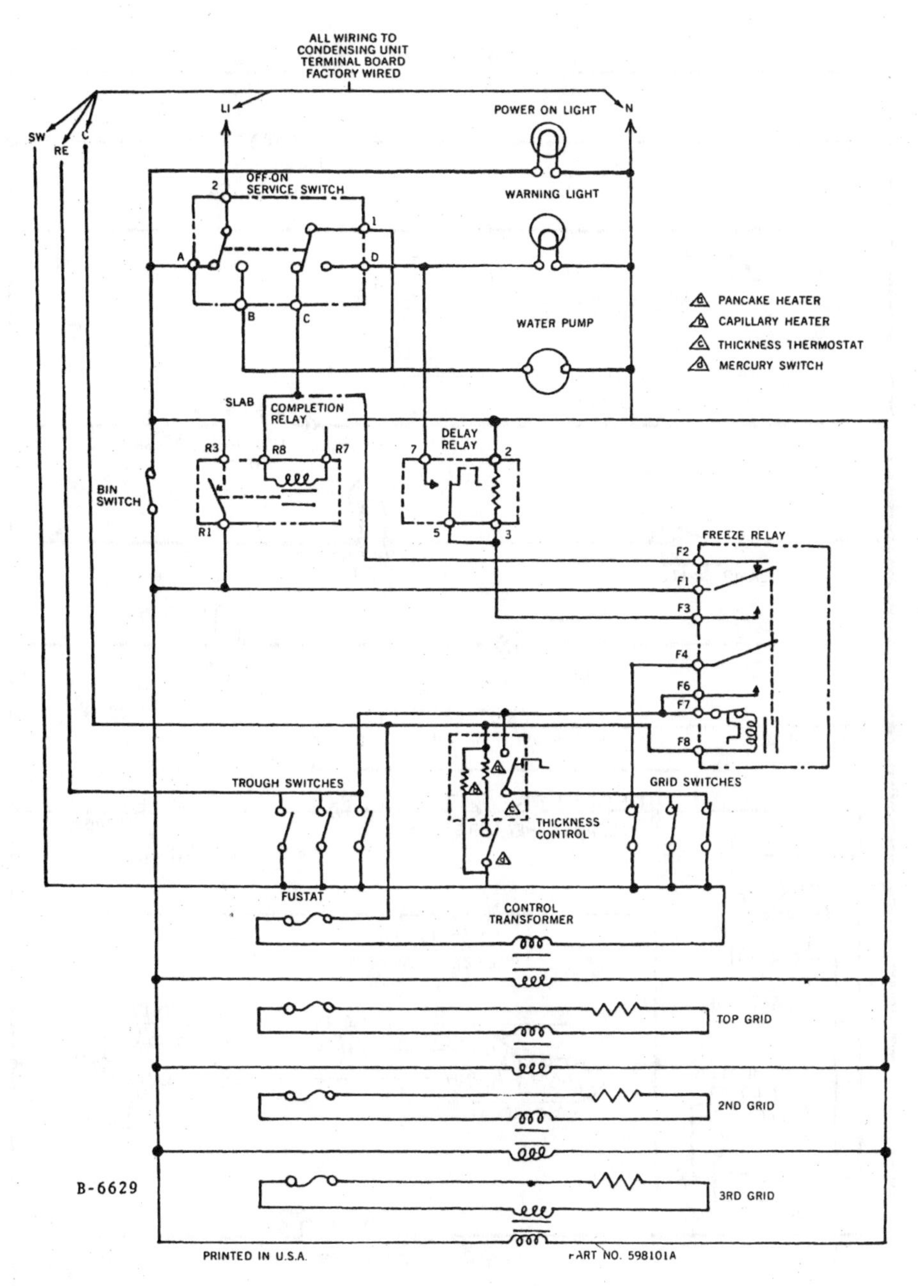

Schematic Diagram MCT-70A&W

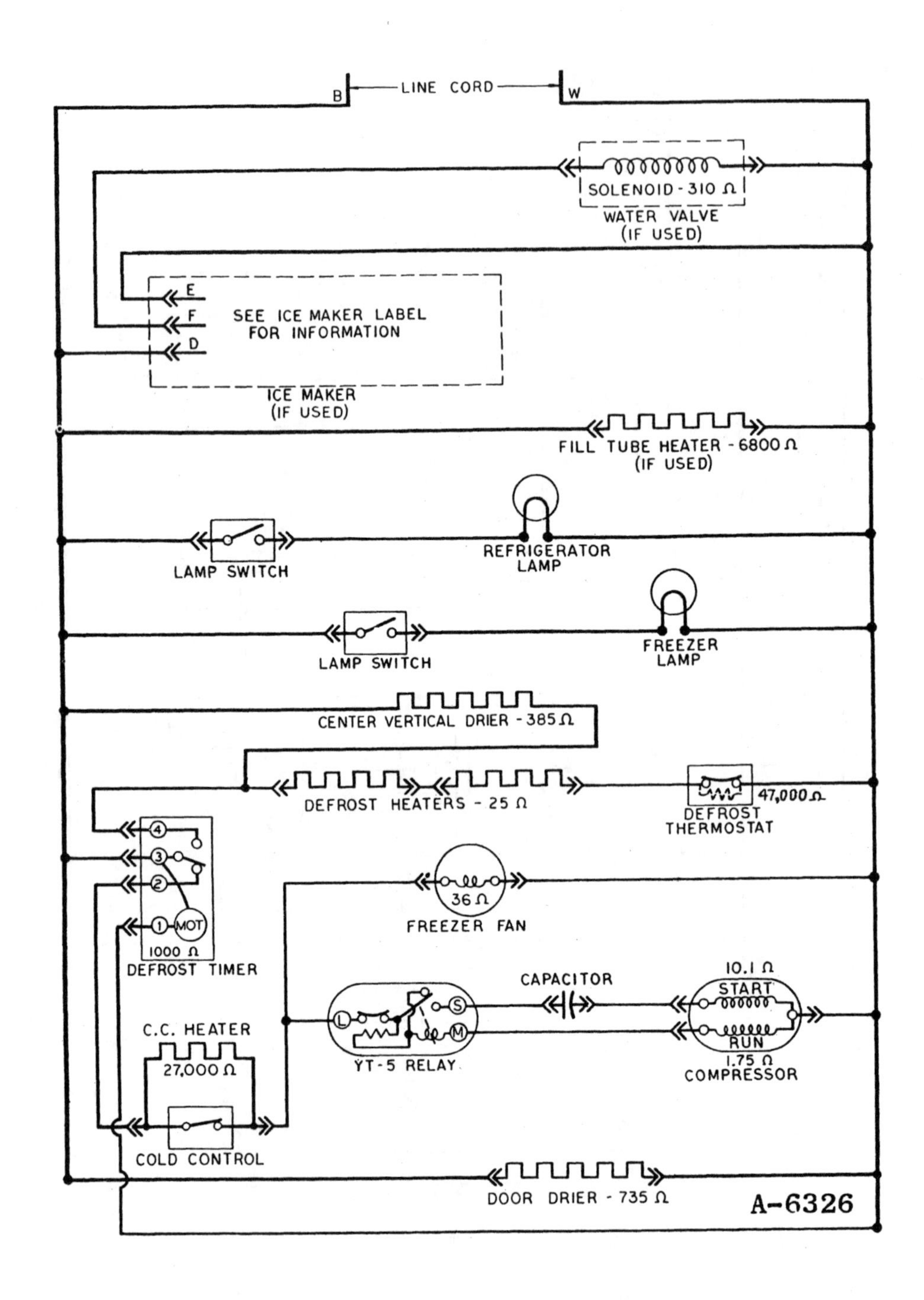
LINE CORD
B
W
SOLENOID - 310 Ω
WATER VALVE
(IF USED)
E
F
D
SEE ICE MAKER LABEL
FOR INFORMATION
ICE MAKER
(IF USED)
FILL TUBE HEATER - 6800 Ω
(IF USED)
LAMP SWITCH
REFRIGERATOR
LAMP
LAMP SWITCH
FREEZER
LAMP
CENTER VERTICAL DRIER - 385 Ω
DEFROST HEATERS - 25 Ω
47,000 Ω
DEFROST
THERMOSTAT
4
3
2
1
MOT
1000 Ω
DEFROST TIMER
36 Ω
FREEZER FAN
CAPACITOR
10.1 Ω
START
RUN
1.75 Ω
COMPRESSOR
L
S
M
YT-5 RELAY
C.C. HEATER
27,000 Ω
COLD CONTROL
DOOR DRIER - 735 Ω
A-6326

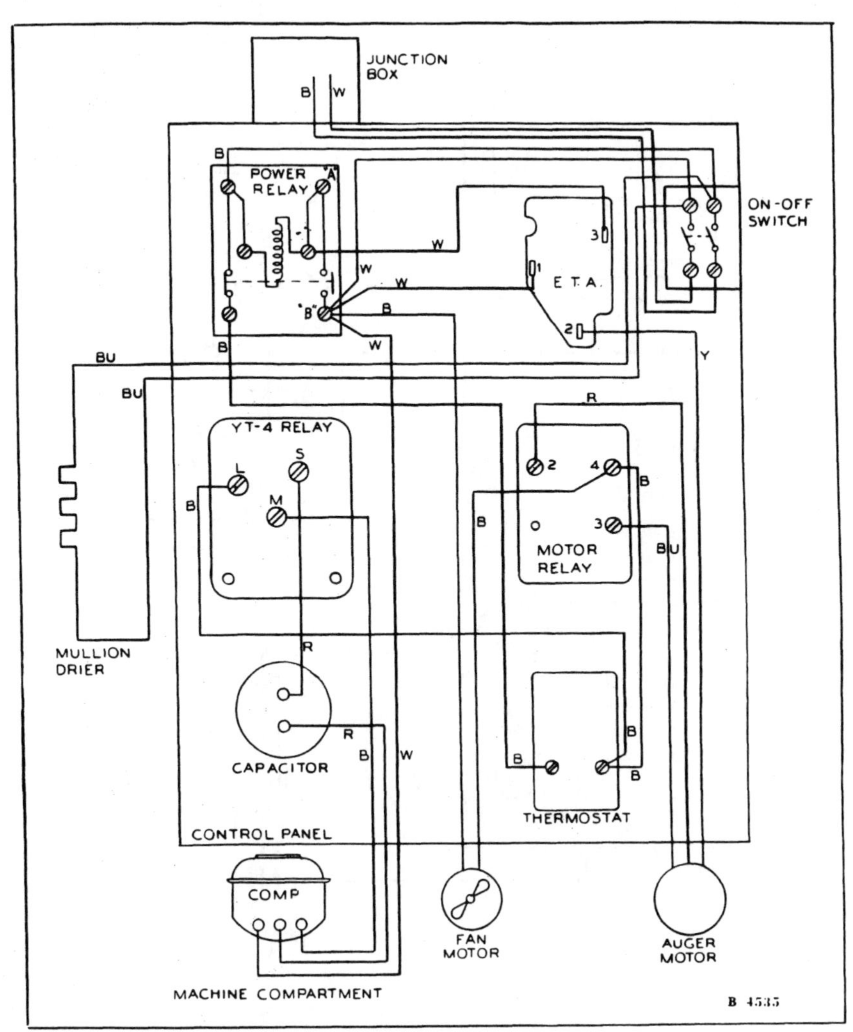

Wiring Diagram

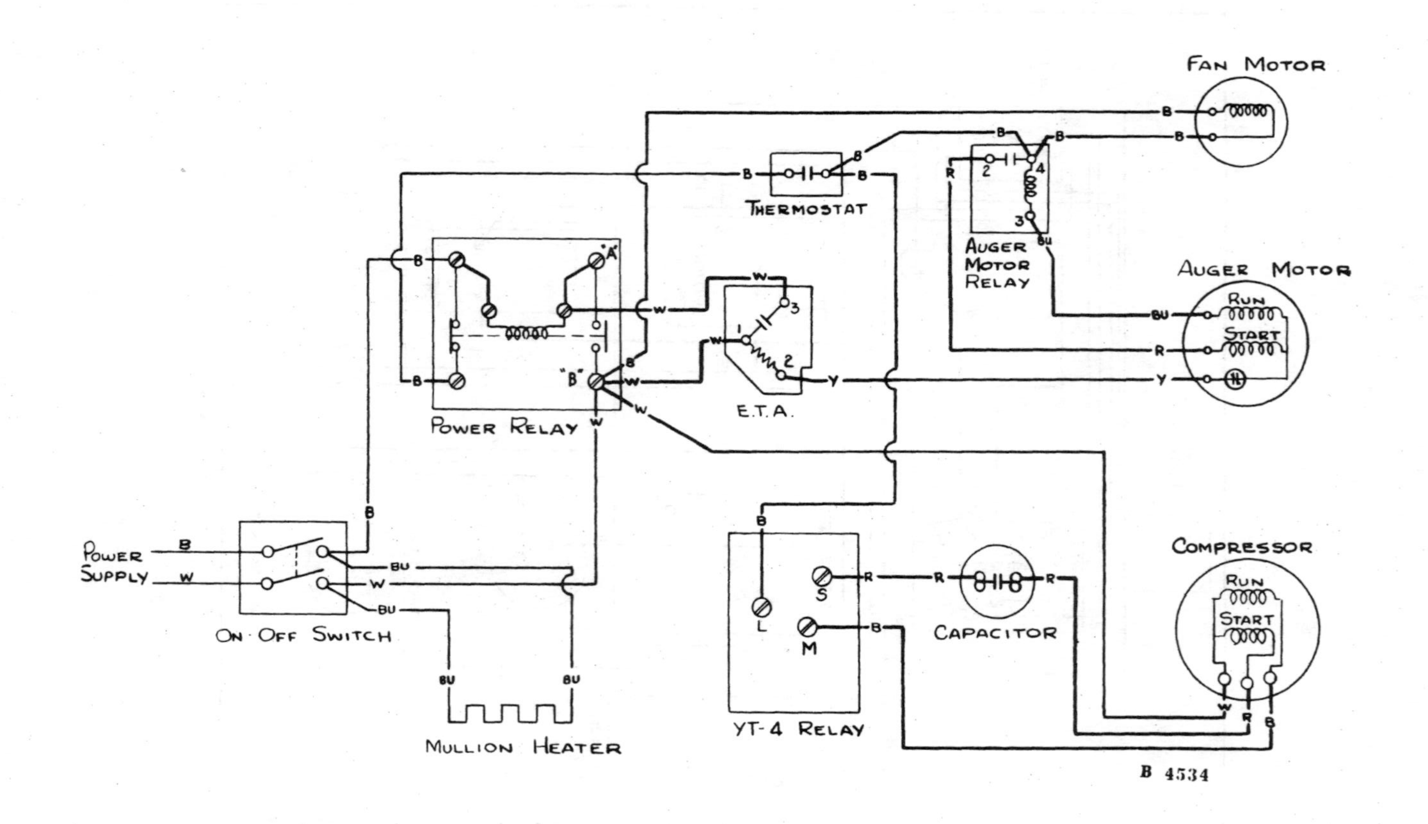
Fan Motor
Thermostat
Auger Motor Relay
Auger Motor
Run
Start
Power Relay
E.T.A.
Power Supply
On-Off Switch
Mullion Heater
YT-4 Relay
Capacitor
Compressor
Run
Start
B 4534

REPRESENTATIVE COPELAND REFRIGERATION WIRING DIAGRAMS

Single Phase Motors

There are four basic types of single phase motors used in hermetic motor compressors: split phase, permanent split capacitor, capacitor start-capacitor run, and capacitor start-induction run. Schematic diagrams of each type are shown in Figures 1, 2, 3 and 4.

A current relay is normally open when de-energized; the coil is wound so that the contacts will close when starting current is being drawn by the motor, but will drop out when the current approaches normal full load conditions. Therefore, the current relay is closed only during the starting cycle.

A potential relay is normally closed when de-energized, and the coil is designed to open the contacts only when sufficient voltage is generated by the start winding. Since the voltage or back-EMF generated by the start winding is proportional to the motor speed, the relay will open only when the motor has started and is approaching normal running speed. The illustrations show the schematic wiring with the motor in operation so the potential relay is in the energized position.

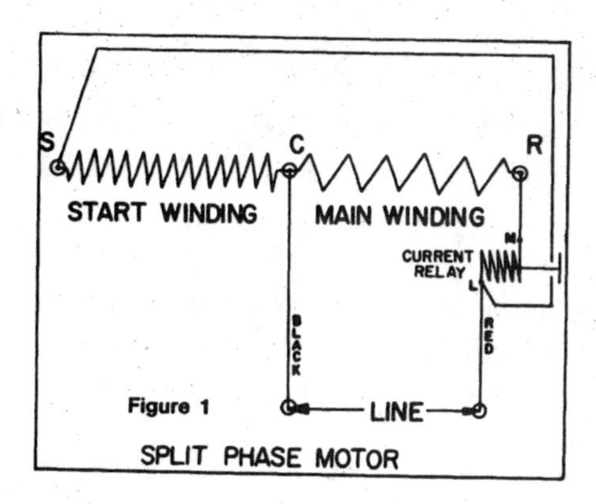

Figure 1
SPLIT PHASE MOTOR

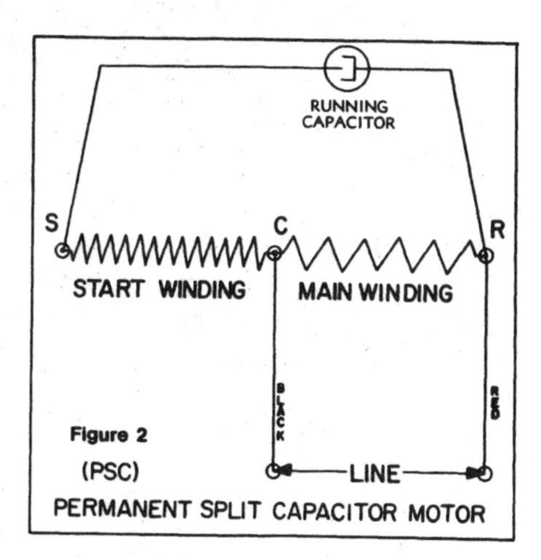

Figure 2
(PSC)
PERMANENT SPLIT CAPACITOR MOTOR

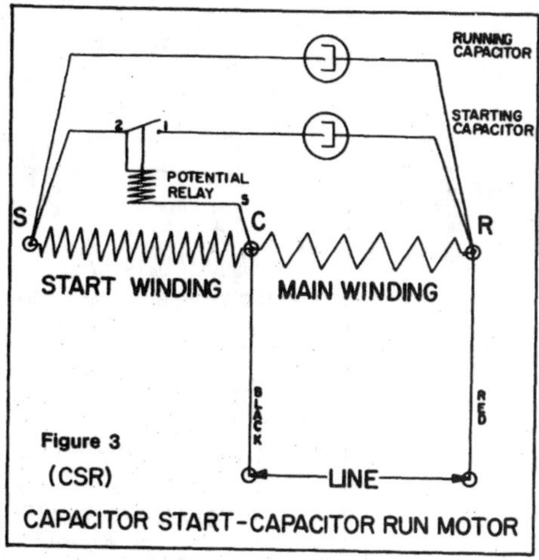

Figure 3
(CSR)
CAPACITOR START-CAPACITOR RUN MOTOR

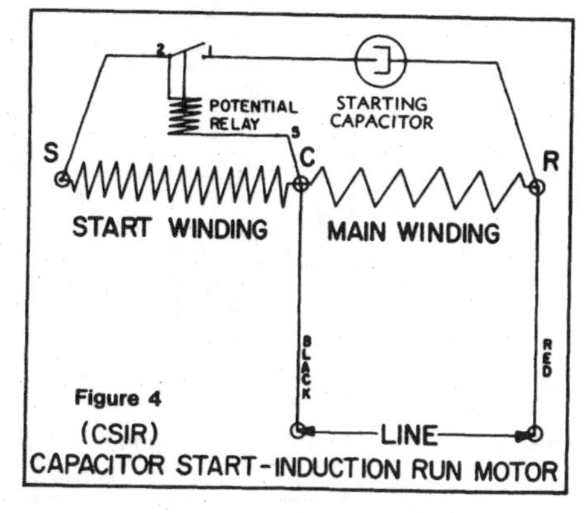

Figure 4
(CSIR)
CAPACITOR START-INDUCTION RUN MOTOR

Run Capacitors

Run capacitors are continuously in the operating circuit, and are normally of the oil filled type. Because of the voltage generated in the motor start winding, the run capacitor has a voltage across its terminals greater than line voltage.

The starting winding of a motor can be damaged by a shorted and grounded running capacitor. This damage usually can be avoided by proper connection of the running capacitor terminals.

The terminal connected to the outer foil (nearest the can) is the one most likely to short to the can and be grounded in the event of a capacitor breakdown. It is identified and marked by most manufacturers of running capacitors. See Figure 5.

From the supply line on a typical 115 or 230 volt circuit, a 115 volt potential exists from the "R" terminal to ground through a possible short in the capacitor. See wiring diagram Figure 6. However, from the "S" or start terminal, a much higher potential, possibly as high as 400 volts, exists because of the EMF generated in the start winding. Therefore, the possibility of capacitor failure is much greater when the identified terminal is connected to the "S" or start terminal.

IMPORTANT: THE IDENTIFIED TERMINAL SHOULD ALWAYS BE CONNECTED TO THE SUPPLY LINE, OR "R" TERMINAL, NEVER TO THE "S" TERMINAL. This applies to PSC as well as capacitor-start, capacitor-run motors.

This section is from Copeland Electrical Handbook, revised edition, December, 1977

If connected in this manner, a shorted and grounded running-capacitor will result in a direct short to ground from the "R" terminal and will blow line fuse No. 1. The motor protector will protect the main winding from excessive temperature.

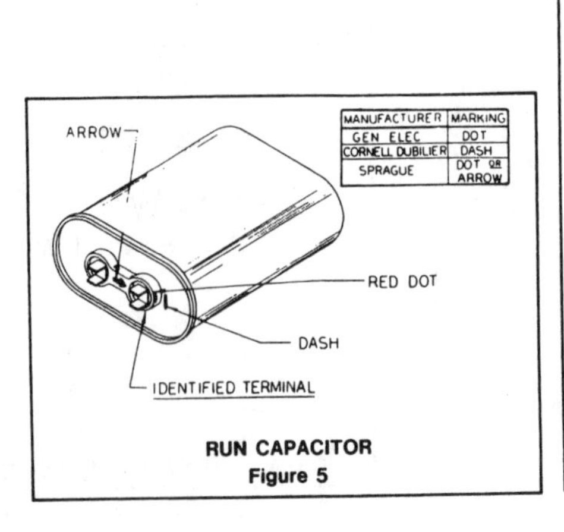

MANUFACTURER	MARKING
GEN ELEC	DOT
CORNELL DUBILIER	DASH
SPRAGUE	DOT OR ARROW

RUN CAPACITOR
Figure 5

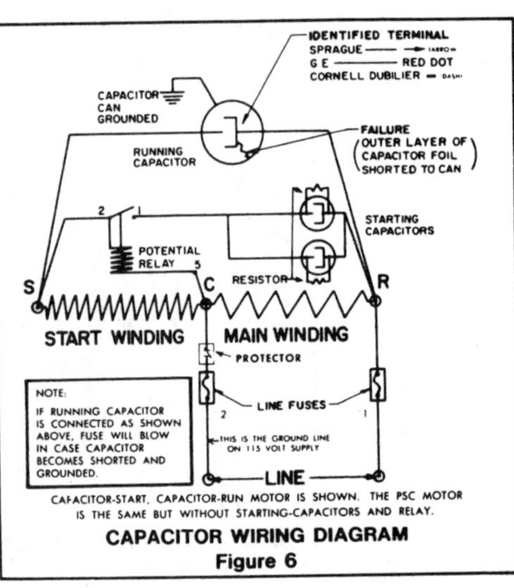

CAPACITOR-START, CAPACITOR-RUN MOTOR IS SHOWN. THE PSC MOTOR IS THE SAME BUT WITHOUT STARTING-CAPACITORS AND RELAY.

CAPACITOR WIRING DIAGRAM
Figure 6

If, however, the shorted and grounded terminal is connected to the start winding terminal "S", current will flow from the supply line through the main winding and through the start winding to ground. Even though the protector may trip, current will continue to flow through the start winding to ground, resulting in a continuing temperature rise and failure of the starting winding.

Start Capacitors

Start capacitors are designed for intermittent service only, and have a high MFD rating. Their construction is of the electrolytic type in order to obtain the high capacity.

All standard Copeland starting-capacitors are supplied with bleed-resistors securely attached and soldered to their terminals as shown in Figure 7.

The use of capacitors without these resistors probably will result in sticking relay contacts and/or erratic relay operation, especially where short cycling is likely to occur.

This is due to the starting capacitor discharging through the relay contacts as they close, following a very short running cycle. The resistor will permit the capacitor charge to bleed down at a much faster rate, preventing arcing and overheating of the relay contacts.

The use of capacitors supplied by Copeland is recommended, but in case of an emergency exchange, a 15,000-18,000 ohm, two watt resistor should be soldered across the terminals of each starting capacitor. Care should be taken to prevent their shorting to the case or other nearby metallic objects.

If sticking contacts are encountered on any starting relay, the first item to check is the starting capacitor resistors. If damaged or not provided, install new resistors, and clean the relay contacts or replace the relay.

Suitable resistors can be obtained from any radio parts wholesaler.

Capacitor Voltage

The voltage rating of a capacitor indicates the nominal voltage at which it is designed to operate. Use of a capacitor at voltages below its rating will do no harm. Run capacitors must not be subjected to voltages exceeding 110% of the nominal rating, and start capacitors must not be subjected to voltages exceeding 130% of

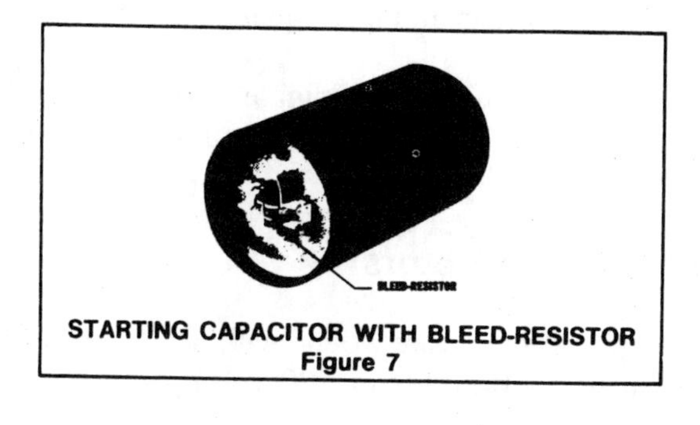

STARTING CAPACITOR WITH BLEED-RESISTOR
Figure 7

the nominal rating. The voltage to which a capacitor is subjected is not line voltage, but is a much higher potential (often, called electromotive force or back EMF) which is generated in the start winding. On a typical 230 volt motor, the generated voltage may be as high as 450 volts, and is determined by the start winding characteristics, the compressor speed, and the applied voltage.

Parallel or Series Capacitor Connections

Capacitors, either start or run, can be connected either in series or parallel to provide the desired characteristics if the voltage and MFD are properly selected. When two capacitors having the same MFD rating are connected in series, the resulting total capacitance will be one half the rated capacitance of a single capacitor. The formula for determining capacitance (MFD) when capacitors are connected in series is as follows:

$$\frac{1}{\text{MFD total}} = \frac{1}{\text{MFD}_1} + \frac{1}{\text{MFD}_2}$$

For example, if a 20 MFD and a 30 MFD capacitor are connected in series, the resultant capacitance will be

$$\frac{1}{\text{MFDt}} = \frac{1}{\text{MFD}_1} + \frac{1}{\text{MFD}_2}$$

$$\frac{1}{\text{MFDt}} = \frac{1}{20} + \frac{1}{30}$$

$$\frac{1}{\text{MFDt}} = \frac{5}{60} = \frac{1}{12}$$

$$\text{MFDt} = 12 \text{ MFD}$$

The voltage rating of similar capacitors connected in series is equal to the sum of the voltage of the two capacitors. However, since the voltage across individual capacitors in series will vary with the rating of the capacitor, for emergency field replacements it is recommended that only capacitors of like voltage and capacitance be connected in series to avoid the possibility of damage due to voltage beyond the capacitor limits.

When capacitors are connected in parallel, their MFD rating is equal to the sum of the individual ratings. The voltage rating is equal to the smallest rating of the individual capacitors.

It is possible to use any combination of single, series, or parallel starting capacitors, with single or parallel running capacitors (running capacitors are seldom used in series).

Dual Voltage Three Phase Motors

Certain Copeland three phase motors are wound with two identical stator windings which are connected in parallel on 208 or 220 volt operation, and in series for

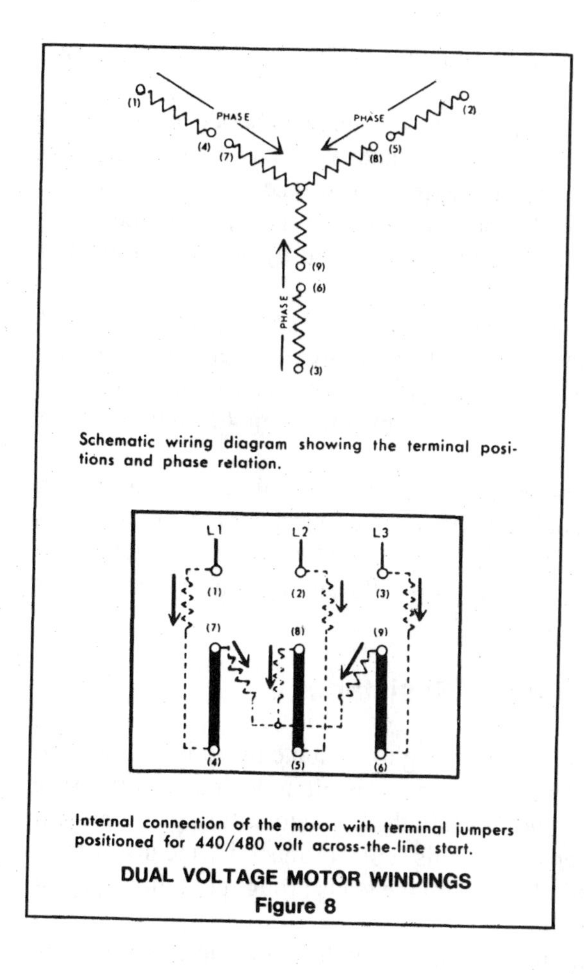

Schematic wiring diagram showing the terminal positions and phase relation.

Internal connection of the motor with terminal jumpers positioned for 440/480 volt across-the-line start.

DUAL VOLTAGE MOTOR WINDINGS
Figure 8

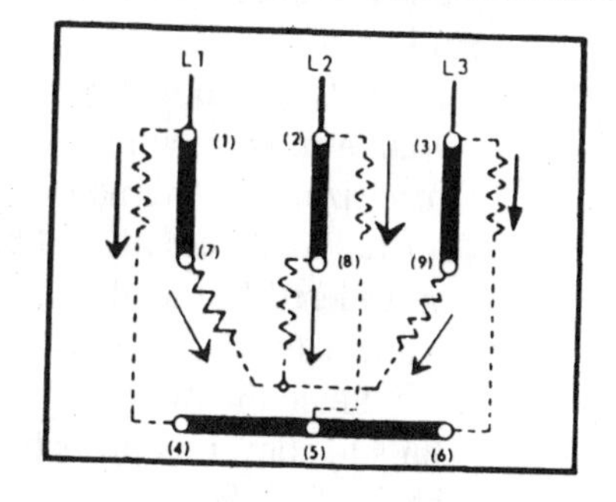

Internal connection of the motor with terminal jumpers positioned for single contactor 208/220 volt across-the-line start.

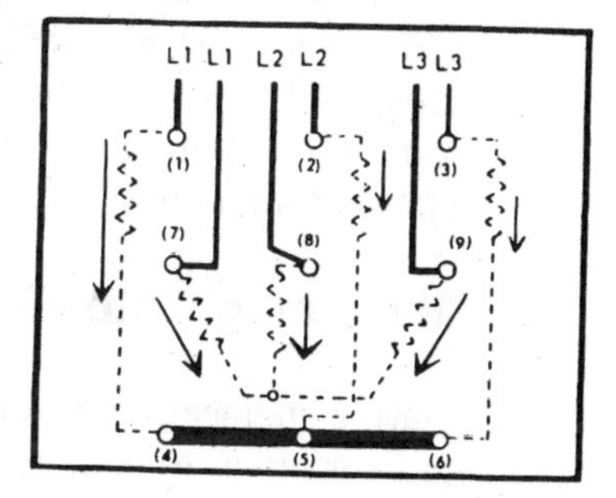

Internal connection of the motor with terminal jumpers positioned for 208/220 volt across-the-line start with two contactors, or part-winding start.

DUAL VOLTAGE MOTOR WINDINGS
Figure 8A

440/480 volt operation. Internal connections of this type of motor are shown in Figures 8 and 8A.

These models have two separate windings with nine leads which must be connected correctly for the voltage of the power supply. If the windings are connected out of phase, or if the jumper bars are not positioned correctly, motor overheating and possible failure can occur.

Fuses and Circuit Breakers

On air conditioners having motor compressors with PSC motors, it is possible that nuisance tripping of household type circuit breakers may occur. PSC motors have very low starting torque, and if pressures are not equalized at start up, the motor may require several seconds to start and accelerate.

This is most apt to occur where a short cycle of the compressor can be caused by the thermostat making contact prematurely due to shock or vibration. Typically this can occur when the thermostat is wall mounted and can be jarred by the slamming of a door.

U.L. and most electrical inspection agencies now require that hermetic type refrigeration motor compressors must comply with the National Electric Code maximum fuse sizing requirement. This establishes the maximum fuse size at 225% of the motor full load current, and by definition the motor compressor nameplate amperage is considered full load current, unless this rating is superseded by another on the unit nameplate.

Since the motor protector may take up to 17 seconds to trip if the compressor fails to start, it is probable that a standard type fuse or circuit breaker sized on the basis of 225% of full load current may break the circuit prior to the compressor protector trip, since locked rotor current of the motor may be from 400% to 500% of nameplate amperage.

To avoid nuisance tripping, Copeland recommends that air conditioners with PSC motors be installed with branch circuit fuses or circuit breakers sized as closely as possible to the 225% maximum limitation, and the fuse or circuit breaker to be of the time delay type with a capability of withstanding motor locked rotor current for a minimum of 17 seconds.

Measuring Motor Temperatures by Resistance

During engineering development or in attempting to evaluate the ability of a compressor motor to operate under severe conditions, it is often desirable to measure motor temperatures under extreme conditions. In the laboratory during compressor development, temperature measurements of this type are made by means of thermocouples installed in the motor windings, with the leads brought out through gasketed surfaces.

In evaluating a production compressor in a system, thermocouple readings are not usually feasible, and the only means of checking motor temperatures is by checking resistance values.

The resistance through the motor windings varies in a fixed ratio to a change in temperature. By establishing a base resistance value at a given base temperature (normally room temperature after storage for a prolonged period without operation), any change in the temperature of the motor windings can be calculated by determining the change in winding resistance. The resistance method, since it reads the overall resistance of the winding, is an average reading and does not accurately reflect the temperature that might exist at hot spots. In some cases there may be hot spots that can reach temperatures of 40°F or more above the average reading, so resistance readings must be evaluated very conservatively.

Motors may be wound with either copper or aluminum wire, and it is necessary to know the motor construction since the resistance change varies depending on the type of wire. The basic formula for calculating winding temperature is as follows:

$$T_h = (R_h/R_c)(T_c + K) - K$$

T_h = Temperature hot

T_c = Temperature cold
R_h = Resistance hot, ohms
R_c = Resistance cold, ohms
K copper = 390.1 for °F; 234.5 for °C
K aluminum = 365.8 for °F, 221.0 for °C

Potential Relay Nomenclature

Both General Electric and RBM have somewhat complicated numbering systems for potential relays of necessity since there are so many different features to be described. Without a complete catalog, the numerical designations are not self-explanatory, but a basic understanding of the nomenclature is helpful in identifying potential replacements in an emergency situation.

RBM Numbering System

A typical RBM potential relay number is as follows:
128146-1653DK

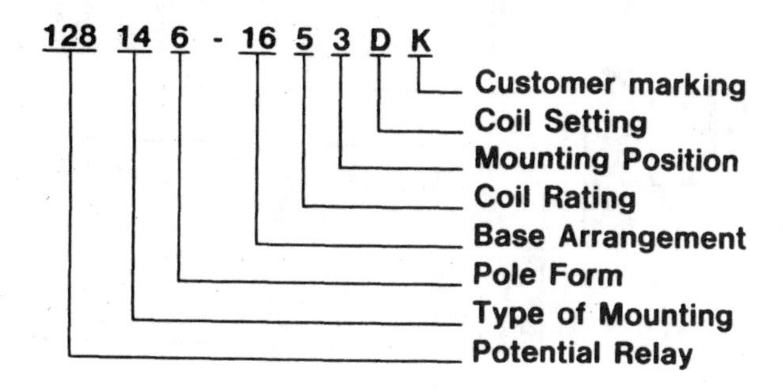

G.E. Numbering System

A typical G.E. potential relay number is as follows:
3ARR3CT6D5

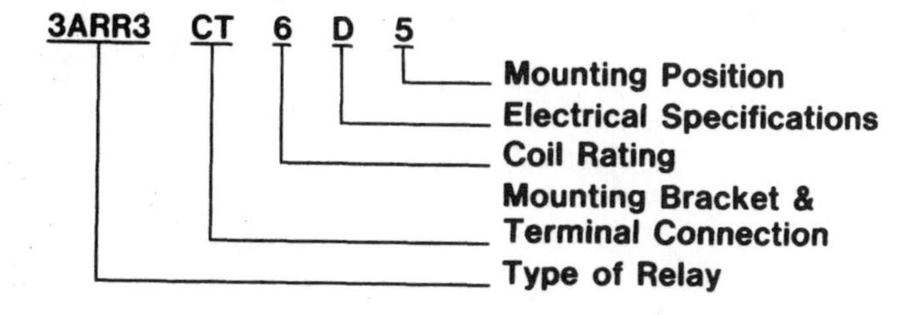

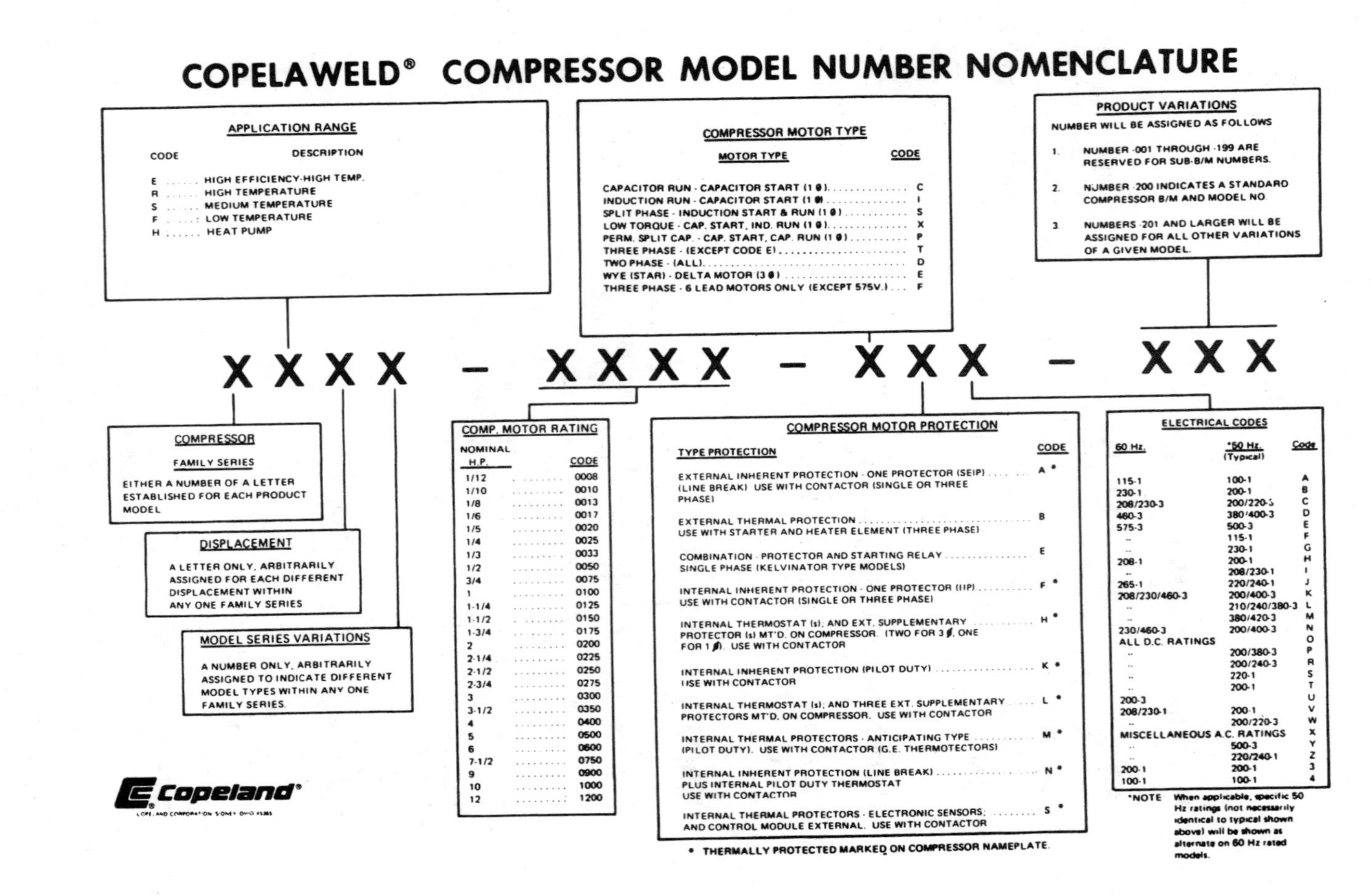

COPELAWELD® COMPRESSOR MODEL NUMBER NOMENCLATURE

X X X X – X X X X – X X X – X X X

APPLICATION RANGE

CODE	DESCRIPTION
E	HIGH EFFICIENCY-HIGH TEMP.
R	HIGH TEMPERATURE
S	MEDIUM TEMPERATURE
F	LOW TEMPERATURE
H	HEAT PUMP

COMPRESSOR

FAMILY SERIES

EITHER A NUMBER OF A LETTER ESTABLISHED FOR EACH PRODUCT MODEL

DISPLACEMENT

A LETTER ONLY, ARBITRARILY ASSIGNED FOR EACH DIFFERENT DISPLACEMENT WITHIN ANY ONE FAMILY SERIES

MODEL SERIES VARIATIONS

A NUMBER ONLY, ARBITRARILY ASSIGNED TO INDICATE DIFFERENT MODEL TYPES WITHIN ANY ONE FAMILY SERIES.

COMP. MOTOR RATING

NOMINAL H.P.	CODE
1/12	0008
1/10	0010
1/8	0013
1/6	0017
1/5	0020
1/4	0025
1/3	0033
1/2	0050
3/4	0075
1	0100
1-1/4	0125
1-1/2	0150
1-3/4	0175
2	0200
2-1/4	0225
2-1/2	0250
2-3/4	0275
3	0300
3-1/2	0350
4	0400
5	0500
6	0600
7-1/2	0750
9	0900
10	1000
12	1200

COMPRESSOR MOTOR TYPE

MOTOR TYPE	CODE
CAPACITOR RUN - CAPACITOR START (1 ∅)	C
INDUCTION RUN - CAPACITOR START (1 ∅)	I
SPLIT PHASE - INDUCTION START & RUN (1 ∅)	S
LOW TORQUE - CAP. START, IND. RUN (1 ∅)	X
PERM. SPLIT CAP. - CAP. START, CAP. RUN (1 ∅)	P
THREE PHASE - (EXCEPT CODE E)	T
TWO PHASE - (ALL)	D
WYE (STAR) - DELTA MOTOR (3 ∅)	E
THREE PHASE - 6 LEAD MOTORS ONLY (EXCEPT 575V.)	F

COMPRESSOR MOTOR PROTECTION

TYPE PROTECTION	CODE
EXTERNAL INHERENT PROTECTION - ONE PROTECTOR (SEIP) (LINE BREAK) USE WITH CONTACTOR (SINGLE OR THREE PHASE)	A *
EXTERNAL THERMAL PROTECTION USE WITH STARTER AND HEATER ELEMENT (THREE PHASE)	B
COMBINATION - PROTECTOR AND STARTING RELAY SINGLE PHASE (KELVINATOR TYPE MODELS)	E
INTERNAL INHERENT PROTECTION - ONE PROTECTOR (IIP) USE WITH CONTACTOR (SINGLE OR THREE PHASE)	F *
INTERNAL THERMOSTAT (s); AND EXT. SUPPLEMENTARY PROTECTOR (s) MT'D. ON COMPRESSOR. (TWO FOR 3 ∅, ONE FOR 1 ∅). USE WITH CONTACTOR	H *
INTERNAL INHERENT PROTECTION (PILOT DUTY) USE WITH CONTACTOR	K *
INTERNAL THERMOSTAT (s); AND THREE EXT. SUPPLEMENTARY PROTECTORS MT'D. ON COMPRESSOR. USE WITH CONTACTOR	L *
INTERNAL THERMAL PROTECTORS - ANTICIPATING TYPE (PILOT DUTY). USE WITH CONTACTOR (G.E. THERMOTECTORS)	M *
INTERNAL INHERENT PROTECTION (LINE BREAK) PLUS INTERNAL PILOT DUTY THERMOSTAT USE WITH CONTACTOR	N *
INTERNAL THERMAL PROTECTORS - ELECTRONIC SENSORS; AND CONTROL MODULE EXTERNAL. USE WITH CONTACTOR	S *

* THERMALLY PROTECTED MARKED ON COMPRESSOR NAMEPLATE.

PRODUCT VARIATIONS

NUMBER WILL BE ASSIGNED AS FOLLOWS

1. NUMBER -001 THROUGH -199 ARE RESERVED FOR SUB-B/M NUMBERS.
2. NUMBER -200 INDICATES A STANDARD COMPRESSOR B/M AND MODEL NO.
3. NUMBERS -201 AND LARGER WILL BE ASSIGNED FOR ALL OTHER VARIATIONS OF A GIVEN MODEL.

ELECTRICAL CODES

60 Hz.	*50 Hz. (Typical)	Code
115-1	100-1	A
230-1	200-1	B
208/230-3	200/220-3	C
460-3	380/400-3	D
575-3	500-3	E
--	115-1	F
--	230-1	G
208-1	200-1	H
--	208/230-1	I
265-1	220/240-1	J
208/230/460-3	200/400-3	K
--	210/240/380-3	L
--	380/420-3	M
230/460-3	200/400-3	N
ALL D.C. RATINGS		O
--	200/380-3	P
--	200/240-3	R
--	220-1	S
--	200-1	T
200-3		U
208/230-1	200-1	V
--	200/220-3	W
MISCELLANEOUS A.C. RATINGS		X
--	500-3	Y
--	220/240-1	Z
200-1	200-1	3
100-1	100-1	4

*NOTE When applicable, specific 50 Hz ratings (not necessarily identical to typical shown above) will be shown as alternate on 60 Hz rated models.

Copeland®

COPELAND CORPORATION SIDNEY, OHIO 45365

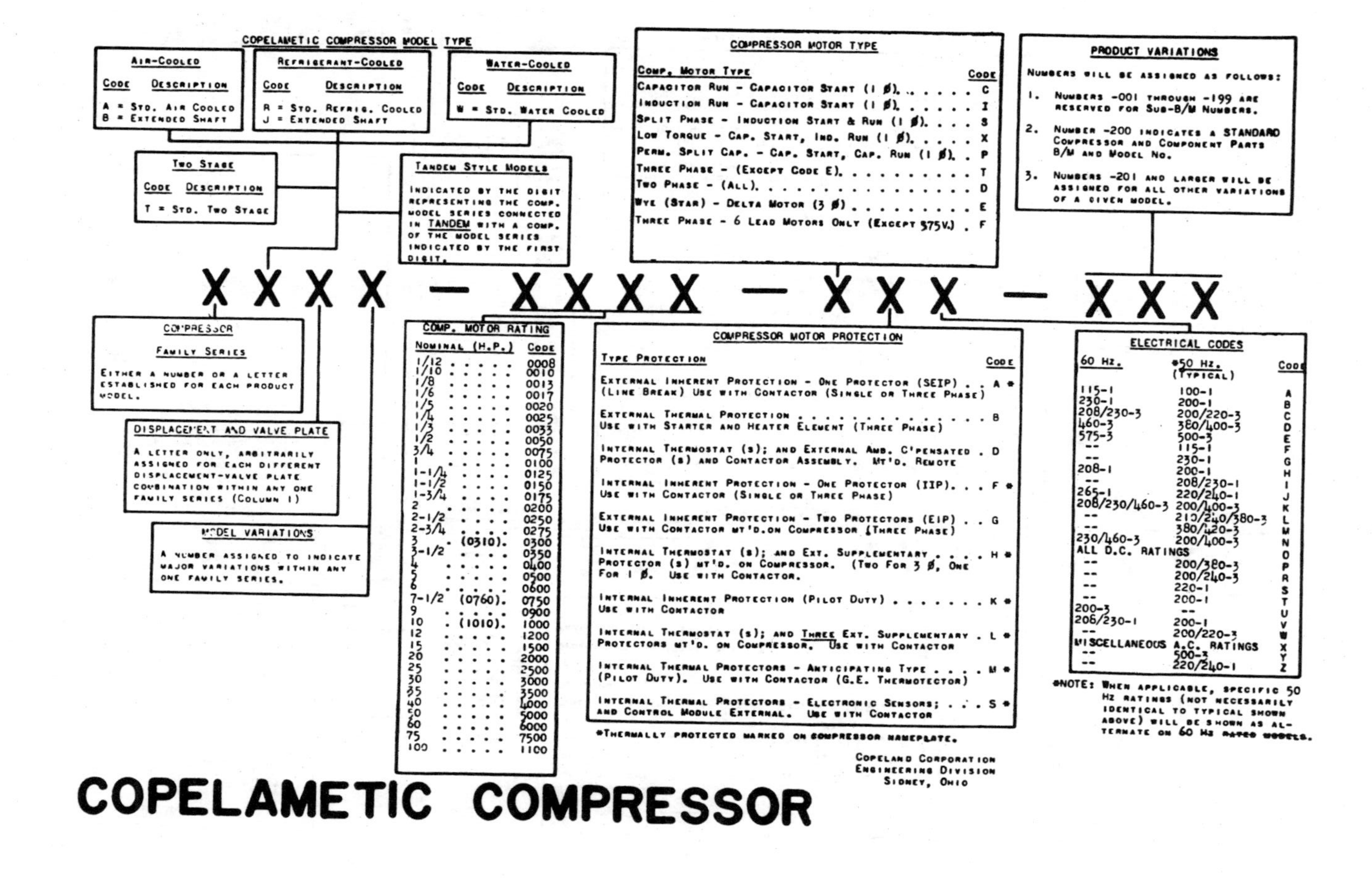
COPELAMETIC COMPRESSOR MODEL TYPE
Air-Cooled
Code Description
A = Std. Air Cooled
B = Extended Shaft
Refrigerant-Cooled
Code Description
R = Std. Refrig. Cooled
J = Extended Shaft
Water-Cooled
Code Description
W = Std. Water Cooled
Two Stage
Code Description
T = Std. Two Stage
Tandem Style Models
Indicated by the digit representing the comp. model series connected in TANDEM with a comp. of the model series indicated by the first digit.
COMPRESSOR MOTOR TYPE
Comp. Motor Type Code
Capacitor Run - Capacitor Start (1 ø). C
Induction Run - Capacitor Start (1 ø). I
Split Phase - Induction Start & Run (1 ø). . . . S
Low Torque - Cap. Start, Ind. Run (1 ø). . . . X
Perm. Split Cap. - Cap. Start, Cap. Run (1 ø). . P
Three Phase - (Except Code E). . . . . T
Two Phase - (All). . . . . D
Wye (Star) - Delta Motor (3 ø) . . . . E
Three Phase - 6 Lead Motors Only (Except 575V.) . F
PRODUCT VARIATIONS
Numbers will be assigned as follows:
1. Numbers -001 through -199 are reserved for Sub-B/M Numbers.
2. Number -200 indicates a STANDARD Compressor and Component Parts B/M and Model No.
3. Numbers -201 and larger will be assigned for all other variations of a given model.
X X X X — X X X X — X X X — X X X
Compressor Family Series
Either a number or a letter established for each product model.
DISPLACEMENT AND VALVE PLATE
A letter only, arbitrarily assigned for each different displacement-valve plate combination within any one family series (Column 1)
MODEL VARIATIONS
A number assigned to indicate major variations within any one family series.
COMP. MOTOR RATING
Nominal (H.P.) Code
1/12 0008
1/10 0010
1/8 0013
1/6 0017
1/5 0020
1/4 0025
1/3 0033
1/2 0050
3/4 0075
1 0100
1-1/4 0125
1-1/2 0150
1-3/4 0175
2 0200
2-1/2 0250
2-3/4 0275
3 (0310). 0300
3-1/2 0350
4 0400
5 0500
6 0600
7-1/2 (0760). 0750
9 0900
10 (1010). 1000
12 1200
15 1500
20 2000
25 2500
30 3000
35 3500
40 4000
50 5000
60 6000
75 7500
100 1100
COMPRESSOR MOTOR PROTECTION
Type Protection Code
External Inherent Protection - One Protector (SEIP) (Line Break) Use with Contactor (Single or Three Phase) . . A *
External Thermal Protection. Use with Starter and Heater Element (Three Phase) . . B
Internal Thermostat (s); and External Amb. C'pensated Protector (s) and Contactor Assembly. Mt'd. Remote . D
Internal Inherent Protection - One Protector (IIP). Use with Contactor (Single or Three Phase) . . F *
External Inherent Protection - Two Protectors (EIP) Use with Contactor mt'd. on Compressor (Three Phase) . . G
Internal Thermostat (s); and Ext. Supplementary Protector (s) mt'd. on Compressor. (Two for 3 ø, One for 1 ø. Use with Contactor. . . . H *
Internal Inherent Protection (Pilot Duty) Use with Contactor . . . . . . K *
Internal Thermostat (s); and Three Ext. Supplementary Protectors mt'd. on Compressor. Use with Contactor . L *
Internal Thermal Protectors - Anticipating Type (Pilot Duty). Use with Contactor (G.E. Thermotector) . . . . M *
Internal Thermal Protectors - Electronic Sensors; and Control Module External. Use with Contactor . . . S *
*Thermally protected marked on compressor nameplate.
ELECTRICAL CODES
60 Hz. *50 Hz. (Typical) Code
115-1 100-1 A
230-1 200-1 B
208/230-3 200/220-3 C
460-3 380/400-3 D
575-3 500-3 E
-- 115-1 F
-- 230-1 G
208-1 200-1 H
-- 208/230-1 I
265-1 220/240-1 J
208/230/460-3 200/400-3 K
-- 210/240/380-3 L
-- 380/420-3 M
230/460-3 200/400-3 N
ALL D.C. RATINGS O
-- 200/380-3 P
-- 200/240-3 R
-- 220-1 S
-- 200-1 T
200-3 -- U
208/230-1 200-1 V
-- 200/220-3 W
MISCELLANEOUS A.C. RATINGS X
-- 500-3 Y
-- 220/240-1 Z
*NOTE: When applicable, specific 50 Hz ratings (not necessarily identical to typical shown above) will be shown as alternate on 60 Hz rated models.
Copeland Corporation
Engineering Division
Sidney, Ohio
COPELAMETIC COMPRESSOR

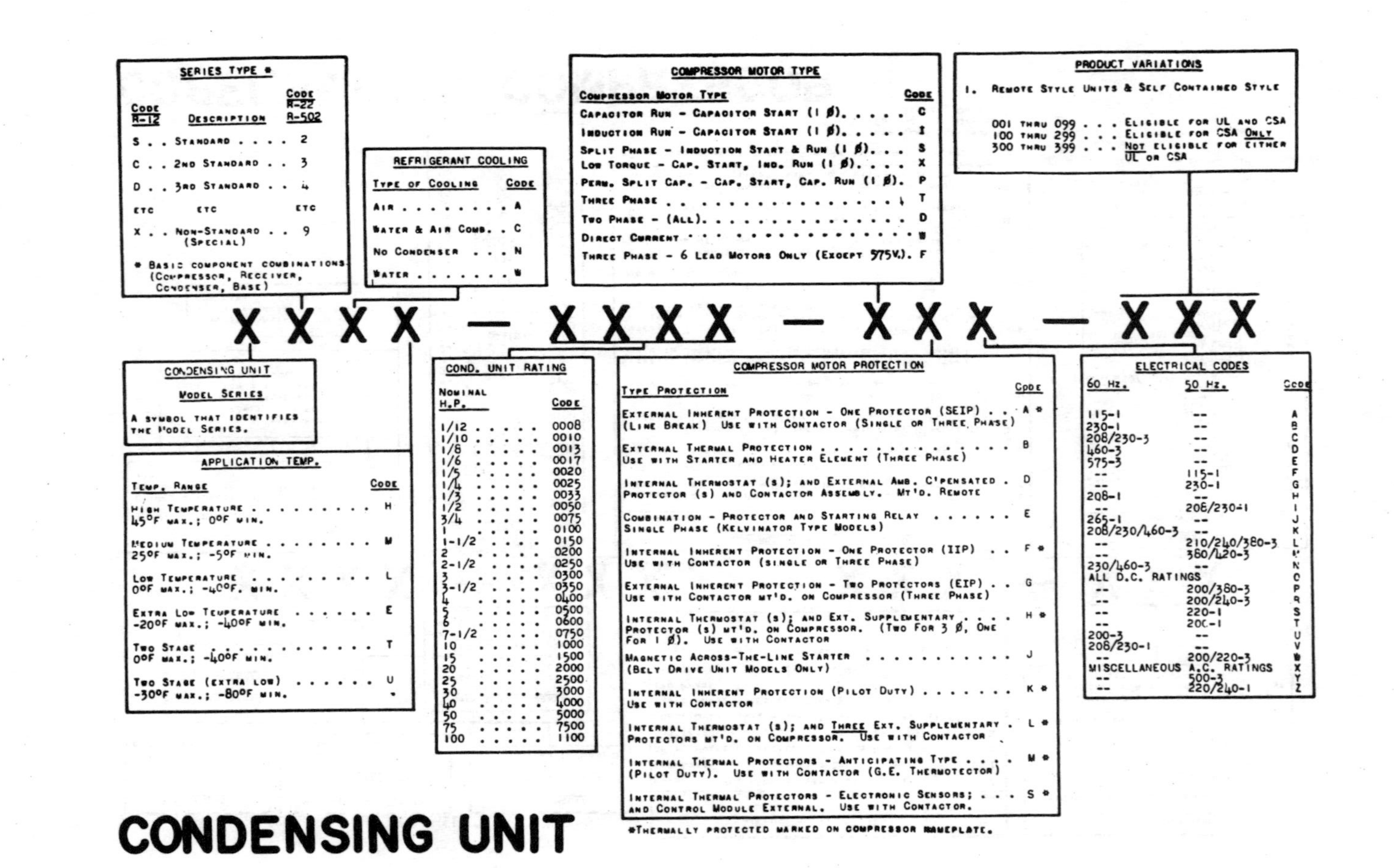

CONDENSING UNIT

INDEX TO ELECTRICAL WIRING DIAGRAMS

COPELAWELD 2 POLE MOTOR-COMPRESSOR

WIRING DIAGRAMS

Motor-Compressor	Electrical Characteristics					
	PA•,1A•, CA•	PF•,CF•	SA•,1A•, CA•	TF•	TH•	TS•
		Current Relay				
J	573		1522			
R	1520	1521	1522	0607		
S	1520	1521		0607		
Y		529		0607		
PR					553	
BRG-0900						633
BRK-1200						632

•May be any letter depending on electrical characteristics.

COPELAMETIC WIRING DIAGRAMS

Motor-Compressor	Electrical Characteristics							
	1A•,SA• Current Relay	1A•, CA• Poten-tial Relay	TA•	TFC	TFD TFE	CF•	THC	THD THE
H	1504	1502						
K		1502	513					
EA		1502	513					
ER		1519	513					
3A		1502	513					
3R		1502	517					
LA		1501	517					
NR				1064	1063	1070		
MR				1064	1063	1070		
9R				1064	1063	1070	503	504

•May be any letter depending on electrical characteristics.

COPELAMETIC WIRING DIAGRAMS

Motor-Compressor	Electrical Characteristics						
	TDK	TMK	TDE	TSK 115V.	TSK 230V.	TSE 115V.	TSE 230V.
4R,6R,-2500,3500,4000		525		610	608	613	612
4R,6R-0750,1000,1500, 2000,3000	580		351	618	616	636	635
	FSM FSR 115V.	FSM 230V.	FSR 230V.	EML	EMM		
4R,6R	613	612	622	591	594		

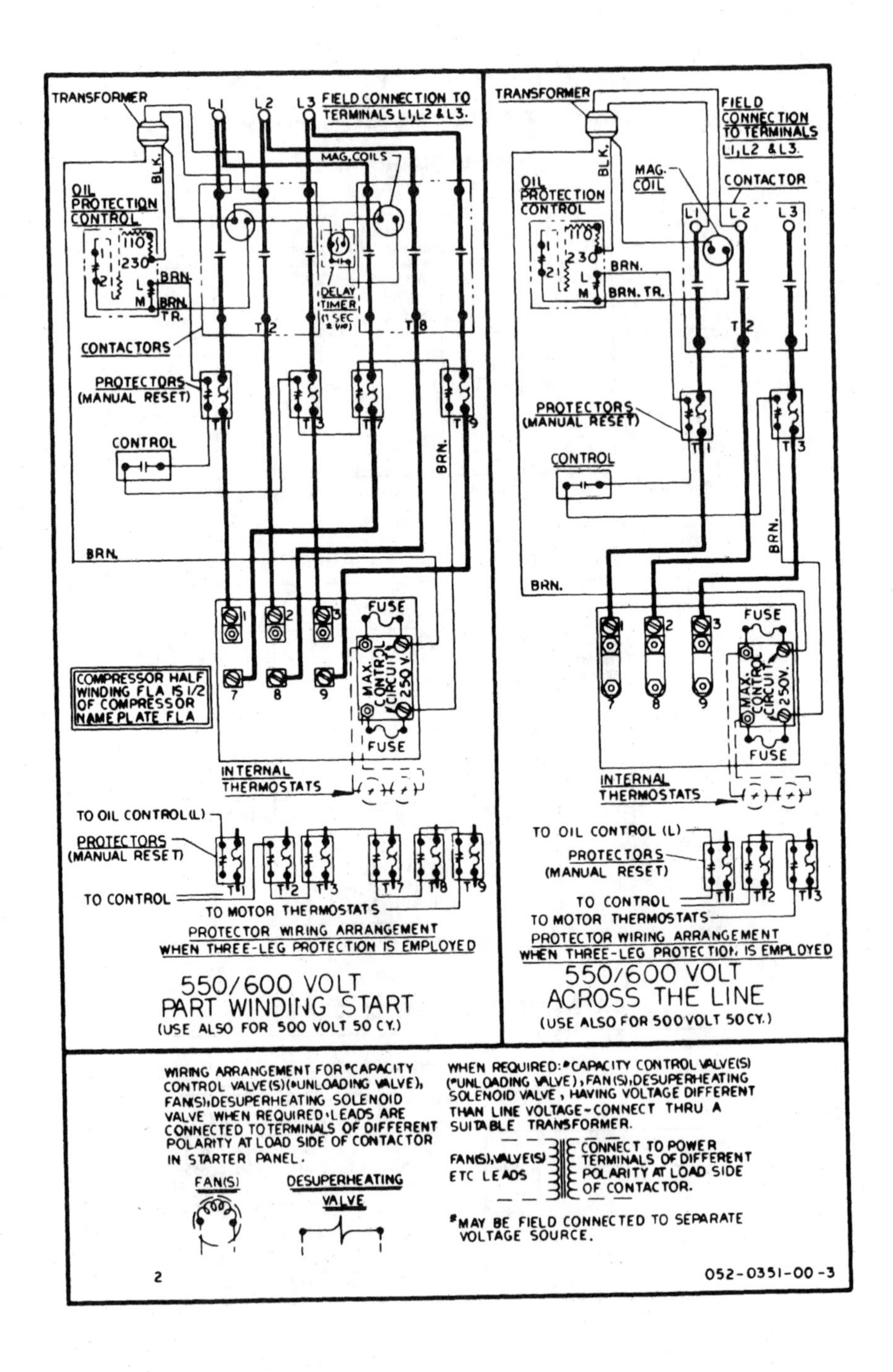

TRANSFORMER
L1
L2
L3
FIELD CONNECTION TO TERMINALS L1, L2 & L3.
MAG. COILS
BLK.
OIL PROTECTION CONTROL
110
230
BRN.
BRN. TR.
DELAY TIMER (1 SEC)
CONTACTORS
PROTECTORS (MANUAL RESET)
CONTROL
BRN.
FUSE
MAX. CONTROL CIRCUIT 250 V.
COMPRESSOR HALF WINDING FLA IS 1/2 OF COMPRESSOR NAME PLATE FLA
INTERNAL THERMOSTATS
TO OIL CONTROL (L)
PROTECTORS (MANUAL RESET)
TO CONTROL
TO MOTOR THERMOSTATS
PROTECTOR WIRING ARRANGEMENT WHEN THREE-LEG PROTECTION IS EMPLOYED
550/600 VOLT PART WINDING START
(USE ALSO FOR 500 VOLT 50 CY.)
TRANSFORMER
FIELD CONNECTION TO TERMINALS L1, L2 & L3.
MAG. COIL
CONTACTOR
OIL PROTECTION CONTROL
BRN.
BRN. TR.
PROTECTORS (MANUAL RESET)
CONTROL
FUSE
MAX. CONTROL CIRCUIT 250V.
INTERNAL THERMOSTATS
TO OIL CONTROL (L)
PROTECTORS (MANUAL RESET)
TO CONTROL
TO MOTOR THERMOSTATS
PROTECTOR WIRING ARRANGEMENT WHEN THREE-LEG PROTECTION IS EMPLOYED
550/600 VOLT ACROSS THE LINE
(USE ALSO FOR 500 VOLT 50 CY.)
WIRING ARRANGEMENT FOR *CAPACITY CONTROL VALVE(S) (*UNLOADING VALVE), FAN(S), DESUPERHEATING SOLENOID VALVE WHEN REQUIRED: LEADS ARE CONNECTED TO TERMINALS OF DIFFERENT POLARITY AT LOAD SIDE OF CONTACTOR IN STARTER PANEL.
FAN(S)
DESUPERHEATING VALVE
WHEN REQUIRED: *CAPACITY CONTROL VALVE(S) (*UNLOADING VALVE), FAN(S), DESUPERHEATING SOLENOID VALVE, HAVING VOLTAGE DIFFERENT THAN LINE VOLTAGE-CONNECT THRU A SUITABLE TRANSFORMER.
FAN(S), VALVE(S) ETC LEADS
CONNECT TO POWER TERMINALS OF DIFFERENT POLARITY AT LOAD SIDE OF CONTACTOR.
*MAY BE FIELD CONNECTED TO SEPARATE VOLTAGE SOURCE.
2
052-0351-00-3

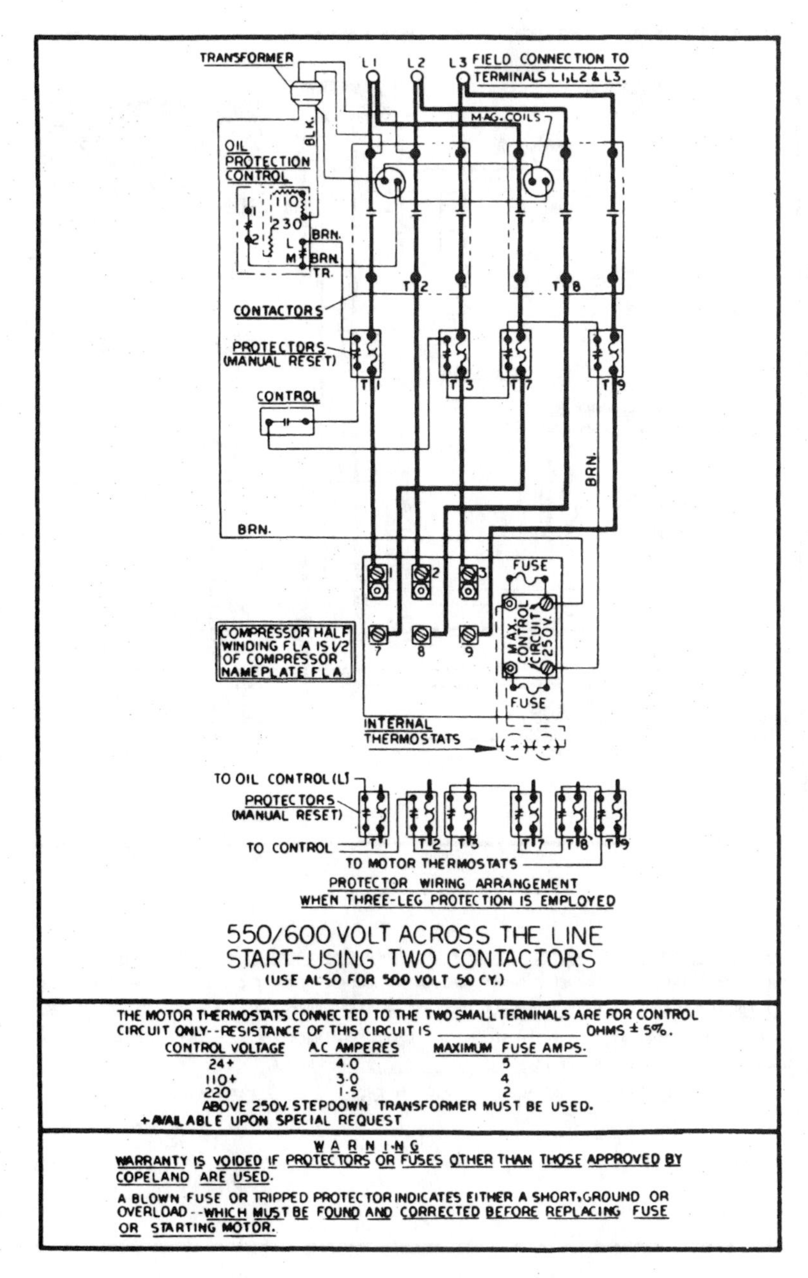
TRANSFORMER
L1 L2 L3 FIELD CONNECTION TO TERMINALS L1, L2 & L3.
MAG. COILS
BLK.
OIL PROTECTION CONTROL
110
230
BRN.
L
M
BRN.
TR.
T2
T8
CONTACTORS
PROTECTORS (MANUAL RESET)
T1 T3 T7 T9
CONTROL
BRN.
BRN.
FUSE
1 2 3
7 8 9
COMPRESSOR HALF WINDING FLA IS 1/2 OF COMPRESSOR NAMEPLATE FLA
MAX. CONTROL CIRCUIT 250V.
FUSE
INTERNAL THERMOSTATS
TO OIL CONTROL (L)
PROTECTORS (MANUAL RESET)
TO CONTROL
T1 T2 T3 T7 T8 T9
TO MOTOR THERMOSTATS
PROTECTOR WIRING ARRANGEMENT WHEN THREE-LEG PROTECTION IS EMPLOYED
550/600 VOLT ACROSS THE LINE START-USING TWO CONTACTORS
(USE ALSO FOR 500 VOLT 50 CY.)
THE MOTOR THERMOSTATS CONNECTED TO THE TWO SMALL TERMINALS ARE FOR CONTROL CIRCUIT ONLY--RESISTANCE OF THIS CIRCUIT IS ______________ OHMS ± 5%.
CONTROL VOLTAGE | A.C AMPERES | MAXIMUM FUSE AMPS.
24+ | 4.0 | 5
110+ | 3.0 | 4
220 | 1.5 | 2
ABOVE 250V. STEPDOWN TRANSFORMER MUST BE USED.
+AVAILABLE UPON SPECIAL REQUEST
WARNING
WARRANTY IS VOIDED IF PROTECTORS OR FUSES OTHER THAN THOSE APPROVED BY COPELAND ARE USED.
A BLOWN FUSE OR TRIPPED PROTECTOR INDICATES EITHER A SHORT, GROUND OR OVERLOAD--WHICH MUST BE FOUND AND CORRECTED BEFORE REPLACING FUSE OR STARTING MOTOR.

No. 351

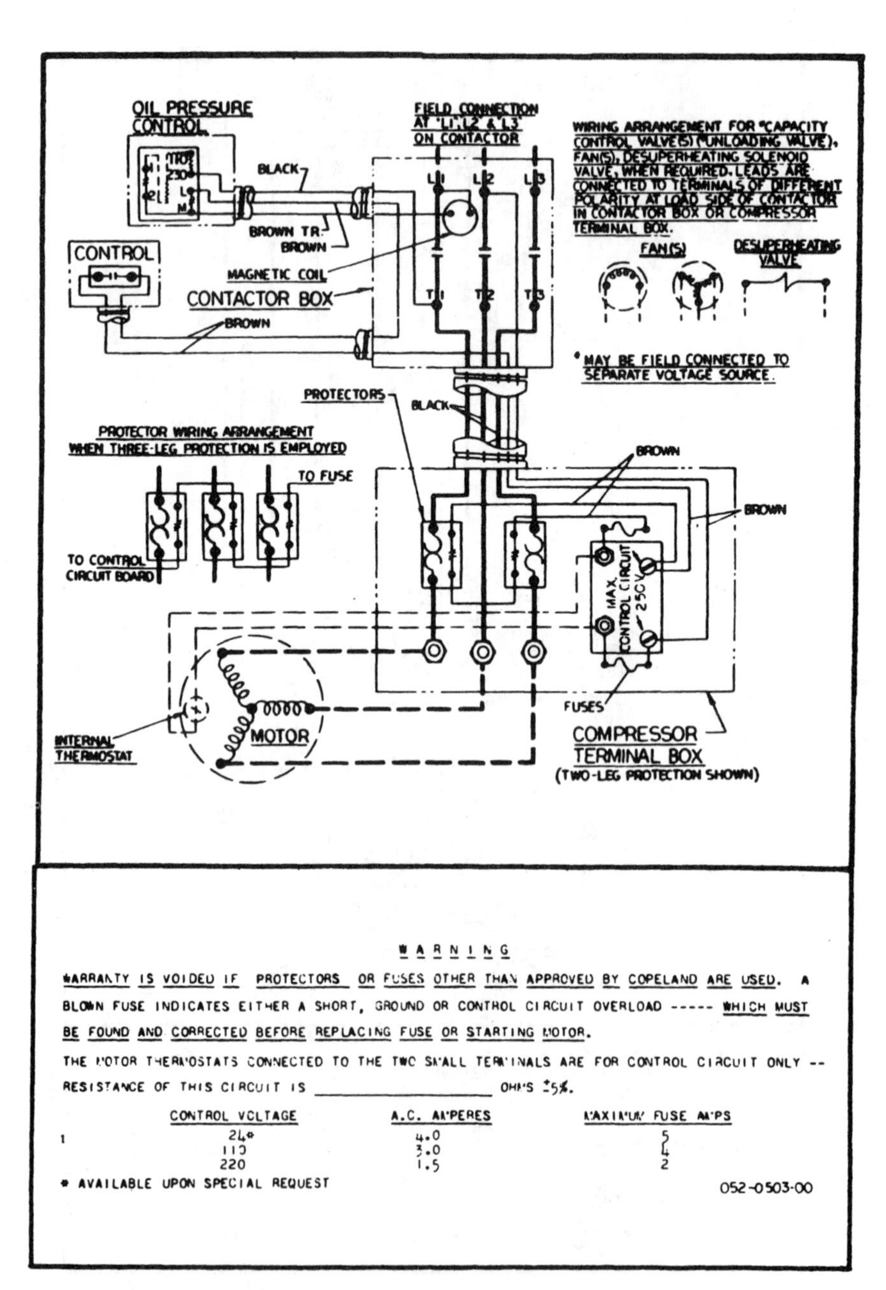

WARNING

WARRANTY IS VOIDED IF PROTECTORS OR FUSES OTHER THAN APPROVED BY COPELAND ARE USED. A BLOWN FUSE INDICATES EITHER A SHORT, GROUND OR CONTROL CIRCUIT OVERLOAD ----- WHICH MUST BE FOUND AND CORRECTED BEFORE REPLACING FUSE OR STARTING MOTOR.

THE MOTOR THERMOSTATS CONNECTED TO THE TWO SMALL TERMINALS ARE FOR CONTROL CIRCUIT ONLY -- RESISTANCE OF THIS CIRCUIT IS ______________ OHMS ±5%.

CONTROL VOLTAGE	A.C. AMPERES	MAXIMUM FUSE AMPS
24*	4.0	5
110	3.0	4
220	1.5	2

* AVAILABLE UPON SPECIAL REQUEST

052-0503-00

Wiring Diagram No. 503

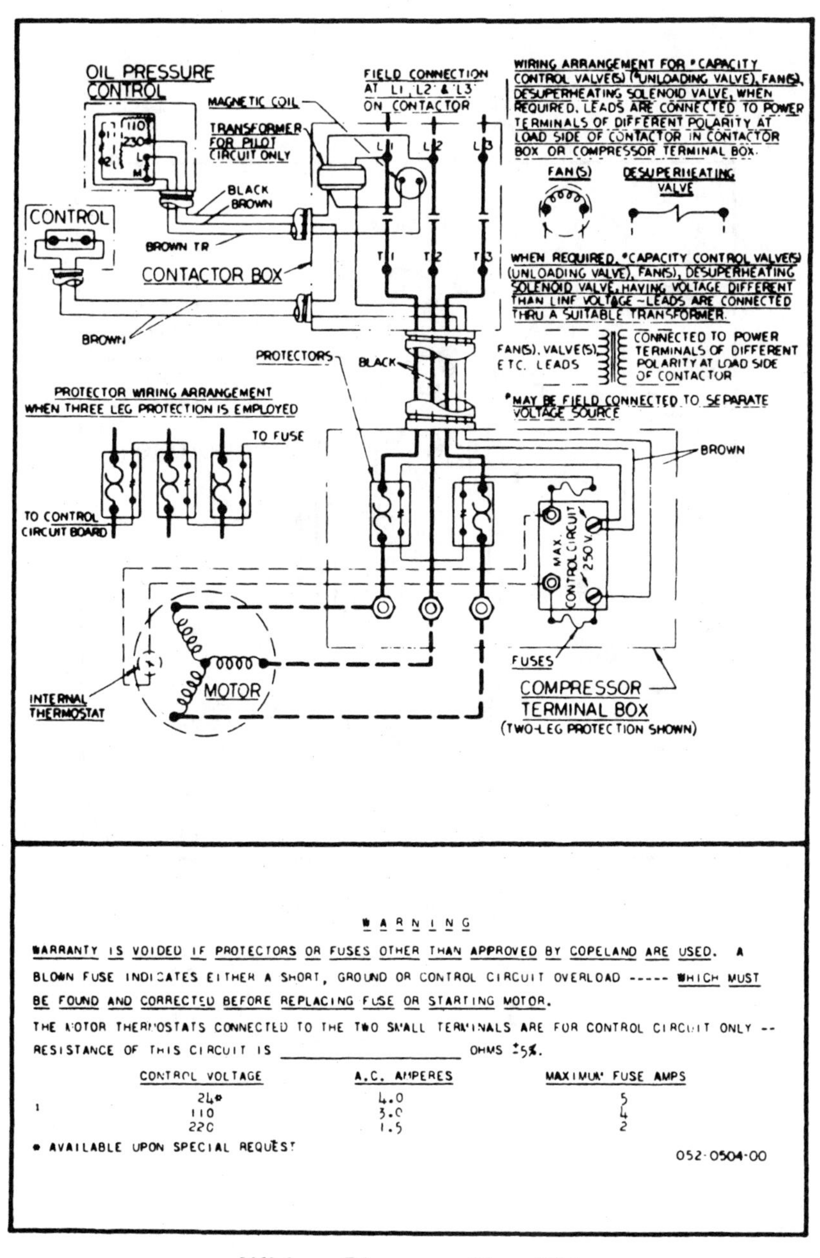

Wiring Diagram No. 504

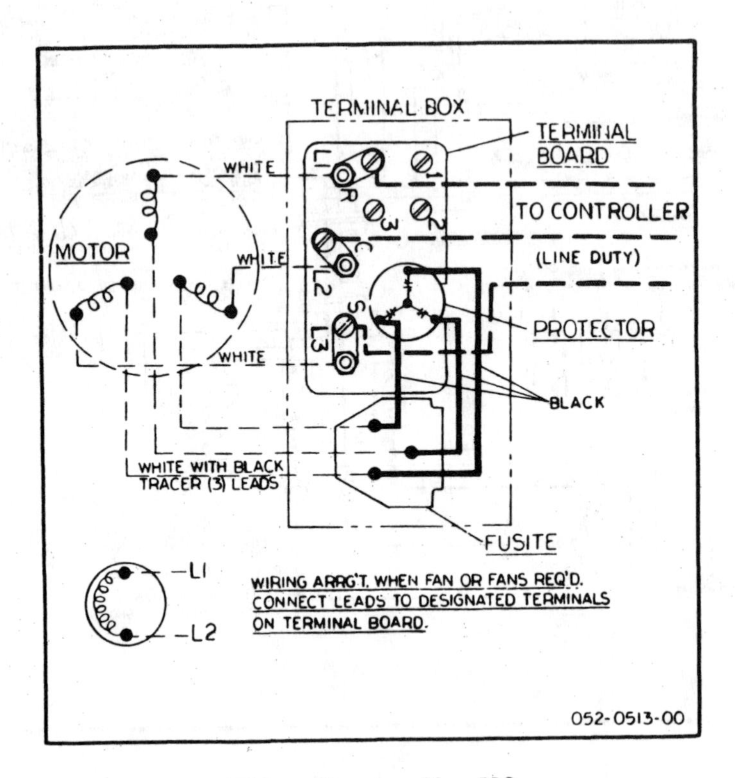

Wiring Diagram No. 513

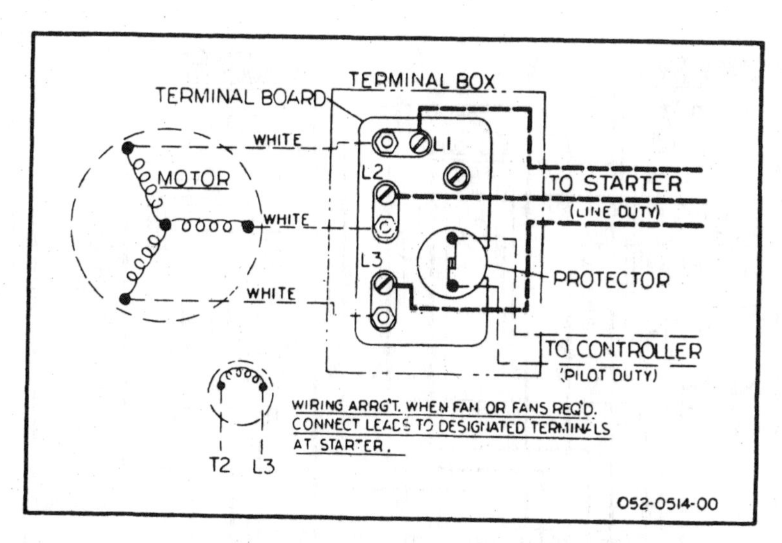

Wiring Diagram No. 514

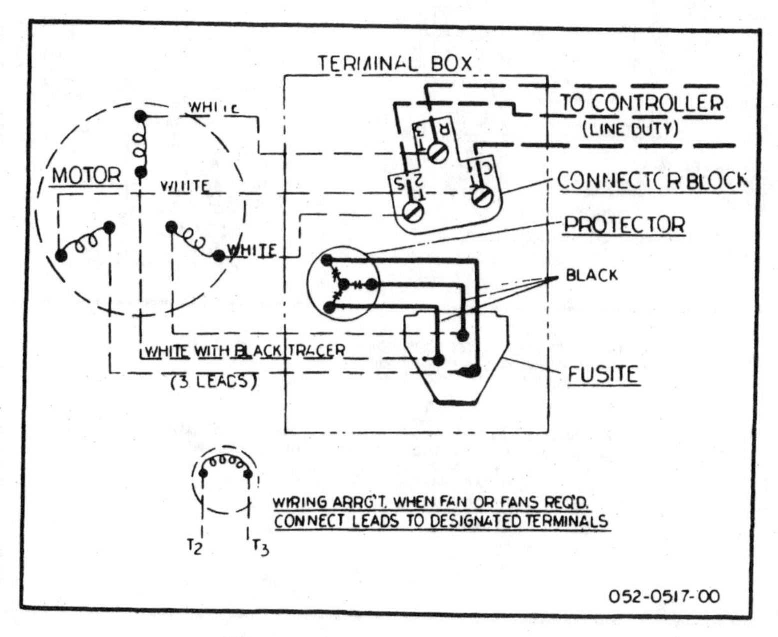

Wiring Diagram No. 517

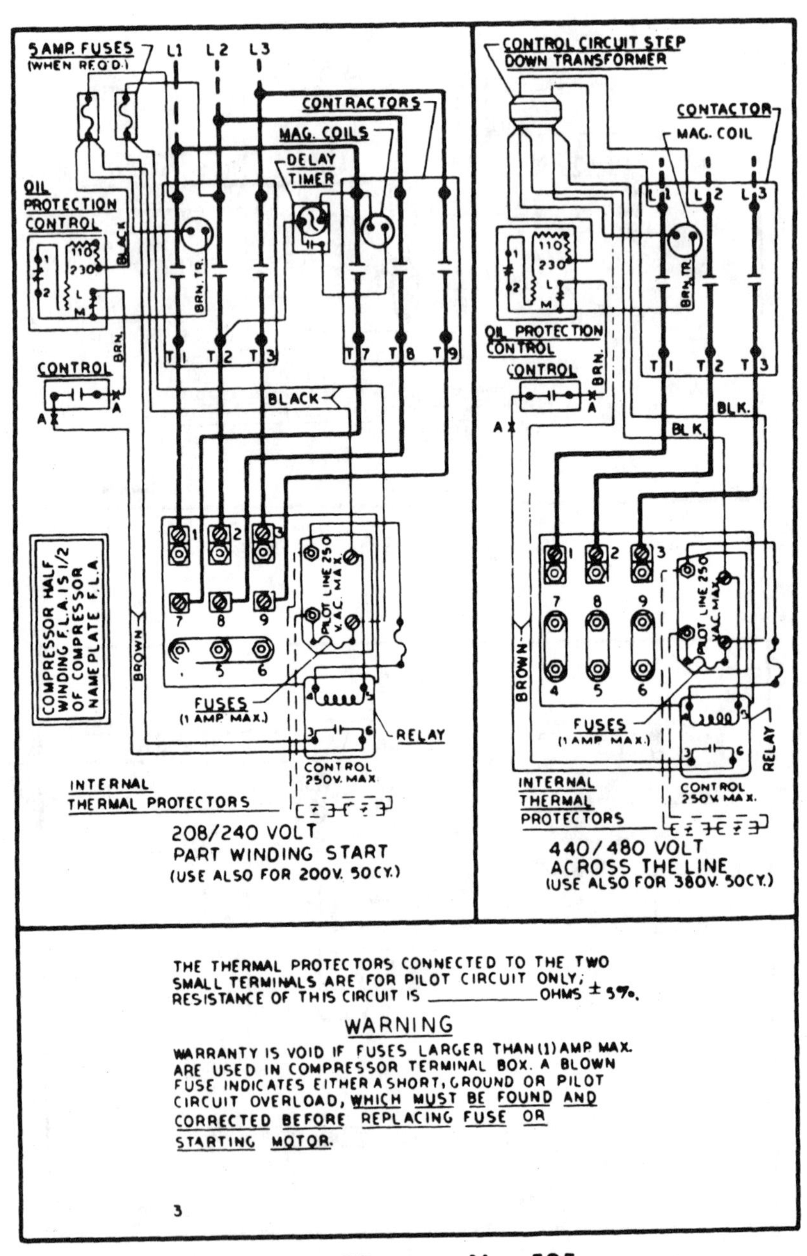

Wiring Diagram No. 525

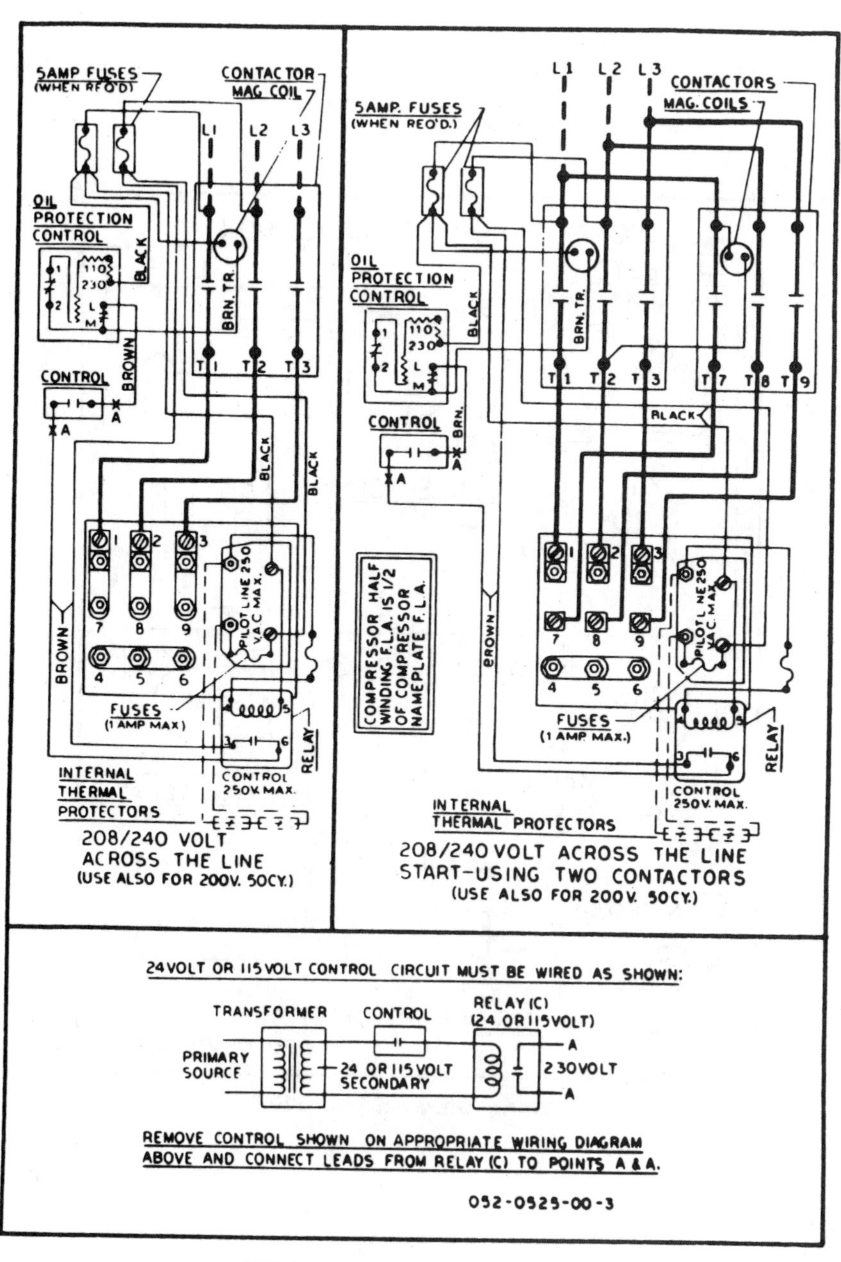

Wiring Diagram No. 525

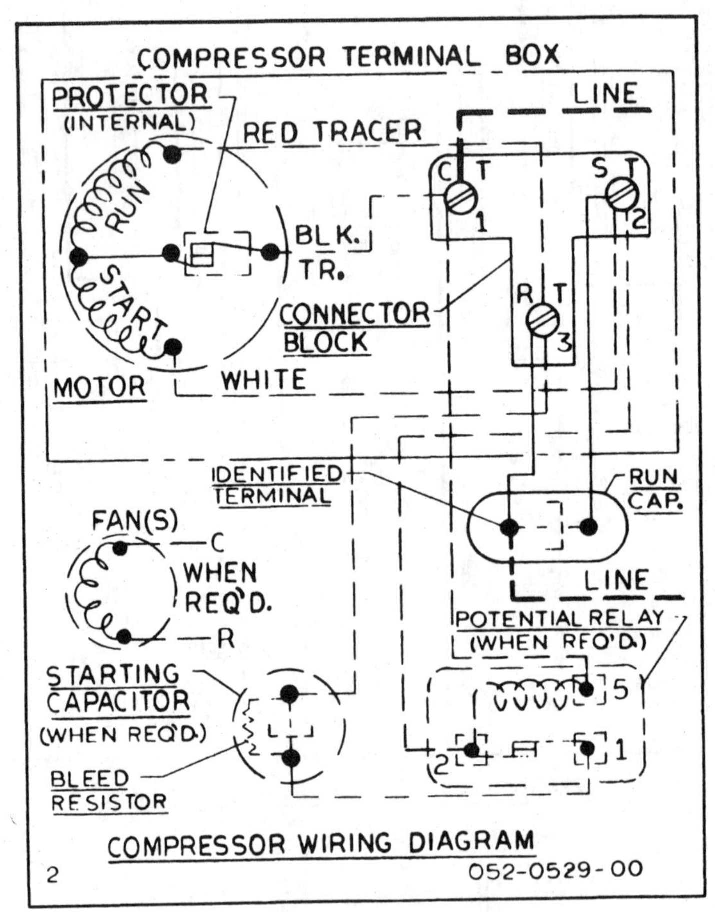

Wiring Diagram No. 529

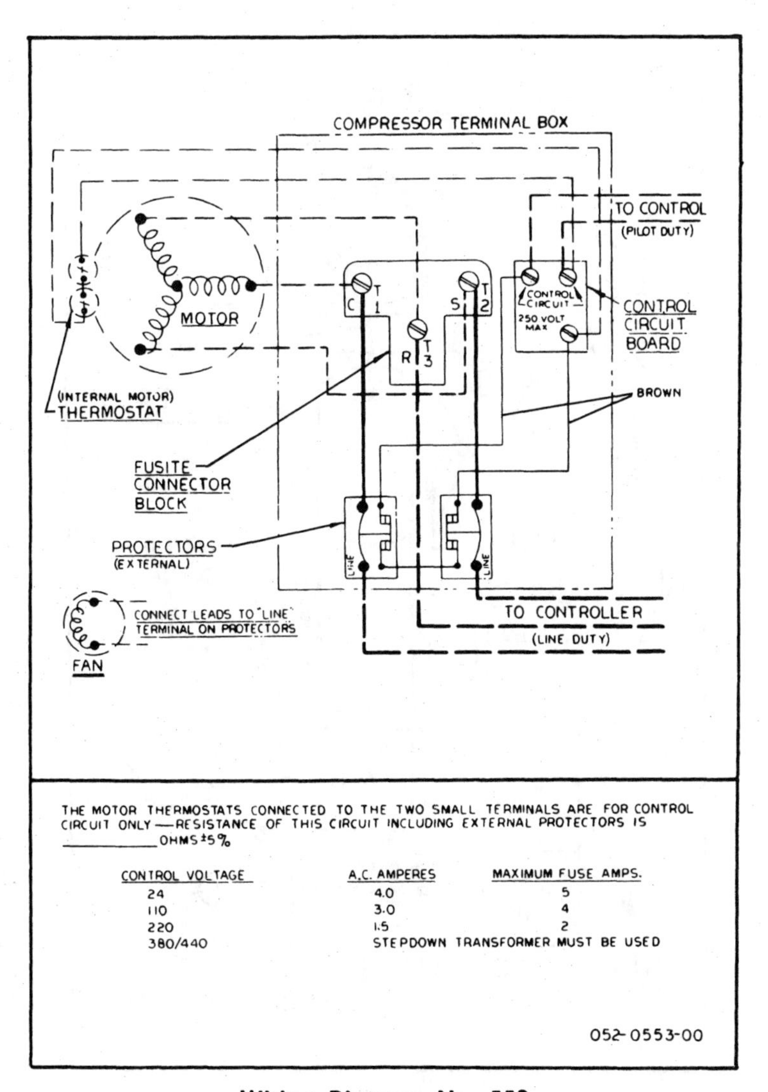

THE MOTOR THERMOSTATS CONNECTED TO THE TWO SMALL TERMINALS ARE FOR CONTROL CIRCUIT ONLY —RESISTANCE OF THIS CIRCUIT INCLUDING EXTERNAL PROTECTORS IS ________ OHMS ±5%

CONTROL VOLTAGE	A.C. AMPERES	MAXIMUM FUSE AMPS.
24	4.0	5
110	3.0	4
220	1.5	2
380/440	STEPDOWN TRANSFORMER MUST BE USED	

Wiring Diagram No. 553

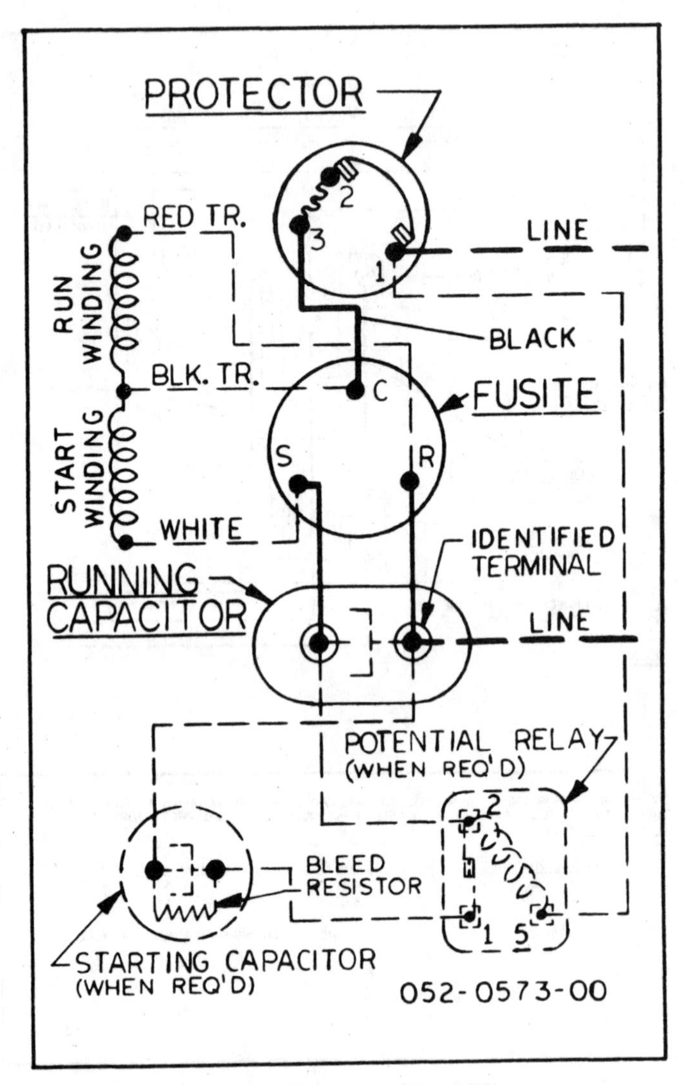

Wiring Diagram No. 573

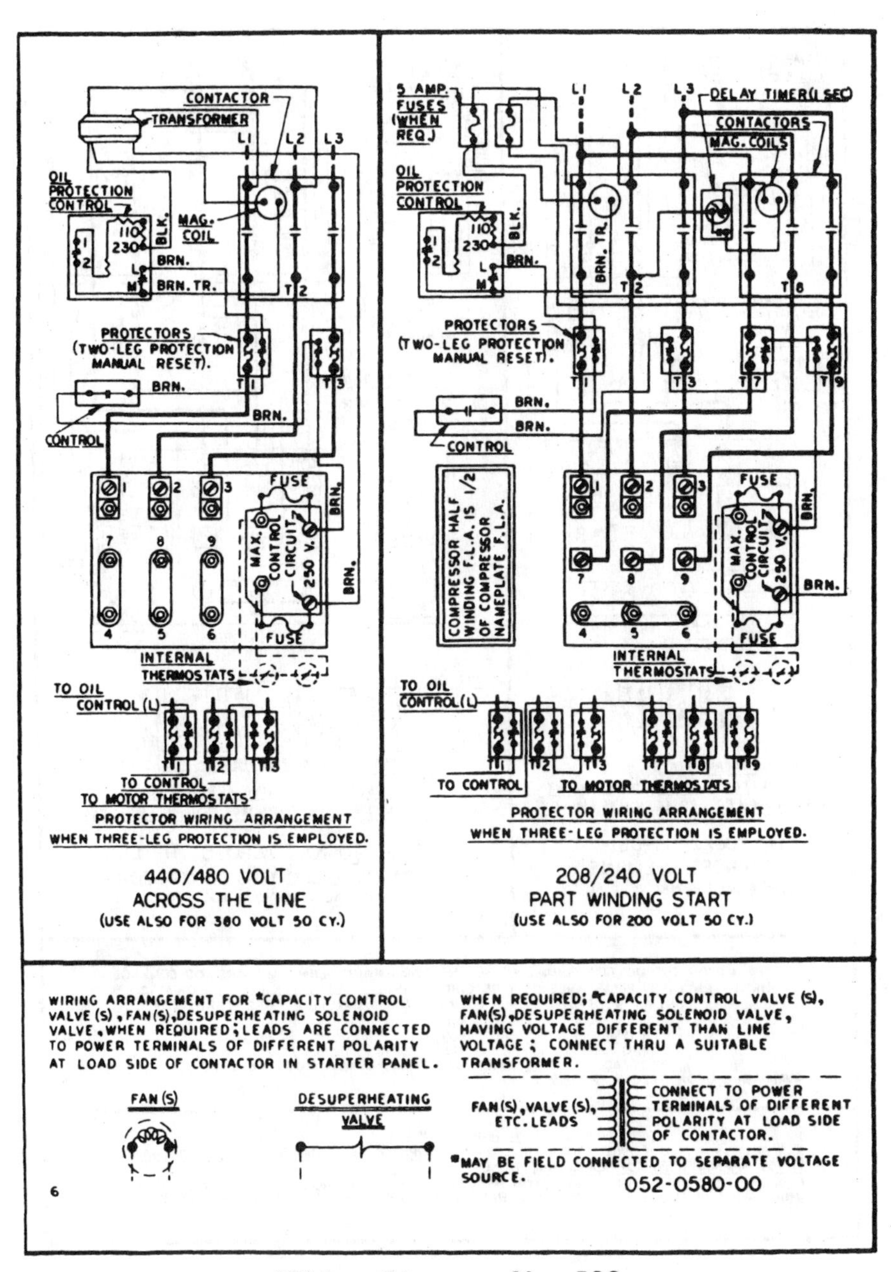

Wiring Diagram No. 580

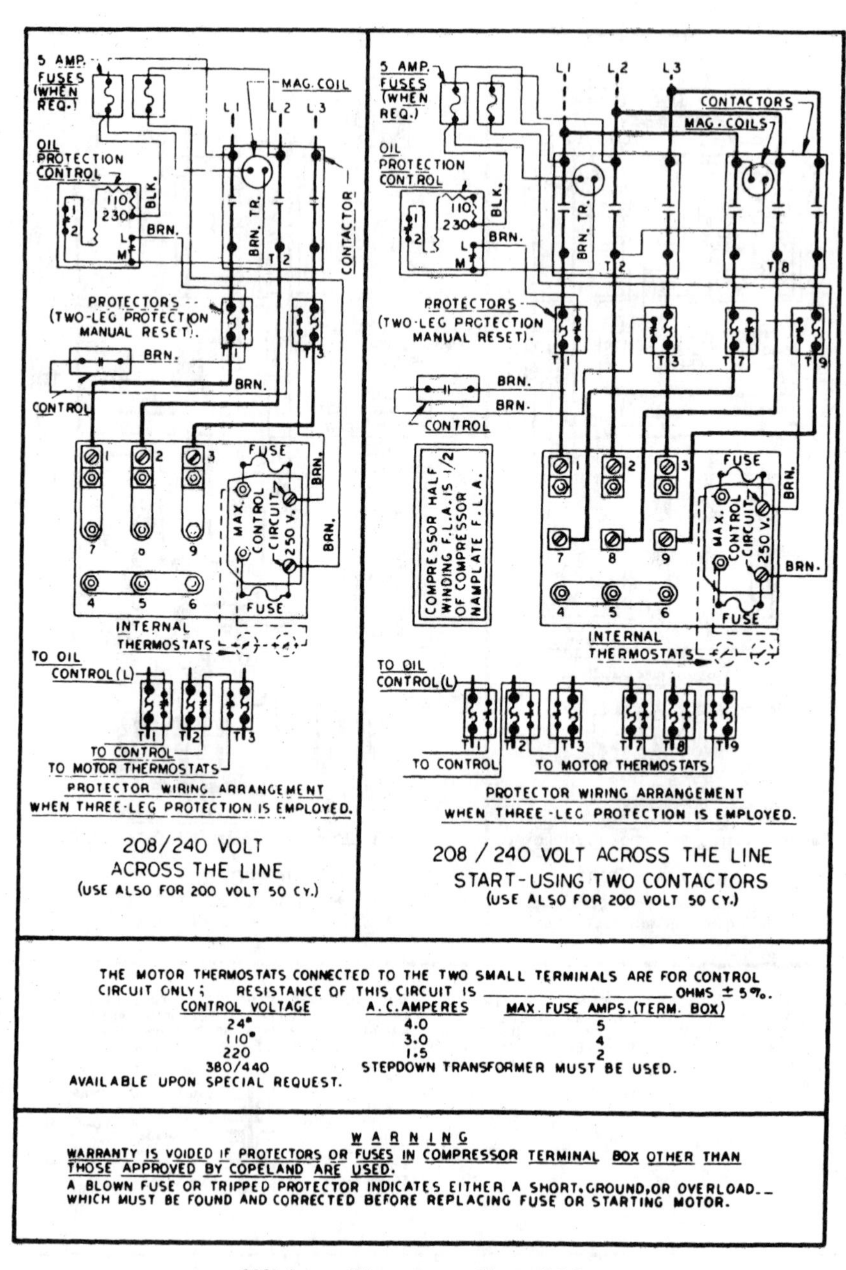

Wiring Diagram No. 580

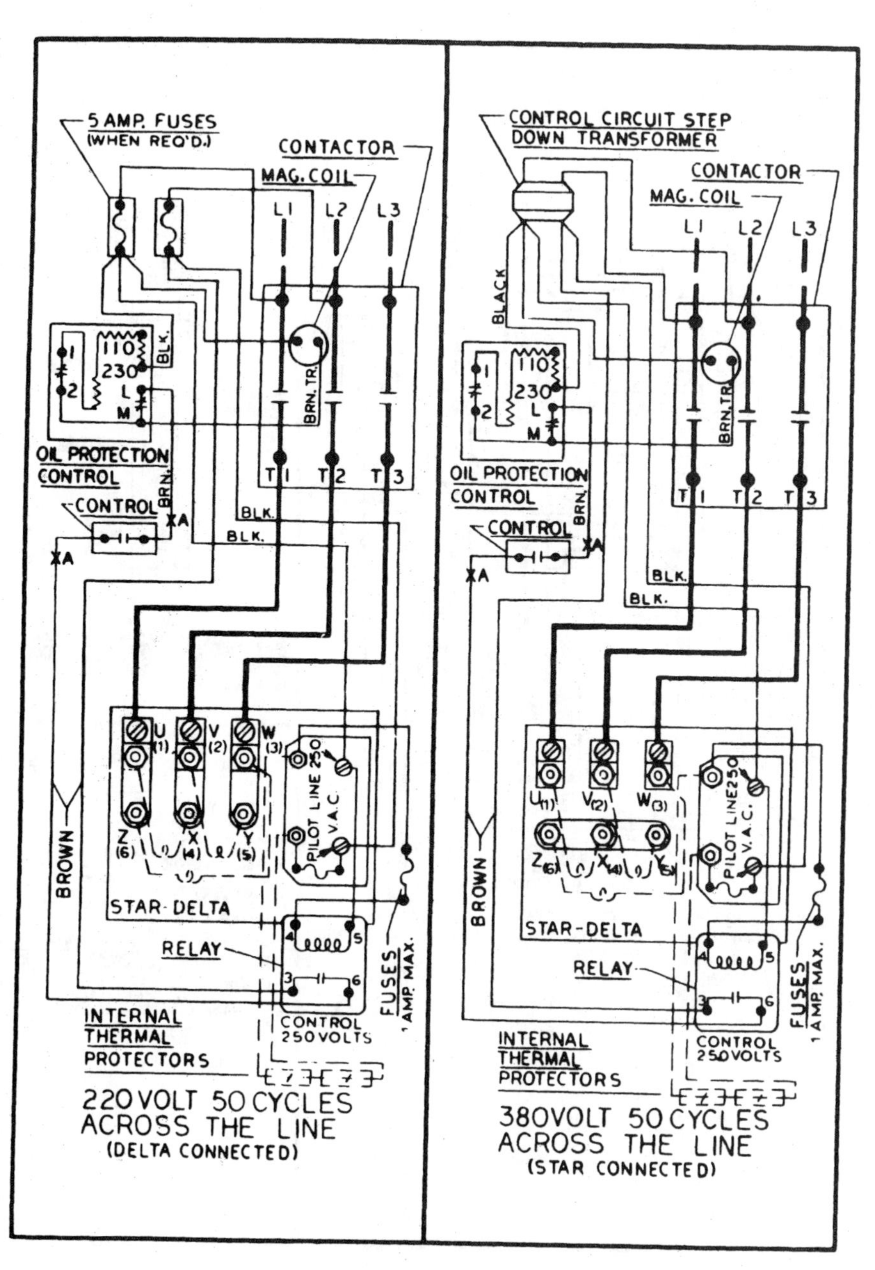

Wiring Diagram No. 591 (Part 1)

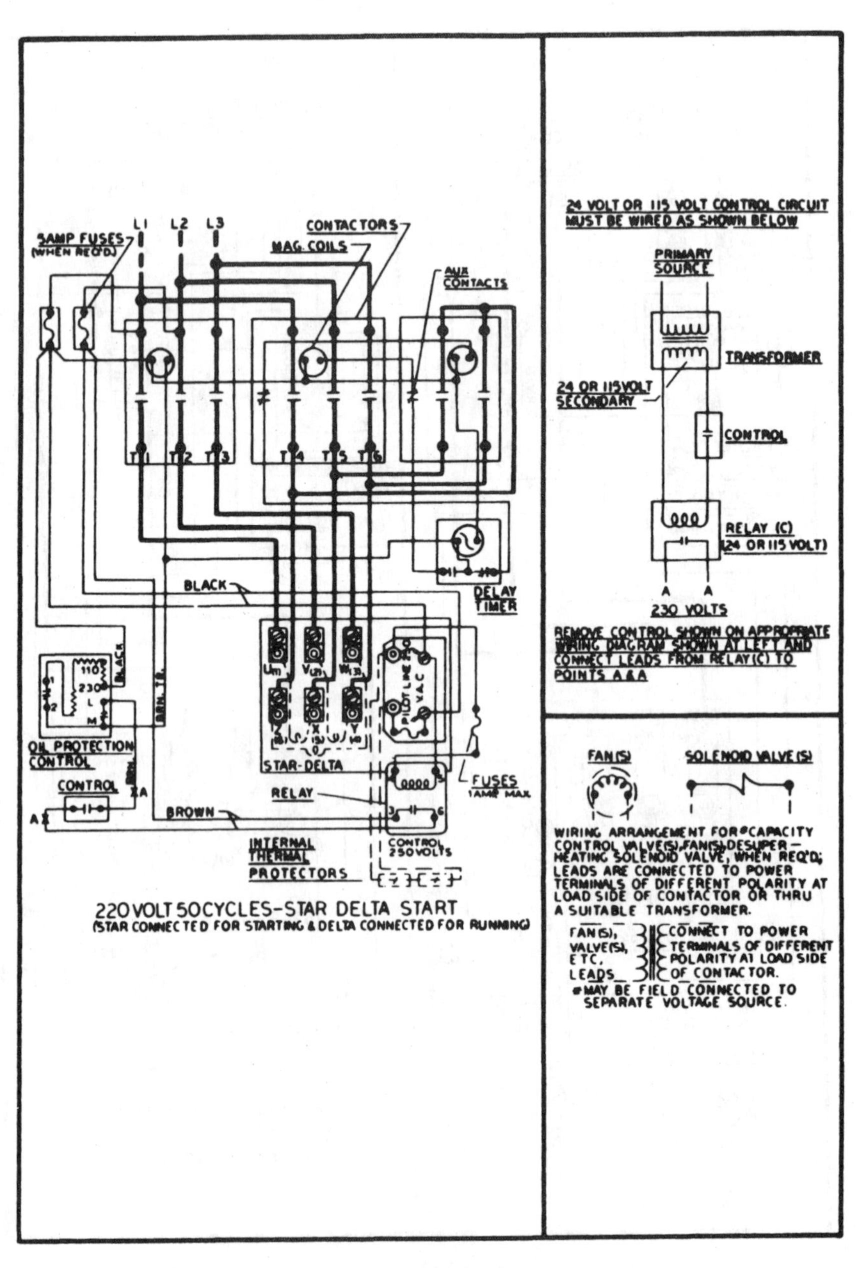

Wiring Diagram No. 591 (Part 2)

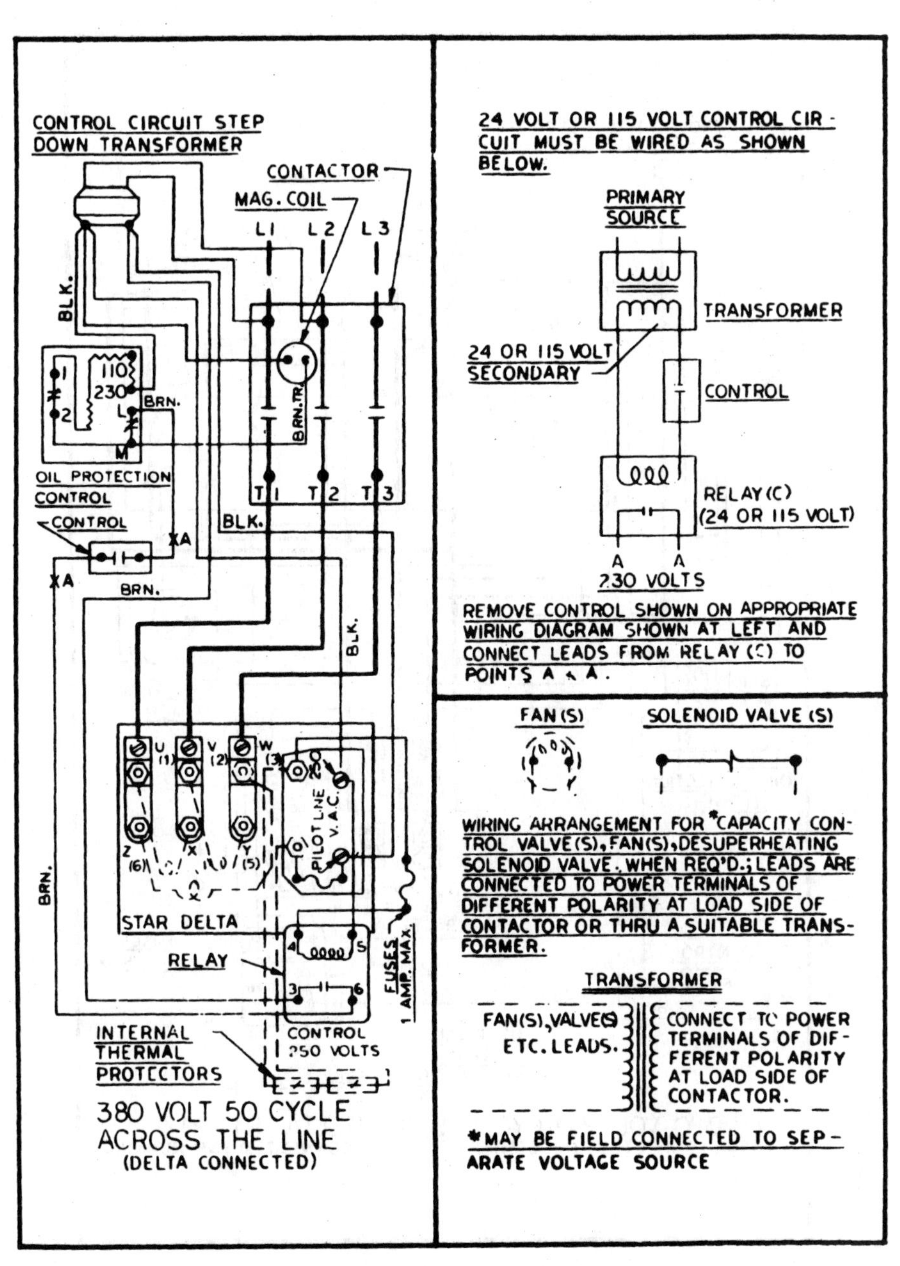

Wiring Diagram No. 594 (Part 1)

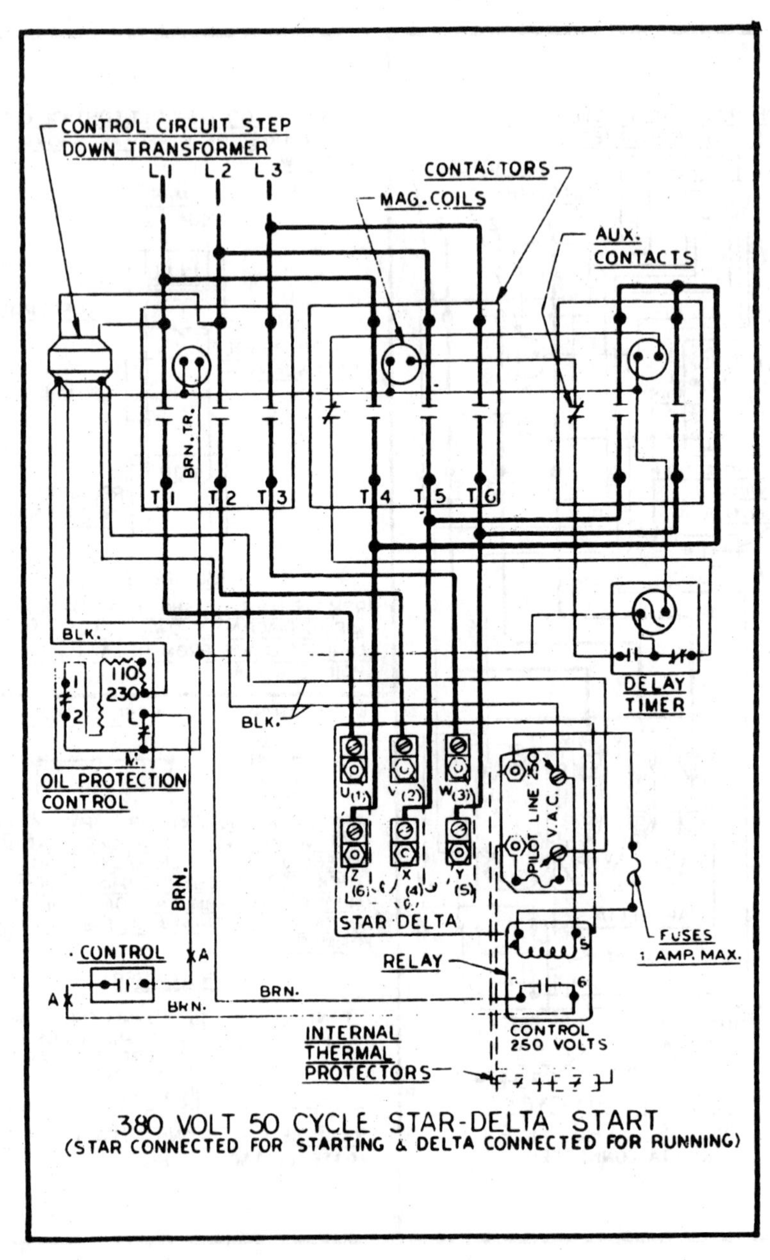

Wiring Diagram No. 594 (Part 2)

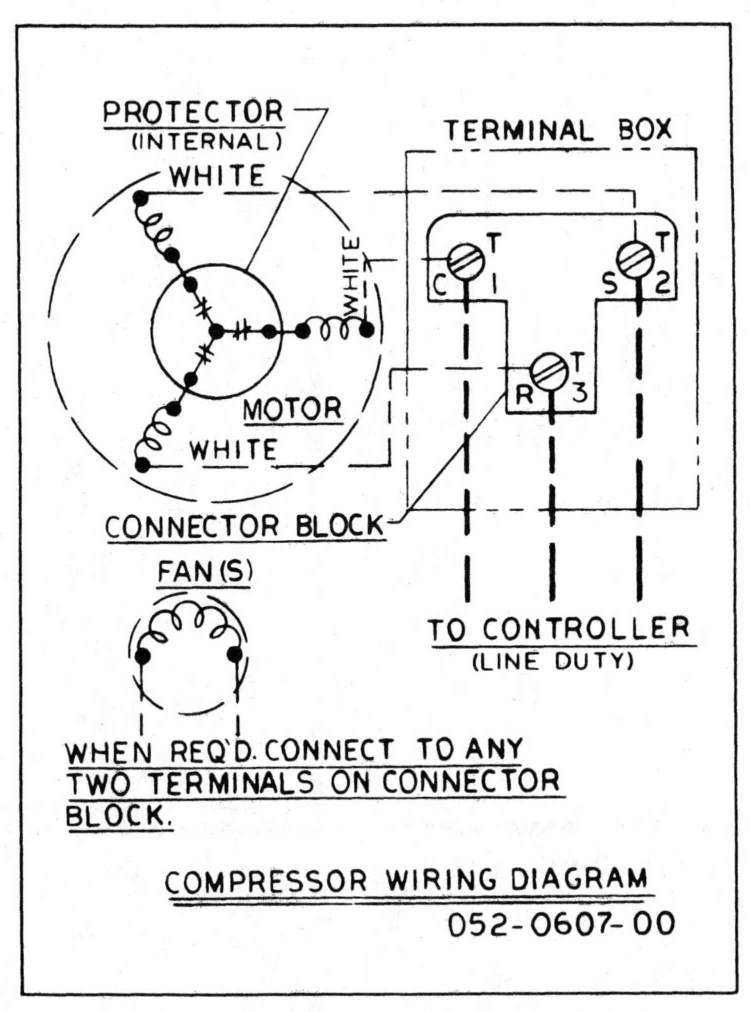

Wiring Diagram No. 607

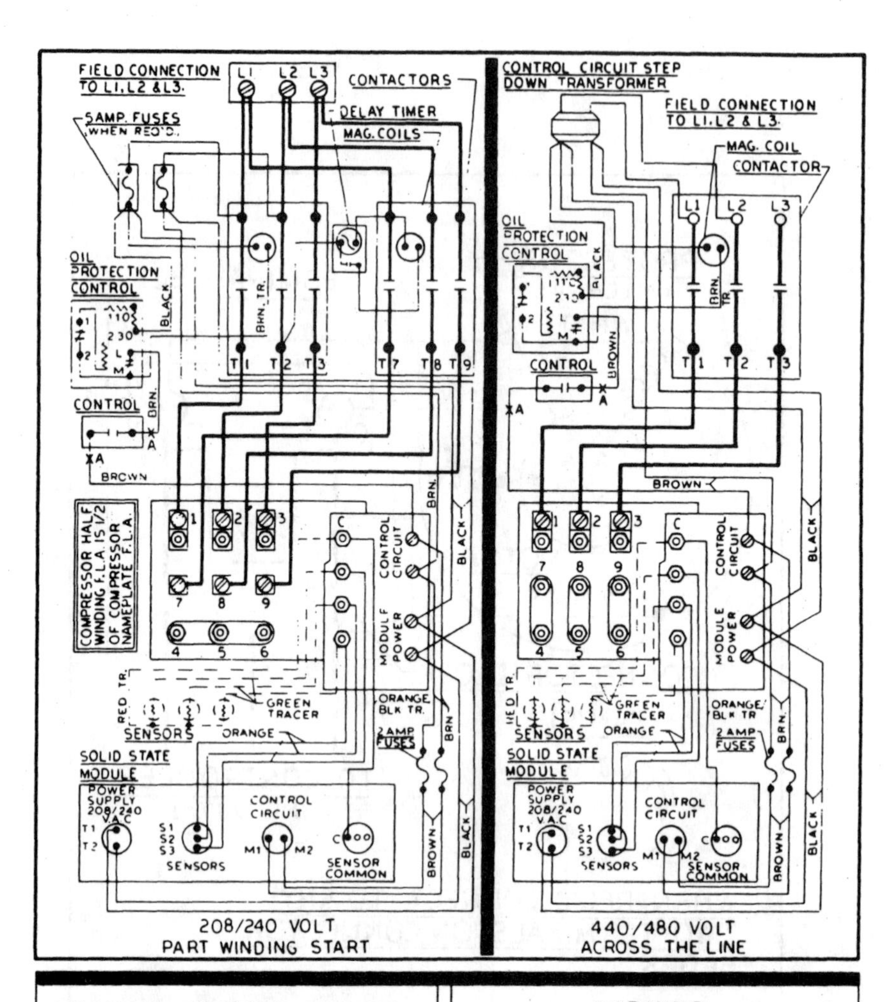

WARNING—TO AVOID DAMAGE

USE OHMMETER ONLY (3 VOLT MAX.) TO CHECK SENSOR RESISTANCE. DO NOT SHORT ACROSS THE TERMINALS. RESISTANCE ACROSS EACH MOTOR SENSOR SHOULD BE APPROXIMATELY 75 OHMS (COLD) TO 125 OHMS (HOT) WITH SENSORS DISCONNECTED FROM MODULE.

1

WARNING

WARRANTY IS VOIDED IF MODULE OR FUSES IN COMPRESSOR TERMINAL BOX OTHER THAN THOSE APPROVED BY COPELAND ARE USED.

A BLOWN FUSE OR TRIPPED MODULE INDICATES EITHER A SHORT, GROUND OR OVERLOAD WHICH MUST BE FOUND AND CORRECTED BEFORE REPLACING FUSE OR STARTING MOTOR.

Solid State Wiring Diagram #608
220 V. Robertshaw Control Circuit

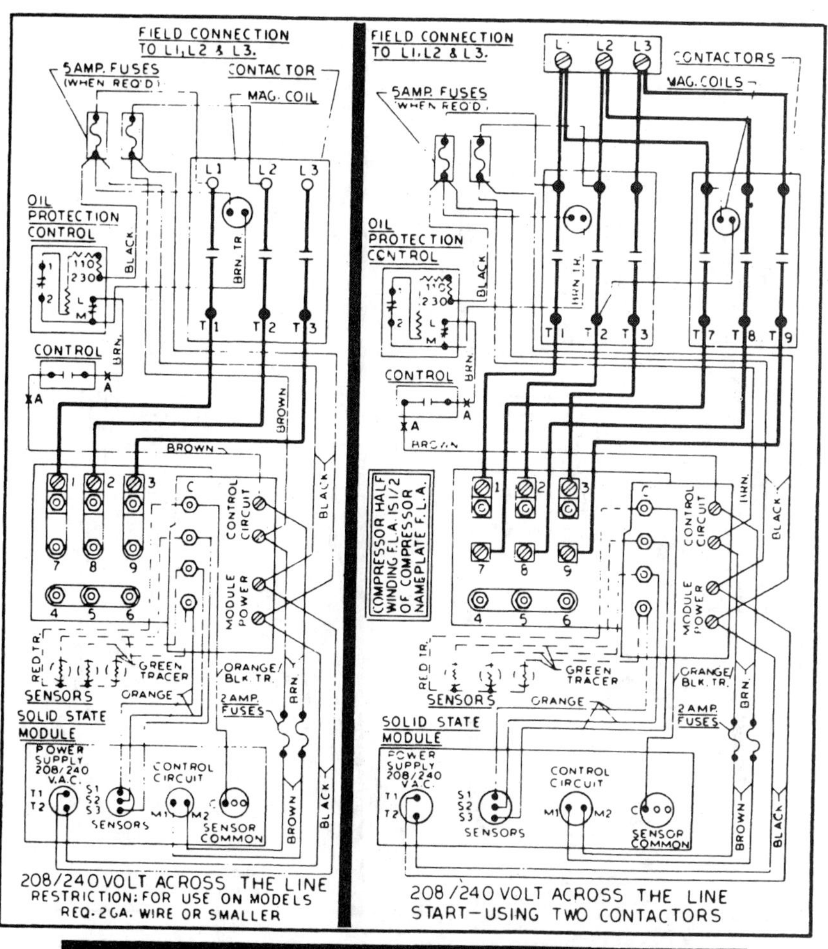

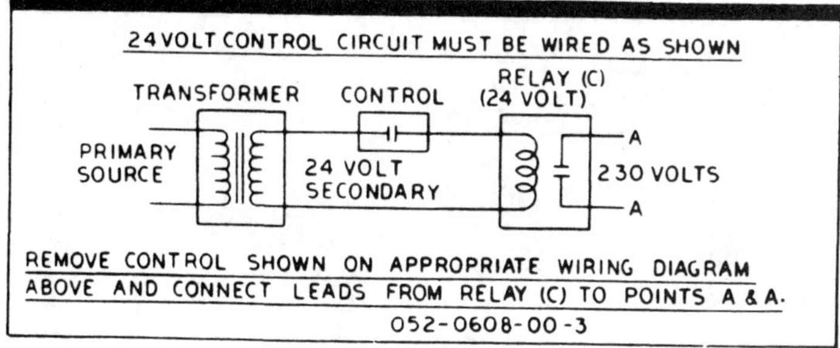

Solid State Wiring Diagram #608
220 V. Robertshaw Control Circuit

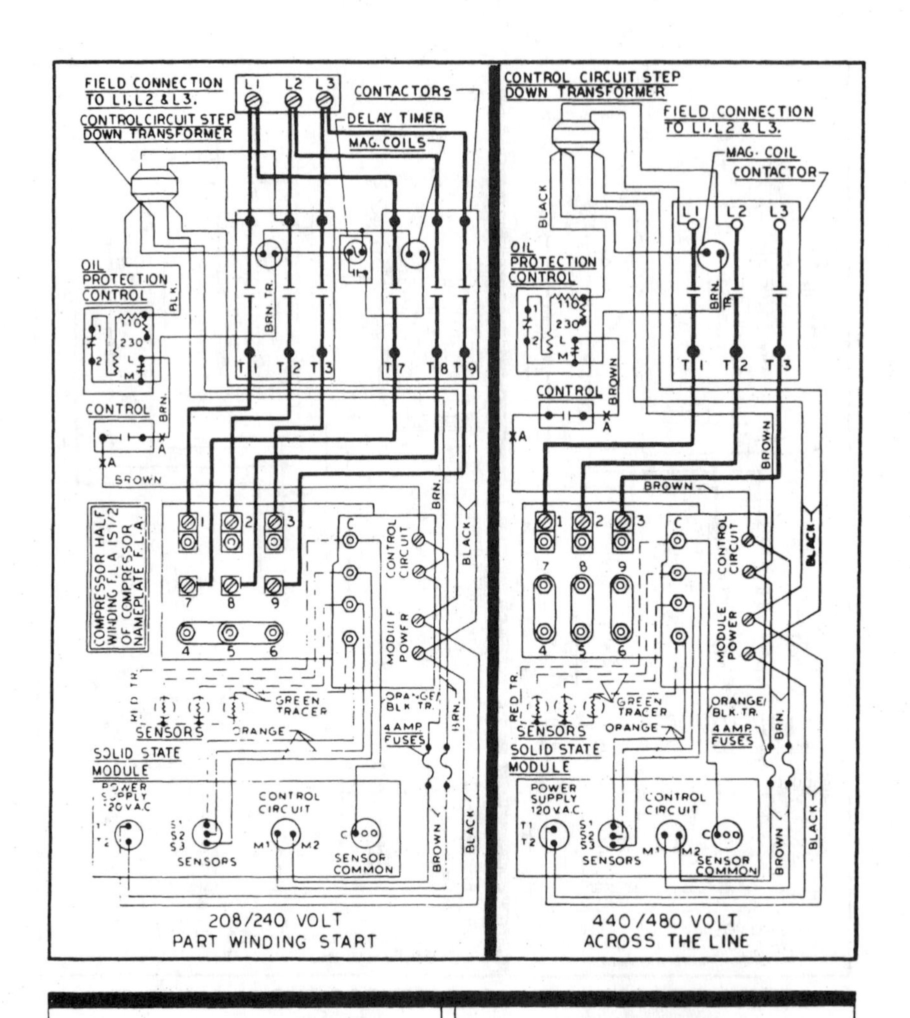

WARNING—TO AVOID DAMAGE

USE OHMMETER ONLY (3 VOLT MAX.) TO CHECK SENSOR RESISTANCE. DO NOT SHORT ACROSS THE TERMINALS. RESISTANCE ACROSS EACH MOTOR SENSOR SHOULD BE APPROXIMATELY 75 OHMS (COLD) TO 125 OHMS (HOT) WITH SENSORS DISCONNECTED FROM MODULE.

WARNING

WARRANTY IS VOIDED IF MODULE OR FUSES IN COMPRESSOR TERMINAL BOX OTHER THAN THOSE APPROVED BY COPELAND ARE USED.

A BLOWN FUSE OR TRIPPED MODULE INDICATES EITHER A SHORT, GROUND OR OVERLOAD WHICH MUST BE FOUND AND CORRECTED BEFORE REPLACING FUSE OR STARTING MOTOR.

Solid State Wiring Diagram #610
120 V. Robertshaw Control Circuit

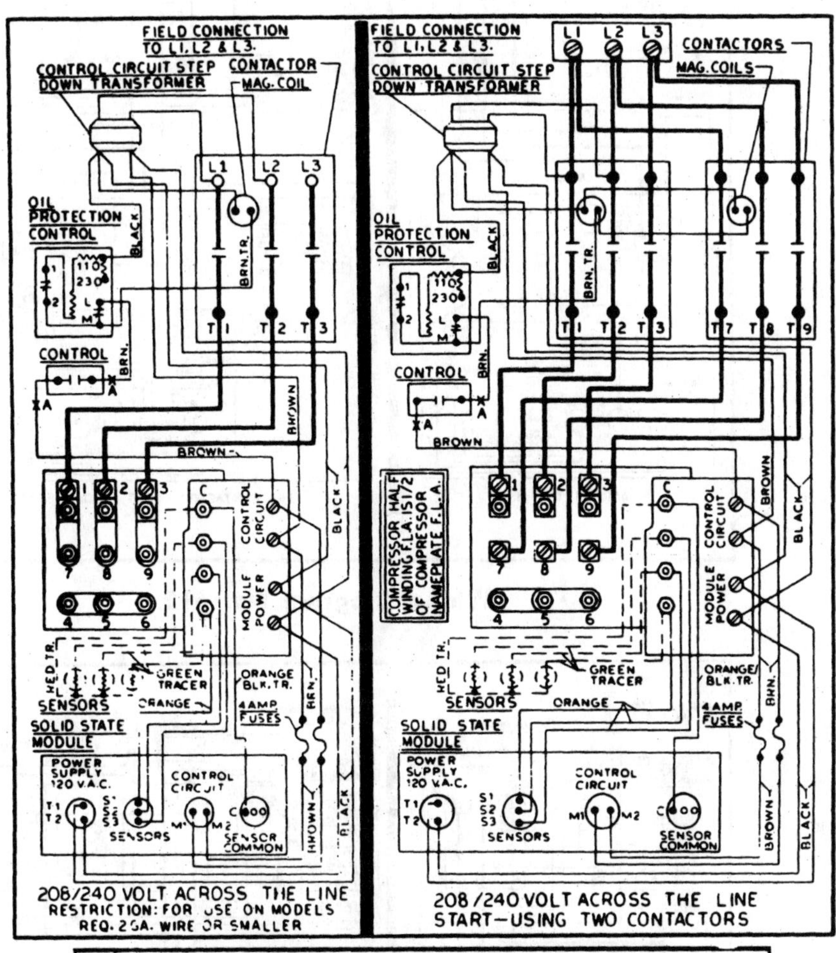

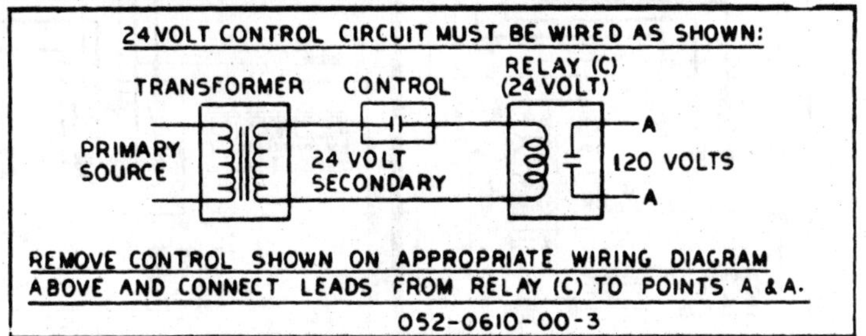

Solid State Wiring Diagram #610
120 V. Robertshaw Control Circuit

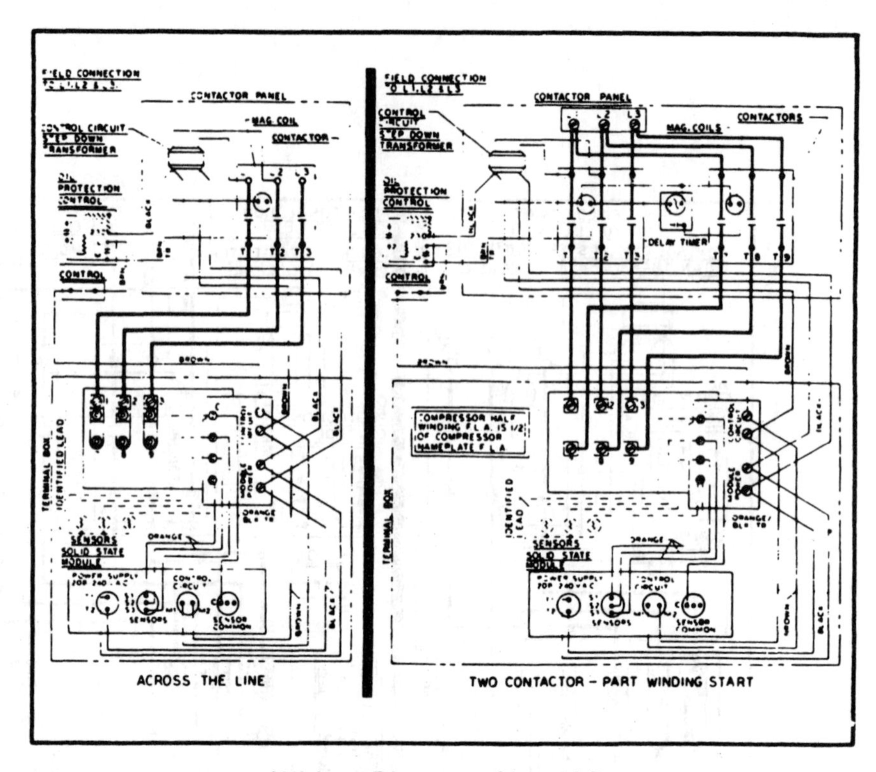

Wiring Diagram No. 612

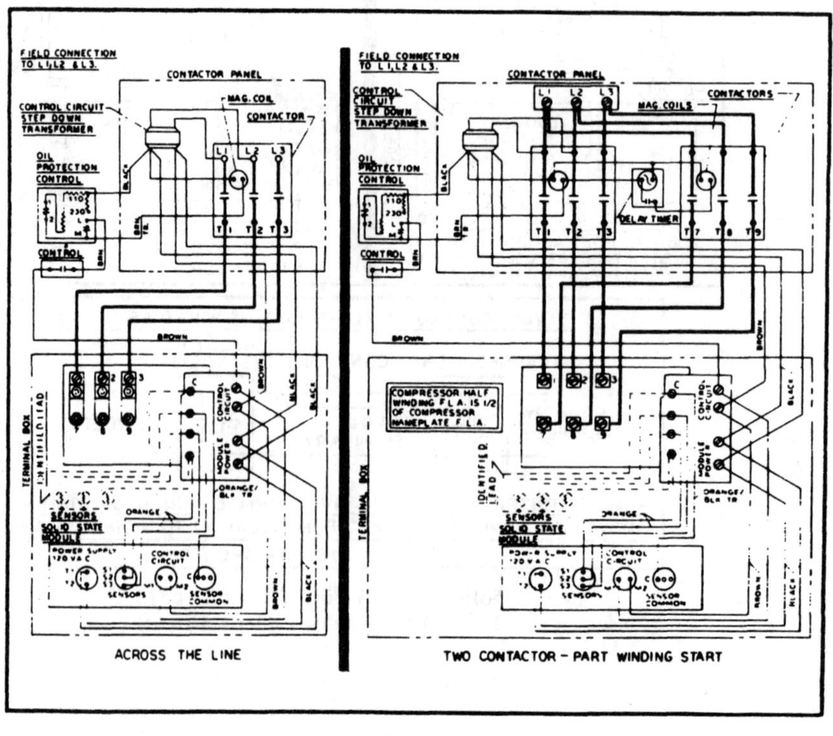

Wiring Diagram No. 613

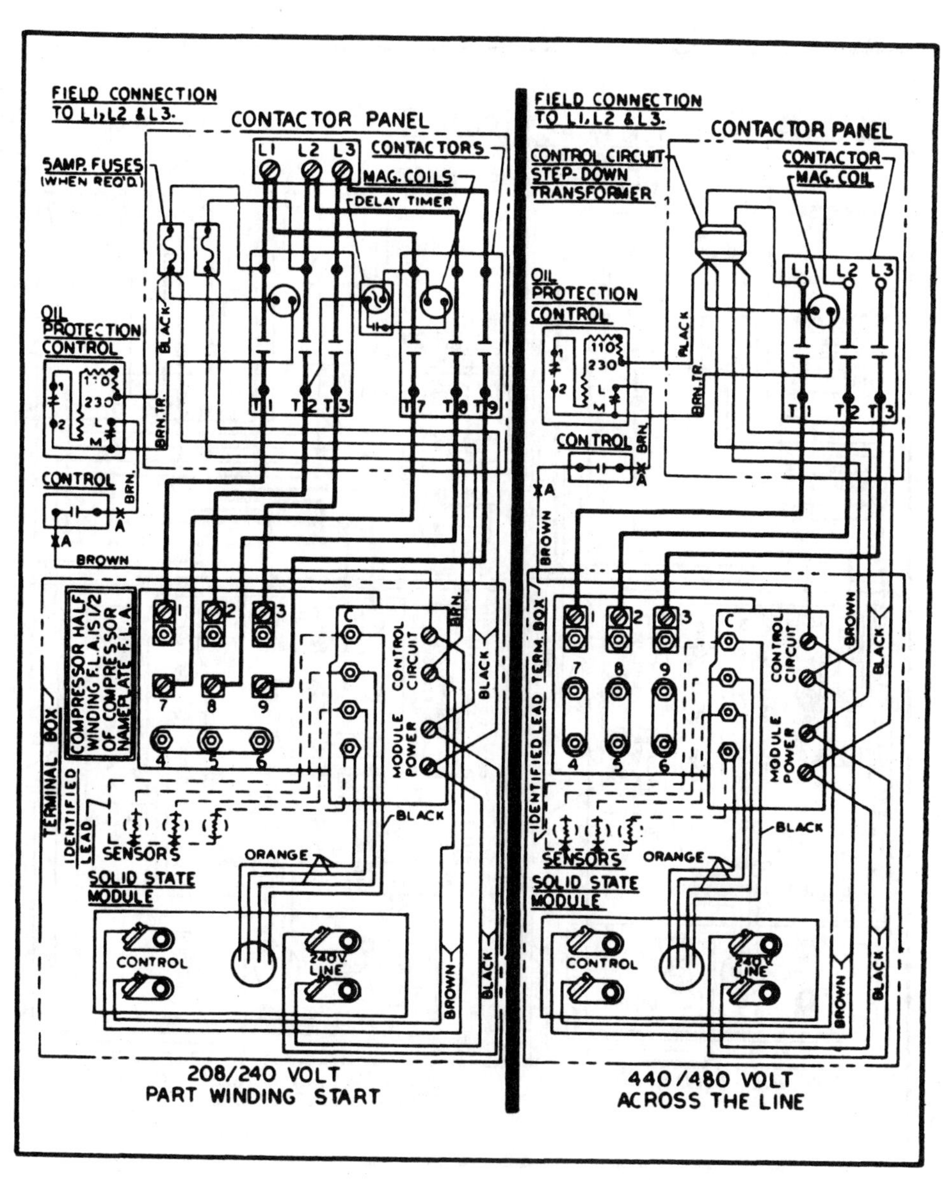

Wiring Diagram No. 616

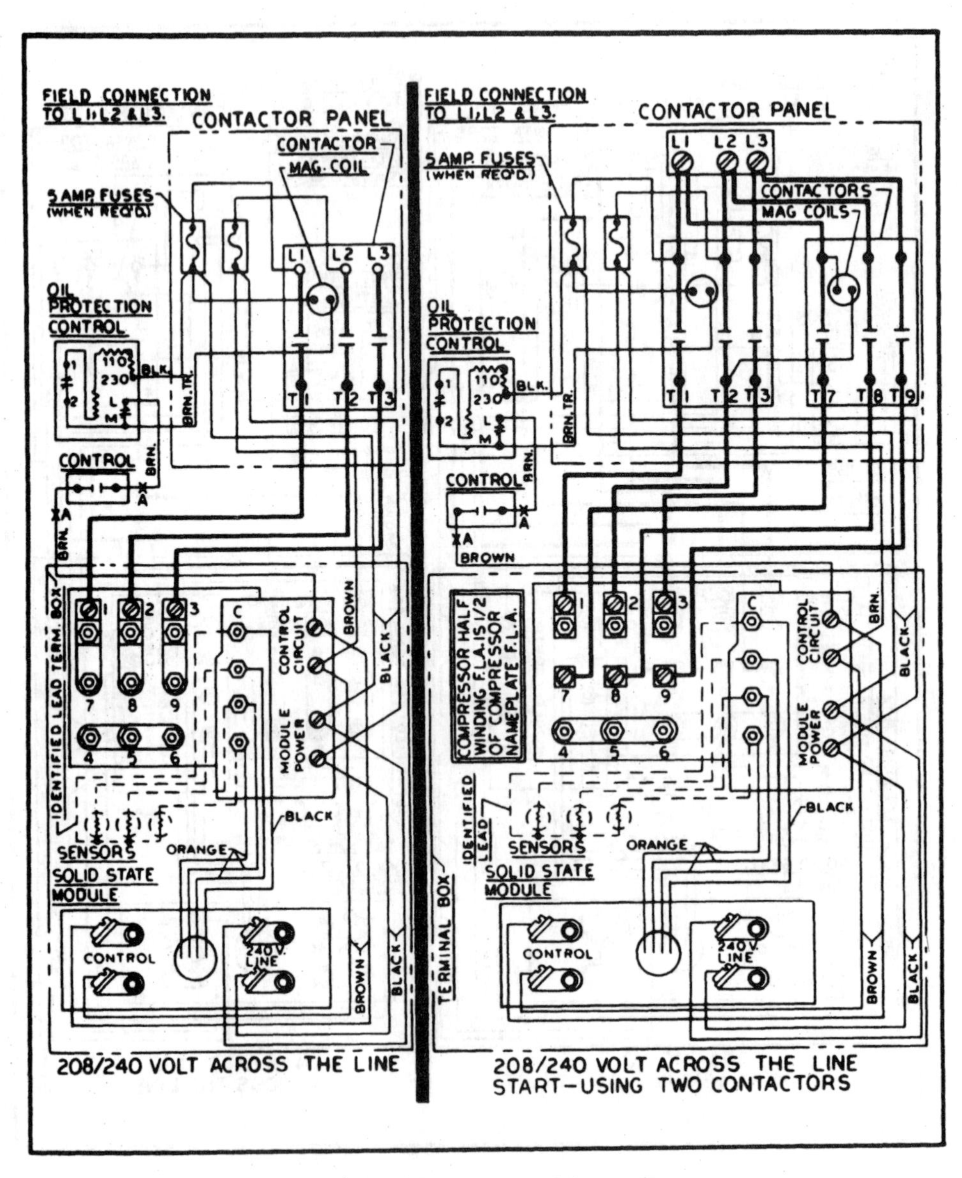

Wiring Diagram No. 616

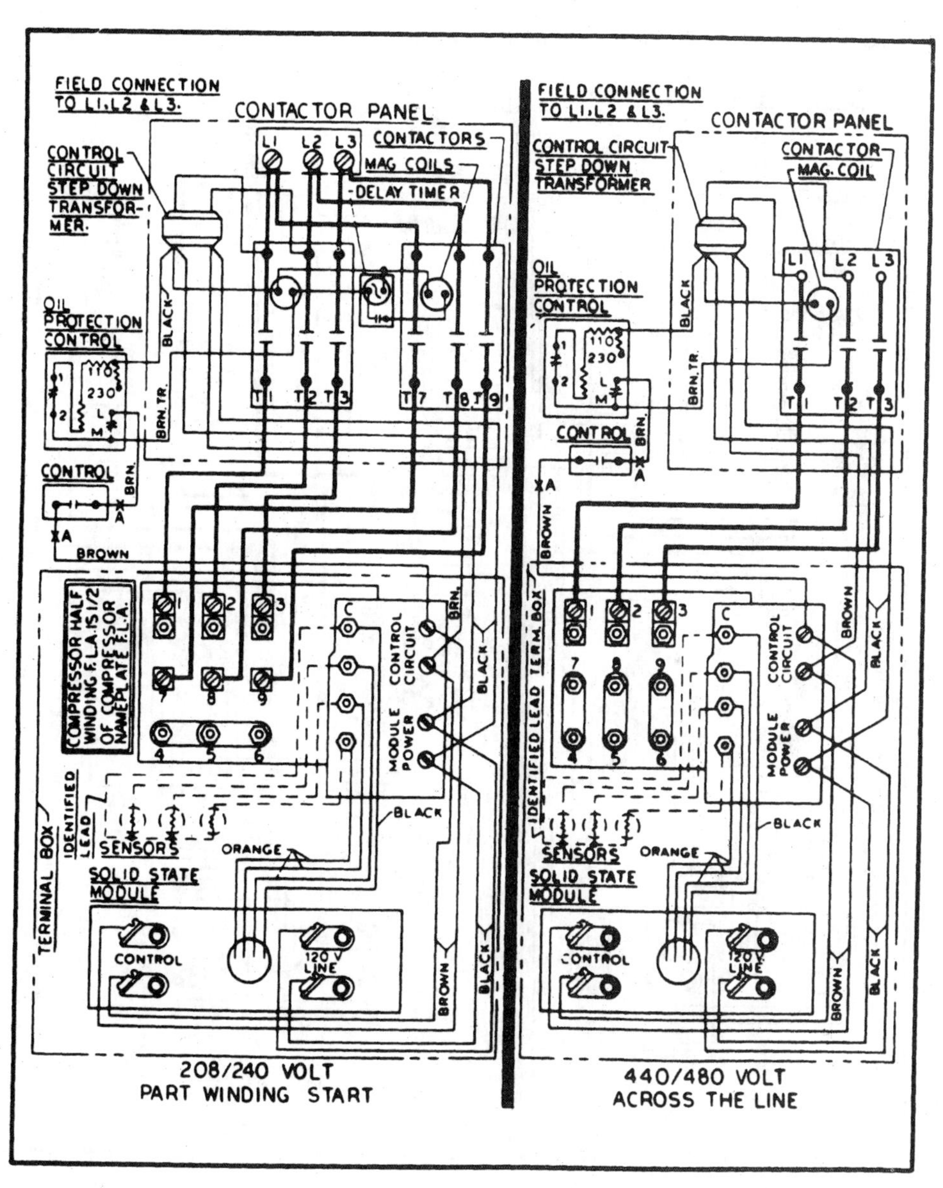

Wiring Diagram No. 618

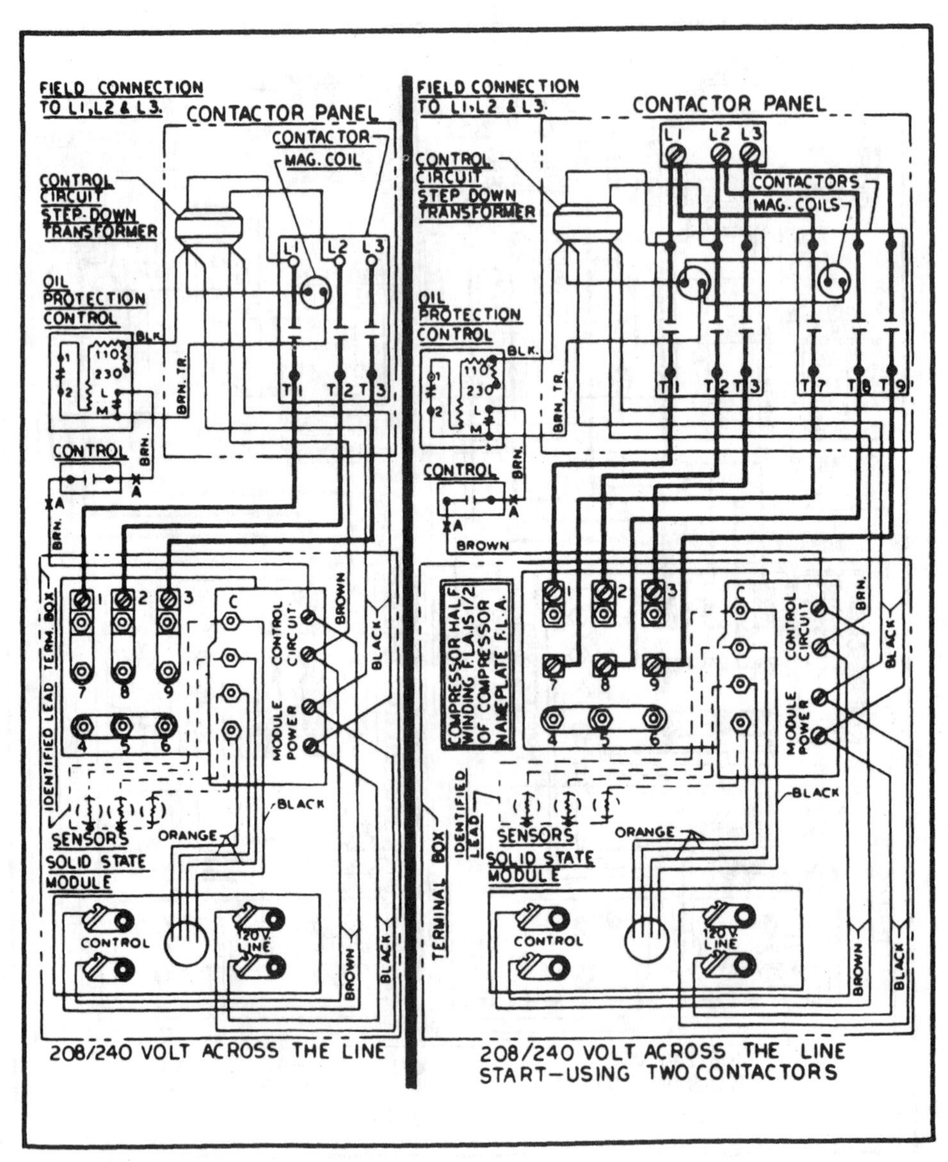

Wiring Diagram No. 618

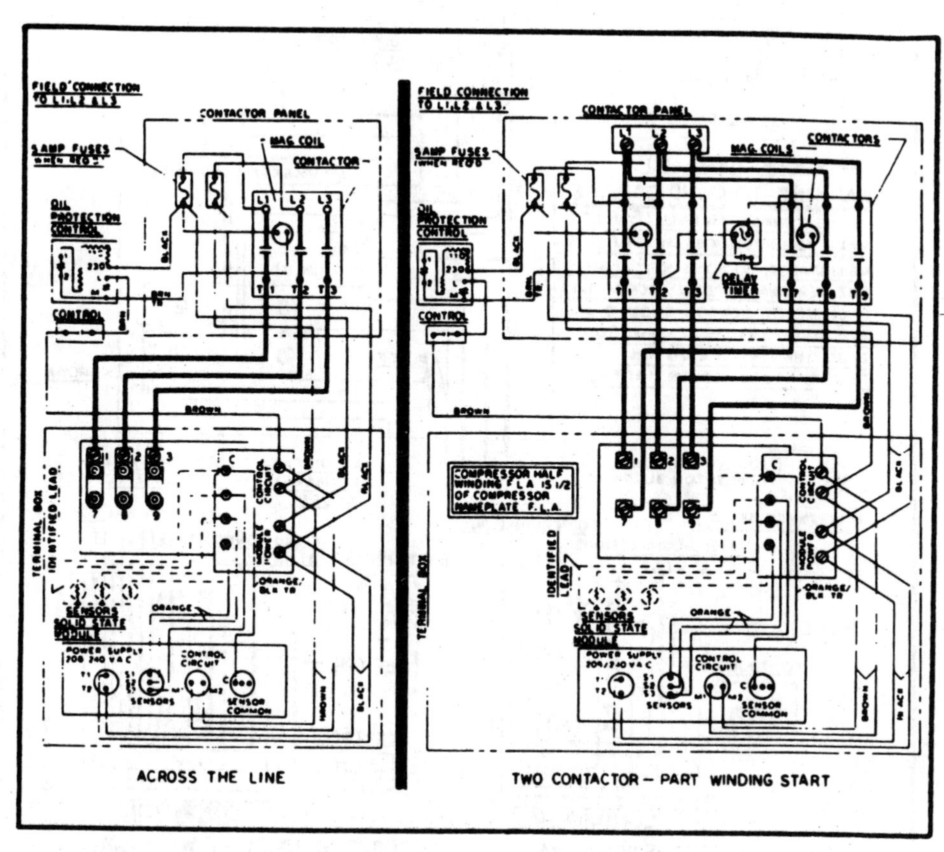

Wiring Diagram No. 622

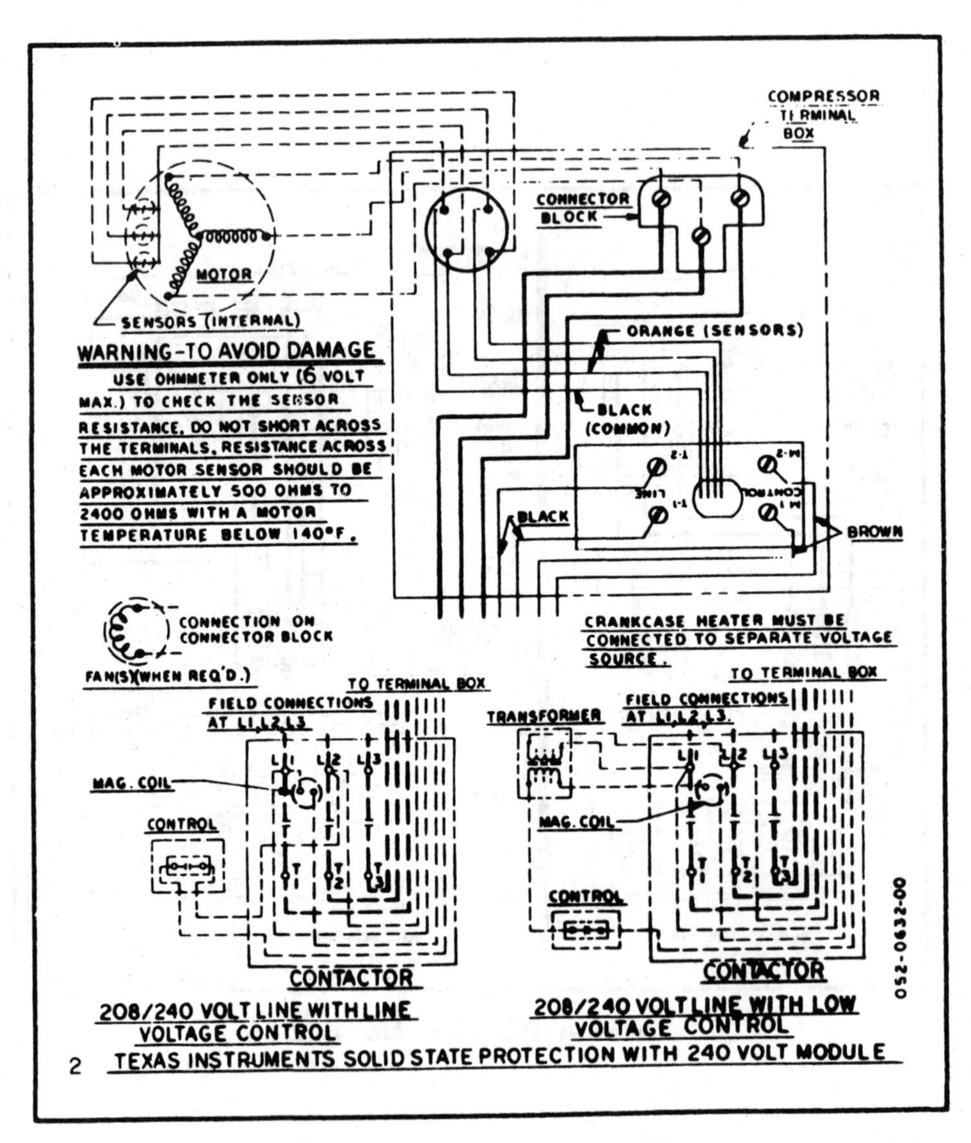

Wiring Diagram No. 632

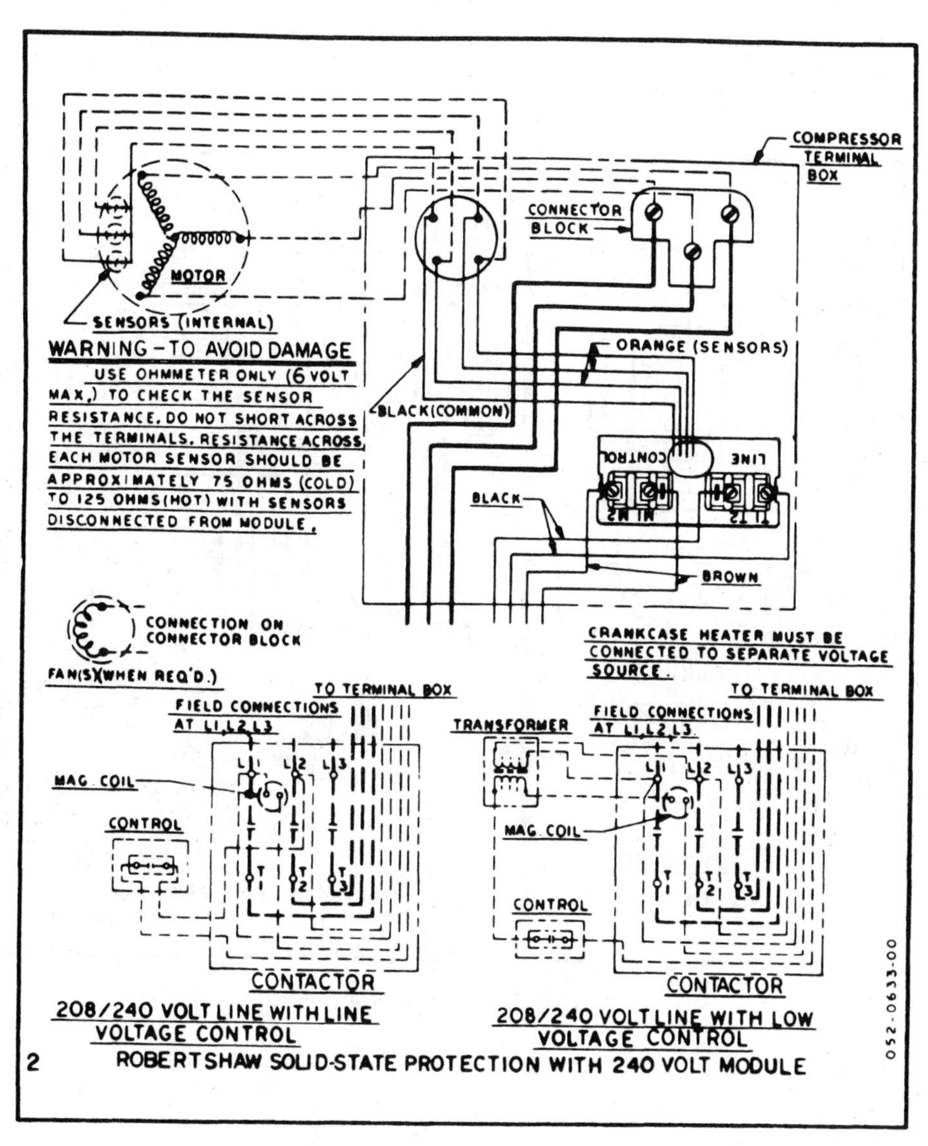

Wiring Diagram No. 633

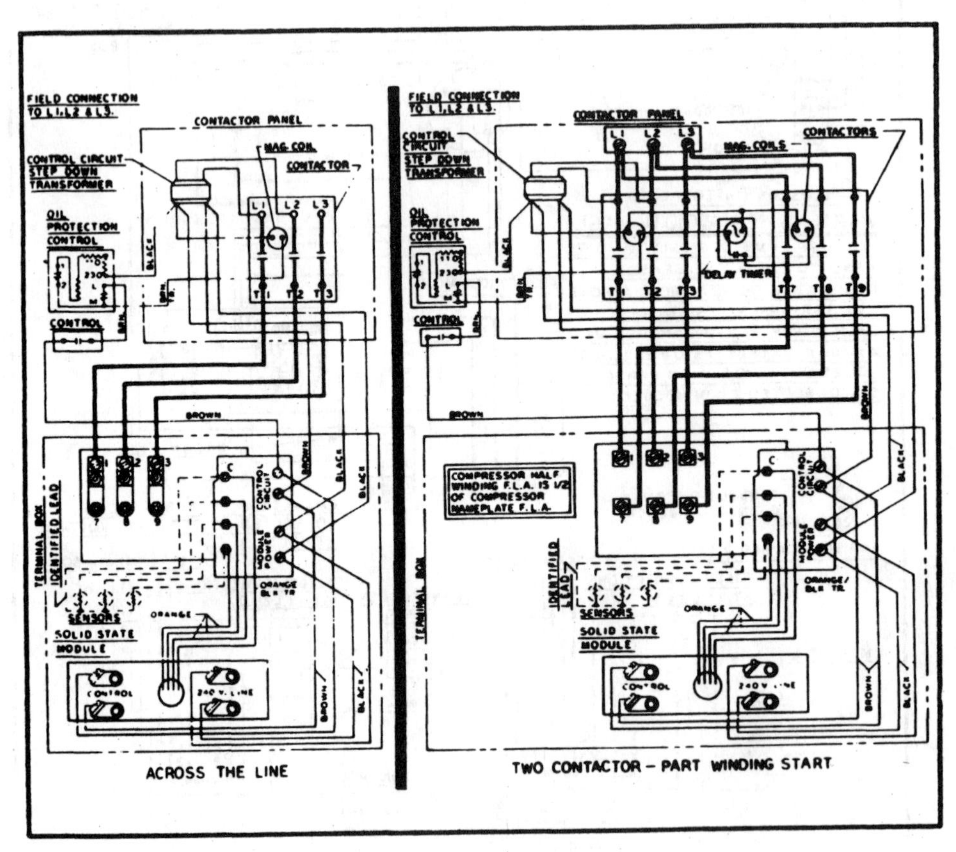

Wiring Diagram No. 635

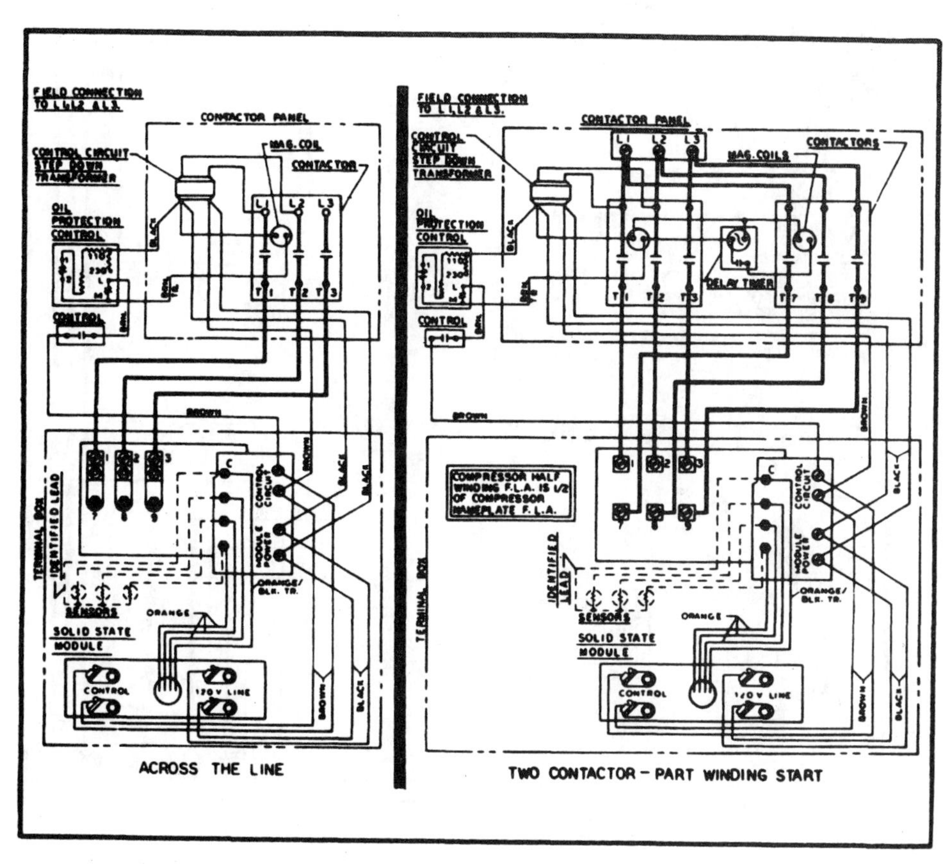

Wiring Diagram No. 636

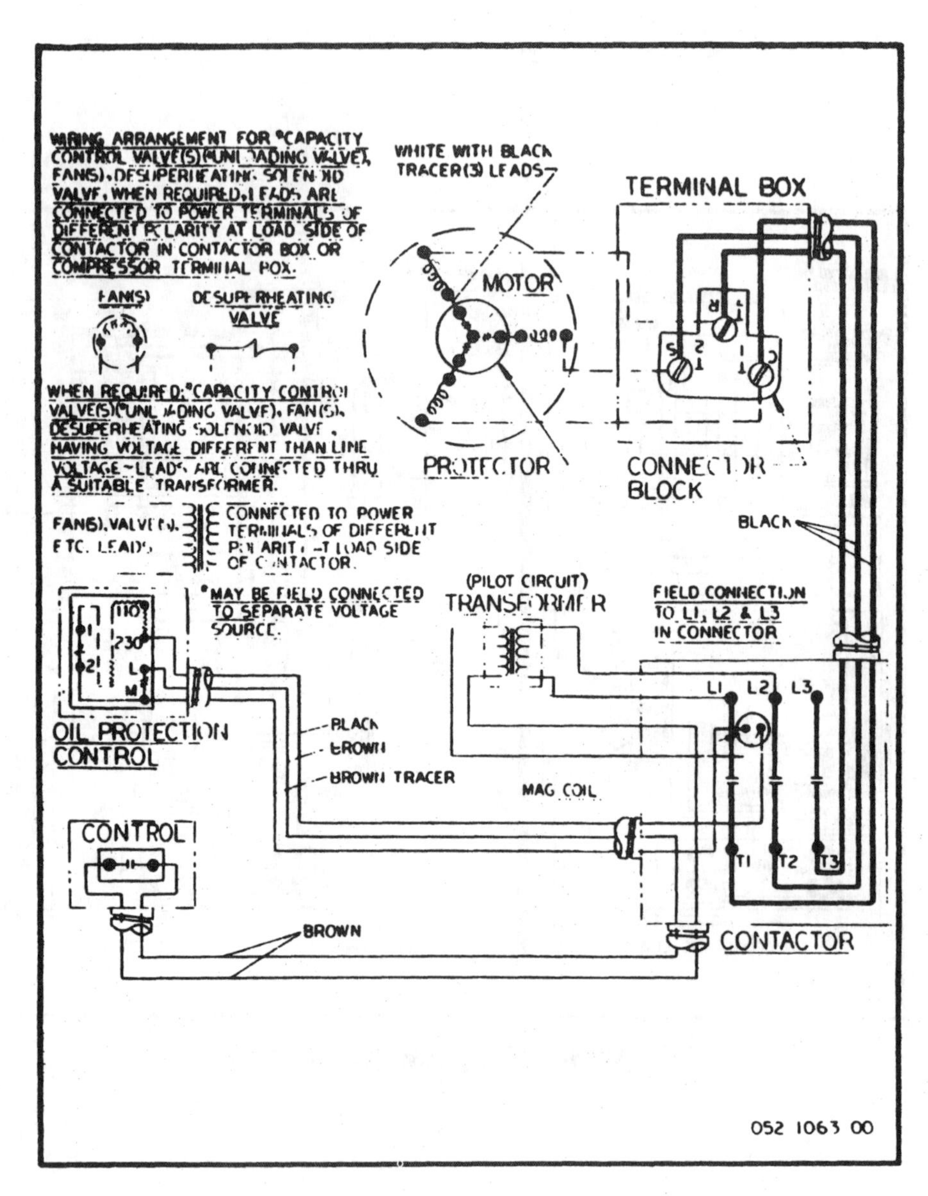

Wiring Diagram No. 1063

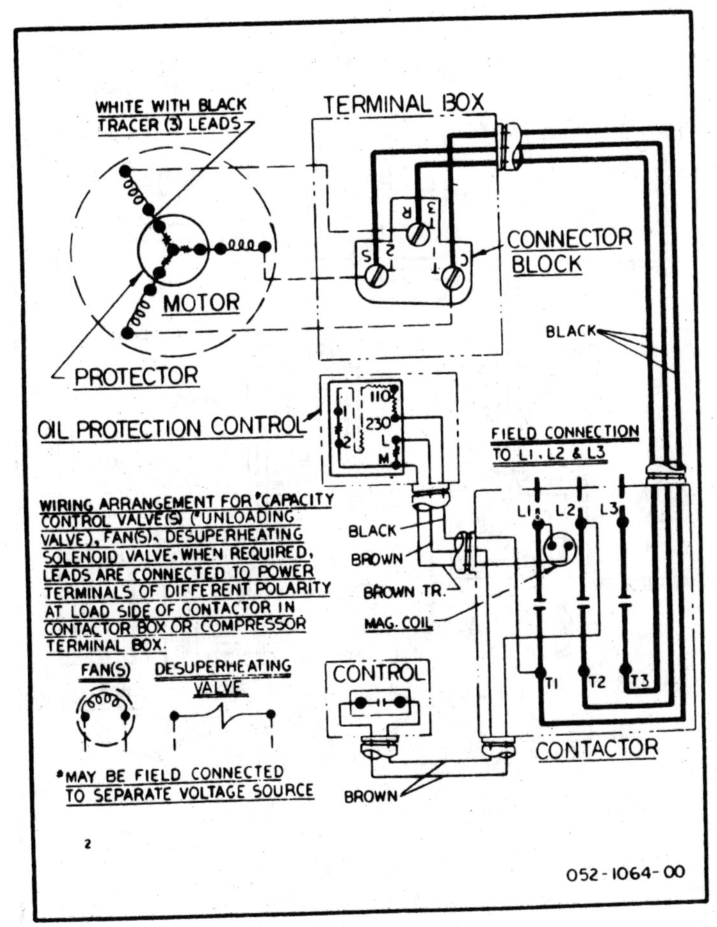

Wiring Diagram No. 1064

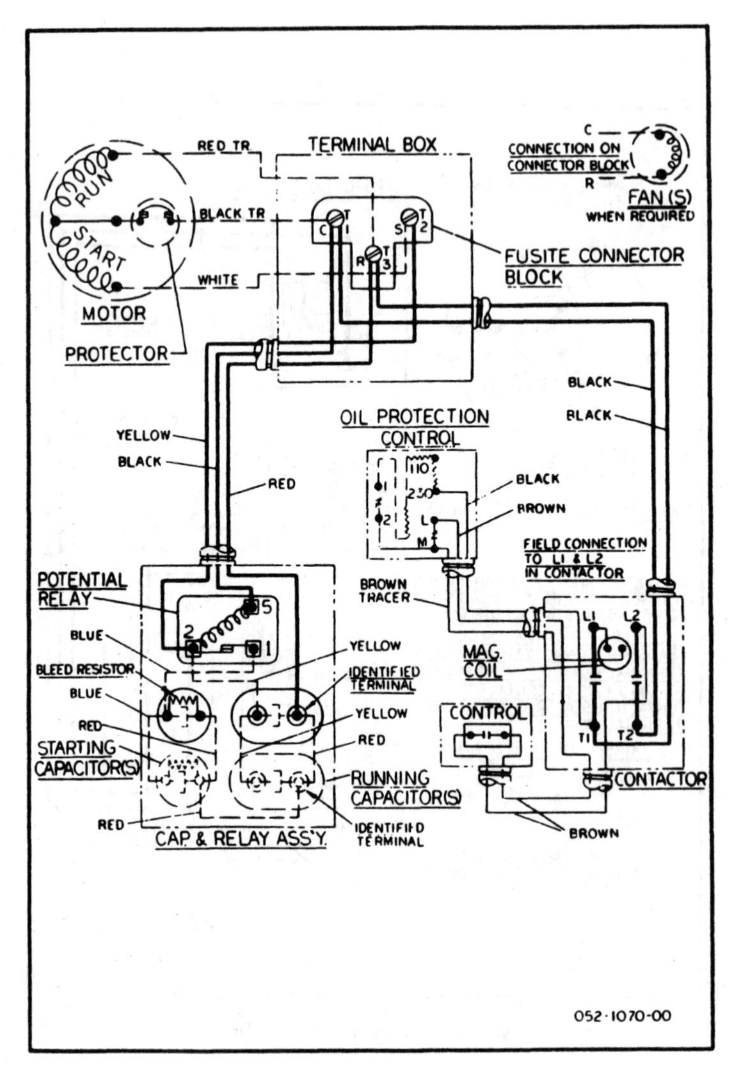

Wiring Diagram No. 1070

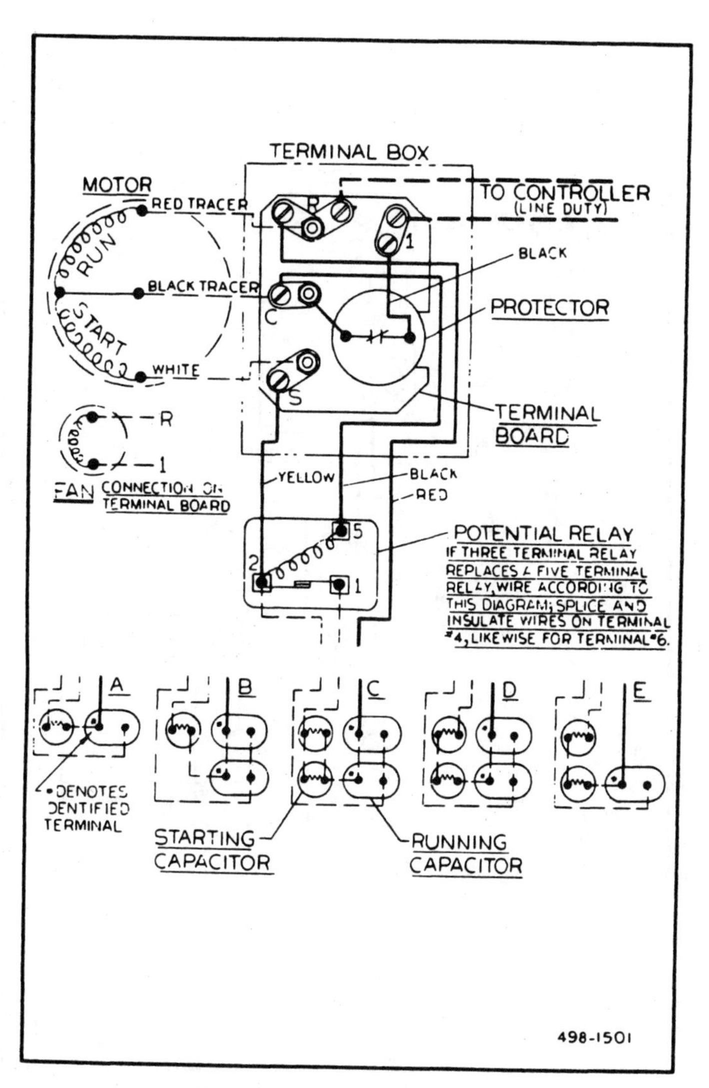

Wiring Diagram No. 1501

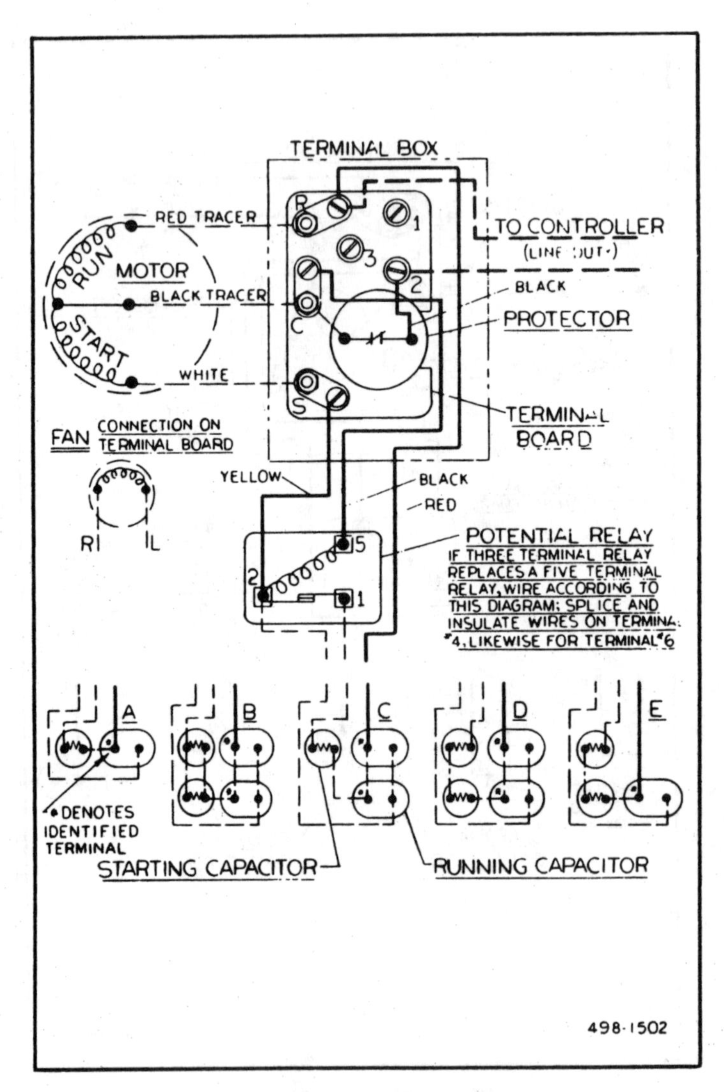

Wiring Diagram No. 1502

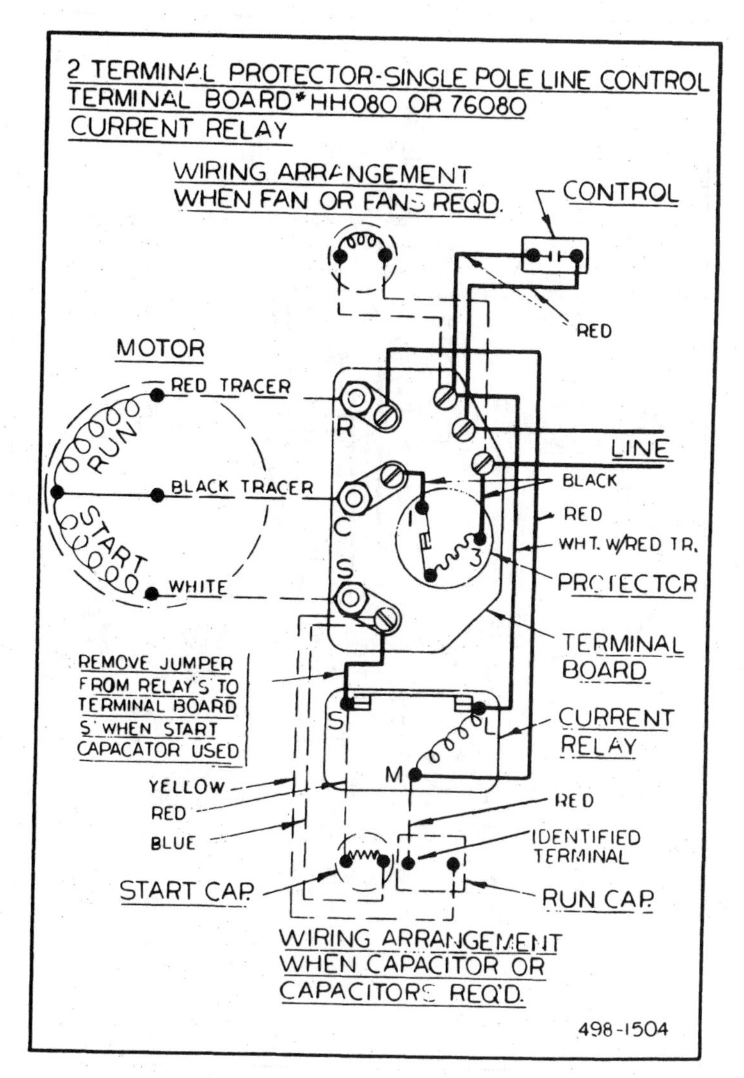

Wiring Diagram No. 1504

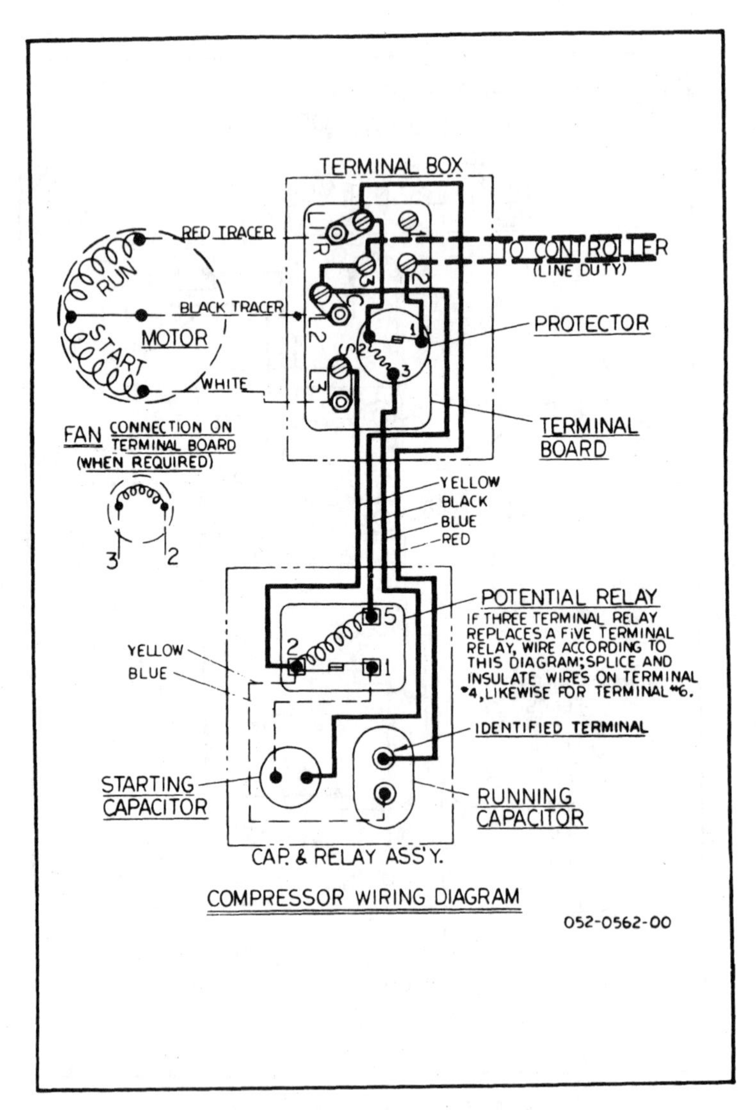

Wiring Diagram No. 1519

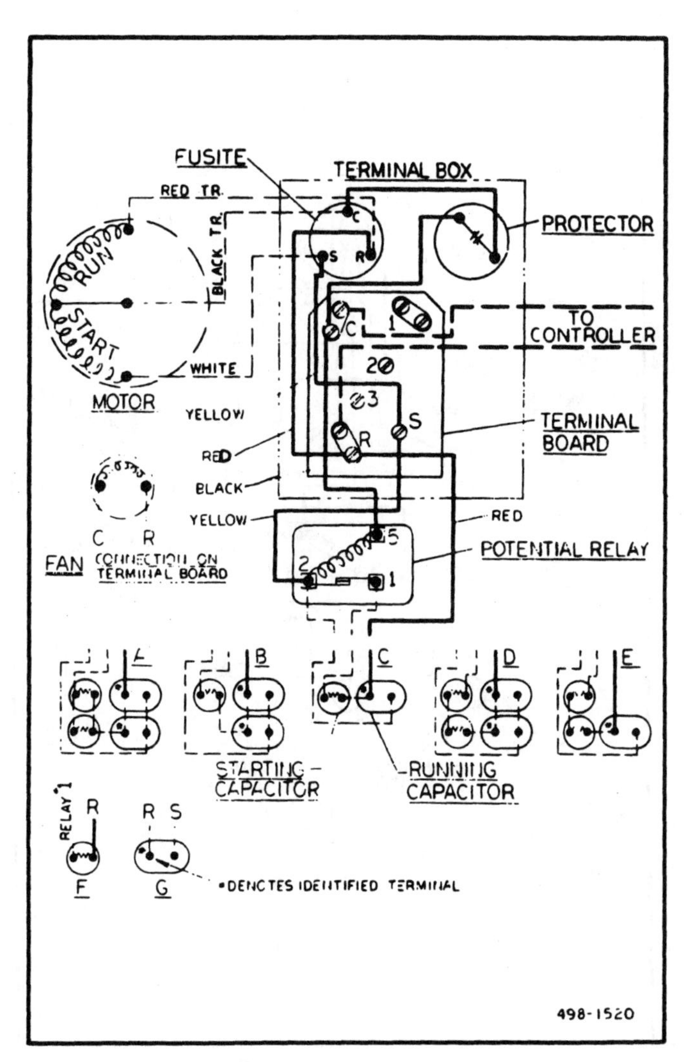

Wiring Diagram No. 1520

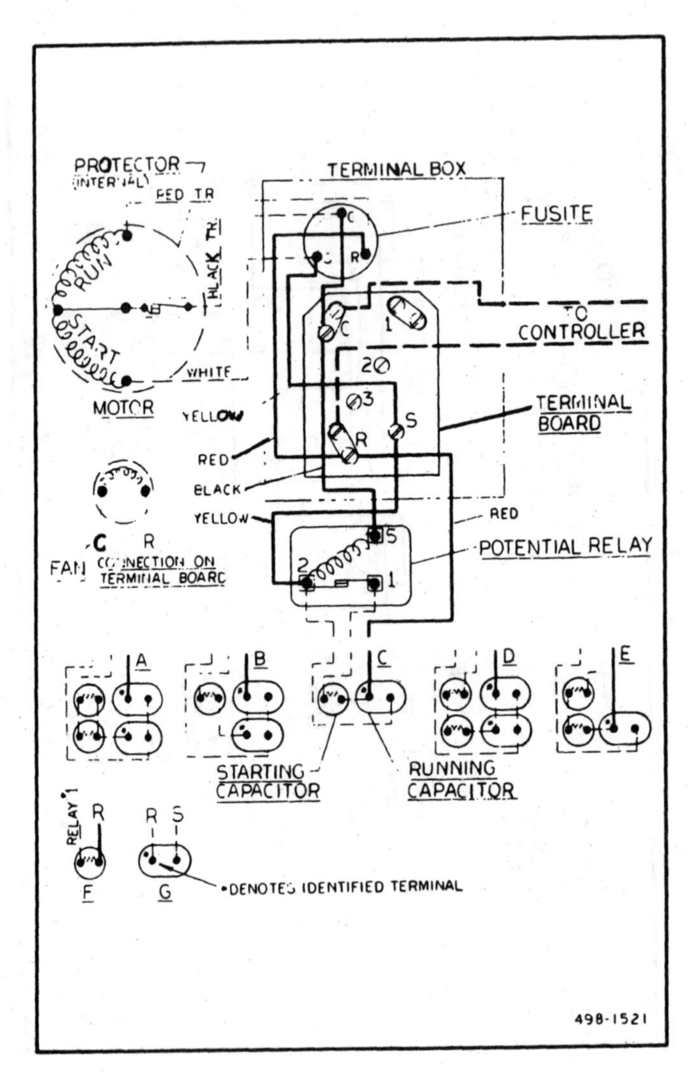

Wiring Diagram No. 1521

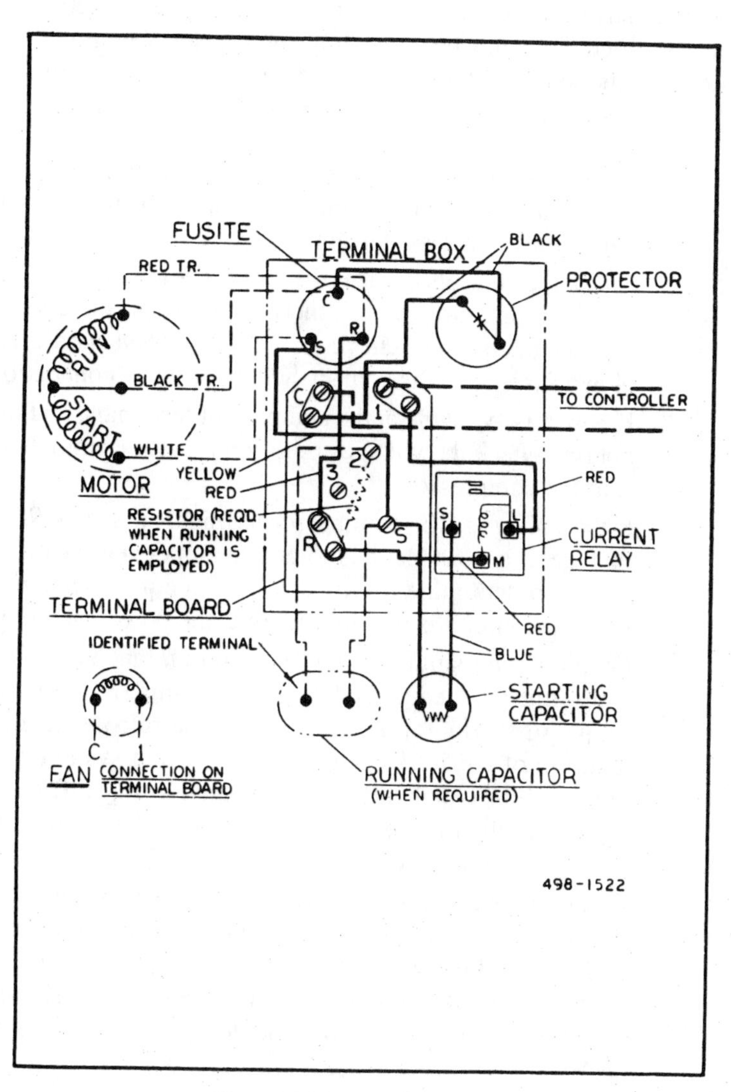

Wiring Diagram No. 1522

Electrical Trouble Shooting

(A) If The Compressor Will Not Run

If a motor compressor fails to start and run properly, it is important that the compressor be tested to determine its condition. It is possible that external electrical components may be defective, the protector may be open, a safety device may be tripped, or other conditions may be preventing compressor operation. If the motor compressor is not the source of the malfunction, replacing the compressor will only result in the unnecessary expenditure of time and money, while the basic problem remains.

1. If there is no voltage at the compressor terminals, follow the wiring diagram and check back from the compressor to the power supply to find where the circuit is interrupted.

   Check the controls to see if the contact points are closed (low pressure control, high pressure control, thermostat, oil pressure safety control, etc.). If a contactor is used, check to see if the contacts are closed. Check for a blown fuse, open disconnect switch, or loose connection.
2. If power is available at the compressor terminals, and the compressor does not run, check the voltage at the compressor terminals while attempting to start the compressor.

   If the voltage at the compressor terminals is below 90% of the nameplate voltage, it is possible the motor may not develop sufficient torque to start. Check to determine if wire sizes are adequate, electrical connections are loose, the circuit is overloaded, or if the power supply is adequate.
3. On units with single phase PSC motors, the suction and discharge pressures must be equalized before starting because of the low starting torque of the motor. Any change in the refrigerant metering device, the addition of a drier, or other changes in the system components may delay pressure equalization and create starting difficulties. If PSC motor starting problems are being encountered, the addition of a capacitor start kit is recommended.
4. On single phase compressors, a defective capacitor or relay may prevent the compressor starting. If the compressor attempts to start, but is unable to do so, or if there is a humming sound, check the relay to see if the relay contacts are damaged or fused. The relay points should be closed during the initial starting cycle, but should open as the compressor comes up to speed.

   Remove the wires from the starting relay and capacitors. Use a high voltage ohmmeter to check for continuity through the relay coil. Replace the relay if there is no continuity. Use an ohmmeter to check across the relay contacts. Potential relay contacts are normally closed when the relay is not energized; current relay contacts are normally open. If either gives an incorrect reading, replace the relay.

   Replace any capacitor found to be bulging, leaking, or damaged.

Make sure capacitors are discharged before checking. Check for continuity between each capacitor terminal and the case. Continuity indicates a short, and the capacitor should be replaced.

Substitute "a known to be good" start capacitor if available. If compressor then starts and runs properly, replace the original start capacitor. On PSC motors, substitute "a known to be good" run capacitor if available. If compressor then starts and runs properly, replace the original run capacitor.

If a capacitor tester is not available, use an ohmmeter to check run and start capacitors for shorts or open circuits. Use an ohmmeter set to its highest resistance scale, and connect prods to capacitor terminals.

a. With a good capacitor, the indicator should first move to zero, and then gradually increase to infinity.
b. If there is no movement of the ohmmeter indicator, an open circuit is indicated.
c. If the ohmmeter indicator moves to zero, and remains there or on a low resistance reading, a short circuit is indicated.
Replace defective capacitors.

5. If the correct voltage is available at the compressor terminals, and no current is drawn, remove all wires from the terminals and check for continuity through the motor windings. On single phase motor compressors, check for continuity from terminals C to R, and C to S. On three phase compressors, check for continuity between the terminals for connections to phases 1 and 2, 2 and 3, and 1 and 3. On compressors with line break inherent protectors, an open overload protector can cause a lack of continuity. If the compressor is warm, wait one hour for compressor to cool and recheck. If continuity cannot be established through all motor windings, the compressor should be replaced.

   Check the motor for ground by means of a continuity check between the common terminal and the compressor shell. If there is a ground, replace the compressor.

6. If the compressor has an external protector, check for continuity through the protector or protectors.

   All external and internal inherent protectors on Copelametic compressors can be replaced in the field. On larger compressors with thermostats, thermotectors, or solid state sensors in the motor windings (D, H, S protection), the internal protective devices cannot be replaced and the stator or compressor must be changed if the internal protectors are defective or damaged.

(B) If The Compressor Starts But Trips Repeatedly On The Overload Protector

1. Check the compressor suction and discharge pressures while the compressor is operating. Be sure the pressure are within the limitations of the compressor. If pressures are excessive, it may be necessary to clean the condenser, purge air from the system, add a crankcase pressure regulating

valve, modify the system control, or take such other action as may be necessary to avoid excessive pressures.

An excessively low suction pressure may indicate a loss of charge, and a suction cooled motor compressor may not be getting enough refrigerant vapor across the motor for proper cooling.

On units with no service gauge ports where pressures cannot be checked, check condenser to be sure it is clean and fan is running. Excessive temperatures on suction and discharge line may also indicate abnormal operating conditions.

2. Check the line voltage at the motor terminals while the compressor is operating. The voltage should be within 10% of the nameplate voltage rating. If outside those limits, the voltage supply must be brought within the proper range, or a motor compressor with different electrical characteristics must be used.
3. Check the amperage drawn while the compressor is operating. Under normal operating conditions, the amperage drawn will seldom exceed 100% of the nameplate amperage and should never exceed 120% of the nameplate amperage. High amperage can be caused by low voltage, high head pressure, high suction pressure, low oil level, compressor mechanical damage, defective running capacitors, or a defective starting relay.

   On three phase compressors, check amperage in each line. One or two high amperage legs on a three phase motor indicates an unbalanced voltage supply, or a winding imbalance. If all three legs are not drawing approximately equal amperage, temporarily switch the leads to the motor to determine if the high leg stays with the line or stays with the terminal. If the high amperage reading stays with the line, the problem is in the line voltage supply. If the high amperage reading stays with the terminal, the problem is in the motor.

   If the amperage is sufficiently unbalanced to cause a protector trip, and the voltage supply is unbalanced, check with the power company to see if the condition can be corrected. If the voltage supply is balanced, indicating a defective motor phase, replace the compressor.
4. Check for a defective running capacitor or starting relay in the same manner described in the previous section.
5. Check the wiring against the wiring diagram in the terminal box. On dual voltage motors, check the location of the terminal jumper bars to be sure phases are properly connected.
6. Overheating of the cylinders and head can be caused by a leaking valve plate. To check, close the suction service valve and pump the compressor into a vacuum. Stop the compressor and crack the suction valve to allow the pressure on the suction gauge to build up to 0 psig. Again close the valve. If the pressure on the gauge continues to increase steadily, the valve plate is leaking. Remove the head and check the valve plate, replace if necessary.
7. If all operating conditions are normal, the voltage supply at the compressor terminals balanced and within limits, the compressor crankcase tem-

perature within normal limits, and the amperage drawn within the specified range, the motor protector may be defective and should be replaced.

If the operating conditions are normal and the compressor is running excessively hot for no observable reason, or if the amperage drawn is above the normal range and sufficient to repeatedly trip the protector, the compressor has internal damage and should be replaced.

(C) If The Compressor Runs But Will Not Refrigerate

1. Check the refrigerant charge, and the operating pressures. Correct any abnormal operating conditions.
2. If the suction pressure is high, and the evaporator and condenser are functioning normally, check the compressor amperage draw. An amperage draw near or above the nameplate rating indicates normal compressor operation, and it is possible the compressor or unit may have damaged valves or does not have sufficient capacity for the application.

   An amperage draw considerably below the nameplate rating may indicate a broken suction reed or broken connecting rod in the compressor. Check the pistons and valve plate on an accessible compressor. If no other reason for lack of capacity can be found, replace a welded compressor.

REPRESENTATIVE RHEEM AIR CONDITIONING WIRING DIAGRAMS

(-)PKA 10 SEER AND (-)PLA 11 SEER — SINGLE PHASE REMOTE HEAT PUMPS TIME/TEMPERATURE DEFROST CONTROL

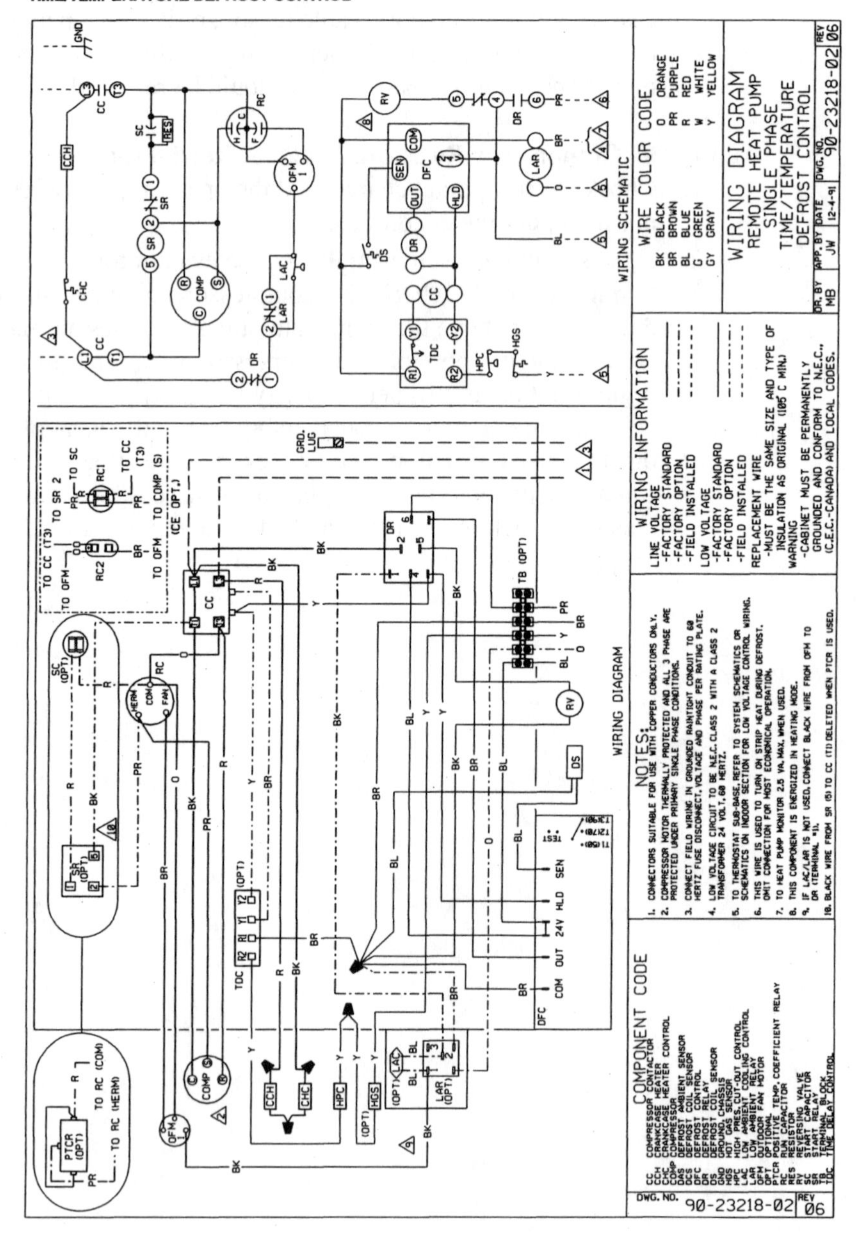

(-)PKA 10 SEER — THREE PHASE REMOTE HEAT PUMP
TIME/TEMPERATURE DEFROST CONTROL

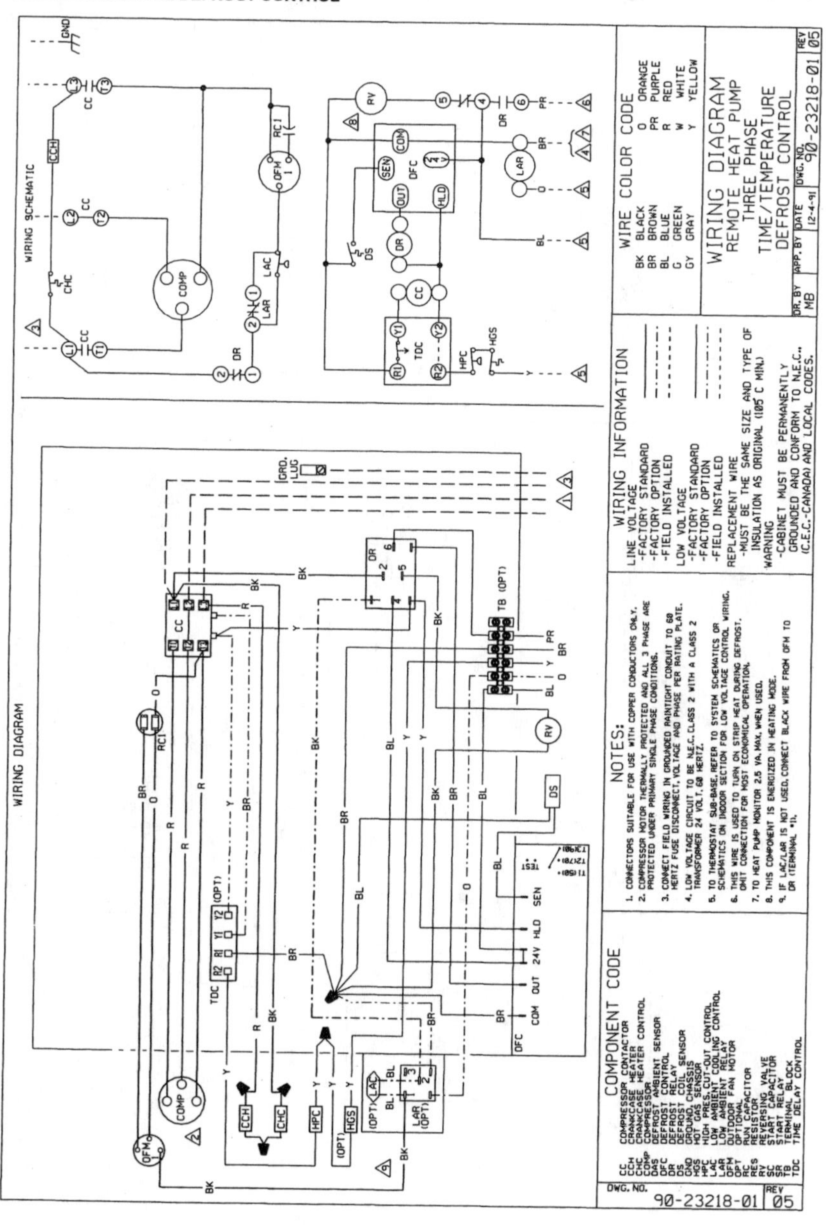

(-)PKA 10 SEER AND (-)PLA 11 SEER — SINGLE PHASE REMOTE HEAT PUMPS
DEMAND (ESSEX) DEFROST CONTROL

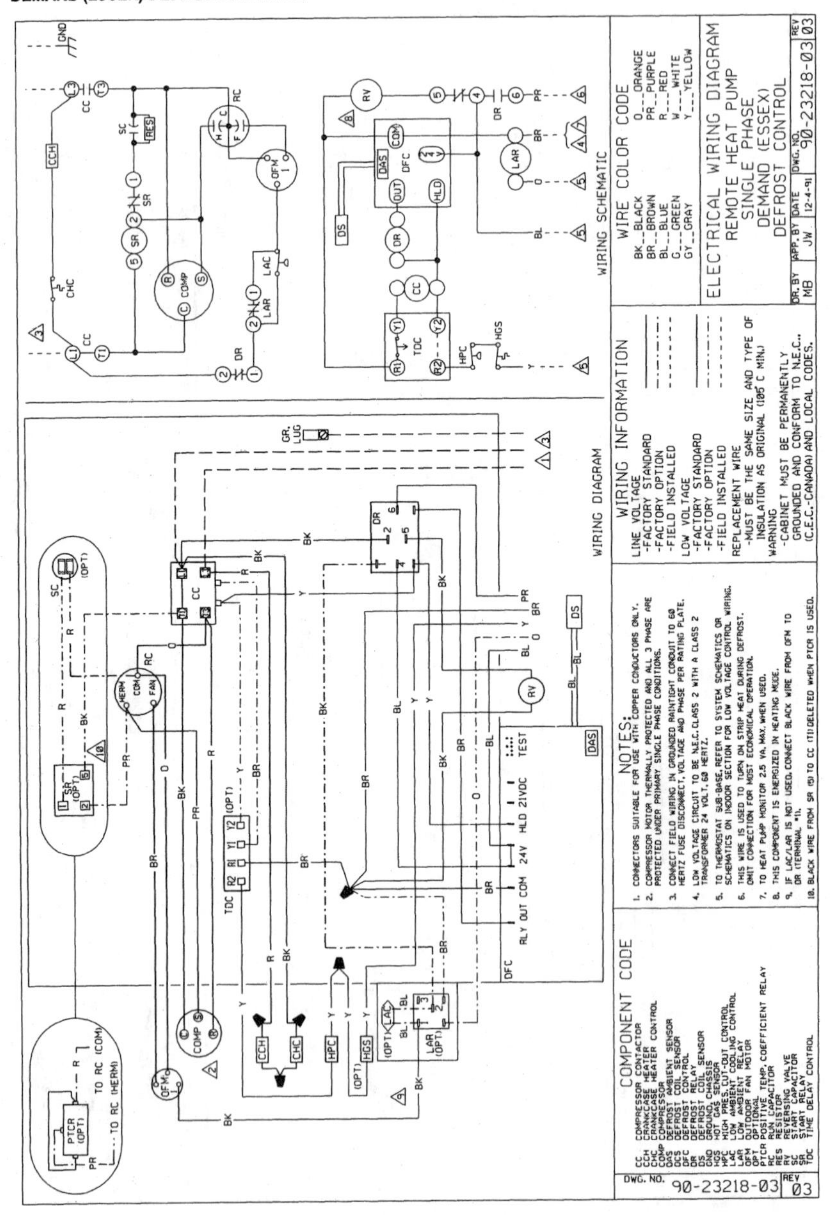

(-)PKA 10 SEER — THREE PHASE REMOTE HEAT PUMP DEMAND (ESSEX) DEFROST CONTROL

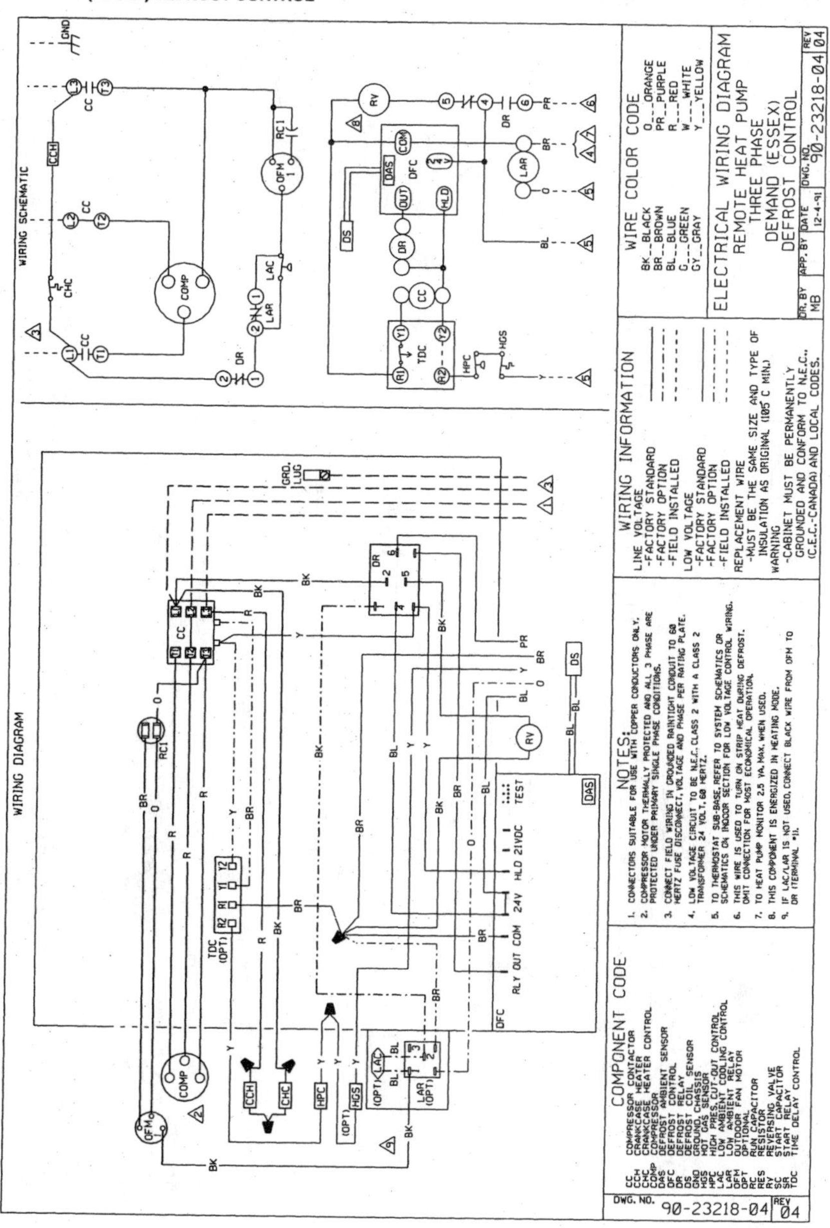

(-)PMA 12 SEER AND (-)PGB 13 SEER — SINGLE PHASE REMOTE HEAT PUMPS
DEMAND DEFROST CONTROL (RANCO DD1)

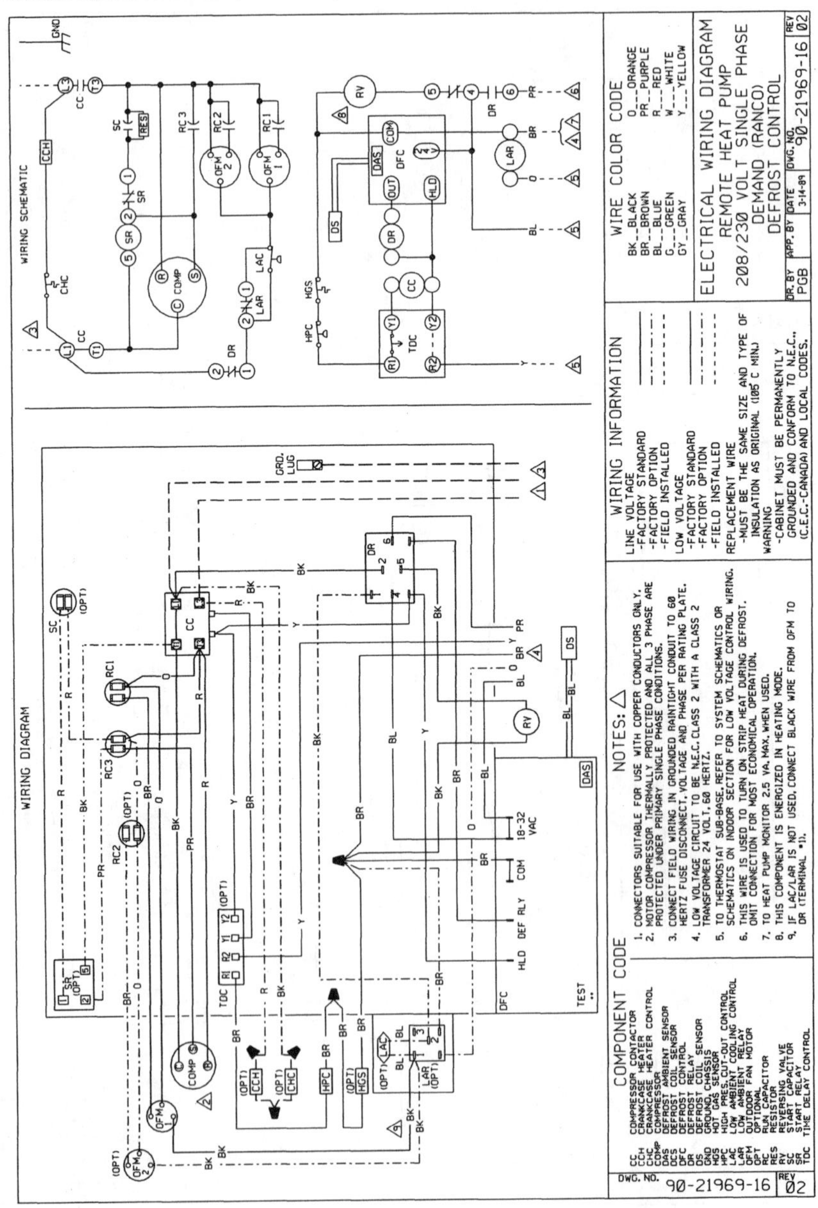

(-)PMA 12 SEER AND (-)PGB 13 SEER — THREE PHASE REMOTE HEAT PUMPS
DEMAND DEFROST CONTROL (RANCO DD1)

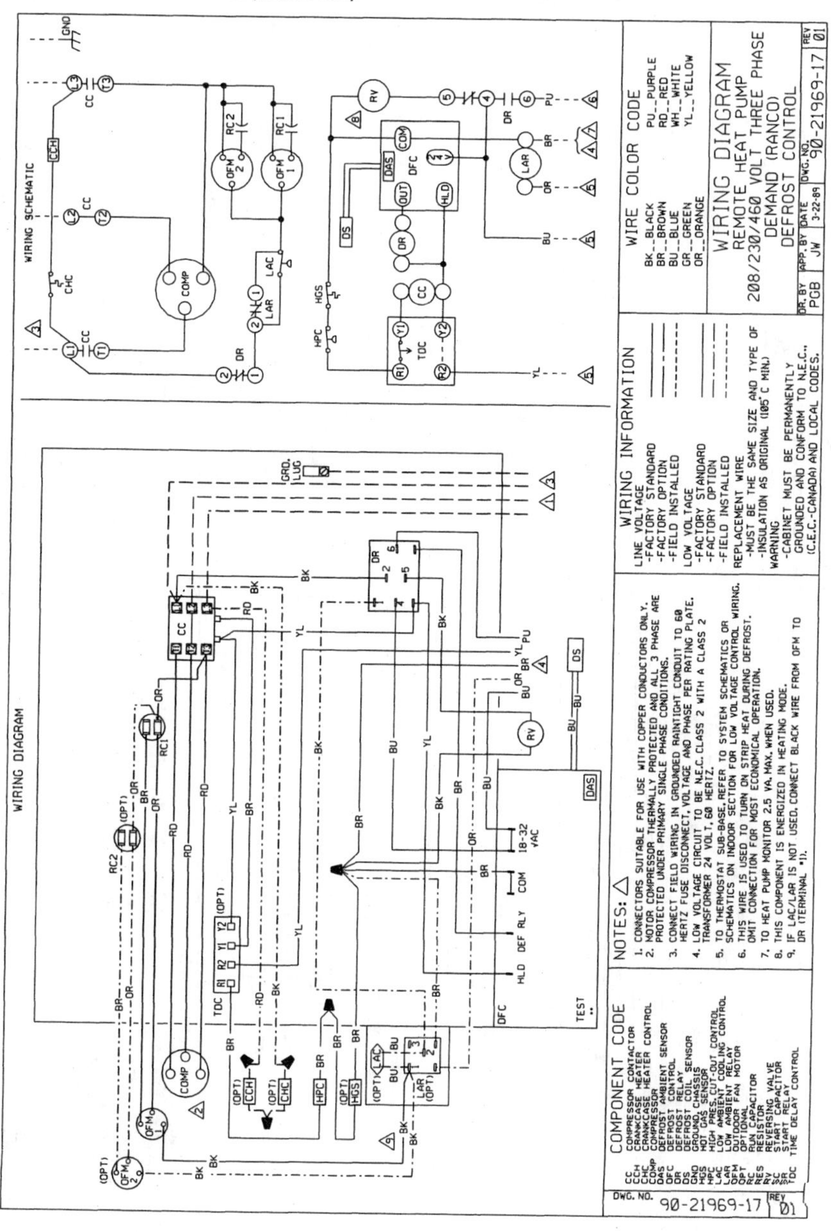

ALL MODELS — SINGLE PHASE REMOTE HEAT PUMP DEMAND DEFROST CONTROL (RANCO DDL)

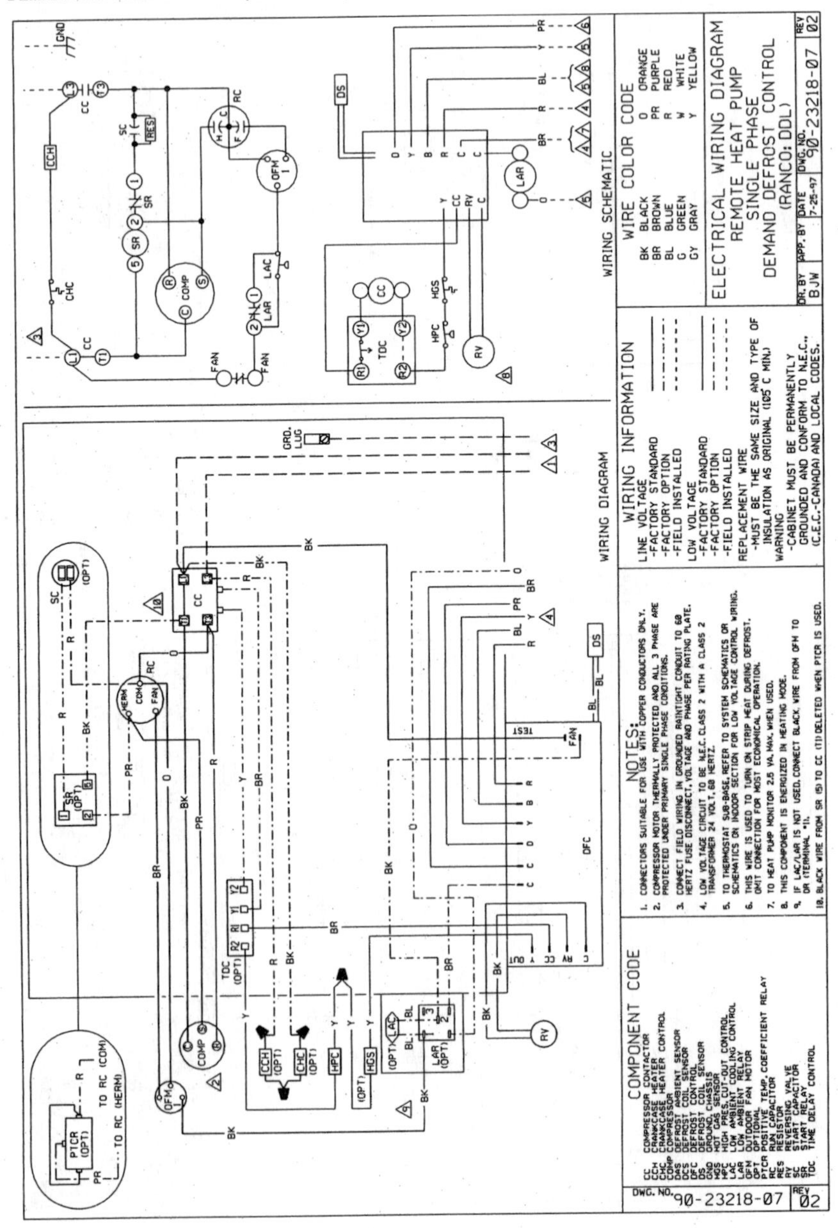

ALL MODELS — THREE PHASE — 208/230 VOLT REMOTE HEAT PUMP DEMAND DEFROST CONTROL (RANCO DDL)

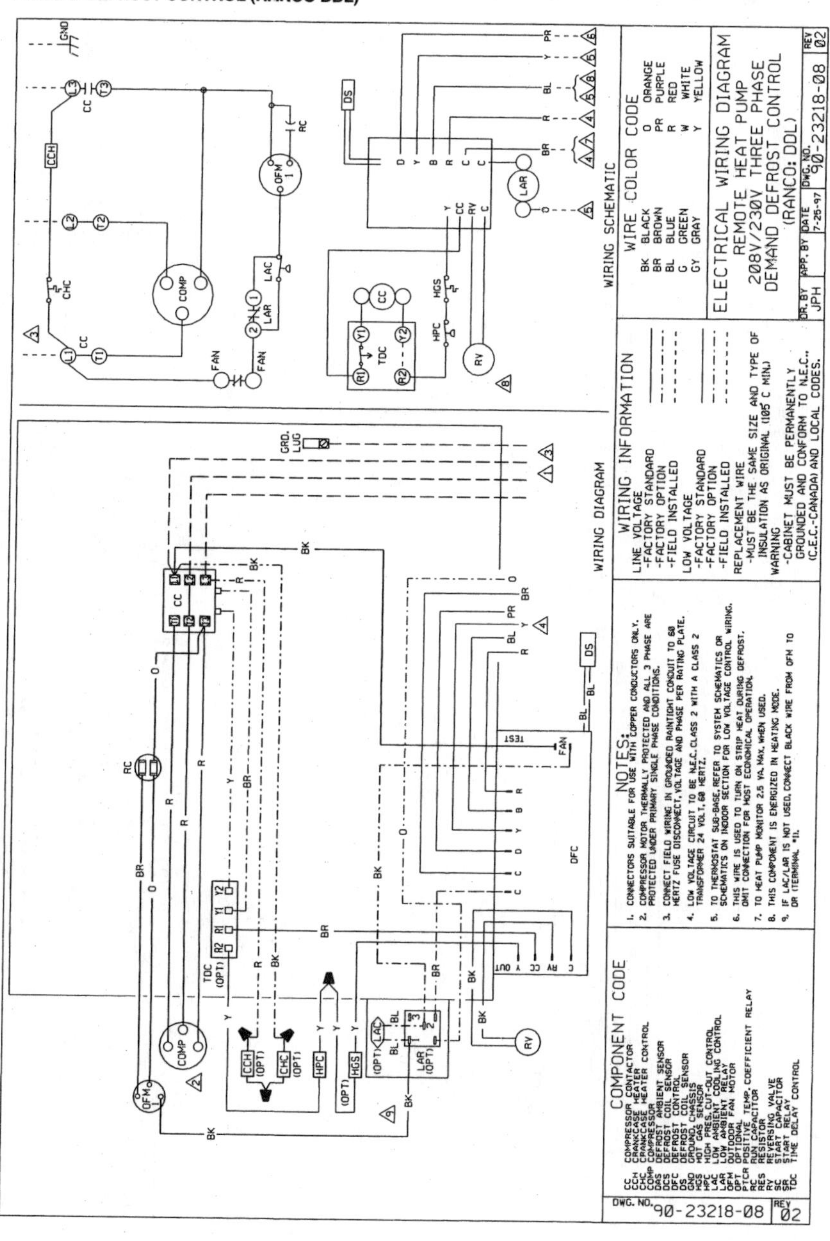

ALL MODELS — THREE PHASE — 480/575 VOLT REMOTE HEAT PUMP DEMAND DEFROST CONTROL (RANCO DDL)

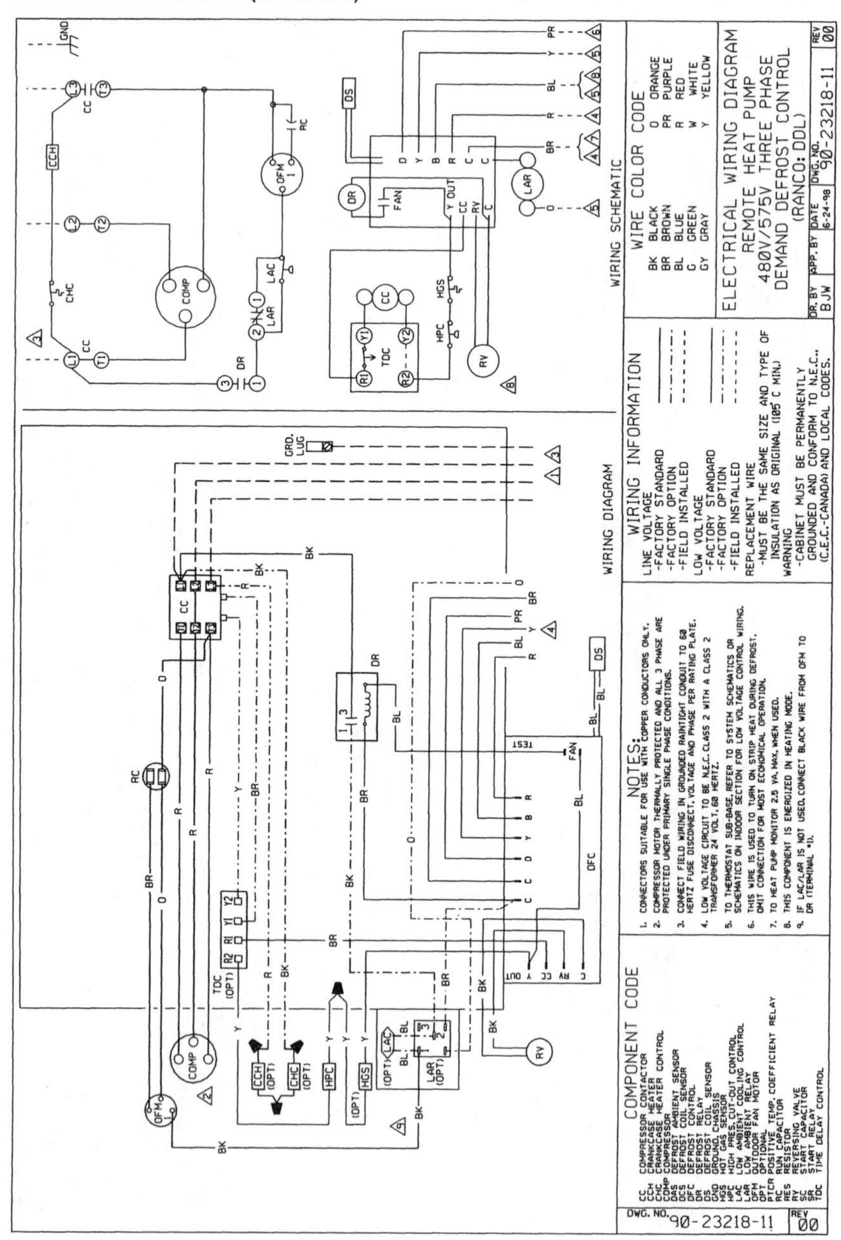

BASIC FIELD WIRING DIAGRAM — TIME-TEMP DEFROST CONTROL

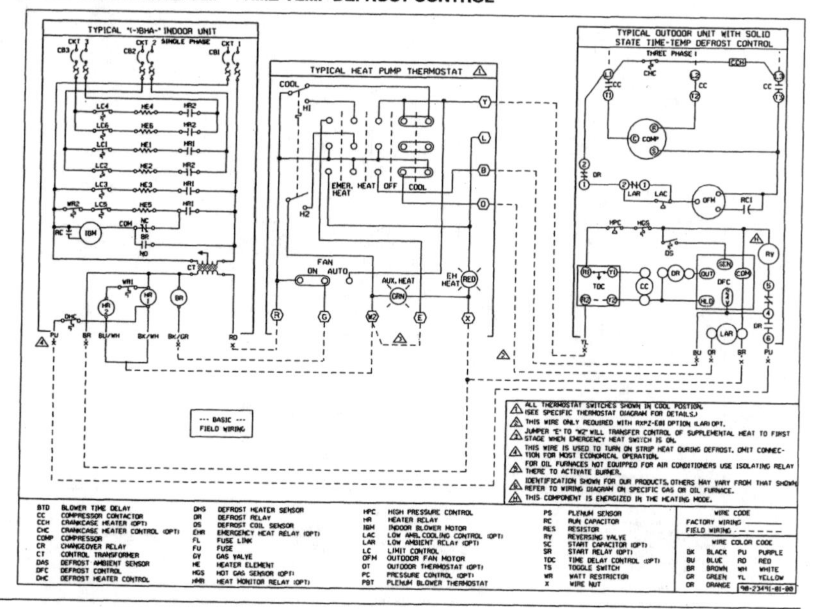

BASIC FIELD WIRING DIAGRAM — RANCO DDL DEMAND DEFROST CONTROL

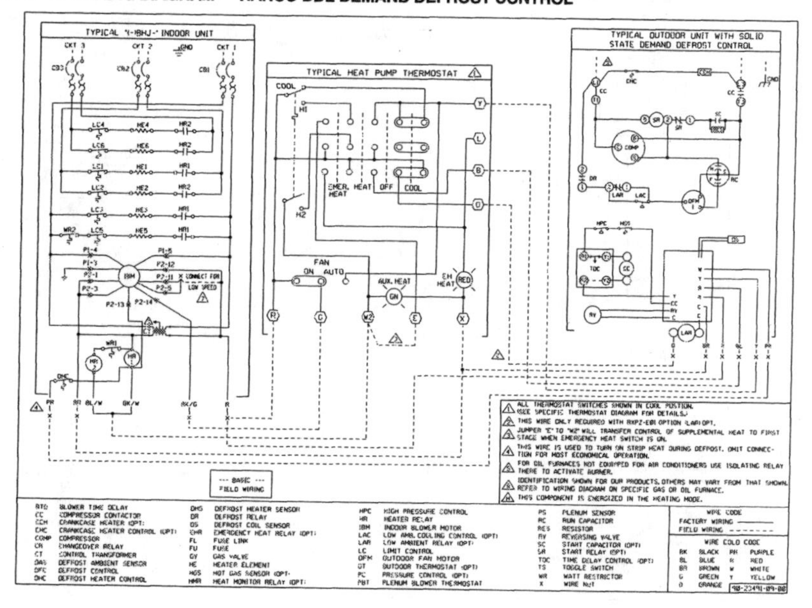

RXPF-E01 FOSSIL FUEL KIT WIRING DIAGRAM

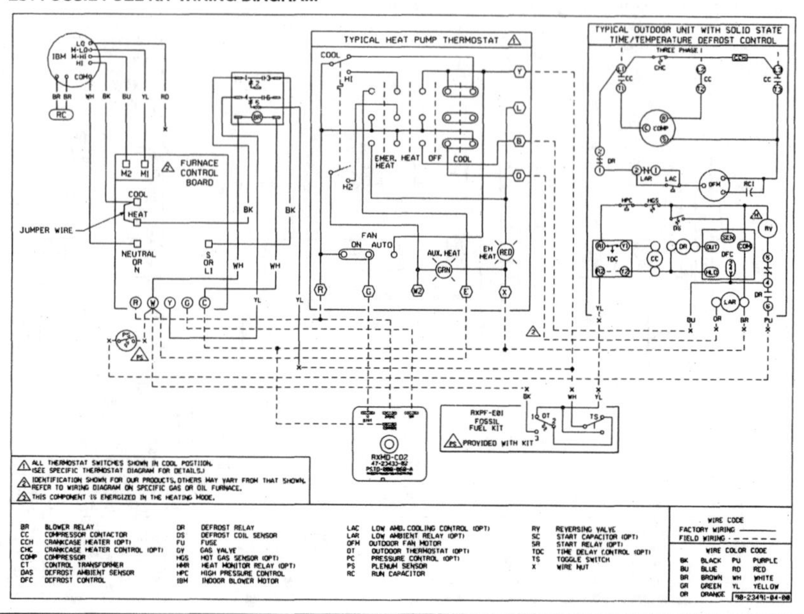

RXPF-F01 FOSSIL FUEL KIT WIRING DIAGRAM

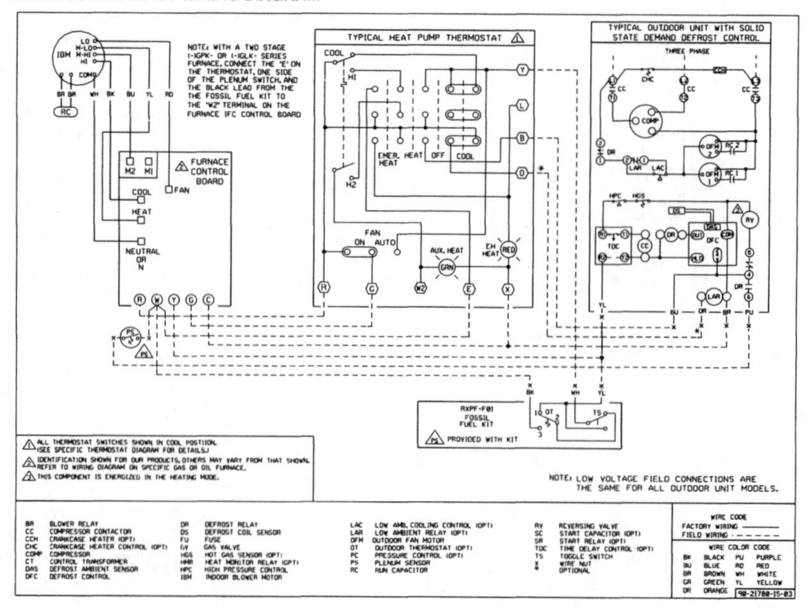

6

Safety Procedures

INTRODUCTION:

The purpose of this list of safety procedures is to provide general safety precautions to be used by persons who own, install, operate, or maintain air conditioning and refrigeration equipment.

This list should not be used to replace instructions that are provided by the equipment manufacturer. Therefore, anyone attempting to work on air conditioning and refrigeration equipment should be thoroughly familiar with the specific instructions for that particular unit.

The following is a list of the criteria that have been used here to indicate the intensity of the hazard:

Danger: This means that there is an immediate hazard that will result in severe personal injury or death.

Warning: This means that hazards or unsafe practices could result in severe personal injury or death.

Caution: This means that potential hazards or unsafe practices could result in minor personal injury.

Safety Instructions: These are general instructions that are necessary for safe working practices.

PERSONAL PROTECTION:

Warning:

1. Do not touch electrical wiring connections with wet hands.
2. Do not touch electrical equipment while standing on a wet surface or wearing wet shoes.
3. Do wear a hard hat or other head protection when there is a possibility of falling objects.

Caution:

1. Do wear safety glasses equipped with side shields when working in manufacturing plants or at construction sites.
2. Do wear gloves when handling system components after a compressor motor burnout. The refrigerant and oil contain acid that can result in acid burns to the skin.
3. Do wear goggles and gloves when handling chemicals; when welding, cutting, grinding, or brazing, or when in an area where these operations are performed.
4. Do wear gloves and other protective clothing when working with sheet metal.
5. Do wear safety shoes when working around or lifting heavy objects.
6. Do wear protective clothing when arc welding to protect from serious burns.
7. Do wear hearing protection when working in areas where sound levels are greater than 90 db.
8. Do not wear rings, jewelry, loose clothing, long ties, or gloves while working around moving belts and machinery.
9. Do not wear rings or watch while working on electrical equipment.

Safety Instructions:

1. Keep your work area clean of debris and free of liquid spills on the floor.
2. Do not continue working if you become ill. An ill person is less observant and is therefore more subject to accidents.

RIGGING (USE OF CRANES):

Danger:

1. Never use cranes under power lines. They may come in contact with the lines, causing a high voltage electrical short.

Warning:

1. Do check for the center of gravity before hoisting heavy equipment.
2. Do check for any specific hoisting instructions by the manufacturer before hoisting equipment.

3. Do check the component and assembly weights before assembling equipment to make certain the crane can lift the unit safely.
4. Do use only approved methods and rigging equipment.
5. Do not use eyebolt holes to hoist an entire assembly.
6. Do not move a loaded hoist, crane, or chain fall until the path is clear.
7. Do not use faulty rigging equipment.

Caution:

1. Do not use rigging equipment when there is a possibility of slipping or losing your balance.
2. Do use platforms or catwalks to cross over a machine. Never climb over a machine.
3. Do not use ladders that are too straight up or have too much slope from top to bottom.

Safety Instructions:

1. Do use lifting lugs according to hoisting instructions.
2. Do keep aware of where fellow workmen are at all times when hoisting equipment.
3. Do post signs indicating that heavy objects are being hoisted.

STORING AND HANDLING REFRIGERANT CYLINDERS:

Warning:

1. Do not heat a refrigerant cylinder with an open flame. When necessary to heat cylinder, use warm water.
2. Do not store refrigerant cylinders in direct sunlight.
3. Do not store refrigerant cylinders where the surrounding temperature may exceed the relief valve setting.
4. Do not reuse disposable refrigerant cylinders. It is dangerous and illegal.
5. Do not attempt to burn or incinerate refrigerant cylinders.
6. Do not alter the safety devices on a refrigerant cylinder.
7. Do not force connections.
8. Do not overfill refillable refrigerant cylinders.
9. Do open cylinder valves slowly to prevent rapid overpressurizing of the system.
10. Do use a proper wrench when opening and closing refrigerant cylinder valves.

Caution:

1. Do not alter refrigerant cylinders.
2. Do not drop, dent, or abuse refrigerant cylinders.

3. Do not charge a cylinder with a refrigerant different from the color coding.
4. Do not entirely depend on the color of a refrigerant cylinder for refrigerant identification.
5. Do not overcharge a rechargeable refrigerant cylinder.
6. Do replace cylinder caps when cylinders are not in use.
7. Do avoid pressure surges when transferring refrigerant from one cylinder to another.
8. Do periodically inspect all hoses, fittings, and charging manifolds and replace them when needed.
9. Do secure all refrigerant cylinders to prevent damage. The following procedure is recommended:
   a. *Large cylinder*: Lay on its side and prevent rolling by using chocks.
   b. *Small cylinder*: Store in an upright position and secure with a strap or chain.

LEAK TESTING AND PRESSURE TESTING SYSTEMS:

Danger:

1. Do not use oxygen for pressurizing a refrigeration system. Oxygen and oil combine to cause an explosion.
2. Do not use full cylinder pressure when pressurizing a system with nitrogen.
3. Do not exceed the specified system test pressures when pressurizing a system.
4. Do not pressurize a system with nitrogen before putting in the refrigerant when leak testing. The nitrogen may have a greater pressure than the refrigerant cylinder can withstand.
5. Do use nitrogen when pressurizing a system above refrigerant pressures.
6. Do use a gauge-equipped regulator when pressurizing a system with nitrogen.
7. Do disconnect the nitrogen cylinder from the system when the system is pressurized.

REFRIGERANTS:

Warning:

1. Do not enter an enclosed area after a refrigerant leak has occurred without thoroughly ventilating the space. Use the buddy system or an approved air pack or both.
2. Do not allow liquid refrigerant to come into contact with the skin or eyes. Immediately wash the skin with soap and water. Immediately flush the eyes with water and consult a physician.

3. Do not breathe the fumes given off by a leak detector or open flame. The fumes are likely to be phosgene, a deadly gas.
4. Do use safety goggles.
5. Do use gloves when working with liquid refrigerants.

Caution:

1. Do not weld or cut a refrigerant line or vessel until all the refrigerant has been removed.
2. Do not use an open flame in a space containing refrigerant vapor. Properly ventilate the area before entering.
3. Do not smoke in a space filled with refrigerant vapor.

Safety Instructions:

1. Avoid heavy concentrations of refrigerant within an enclosed area. A refrigerant can displace enough oxygen to cause suffocation.
2. Do not allow heating devices, such as gas flames or electric elements, to operate in an area filled with refrigerant vapor. Heat can decompose the refrigerant into hazardous substances such as hydrochloric acid, hydrofluoric acid, and phosgene gas.
3. If a strong, irritating odor is detected, warn other persons and immediately leave the area. Report the problem to the proper persons.

RECIPROCATING COMPRESSORS:

Danger:

1. Do not work on electrical wiring until all electrical power is off.
2. Do not take ohmmeter measurements with the electrical power on.

Warning:

1. Do not use a hermetic compressor for a vacuum pump. The winding can short out or a terminal may blow out, causing serious injury.
2. Do not use a welding torch when removing a compressor from the refrigerant system. The oil could catch fire and cause severe burns.
3. Do not purge refrigerant from the system through a cut, loosened connection, or broken pipe. Use the gauge manifold so that the rate of purging can be controlled.
4. Do not apply voltage to a compressor motor while the terminal box cover is off.
5. Do not attempt to loosen or remove bolts from the compressor while it is under pressure. Pump the system down to 0 to 2 psig.
6. Do not attempt to operate a compressor with the suction and discharge service valves closed.
7. Do open, tag, and lock all electrical disconnect switches while servicing the electrical circuits and connections.

Caution:

1. Do valve off all the compressors in a multiple-compressor system before attempting to service any circuit. Otherwise, the oil equalizer connection will prevent purging pressure from a single compressor.

AIR HANDLING EQUIPMENT:

Danger:

1. Do not enter an enclosed fan cabinet while the unit is running.
2. Do not reach into a unit or fan cabinet while the unit is running.
3. Do not work on a fan or fan motor until the electrical disconnect switch is open, tagged, locked, and the fuses removed.
4. Do not work on electric circuits, heating elements, or connections until the electrical disconnect switch is open, tagged, locked, and the fuses removed.

Warning:

1. Do not operate belt-driven equipment without the belt guards in place.
2. Do not service air control dampers until the operators are disconnected.
3. Do not handle access covers in high winds without sufficient help.
4. Do not pressurize a coil with a liquid refrigerant for leak testing.
5. Do not steam-clean coils until all personnel are clear of the area.
6. Do insure there is adequate ventilation when welding or cutting inside an air handling unit.
7. Do be sure that rooftop units are properly grounded.

Caution:

1. Do not work on fans without securing the sheave with a rope or strap to prevent fan freewheeling.
2. Do not exceed the specified test pressure when pressurizing a system.
3. Do protect all flammable material when welding or cutting inside an air handling unit.

Safety Instructions:

1. When operating and servicing air handling equipment, use good judgment and safe working practices to prevent damage to equipment and personal injury or property damage.

OXYACETYLENE WELDING AND CUTTING:

Danger:

1. Do not use oxygen for any purpose except welding and cutting.
2. Do not use oxygen to pressurize a refrigerant system.

Warning:

1. Do not store oxygen cylinders near oil and/or grease.
2. Do not store oxygen cylinders near a combustible material.
3. Do not use oily hands or gloves to handle oxygen cylinders.
4. Do not weld or cut in an atmosphere filled with refrigerant vapor.
5. Do not weld or cut near combustible materials.
6. Do not weld or cut lines or pressure vessels until they have been properly evacuated.
7. Do not weld or cut unless adequate ventilation is provided.
8. Do use an approved breathing system and a buddy when necessary to weld or cut in an unventilated area.
9. Do wear goggles and welding gloves when welding or cutting.

Caution:

1. Do not store oxygen and acetylene cylinders next to each other.
2. Do not store oxygen and acetylene cylinders near a heat source.
3. Do not block passageways, stairways, or ladders with welding equipment.
4. Do store oxygen and acetylene cylinders strapped or chained in an upright position.
5. Do wear suitable protective clothing when welding or cutting.

Safety Instructions:

1. Do not use damaged or worn hoses.
2. Do not use connectors other than those made specifically for oxyacetylene welding and cutting equipment.
3. Do not stand in front of the regulator when opening a cylinder valve.
4. Do not ignite torches with anything other than a friction lighter.
5. Do observe the color coding of pipelines, cylinders, and hoses.
6. Do crack the cylinder valve before attaching the pressure regulator.
7. Do release tension on the regulator adjusting screw before opening the cylinder valve.
8. Do inspect the equipment for leaking shut-off valves and connections before use.

REFRIGERATION AND AIR CONDITIONING MACHINERY (GENERAL):

Warning:

1. Do not attempt to siphon refrigerants or other chemicals by mouth.
2. Do not attempt to weld, cut, or remove fittings while the system is pressurized.

Caution:

1. Do not attempt to bend or step on pressurized refrigerant lines.
2. Do not weld or cut in an area containing refrigerant.
3. Do not loosen a packing gland nut before making sure there are plenty of threads engaged to prevent a blowout.
4. Do use replacement parts that meet the requirements of the equipment.
5. Do tag and valve off refrigerant, water, and steam lines before opening them.
6. Do periodically inspect fittings, piping, and valves for corrosion, leaks, damage, or rust.

Safety Instructions:

1. Do be sure that all shipping bolts and plugs have been removed before initial start-up.
2. Do periodically check all refrigerant and oil sight glasses for cracks.
3. Do use an ice-melting spray to remove ice from sight glasses.
4. Do not chip ice from sight glasses.

CENTRIFUGAL LIQUID CHILLERS (HEAT EXCHANGERS):

Danger:

1. Do not exceed the specified test pressures when pressurizing a system.
2. Do not use oxygen to pressurize, purge, or leak-test a refrigerant system.
3. Do not make any pressure-relieving device inoperative.
4. Do not operate any machine without the proper pressure relief devices installed and functioning.

Warning:

1. Do not attempt to repair any pressure relief device.
2. Do not make any pressure-relieving device inoperative.
3. Do not vent pressure-relieving devices inside a building.
4. Do not open a refrigerant system that is under pressure.
5. Do inspect all pressure relief devices at least once a year.
6. Do check the type of refrigerant before charging a system.
7. Do properly ventilate an area that contains refrigerant vapor before attempting to make repairs.

Caution:

1. Do not loosen water box cover bolts without completely draining the water box.
2. Do not install shut-off valves on both sides of a component in such a way that liquid might be trapped.
3. Do install a drain in the vent line near each pressure relief device.

CENTRIFUGAL LIQUID CHILLERS (ELECTRICAL CIRCUITS AND CONTROLS):

Danger:

1. Do not attempt to check voltage without proper instructions.
2. Do not attempt to take high voltage measurements (600 volts or more) with hand-held instruments.
3. Do use current and potential transformers when making high-voltage measurements.

Warning:

1. Do not work on electrical circuits until they are opened, tagged, and locked.
2. Do not attempt to stop a unit by opening a knife switch.
3. Do not attempt to check the resistance on an energized circuit.
4. Do not remove terminal box covers while the compressor is running.
5. Do not handle capacitors until they have been properly discharged.
6. Do ground all electrical equipment.
7. Do use a ground-fault current interrupter when using hand tools.

Caution:

1. Do not bypass or block electrical circuit interlocks.
2. Do not check an electrical circuit until you are certain that the electric power is off in all circuits.
3. Do use insulated fuse pullers on cartridge fuses.

Safety Instructions:

1. Do not store tools or parts in control and starter boxes.
2. Do not use meters for purposes other than those for which they were designed.
3. Do periodically inspect electrical cords and replace or repair any damaged cords, plugs, or connections.

CENTRIFUGAL LIQUID CHILLERS (COUPLINGS):

Danger:

1. Do not remove guards until all rotating parts have completely stopped.
2. Do not start a unit until all bolts are tight.
3. Do be sure that all tools and materials have been removed before starting a unit.

Warning:

1. Do not stand beside rotating couplings.
2. Do not operate a beltdrive unit without the guards in place.

Caution:

1. Do not start a unit until all wrenches, dial indicators, etc., have been removed.
2. Do tighten all coupling bolts twice to make certain that all are tight.
3. Do periodically check couplings for proper lubrication and alignment.

CENTRIFUGAL LIQUID CHILLERS (TURBINES):

Danger:

1. Do not start a unit until all superheated steam connections have been checked.
2. Do not stand in line with turbine trip valve handles.
3. Do familiarize yourself with the proper start-up and operation procedures before starting the unit.

Warning:

1. Do not open drain lines on a turbine while the unit is under a vacuum.
2. Do not block open a steam governor.
3. Do not block a steam trip handle.
4. Do not block a pressure relief valve.
5. Do not operate turbines at speeds greater than their designed speed.
6. Do install a relief valve between the turbine exhaust flange and the first shut-off valve to protect against rupturing the turbine casing.
7. Do install the proper blinds or blank flanges in the inlet and exhaust lines before dismantling a turbine.

Caution:

1. Do not touch a hot turbine or components without the proper protective clothing.
2. Do keep the area clean of oil spills and oily rags.

Safety Instructions:

1. Do not attempt to operate a turbine until all preoperative safety and control devices have been checked.
2. Do check and adjust turbine safeties and speed trip devices annually.

ABSORPTION LIQUID CHILLERS:

Danger:

1. Do not exceed the specified test pressures when pressurizing a refrigerant system.
2. Do not use oxygen to purge lines or to pressurize a refrigeration system.

Warning:

1. Do not attempt to remove fittings or components while the system is under pressure.
2. Do not siphon lithium bromide by mouth.
3. Do not flame-cut a purge chamber until all the hydrogen has been evacuated.
4. Do not service an electrical circuit until the switch is opened, tagged, and locked.
5. Do wear goggles and the proper protective clothing when handling inhibitor, octyl alcohol, lithium hydroxide, hydrobromic acid, and lithium bromide.
6. Do immediately wash any chemical spills from the skin with soap and water.
7. Do immediately flush your eyes with water and consult a physician if your eyes have been exposed to chemicals.
8. Do properly ventilate the area when welding or cutting to remove any noxious fumes.

Caution:

1. Do not loosen cover bolts before draining water boxes.
2. Do keep the floor clean of spills and debris.
3. Do tag and valve off water, steam, and brine lines before opening them.

7

Troubleshooting Charts

(1) NUMBER @ TOP IS MOST COMMON

Electric Heat

Condition	Possible Cause	Corrective Action	Reference
Unit will not run	1. Blown fuse	1. Replace fuse and correct cause	3–1
	2. Burned transformer	2. Replace transformer and correct cause	48–1; 48–1A
	3. Thermostat not calling for heat	3. Set thermostat	8–1; 8–1A, B
	4. Defective thermostat	4. Replace thermostat	8–1; 8–1A, B
	5. Defective heating relay	5. Replace relay	55–1
Fan will not run	1. Burned fan motor	1. Repair or replace fan motor	47–1; 47–1A, B, C, D, E, F
	2. Broken fan belt	2. Replace fan belt	18-1; 18–1B
	3. Burned contacts in fan relay	3. Replace fan relay	51–1
	4. Defective fan control	4. Replace fan control	56–1; 56–1A, B
	5. Defective wiring or connections	5. Repair wiring or connections	10–1
Fan motor hums but will not start	1. Defective fan motor bearings	1. Replace bearings or fan motor	47–1; 47–1E, F
	2. Defective fan motor starting switch	2. Repair starting switch or replace motor	47–1; 47–1D, F
	3. Defective starting capacitor	3. Replace capacitor	13–1; 13–1A
	4. Burned start windings in motor	4. Repair or replace motor	47–1; 47–1A, B, C
	5. Defective blower bearings	5. Replace bearings	60–1; 60–1B
	6. Loose wiring connections in motor starting circuit	6. Repair wiring	10–1
Fan motor cycles	1. Defective motor bearings	1. Replace bearings or motor	47–1; 47–1E, F
	2. Defective blower bearings	2. Replace blower bearings	60–1; 60–1B
	3. Defective run capacitor	3. Replace capacitor	13–1; 13–1B
	4. Defective fan control	4. Replace fan control	56–1; 56–1A, B
	5. Defective fan relay	5. Replace relay	51–1
	6. Defective motor windings	6. Repair or replace motor	47–1; 47–1A, B, C
Fan blows cold air	1. Defective heat sequencing relays	1. Replace relays	57–1; 57–1A, B, C
	2. Burned heat elements	2. Replace elements	58–1; 58–1A, B
	3. Loose wiring connections	3. Repair wiring	10–1

Electric Heat			
Condition	Possible Cause	Corrective Action	Reference
Fan blows cold air—cont'd.	4. Defective thermostat	4. Replace thermostat	8–1; 8–1A, B
	5. Fan set to "on" position	5. Set to "auto" position	8–1; 8–1D
	6. Defective fan control	6. Replace fan control	56–1; 56–1A, B
Not enough heat	1. Dirty air filters	1. Clean or replace filters	19–1; 19–1B; 56–1; 56–1A
	2. Unit too small	2. Install more elements	53–1
	3. Too little air flow through furnace	3. Increase air flow; remove restrictions	59–1
	4. Thermostat heat anticipator not properly set	4. Set heat anticipator	8–1; 8–1B
	5. Defective fan motor	5. Repair or replace fan motor	47–1; 47–1A, B, C, D, E, F
	6. Air conditioning evaporator dirty	6. Clean evaporator	19–1; 19–1C
	7. Thermostat not properly located	7. Relocate thermostat	8–1; 8–1A, B
	8. Thermostat set too low	8. Set thermostat	8–1; 8–1A, B
	9. Thermostat out of calibration	9. Calibrate thermostat	8–1; 8–1A, B
	10. Low voltage	10. Correct cause	12–1
	11. Air ducts not insulated	11. Insulate ducts	52–1
	12. Burned elements	12. Replace elements	58–1; 58–1A, B
	13. Defective heat sequencing relays	13. Replace relays	57–1; 57–1A, B, C
	14. Defective thermostat	14. Replace thermostat	8–1; 8–1A, B
	15. Defective element limits	15. Replace limits	58–1; 58–1A, B
	16. Outdoor thermostat set too low	16. Reset thermostat	8–1; 8–1E
Too much heat	1. Unit too large	1. Reduce Btu input	61–1
	2. Thermostat heat anticipator not properly set	2. Set heat anticipator	8–1; 8–1B
	3. Thermostat not properly located	3. Relocate thermostat	8–1; 8–1A, B
	4. Thermostat set too high	4. Set thermostat	8–1; 8–1A, B
	5. Thermostat out of calibration	5. Calibrate thermostat	8–1; 8–1A, B
High humidity in building	1. Humidity due to cooking	1. Vent cookstove	62–1
	2. Humidity due to bathing	2. Vent bathroom	62–1
	3. Humidity due to rain	3. Increase temperature rise through furnace	59–1

Electric Heat

Condition	Possible Cause	Corrective Action	Reference
Blown element limits	1. Shorted heating element	1. Replace element and correct cause	58–1
	2. Dirty filters	2. Clean or replace filters	19–1; 19–1B; 56–1; 56–1A
	3. Dirty blower	3. Clean blower	60–1; 60–1A
	4. Broken or slipping fan belt	4. Adjust or replace belt	18–1; 18–1B
	5. High or low voltage	5. Notify power company	12–1
	6. Defective blower motor	6. Replace motor	47–1; 47–1A, B, C, D, E, F
	7. Not enough air through furnace	7. Remove restriction	14–1B; 19–1; 47–1; 47–1A, B, C, D, E, F; 56–1; 56–1A; 59–1
	8. Loose electrical connections	8. Repair connections	10–1
High operating costs	1. Unit too small	1. Increase number of elements	53–1
	2. Dirty air filters	2. Clean or replace filters	19–1; 19–1B; 56–1; 56–1A
	3. Dirty air conditioning evaporator	3. Clean evaporator	19–1; 19–1C; 56–1; 56–1A
	4. Air ducts not insulated	4. Insulate ducts	52–1
	5. Thermostat in wrong location	5. Relocate thermostat	8–1; 8–1A, B
	6. Dirty blower	6. Clean blower	60–1; 60–1A
	7. Defective thermostat	7. Replace thermostat	8–1; 8–1A, B
	8. Fan belt slipping	8. Replace or adjust fan belt	18–1; 18–1B
	9. Low or high voltage	9. Notify power company	12–1
	10. Thermostat setting too high	10. Set to lower setting	8–1; 8–1A, B

Gas Heat			
Condition	Possible Cause	Corrective Action	Reference
Unit will not run	1. Blown fuse	1. Replace fuse and correct cause	3–1
	2. Burned transformer	2. Replace transformer and correct cause	48–1; 48–1A
	3. Pilot out	3. See entry "Pilot not burning properly or out"	
	4. Thermostat not calling for heat	4. Set thermostat	8–1; 8–1A, B
	5. Defective wiring or connections	5. Repair wiring or connections	10–1
Fan will not run	1. Burned fan motor	1. Repair or replace motor	47–1; 47–1A, B, C, D, E, F
	2. Broken fan belt	2. Replace fan belt	18–1; 18–1B
	3. Burned contacts in fan relay	3. Replace fan relay	51–1
	4. Defective fan control	4. Replace fan control	56–1; 56–1A, B
	5. Defective wiring or connections	5. Repair wiring or connections	10–1
Fan motor hums, but will not start	1. Defective bearings in fan motor	1. Replace bearings or fan motor	47–1; 47–1E, F
	2. Defective starting switch in fan motor	2. Repair starting switch or replace motor	47–1; 47–1D, F
	3. Defective starting capacitor	3. Replace capacitor	18–1; 13–1A
	4. Burned start winding in motor	4. Repair or replace motor	47–1; 47–1A, B, C
	5. Defective blower bearings	5. Replace bearings	60–1; 60–1B
	6. Loose wiring connections in motor starting circuit	6. Repair wiring	10–1
Fan motor cycles	1. Defective motor bearings	1. Replace bearings or fan motor	47–1; 47–1E, F
	2. Defective blower bearings	2. Replace blower bearings	60–1; 60–1B
	3. Defective run capacitor	3. Replace run capacitor	13–1; 13–1B
	4. Defective fan control	4. Replace fan control	56–1; 56–1A, B
	5. Return air too cool	5. Allow air to warm	56–1; 56–1B
	6. Fan control differential too close	6. Adjust fan control	58–1; 56–1A, B
	7. Fan control "off" setting too high	7. Adjust fan control	56–1; 56–1A, B
	8. Too much air flow through furnace	8. Reduce air flow	56–1; 56–1B; 59–1
	9. Defective motor windings	9. Repair or replace motor	47–1; 47–1A, B, C
	10. Fan control "on" setting too low	10. Adjust control	56–1; 56–1A, B

Gas Heat

Condition	Possible Cause	Corrective Action	Reference
Pilot not burning properly or out	1. Faulty thermocouple	1. Replace thermocouple	63–1; 63–1A
	2. Dirty or corroded thermocouple connection	2. Clean connection	63–1; 63–1A
	3. Gas supply turned off	3. Restore gas supply	
	4. Pilot burner orifice dirty	4. Clean orifice	63–1; 63–1B
	5. Thermocouple not installed in flame properly	5. Properly install thermocouple	63–1; 63–1A
	6. Drafts affecting pilot flame	6. Shield pilot from drafts	63–1; 63–1B
	7. Defective pilot safety device	7. Replace pilot safety device	63–1; 63–1C
Fan cycles while main burner stays on	1. Wrong size orifices in burners	1. Replace orifices with proper size	64–1; 64–1A, C
	2. Low manifold gas pressure	2. Increase gas pressure	67–1
	3. Too much air flowing through furnace	3. Reduce air flow	56–1; 56–1B
	4. Too cold return air	4. Allow air to warm	56–1; 56–1B
	5. Electrical or motor problems	5. See previous entry "Fan motor cycles"	10–1; 47–1; 47–1A, B, C, D, E
Main burner cycles while blower stays on	1. Dirty air filters	1. Clean or replace filters	19–1; 19–1B
	2. Wrong size orifices in burners	2. Replace orifices with proper size	64–1; 64–1A, C
	3. High manifold gas pressure	3. Reduce gas pressure	67–1
	4. Faulty limit control	4. Replace limit control	70–1
	5. Too little air flow through furnace	5. Increase air flow; clear restrictions	19–1B, C, D, E; 59–1; 60–1; 60–1A, B
Not enough heat	1. Dirty air filters	1. Clean or replace filters	19–1; 19–1B
	2. Wrong size orifices in burners	2. Replace orifices with proper size	64–1; 64–1A, C
	3. Low manifold gas pressure	3. Increase gas pressure	67–1
	4. Too little air flow through furnace	4. Increase air flow; clear restrictions	19–1; 19–B; 59–1; 60–1; 60–1A, B
	5. Thermostat heat anticipator not properly set	5. Set heat anticipator	8–1; 8–1B
	6. Defective fan motor	6. Replace or repair fan motor	47–1; 47–1A, B, C, D, E, F
	7. Unit too small	7. Replace unit with proper size	53–1; 61–1

Gas Heat			
Condition	Possible Cause	Corrective Action	Reference
Not enough heat—cont'd.	8. Air conditioning evaporator dirty	8. Clean evaporator	19–1; 19–1C
	9. Thermostat not properly located	9. Move thermostat	8–1; 8–1A, B
	10. Thermostat set too low	10. Set thermostat	8–1; 8–1A, B
	11. Thermostat out of calibration	11. Calibrate thermostat	8–1; 8–1A, B
Too much heat	1. Unit too large	1. Reduce Btu input; replace unit	61–1
	2. Thermostat heat anticipator not properly set	2. Set heat anticipator	8–1; 8–1B
	3. Thermostat not properly located	3. Move thermostat	8–1; 8–1A, B
	4. Thermostat set too high	4. Set thermostat	8–1; 8–1A, B
	5. Thermostat out of calibration	5. Calibrate thermostat	8–1; 8–1A, B
Pilot burning; main gas valve will not operate	1. Blown fuse	1. Replace fuse and check for cause	3–1
	2. Defective gas valve	2. Replace gas valve	66–1
	3. Burned transformer	3. Replace transformer and check for cause	48–1; 48–1A
	4. Burned thermostat heat anticipator	4. Replace thermostat	8–1; 8–1B
	5. Bad thermostat	5. Replace thermostat	8–1; 8–1A, B
	6. Bad electrical connections	6. Repair connections	10–1
	7. Broken thermostat wire	7. Repair broken wire	10–1
Delayed ignition of main burner	1. Poor flame travel to the burner	1. Correct flame travel	65–1; 65–1A
	2. Poor flame distribution over the burner	2. Correct flame distribution	65–1; 65–1B
	3. Low manifold gas pressure	3. Adjust gas pressure	67–1
	4. Defective step-opening regulator	4. Adjust or replace regulator	67–1; 67–1A
Roll-out on main burner ignition	1. Restricted heat exchanger	1. Clear restrictions	68–1; 68–1A
	2. Quick opening main gas valve	2. Install surge arrestor	66–1; 66–1A
Flame flashback	1. Low manifold gas pressure	1. Adjust manifold gas pressure	67–1
	2. Extremely small main burner flame	2. Adjust primary air	65–1; 65–1C

Gas Heat			
Condition	Possible Cause	Corrective Action	Reference
Flame flashback—cont'd.	3. Distorted burner or carry-over wing slots	3. Repair burner or carry-over wing slots	65–1; 65–1A
	4. Defective main burner orifice	4. Replace orifices	64–1; 64–1A, C
	5. Orifice misaligned	5. Replace orifices	64–1; 64–1B
	6. Erratic gas valve operation	6. Replace gas valve	66–1
	7. Dirty burner	7. Clean burners	65–1; 65–1B
	8. Improper gas-air mixture	8. Be sure that proper gas is being used	64–1; 64–1B
	9. Unstable gas supply pressure	9. Install a two-stage pressure regulator	67–1; 67–1B
Resonance (loud, rumbling noise)	1. Excess primary air to main burner	1. Adjust primary air	65–1; 65–1E
	2. Defective orifice spud	2. Replace orifice spud	64–1; 64–1B
	3. Dirty orifice spuds	3. Clean orifices	64–1; 64–1B
Yellow flame	1. Too little primary air	1. Adjust primary air	65–1; 65–1C
	2. Dirty orifice spud	2. Clean spuds	
	3. Orifice spuds misaligned	3. Align orifice spuds	64–1; 64–1B
	4. Restricted heat exchanger	4. Clean heat exchanger	68–1; 68–1A
	5. Poor vent operation	5. Correct venting	69–1
Floating main burner flame	1. Air blowing into heat exchanger	1. Check for defective heat exchanger	68–1; 68–1B
	2. Restricted heat exchanger	2. Clean heat exchanger	68–1; 68–1A
	3. Negative pressure in furnace room	3. Increase air supply to room	69–1
Main burner flame too large	1. Orifices too large	1. Replace orifices	64–1; 64–1A
	2. Excessive manifold gas pressure	2. Adjust pressure regulator	67–1
	3. Defective gas pressure regulator	3. Replace regulator	67–1
	4. Wrong type of gas being used	4. Install changeover kit	64–1
Main burner flame too small	1. Dirty orifice spuds	1. Clean orifice spuds	64–1; 64–1B
	2. Low manifold gas pressure	2. Adjust pressure regulator	67–1
	3. Orifices too small	3. Replace orifices	64–1; 64–1C
	4. Wrong type of gas being used	4. Install changeover kit	64–1
Odor in building	1. Vent not operating properly	1. Correct venting problem	69–1
	2. Poor ventilation	2. Check flame conditions and correct	69–1

Gas Heat

Condition	Possible Cause	Corrective Action	Reference
High operating costs	1. Unit too small	1. Install proper size unit	53–1; 61–1
	2. Dirty air filters	2. Clean or replace filters	19–1; 19–1B; 56–1A
	3. Dirty air conditioning evaporator	3. Clean evaporator	19–1; 19–1C
	4. Air ducts not insulated	4. Insulate ducts	52–1
	5. Thermostat in wrong location	5. Relocate thermostat	8–1; 8–1A, B
	6. Dirty blower	6. Clean blower	60–1; 60–1A

Heat Pump (Cooling Cycle)

Condition	Possible Cause	Corrective Action	Reference
No cooling, but compressor runs continuously	1. Defective compressor valves	1. Replace valves and valve plate or compressor	4–3; 4–3C
	2. Shortage of refrigerant	2. Repair leak and recharge	19–1; 19–1A
	3. Defective reversing valve	3. Replace reversing valve	92–1
	4. Air or noncondensables	4. Remove noncondensables	18–1; 18–1G
	5. Wrong superheat setting on indoor expansion valve	5. Adjust superheat setting	20–1
	6. Loose thermal bulb on indoor expansion valve	6. Tighten thermal bulb	20–1; 20–1A
	7. Dirty indoor coil	7. Clean coil	19–1; 19–1C
	8. Dirty indoor filters	8. Clean or replace filters	19–1; 19–1B
	9. Indoor blower belt slipping	9. Replace or adjust belt	18–1; 18–1B
	10. Restriction in refrigerant system	10. Locate and remove restriction	22–1; 22–1A, B, C, D, E
Too much cooling: compressor runs continuously	1. Faulty wiring	1. Repair wiring	10–1
	2. Faulty thermostat	2. Replace thermostat	18–1; 18–1A, B
	3. Wrong thermostat location	3. Relocate thermostat	8–1; 8–1A, B
Liquid refrigerant flooding compressor (TXV system)	1. Wrong superheat setting on indoor expansion valve	1. Adjust superheat	20–1
	2. Loose thermal bulb on indoor expansion valve	2. Tighten thermal bulb	20–1; 20–1A
	3. Faulty indoor expansion valve	3. Replace expansion valve	20–1; 20–1E
	4. Defective indoor check valve	4. Replace check valve	83–1
	5. Refrigerant overcharge	5. Remove overcharge	18–1; 18–1F
Liquid refrigerant flooding compressor (capillary tube system)	1. Refrigerant overcharge	1. Remove overcharge	18–1; 18–1F
	2. High head pressure	2. See entry, "High head pressure"	
	3. Dirty indoor filter	3. Clean or replace filter	19–1; 19–1B
	4. Dirty indoor coil	4. Clean coil	19–1; 19–1C
	5. Indoor blower belt slipping	5. Replace or adjust belt	18–1; 18–1B
	6. Indoor check valve defective	6. Replace check valve	83–1

Heat Pump (Heating Cycle)

Condition	Possible Cause	Corrective Action	Reference
No heating, but compressor runs continuously	1. Refrigerant shortage	1. Repair leak and recharge	19–1; 19–1A
	2. Compressor valves defective	2. Replace valves and valve plate or compressor	4–3; 4–3C
	3. Leaking reversing valve	3. Replace reversing valve	92–1
	4. Defective defrost control, time clock, or relay	4. Replace defrost control, time clock, or relay	93–1
Too much heat; compressor runs continuously	1. Faulty wiring	1. Repair wiring	10–1
	2. Faulty thermostat	2. Replace thermostat	8–1; 8–1A, B
	3. Wrong thermostat location	3. Relocate thermostat	8–1; 8–1B
Compressor cycles on low pressure control at end of defrost cycle	1. Defective reversing valve	1. Replace reversing valve	92–1
	2. Defective power element on indoor expansion valve	2. Replace power element	20–1
	3. Shortage of refrigerant	3. Repair leak and recharge	19–1; 19–1A
Unit runs in cooling cycle but pumps down in cooling cycle	1. Faulty outdoor expansion valve	1. Clean or replace expansion valve	20–1; 20–1A, B, C, D, E, F
	2. Defective power element on outdoor expansion valve	2. Replace power element	20–1
	3. Defective reversing valve	3. Replace reversing valve	92–1
	4. Dirty outdoor coil	4. Clean coil	18–1; 18–1B
	5. Belt slipping on outdoor blower	5. Replace or adjust belt	18–1; 18–1B
	6. Defective indoor check valve	6. Replace check valve	83–1
	7. Restriction in refrigerant circuit	7. Locate and remove restriction	22–1; 22–1A, B, C, D, E
Defrost cycle will not terminate	1. Shortage of refrigerant	1. Repair leak and recharge	19–1; 19–1A
	2. Defrost control out of adjustment	2. Adjust control	93–1
	3. Defective defrost control, time clock, or relay	3. Replace defrost control, time clock, or relay	93–1
	4. Defective reversing valve	4. Replace reversing valve	92–1
	5. Defective compressor valves	5. Replace valves and valve plate or compressor	4–3; 4–3C
	6. Faulty electrical wiring	6. Repair wiring	10–1

Heat Pump (Heating Cycle)

Condition	Possible Cause	Corrective Action	Reference
Defrost cycle initiates without ice on coil	1. Shortage of refrigerant	1. Repair leak and recharge	19–1; 19–1A
	2. Defrost control out of adjustment	2. Adjust control	93–1
	3. Defective defrost control, time clock, or relay	3. Replace defrost control, time clock, or relay	93–1
	4. Defrost control sensing element not making proper contact	4. Improve contact	93–1
	5. Outdoor coil dirty	5. Clean coil	18–1; 18–1B
	6. Outdoor fan belt slipping	6. Replace belt or adjust	18–1; 18–1B
Reversing valve will not shift	1. Defective reversing valve	1. Replace reversing valve	92–1
	2. Defective compressor valves	2. Replace valves and valve plate or compressor	4–3; 4–3C
	3. Faulty fan relay on either indoor or outdoor section	3. Replace relay	51–1; 54–1
	4. Burned transformer	4. Replace transformer	48–1; 48–1A
Indoor blower off with auxiliary heat on	1. Defective indoor fan relay	1. Replace fan relay	51–1
	2. Defective indoor fan motor	2. Repair or replace motor	47–1; 47–1A, B, C, D, E, F
	3. Faulty wiring or loose terminals	3. Repair wiring or terminals	10–1
	4. Faulty thermostat	4. Replace thermostat	8–1; 8–1A, B
Outdoor blower runs during defrost cycle	1. Faulty outdoor fan relay	1. Replace fan relay	54–1
Compressor short cycles on defrost control	1. Shortage of refrigerant	1. Repair leak and recharge	19–1; 19–1A
	2. Defrost control out of adjustment	2. Adjust defrost control	93–1
	3. Defective defrost control, time clock, or relay	3. Replace defrost control, time clock, or relay	93–1
	4. Defective power element on outdoor expansion valve	4. Replace power element	20–1
	5. Fan belt slipping on outdoor blower	5. Replace or adjust belt	18–1; 18–1B

Heat Pump (Heating Cycle)

Condition	Possible Cause	Corrective Action	Reference
Excessive ice build-up on indoor coil	1. Defective defrost relay	1. Replace defrost relay	96–1
	2. Defective compressor valves	2. Replace valves and valve plate or compressor	4–3; 4–3C
	3. Shortage of refrigerant	3. Repair leak and recharge	19–1; 19–1A
	4. Defrost control out of adjustment	4. Adjust defrost control	93–1
	5. Defrost control sensing element not making proper contact	5. Improve contact	93–1
	6. Defective defrost control, time clock, or relay	6. Replace control, time clock, or relay	93–1
	7. Defective reversing valve	7. Replace reversing valve	92–1
	8. Wrong superheat setting on outdoor expansion valve	8. Adjust superheat	20–1
	9. Defective power element on outdoor expansion valve	9. Replace power element	20–1E
	10. Plugged outdoor expansion valve	10. Clean or replace expansion valve	20–1; 20–1A, B, C, D, E, F
Ice build-up on lower section of outdoor coil	1. Defective defrost relay	1. Replace defrost relay	96–1
	2. Defective compressor valves	2. Replace valves and valve plate or compressor	4–3; 4–3C
	3. Shortage of refrigerant	3. Repair leak and recharge	19–1; 19–1A
	4. Defrost control out of adjustment	4. Adjust defrost control	93–1
	5. Defrost sensing element not making proper contact	5. Improve contact	93–1
	6. Defective reversing valve	6. Replace reversing valve	92–1
	7. Wrong superheat setting on outdoor expansion valve	7. Adjust superheat	20–1
Liquid refrigerant flooding compressor on heating cycle (TXV system)	1. Wrong superheat setting on outdoor expansion valve	1. Adjust superheat	20–1
	2. Outdoor expansion valve thermal bulb not making proper contact	2. Improve contact	20–1; 20–1A

Heat Pump (Heating Cycle)

Condition	Possible Cause	Corrective Action	Reference
Liquid refrigerant flooding compressor on heating cycle (TXV system)—cont'd	3. Outdoor expansion valve stuck open 4. Leaking outdoor check valve	3. Clean or replace expansion valve 4. Replace check valve	20–1; 20–1B 83–1
Liquid refrigerant flooding compressor on heating cycle (capillary tube system)	1. Refrigerant overcharge 2. High head pressure 3. Defective outdoor check valve	1. Remove overcharge 2. See entry "High head pressure" 3. Replace check valve	18–1; 18–1F 83–1
Excessive operating costs	1. Refrigerant shortage 2. Defective reversing valve 3. Defrost control out of adjustment 4. Refrigerant overcharge 5. Dirty indoor or outdoor coil 6. Blower belt slipping on indoor or outdoor blower 7. Dirty indoor air filters 8. Wrong thermostat location 9. Ducts not insulated 10. Wrong size unit 11. Outdoor thermostat not controlling auxiliary heat	1. Repair leak and recharge 2. Replace reversing valve 3. Adjust defrost control 4. Remove overcharge 5. Clean coil 6. Replace or adjust belt 7. Clean or replace filters 8. Relocate thermostat 9. Insulate ducts 10. Replace with proper size 11. Adjust, relocate, or provide shield	19–1; 19–1A 92–1 93–1 18–1; 18–1F 18–1; 18–1B; 19–1; 19–1C 18–1; 18–1B 19–1; 19–1B 8–1; 8–1A, B 52–1 53–1 8–1; 8–1E

Heat Pump (Heating or Cooling Cycle)			
Condition	Possible Cause	Corrective Action	Reference
Compressor hums but will not start	1. Faulty fuse	1. Replace fuse and correct cause	3–1
	2. Faulty wiring	2. Repair wiring	10–1
	3. Loose electrical terminals	3. Repair loose connections	10–1
	4. Compressor overloaded	4. Locate and remove overload	53–1
	5. Faulty starting capacitor	5. Replace capacitor	13–1; 13–1A
	6. Faulty starting relay	6. Replace relay	14–1; 14–1A, B, C, D, E
	7. Burned compressor motor	7. Replace compressor	4–1; 4–2A, B, C, D
	8. Defective compressor bearings	8. Replace bearings or compressor	4–3; 4–3A, B
	9. Stuck compressor	9. Replace compressor	4–3; 4–3A, B
Compressor cycling on overload	1. Low voltage	1. Determine reason and repair	12–1
	2. Loose electrical terminals	2. Repair terminals	10–1
	3. Single-phasing of phase power	3. Replace fuse or repair wiring; or notify power company	3–1; 10–1; 12–1
	4. Defective contactor contacts	4. Replace contacts or contactor	5–1; 5–1B
	5. Defective compressor overload	5. Replace overload	4–2E, F, G
	6. Compressor overloaded	6. Locate and remove overload	53–1
	7. Defective start capacitor	7. Replace capacitor	13–1; 13–1A
	8. Defective run capacitor	8. Replace capacitor	13–1; 13–1B
	9. Defective starting relay	9. Replace starting relay	14–1; 14–1A, B, C, D, E
	10. Refrigerant overcharge	10. Remove overcharge	18–1; 18–1F
	11. Defective compressor bearings	11. Replace bearings or compressor	4–3; 4–3A, B
	12. Air or noncondensables in system (high head pressure)	12. Remove noncondensables from system	18–1; 18–1G
	13. Defective reversing valve	13. Replace reversing valve	92–1

Heat Pump (Heating or Cooling Cycle)

Condition	Possible Cause	Corrective Action	Reference
Compressor off on high pressure control	1. Refrigerant over-charge	1. Remove overcharge	18–1; 18–1F
	2. Control out of adjustment	2. Adjust control	9–1; 9–1B
	3. Defective indoor fan motor	3. Repair or replace motor	4–1; 4–2A, B, C, D; 18–1; 18–1B
	4. Defective outdoor fan motor	4. Repair or replace motor	4–1; 4–2A, B, C, D; 18–1; 18–1B
	5. Defective fan relay on either indoor or outdoor section	5. Repair or replace motor	51–1; 54–1
	6. Too long defrost cycle	6. Replace time clock, defrost relay, or termination thermostat	93–1
	7. Defective reversing valve	7. Replace reversing valve	92–1
	8. Blower belt slipping on indoor or outdoor coil	8. Adjust or replace belt	18–1; 18–1B
	9. Indoor or outdoor coil dirty	9. Clean proper coil	18–1; 18–1B; 19–1; 19–1C
	10. Dirty indoor air filters	10. Replace or clean filters	19–1; 19–1B
	11. Air bypassing indoor or outdoor coil	11. Prevent air bypassing	94–1
	12. Air volume too low over indoor or outdoor coil	12. Increase indoor ductwork or remove restriction from coils	52–1
	13. Auxillary heat strips ahead of indoor coil	13. Locate heat strips downstream of indoor coil	95–1
Compressor cycles on low pressure control	1. Refrigerant shortage	1. Repair leak and recharge	19–1; 19–1A
	2. Low suction pressure	2. Increase load (see Suction Pressure Low)	
	3. Defective expansion valve	3. Repair or replace expansion valve	20–1; 20–A, B, C, D, E, F
	4. Dirty indoor or outdoor coil	4. Clean coil	19–1; 19–1C; 18–1; 18–1B
	5. Slipping blower belt	5. Replace or adjust blower belt	18–1; 18–1B
	6. Dirty air indoor filter	6. Clean or replace filter	19–1; 19–1B
	7. Ductwork restriction	7. Increase ductwork	52–1
	8. Liquid drier or suction strainer restricted	8. Replace drier or strainer	22–1; 22–1A, B, C, D, E
	9. Defrost thermostat element loose or making poor contact	9. Tighten or increase contact	93–1

Heat Pump (Heating or Cooling Cycle)			
Condition	Possible Cause	Corrective Action	Reference
Compressor cycles on low pressure control—cont'd.	10. Air temperature too low for evaporation	10. Relocate unit or provide adequate air temperature	8–1; 8–1A, B
	11. Defrost cycle too long	11. Replace time clock, defrost relay, or termination thermostat	93–1
	12. Defective evaporator fan motor	12. Repair or replace fan motor or relay	18–1; 18–1B, 47–1; 47–1E
Outdoor fan runs, but compressor will not	1. Faulty electrical wiring or loose connections	1. Repair wiring or connections	10–1
	2. Defective starting capacitor	2. Replace starting capacitor	13–1; 13–1A
	3. Defective starting relay	3. Replace starting relay	14–1; 14–1A, B, C, D, E
	4. Defective run capacitor	4. Replace run capacitor	13–1; 13–1B
	5. Shorted or grounded compressor motor	5. Replace compressor	4–1; 4–2A, B, C, D
	6. Stuck compressor	6. Replace compressor	4–3; 4–3A, B
	7. Compressor overloaded	7. Determine and remove overload	4–1; 4–2; 4–3; 4–3A, B; 4–4; 4–4A, B; 53–1
	8. Defective contactor contacts	8. Replace contactor or contacts	5–1; 5–1B
	9. Single-phasing of three-phase power	9. Locate problem and repair or contact power company	3–1; 10–1
	10. Low voltage	10. Locate and correct cause	11–1
Outdoor fan motor will not start	1. Faulty electrical wiring or loose connections	1. Repair wiring or connections	10–1
	2. Defective outdoor fan motor	2. Repair or replace motor	4–1; 4–2A, B, C, D; 18–1; 18–1B; 47–1; 47–1A, B, C, D, E, F
	3. Defective outdoor fan relay	3. Replace fan relay	54–1
	4. Defective defrost control, timer, or relay	4. Replace control, timer, or relay	93–1
Outdoor section does not run	1. No electrical power	1. Inform power company	12–1
	2. Blown fuse	2. Replace fuse and correct fault	3–1
	3. Faulty electrical wiring or loose terminals	3. Repair wiring or terminals	10–1
	4. Compressor overloaded	4. Determine overload and correct	4–1; 4–2; 4–3; 4–3A, B; 4–4; 4–4A, B; 53–1

Heat Pump (Heating or Cooling Cycle)

Condition	Possible Cause	Corrective Action	Reference
Outdoor section does not run—cont'd.	5. Defective transformer	5. Replace transformer	48–1; 48–1A
	6. Burned contactor coil	6. Replace contactor coil	5–1; 5–1A
	7. Compressor overload open	7. Determine cause and correct	4–2E, F, G
	8. High pressure control open	8. Determine cause and correct	9–1; 9–1B
	9. Low pressure control open	9. Determine cause and correct	9–1; 9–1A
	10. Thermostat off	10. Turn thermostat on and set	8–1; 8–1A, B
Indoor blower will not run	1. Blown fuse	1. Replace fuse and correct cause	3–1
	2. Faulty electrical wiring or loose connections	2. Repair wiring or connections	10–1
	3. Burned transformer	3. Replace transformer	48–1; 48–1A
	4. Indoor fan relay defective	4. Replace fan relay	51–1
	5. Faulty indoor fan motor	5. Repair or replace motor	47–1; 47–1A, B, C, D, E, F
	6. Faulty thermostat	6. Replace thermostat	8–1; 8–1A, B
Indoor coil iced over	1. Dirty filters	1. Clean or replace filters	19–1; 19–1B
	2. Dirty coil	2. Clean coil	19–1; 19–1C
	3. Blower fan belt slipping	3. Replace or adjust belt	18–1; 18–1B
	4. Outdoor check valve sticking closed	4. Replace check valve	83–1
	5. Defective indoor expansion valve	5. Clean or replace expansion valve	20–1; 20–1A, B, C, D, E
	6. Low indoor air temperature	6. Increase temperature	18–1; 18–1A, B
	7. Shortage of refrigerant	7. Repair leak and recharge	19–1; 19–1A
Noisy compressor	1. Low oil level in compressor	1. Determine reason for loss of oil and correct. Replace oil.	19–1; 19–1A, B, C, D, E, F; 20–1; 20–1B; 22–1; 22–1A, B, C, D, E; 31–1; 31–1A, B, C; 32–1; 32–1A
	2. Defective suction and/or discharge valves	2. Replace valves and plate or compressor	4–3; 4–3C
	3. Loose hold-down bolts	3. Tighten	25–1; 26–1
	4. Broken internal springs	4. Replace compressor	20–1; 20–1A, B, C, D, E, F
	5. Inoperative check valves	5. Repair or replace check valve	83–1
	6. Loose thermal bulb on indoor expansion valve	6. Tighten thermal bulb	20–1; 20–1

Heat Pump (Heating or Cooling Cycle)			
Condition	Possible Cause	Corrective Action	Reference
Noisy compressor—cont'd.	7. Improper superheat setting on indoor expansion valve	7. Adjust superheat	20–1
	8. Stuck open indoor expansion valve	8. Clean or replace valve	20–1; 20–1E
Compressor loses oil	1. Refrigerant shortage	1. Repair leak and recharge	19–1; 19–1A
	2. Low suction pressure	2. Increase load on evaporator	19–1; 19–1A, B, C, D, E, F
	3. Restriction in refrigerant circuit	3. Remove restriction	22–1; 22–1A, B, C, D, E
	4. Indoor expansion valve stuck open	4. Clean or replace expansion valve	20–1; 20–1B
Unit operates normally in one cycle, but high suction pressure on other cycle	1. Leaking check valve	1. Replace check valve	83–1
	2. Loose thermal bulb on outdoor or indoor expansion valve	2. Tighten thermal bulb	20–1; 20–1A
	3. Leaking reversing valve	3. Replace reversing valve	92–1
	4. Expansion valve stuck open on indoor or outdoor	4. Repair or replace, expansion valve	20–1; 20–1B
Unit pumps down in cool or defrost cycle but operates normally in heat cycle	1. Defective reversing valve	1. Replace reversing valve	92–1
	2. Defective power element on indoor expansion valve	2. Replace power element	20–1; 20–1E
	3. Restriction in refrigerant circuit	3. Locate and remove restriction	22–1; 22–1A, B, C, D, E
	4. Clogged indoor expansion valve	4. Clean or replace expansion valve	20–1; 20–1B, C, D, E
	5. Check valve in outdoor section sticking closed	5. Replace check valve	83–1
Head pressure high	1. Overcharge of refrigerant	1. Remove overcharge	18–1; 18–1F
	2. Air or noncondensables in system	2. Remove noncondensables	18–1; 18–1G
	3. High air temperature supplied to condenser	3. Reduce air temperature	49–1
	4. Dirty indoor or outdoor coil	4. Clean coil	18–1; 18–1B; 19–1; 19–1C
	5. Dirty indoor air filters	5. Clean or replace filters	19–1; 19–1B, C

Heat Pump (Heating or Cooling Cycle)			
Condition	Possible Cause	Corrective Action	Reference
Head pressure high—cont'd.	6. Indoor or outdoor blower belt slipping	6. Replace or adjust blower belt	18–1; 18–1B
	7. Air bypassing indoor or outdoor coil	7. Prevent air bypassing	94–1
Suction pressure high	1. Defective compressor suction valves	1. Replace valves and valve plate or compressor	4–3; 4–3C
	2. High head pressure	2. See previous entry "Head pressure high"	
	3. Excessive load on cooling	3. Determine cause and correct	53–1
	4. Leaking reversing valve	4. Replace reversing valve	92–1
	5. Leaking check valves	5. Replace check valve	83–1
	6. Indoor or outdoor expansion valve stuck open	6. Clean or replace expansion valve	20–1; 20–1B
	7. Loose thermal bulb on indoor or outdoor expansion valve	7. Tighten bulb	20–1; 20–1A
Suction pressure low	1. Shortage of refrigerant	1. Repair leak and recharge	19–1; 19–1A
	2. Blower belt slipping on indoor or outdoor blower	2. Replace belt or adjust	18–1; 18–1B
	3. Dirty Indoor air filters	3. Clean or replace	19–1; 19–1B
	4. Defective check valves	4. Replace check valves	83–1
	5. Restriction in refrigerant circuit	5. Locate and remove restriction	22–1; 22–1A, B, C, D, E
	6. Ductwork small or restricted	6. Repair or replace ductwork	52–1
	7. Defective expansion valve power element on indoor or outdoor coil	7. Replace power element	20–1; 20–1A, B, C, D, E
	8. Clogged indoor or outdoor expansion valve	8. Clean or replace valve	20–1; 20–1E
	9. Wrong superheat setting on indoor or outdoor expansion valve	9. Adjust superheat setting	20–1
	10. Dirty indoor or outdoor coil	10. Clean coil	19–1; 19–1C; 18–1; 18–1B
	11. Bad contactor contacts	11. Replace contactor or contacts	5–1; 5–1B
	12. Low refrigerant charge	12. Repair leak and recharge	19–1; 19–1A

Refrigeration

Condition	Possible Cause	Corrective Action	Reference
Compressor fails to start (no hum)	1. Power failure	1. Contact power company	1–1
	2. Disconnect switch open	2. Close switch and check circuits	2–1
	3. Fuse blown	3. Replace fuse and determine cause	3–1
	4. Burned-out compressor motor	4. Replace	4–1; 4–2A, B, C, D
	5. Inoperative motor starter	5. Repair or replace	5–1; 5–1A, B, C
	6. Control circuit open	6. Locate cause and repair	6–1
	a. Oil failure control	a. Reset and check control	7–1
	b. Overload protector tripped	b. Check overload	4–2E, F, G, H
	c. Thermostat setting too high	c. Set to lower temperature	8–1; 8–1A, C, D, E
	d. Low-pressure control open	d. Reset and check pressures	9–1; 9–1A
	e. High-pressure control open	e. Reset and check pressures	9–1; 9–1B
	7. Loose wiring	7. Repair wiring	10–1
Compressor will not start (hums and trips overload protector)	1. Improperly wired	1. Rewire unit	10–1; 11–1
	2. Low voltage to unit	2. Determine reason and correct	12–1
	3. Bad starting capacitor	3. Determine reason and replace	13–1; 13–1A
	4. Starting relay open	4. Determine reason and replace starting relay	14–1; 14–1A, B, C, D, E
	5. Burned-out compressor motor	5. Replace compressor motor	4–1; 4–2A, B, C, D
	6. Mechanical problems in compressor	6. Replace compressor	4–3; 4–3A, B
	7. Liquid refrigerant in compressor crankcase	7. Install crankcase heater	4–4; 4–4B; 15–1
	8. Bad run capacitor	8. Determine reason and replace capacitor	13–1; 13–1B
	9. Unequalized pressures on PSC motor	9. Allow pressures to equalize or install a hard-start kit	16–1; 17–1
Compressor will run but remains on start winding	1. Improperly wired	1. Rewire unit	10–1; 11–1
	2. Low voltage to unit	2. Determine reason and correct	12–1

Refrigeration

Condition	Possible Cause	Corrective Action	Reference
Compressor will run but remains on start winding—cont'd.	3. Starting relay does not open	3. Determine cause and replace starting relay	14–1; 14–1D, E
	4. Bad running capacitor	4. Determine reason and replace	13–1; 13–1B
	5. High discharge pressure	5. Open compressor discharge service valve; purge possible overcharge of refrigerant	18–1; 18–1A, B, C, D, E, F, G
	6. Open or shorted motor winding	6. Replace compressor	4–2; 4–2A, B
	7. Mechanical trouble in compressor	7. Replace compressor	4–1; 4–3; 4–3A, B, C
	8. Defective overload protector	8. Replace overload protector	4–2; 4–2E, F, G, H
Compressor starts and runs, but short cycles	1. Defective overload protector	1. Replace overload protector	4–2; 4–2E, F, G, H
	2. Low voltage to unit	2. Determine reason and correct	12–1
	3. Defective run capacitor	3. Determine reason and replace	13–1; 13–1B
	4. High discharge pressure	4. Open compressor discharge service valve. Purge possible overcharge of refrigerant. Provide sufficient condenser cooling air to unit	18–1; 18–1A, B, C, D, E, F, G
	5. Suction pressure too low	5. Properly charge system with refrigerant. Increase load on evaporator	19–1; 19–1A, B, C, D, E, F; 22–1; 22–1A, B, C, D, E; 23–1; 24–1
	6. Suction pressure too high	6. Reduce air flow over evaporator. Purge overcharge of refrigerant. Replace compressor valves	
	7. Compressor too hot	7. Properly charge system with refrigerant	19–1; 19–1A
	8. Shorted motor winding	8. Replace compressor	4–2; 4–2A, B
	9. Dirty or iced evaporator	9. Increase air flow over evaporator. Replace broken belt. Replace defective fan motor	19–1; 19–1A, B, C, D, E, F; 22–1; 22–1A, B, C, D, E; 4–4C

Refrigeration			
Condition	Possible Cause	Corrective Action	Reference
Compressor starts and runs, but short cycles—cont'd.	10. Low-pressure control differential set too close	10. Readjust differential	9–1; 9–1A
	11. High pressure control differential set too close	11. Readjust or replace control	9–1; 9–1B
	12. Condenser water regulating valve inoperative	12. Clean and repair or replace	18–1; 18–1C
	13. Condenser water temperature too high	13. Clean and repair water pump, piping spray nozzles, and coil	18–1; 18–1C, D, E
	14. Erratic thermostat	14. Relocate or replace thermostat	8–1; 8–1B, C, D, E
Unit operates excessively	1. Short of refrigerant	1. Repair leak and recharge unit	19–1; 19–1A
	2. Thermostat contacts stuck closed	2. Clean contacts or replace thermostat	8–1; 8–1A, B, C, D, E
	3. Excessive load	3. Check heat load and replace unit accordingly; replace insulation	23–1; 24–1
	4. Evaporator coil iced	4. Defrost unit and check operation	19–1; 19–1A, B, C, D, E, F
	5. Restriction in refrigerant system	5. Locate and remove	22–1; 22–1A, B, C, D, E; 32–1; 32–1A
	6. Dirty condenser	6. Clean condenser	18–1; 18–1A, B, C, D, E, F, G
	7. Restricted air over evaporator	7. Determine cause and correct	19–1; 19–1A, B, C, D, E, F; 22–1; 22–1A, B, C, D, E
	8. Inefficient compressor	8. Check compressor valves and repair	4–1; 4–3C; 4–4; 4–4A, B
Compressor loses oil	1. Traps in hot gas and/or suction lines	1. Reroute lines to provide proper pitch	24–1; 24–1A, B, C, D
	2. Refrigerant velocity too low in risers	2. Resize risers or install oil return traps	33–1; 33–1A, B, C
	3. Shortage of refrigerant	3. Repair leak and recharge	19–1; 19–1A
	4. Liquid refrigerant flooding back to compressor	4. Adjust expansion valve; alter refrigerant charge on capillary tube system	20–1; 20–1A, B, C, E; 21–1
	5. Gas-oil ratio low	5. Add 1 pt of oil for each 10 lb of refrigerant added to the factory charge	24–1; 24–1A, B, C, D

Refrigeration

Condition	Possible Cause	Corrective Action	Reference
Compressor loses oil—cont'd.	6. Plugged expansion valve or strainer	6. Clean or replace	20–1; 20–1A, D, E; 22–1; 22–1E
	7. Compressor short cycling	7. See items under entry "Compressor starts and runs, but short cycles"	
	8. Superheat too high at compressor suction	8. Change location of TXV bulb or adjust superheat to return wet refrigerant to the compressor	20–1; 20–1A, B, D, E; 21–1
Compressor noisy	1. Lack of compressor oil	1. Add oil to correct level	24–1; 24–1A, B, C, D
	2. Tubing rattle	2. Reroute tubing	25–1
	3. Mounting loose	3. Repair mounting	26–1
	4. Oil slugging	4. Adjust oil level or refrigerant charge	4–4; 4–4A
	5. Refrigerant flooding compressor	5. Check expansion valve for leak or oversized orifice	4–4; 4–4B
	6. Dry or scored shaft seal	6. Check oil level	4–3; 4–3A, B, C; 4–4; 4–4A, B
	7. Internal parts of compressor broken or worn	7. Overhaul compressor	4–4; 4–4A, B
	8. Compressor drive coupling loose	8. Tighten coupling and check alignment	27–1
Unit low on capacity	1. Ice or dirt on evaporator	1. Clean coil or defrost	19–1; 19–1A, B, C, D, E, F
	2. Expansion valve stuck or dirty	2. Clean or replace expansion valve	20–1; 20–1A, D, E; 22–1; 22–1E
	3. Improper TXV superheat adjustment	3. Adjust expansion valve	20–1; 20–1A, B, D, E
	4. Wrong size expansion valve	4. Replace valve	20–1; 20–1C, D
	5. Excessive pressure drop in evaporator	5. Adjust expansion valve	20–1; 28–1
	6. Clogged strainer	6. Clean or replace strainer	22–1; 22–1A, B, C, D
	7. Liquid flashing in liquid line	7. Subcool liquid or add refrigerant	31–1; 31–1A, B, C
Space temperature too high	1. Control setting too high	1. Adjust control	29–1
	2. Expansion valve too small	2. Replace valve	20–1; 20–1D, E
	3. Evaporator too small	3. Replace coil	30–1

Refrigeration			
Condition	Possible Cause	Corrective Action	Reference
Space temperature too high—cont'd.	4. Insufficient air circulation	4. Correct circulation	19–1; 19–1A, B, C, D, E, F
	5. Shortage of refrigerant	5. Repair leak and recharge	19–1; 19–1A
	6. Expansion valve plugged	6. Clean or replace	20–1; 20–1D, E
	7. Inefficient compressor	7. Check efficiency	4–3C
	8. Restricted or undersized refrigerant lines	8. Clear restriction or resize lines	22–1; 22–1A, B, C, D, E
	9. Evaporator iced or dirty	9. Clean and defrost evaporator	4–4C; 19–1; 19–1A, B, C, D, E, F
Suction line frosted or sweating	1. Superheat setting too low	1. Adjust superheat setting	20–1; 20–1B, D, E; 22–1E
	2. Expansion valve stuck open	2. Clean or replace valve	20–1; 20–1B, C
	3. Evaporator fan not running	3. Correct problem	19–1; 19–1E, F
	4. Overcharge of refrigerant	4. Correct charge	37–1; 37–1B
Liquid line frosted or sweating	1. Restricted drier or strainer	1. Replace drier or strainer	22–1; 22–1B, C
	2. Liquid line shut-off valve partially closed	2. Open valve	33–1
Hot liquid line	1. Expansion valve open too wide	1. Adjust expansion valve	20–1; 20–1B, C
	2. Refrigerant shortage	2. Repair leak and recharge	19–1; 19–1A
Top of condenser coils cool when unit is operating	1. Refrigerant shortage	1. Repair leak and recharge	19–1; 19–1A
	2. Refrigerant overcharge	2. Remove part of charge	37–1; 37–1B
	3. Inefficient compressor	3. Check efficiency and correct	4–3C
Unit in vacuum—frost on expansion valve only	1. Ice plugging expansion valve orifice	1. Apply hot wet cloth to expansion valve body; an increase in suction pressure indicates moisture; install a new drier	20–1; 22–1E
	2. Expansion valve strainer plugged	2. Clean strainer or replace valve	20–1; 22–1B; 38–1

Refrigeration

Condition	Possible Cause	Corrective Action	Reference
High head pressure	1. Overcharge of refrigerant	1. Remove overcharge	37–1; 37–1B
	2. Air in system	2. Remove air	18–1; 18–1G
	3. Dirty condenser	3. Clean condenser	18–1; 18–1B
	4. Unit in too hot location	4. Relocate unit	
	5. Water-cooled condenser plugged	5. Clean or replace condenser	18–1; 18–1B, D, E
	6. Condenser water too warm	6. Provide sufficient cool water; adjust water regulating valve	18–1; 18–1C, D, E
	7. Cooling water shut off	7. Turn on water	
Low head pressure	1. Shortage of refrigerant	1. Repair leak and recharge	19–1; 19–1A
	2. Cold unit location	2. Provide warm condenser air	34–1
	3. Cold condenser water	3. Adjust water regulating valve or provide less water	18–1; 18–1C, D, E
	4. Inefficient compressor valves	4. Replace leaky valves	4–3C
	5. Leaky oil return valve in oil separator	5. Repair or replace	24–1A
High suction pressure	1. Evaporator overloaded	1. See previous entry "Unit operates excessively"	
	2. Expansion valve stuck open	2. Repair or replace valve	20–1; 20–1B
	3. Expansion valve too large	3. Replace valve	20–1; 20–1C
	4. Leaking compressor suction valves	4. Replace suction valves or compressor	4–3C
	5. Evaporator too large	5. Resize evaporator	35–1; 37–1; 37–1A, B, C
Low suction pressure	1. Shortage of refrigerant	1. Repair leak and recharge	19–1; 19–1A
	2. Evaporator underloaded	2. Clean or defrost evaporator	36–1
	3. Liquid line strainer clogged	3. Clean or replace strainer	22–1; 22–1B
	4. Plugged expansion valve	4. Clean or replace valve	20–1; 22–1E; 38–1A
	5. Lost charge on TXV power assembly	5. Replace power assembly	20–1; 20–1E

Refrigeration			
Condition	Possible Cause	Corrective Action	Reference
Low suction pressure—cont'd.	6. Space temperature too low	6. Adjust or replace thermostat	8–1; 8–1C, D
	7. Expansion valve too small	7. Replace valve	20–1; 20–1D
	8. Excessive pressure drop through evaporator	8. Check for plugged external equalizer	28–1
	9. Oversized compressor	9. Resize compressor	4–5
Loss of compressor oil pressure	1. Loss of compressor oil	1. See previous entry "Compressor loses oil"	
	2. Malfunctioning oil pump	2. Repair or replace oil pump	4–3; 4–3A, B, C
	3. Oil pump inlet screen plugged	3. Clean or replace screen	4–3; 4–3A, B, C
Starting relay burned out	1. Compressor short cycling	1. See previous entry "Compressor starts and runs but short cycles"	
	2. Improper relay mounting	2. Mount relay properly	14–1; 14–1A; B, C, D
	3. Relay vibrating	3. Mount relay in rigid position	14–1; 14–1D
	4. Wrong relay	4. Replace with proper relay	14–1; 14–1A, B, C, D
	5. Wrong running capacitor	5. Replace with proper capacitor	13–1; 13–1B
	6. Excessive line voltage	6. Reduce voltage to a maximum of 10% over motor rating	1–1
	7. Low line voltage	7. Increase voltage to not less than 10% below motor rating	1–1
Starting relay contacts stuck	1. Unit short cycling	1. See previous entry "Compressor starts and runs but short cycles"	
	2. Bad bleed resistor	2. Replace resistor or capacitor	13–1; 13–1A
Starting capacitors burned out	1. Compressor short cycling	1. See previous entry "Compressor starts and runs but short cycles"	
	2. Prolonged operation on starting winding	2. Reduce starting load; increase low voltage	1–1; 14–1A, B, C, D, E
	3. Sticking relay contacts	3. Replace relay	14–1; 14–1A, B, C, D, E
	4. Wrong capacitor	4. Replace with proper capacitor	13–1; 13–1A, B

Refrigeration

Condition	Possible Cause	Corrective Action	Reference
Running capacitors burned out	1. Excessive line voltage	1. Reduce line voltage to not more than 10% over motor rating	1–1
	2. Wrong capacitor	2. Replace with proper capacitor	13–1; 13–1A, B
	3. Light compressor load	3. Check voltage on capacitor and replace capacitor	13–1; 13–1A, B
Evaporator freezes but defrosts while unit is running	1. Moisture in system	1. Remove refrigerant evacuate system; install new drier; recharge system	38–1; 38–1A, B, C
Evaporator coil iced over	1. Automatic defrost control erratic or inoperative	1. Replace control	39–1; 39–1A
	2. Automatic defrost control improperly wired	2. Rewire control	39–1; 39–1B
	3. Defective defrost control thermal element	3. Replace control	39–1; 39–1C
	4. Improperly installed control thermal element	4. Relocate element	39–1; 39–1C
	5. Defrost control termination point too low	5. Replace or adjust control	39–1; 39–1D
	6. Defrost valve solenoid burned out	6. Replace solenoid	39–1; 39–1E
	7. Stuck closed defrost valve	7. Repair or replace valve	39–1; 39–1F
	8. Restricted hot gas bypass line	8. Replace line	39–1; 39–1G
	9. Inoperative freezer compartment door switch	9. Replace switch	39–1; 39–1H
	10. Inoperative freezer compartment fan	10. Clear fan or replace motor	39–1; 39–1I
	11. Freezer defrost element burned out	11. Replace element	39–1; 39–1J
	12. Freezer compartment drain trough or drain pan heater burned out	12. Replace heater	39–1; 39–1K
	13. Freezer compartment drain line plugged	13. Clean drain line	41–1; 41–1A, B, C, D, E; 39–1K

Refrigeration			
Condition	Possible Cause	Corrective Action	Reference
Refrigerator remains in the defrost cycle	1. Defrost control incorrectly wired	1. Rewire defrost control	39–1; 39–1L
	2. Automatic defrost control inoperative	2. Replace defrost control	39–1; 39–1A
	3. Defrost control termination point too high	3. Replace or adjust control	39–1; 39–1M
	4. Defrost solenoid valve stuck open	4. Clean or replace solenoid valve	39–1; 39–1E, N
	5. Room temperature too low (below 55°F or 12.8°C)	5. Relocate unit or provide heat	40–1
Water collects in bottom of refrigerator	1. Drain tube plugged	1. Clean tube	41–1; 41–1A, B, C, D, E
	2. Drain tube frozen	2. Check drain heater element and repair or replace	41–1; 41–1A, B, C, D, E; 39–1C
	3. Split drain trough	3. Replace trough	41–1; 41–1A
	4. Water leakage between trough and cabinet liner	4. Seal with suitable sealer	41–1; 41–1B
	5. Fresh food compartment liner warped	5. Replace liner or seal with suitable sealer	41–1; 41–1C
	6. Evaporator baffle not properly installed	6. Install baffle properly	41–1; 41–1D
	7. Humidiplate not adjusted properly	7. Adjust humidiplate	41–1; 41–1E
	8. Door gasket not sealing properly	8. Adjust door or replace gasket	42–1
Condensation on outside of cabinet	1. Door gaskets leaking	1. Adjust door or replace gasket	42–1
	2. Mullion heater burned out	2. Replace mullion heater	43–1; 43–1A
	3. Wire loose to mullion heater	3. Reconnect wire	43–1; 43–1A
	4. Abnormally high humidity	4. None, (explain reason to owner)	43–1; 43–1B
Water or ice collects in bottom of freezer compartment	1. Drain tube frozen	1. Defrost and repair heater	41–1; 41–1A, B, C, D, E
	2. Drain tube plugged	2. Clean tube	41–1; 41–1A, B, C, D, E
	3. Drain trough heater burned out	3. Replace heater	39–11C; 41–1; 41–1A, B, C, D, E;
	4. Evaporator cover plate mislocated	4. Install properly	41–1; 41–1A, B, C, D, E

Refrigeration

Condition	Possible Cause	Corrective Action	Reference
Fresh food compartment too warm	1. Poor air distribution to food compartment	1. Adjust controls for better air distribution	44–1
	2. Thermostat setting too high	2. Adjust thermostat	8–1; 8–1E
	3. Thermostat control bulb not making good contact	3. Provide good contact	8–1; 8–1E
	4. Bad thermostat	4. Replace thermostat	8–1; 8–1E
Freezer compartment warm	1. Thermostat set warm	1. Adjust thermostat	8–1; 8–1D, E
	2. Bad thermostat	2. Replace thermostat	8–1; 8–1C
	3. Freezer fan motor not running	3. Free blade or replace motor	39–1; 39–1I; 47–1; 47–1A, B, C, E
	4. Evaporator iced up	4. See previous entry "Evaporator coil iced over"	
	5. Light stays on	5. Replace switch or rewire	45–1
	6. Freezer door gaskets not sealing	6. Adjust door or replace gasket	42–1
	7. Freezer compartment door switch erratic	7. Replace switch	39–1; 39–1H
	8. Defective automatic defrost control	8. Replace defrost control	39–1; 39–1A
	9. Defrost valve solenoid burned out	9. Replace solenoid	39–1; 39–1E
	10. Restricted hot gas bypass line	10. Replace hot gas bypass line	39–1; 39–1G
	11. Loose wire at automatic defrost control or solenoid valve	11. Reconnect wire	39–1; 39–1A
	12. Excessive freezer compartment load	12. Advise customer	46–1
	13. Drain trough heater burned out	13. Replace heater	39–1K; 41–1; 41–1A, B, C, D, E
	14. Abnormally low room temperature	14. Relocate cabinet or provide heat	40–1
	15. Packages blocking air distribution	15. Advise customer	44–1
Food compartment too warm	1. System short of refrigerant	1. Repair leak and recharge	19–1; 19–1A
	2. Inefficient compressor	2. Replace compressor	4–1; 4–3C; 4–4; 4–4A, B
	3. Thermostat set too high	3. Adjust thermostat	8–1; 8–1D, E
	4. Dirty condenser	4. Clear obstruction	18–1; 18–1B

Refrigeration

Condition	Possible Cause	Corrective Action	Reference
Food compartment too warm—cont'd.	5. Inoperative condenser fan	5. Replace fan motor	39–1; 39–1I; 47–1; 47–1A, B, C, E, F
	6. Freezer compartment fan inoperative	6. Free blade or replace motor	39–1; 39–1I; 47–1; 47–1A, B, C, E, F
	7. Fresh food compartment fan inoperative	7. Free blade or replace motor	39–1; 39–1I; 47–1; 47–1A, B, C, E, F
	8. Freezer compartment door switch inoperative	8. Replace switch	39–1; 39–1H
	9. Door gasket not sealing	9. Adjust door or replace gasket	42–1
	10. Evaporator baffle not installed properly	10. Install properly	41–1; 41–1D
	11. Shelves covered, restricting air flows	11. Remove covering and advise customer	44–1
	12. Excessive food compartment load	12. Advise customer	46–1
	13. Restricted strainer, filter drier, or capilliary tube	13. Replace and charge unit	22–1; 22–1A, B, C, D, E
	14. Freezer coil iced over	14. See previous entry "Evaporator coil iced over"	
Inoperative defrost circuit	1. Defrost timer motor inoperative	1. Replace timer	39–1; 39–1A
	2. Inoperative defrost heater	2. Replace heater	39–1; 39–1J
	3. Faulty defrost limiter	3. Replace defrost limiter	39–1; 39–1P

Air Conditioning

Condition	Possible Cause	Corrective Action	Reference
Unit will not run	1. Blown power fuse	1. Replace fuse and check for cause	3–1
	2. Thermostat not demanding	2. Turn on thermostat and set temperature	8–1; 8–1A, B
	3. Blown transformer fuse	3. Replace fuse and check for cause	3–1
	4. Burned out transformer	4. Replace transformer	48–1; 48–1A
	5. Faulty wiring or loose connections	5. Repair wiring or connections	10–1
Outdoor unit will not run	1. Blown fuse to outdoor unit	1. Replace fuse and check for cause	3–1
	2. Thermostat set too high	2. Set thermostat	8–1; 8–1A, B
	3. Burned out contactor coil	3. Replace coil	5–1; 5–1A
	4. Burned contactor contacts	4. Replace contacts	5–1; 5–1B
	5. Compressor overload open	5. Determine overload and correct	4–2E, F, G
	6. Off on high-pressure control	6. See entry “High head pressure”	9–1; 9–1B
	7. Off on low-pressure control	7. See entry “Low suction pressure”	9–1; 9–1A
	8. Faulty wiring or connections	8. Repair wiring or connections	10–1
Compressor will not start	1. Defective contactor contacts	1. Replace contacts	5–1; 5–1B
	2. Compressor overload open	2. Determine overload and correct	4–2E, F, G, H
	3. Bad starting capacitor	3. Replace starting capacitor	13–1; 13–1A
	4. Bad starting relay	4. Replace starting relay	14–1; 14–1A, B, C, D, E
	5. Bad run capacitor	5. Replace run capacitor	13–1; 13–1B
	6. Burned out compressor motor	6. Repair motor or replace compressor	4–1; 4–2A, B, C, D
	7. Stuck compressor	7. Replace compressor	4–3; 4–3A, B
Outdoor fan motor will not start	1. Faulty wiring or loose connections	1. Repair wiring or connections	10–1
	2. Burned out fan motor	2. Replace fan motor	4–2; 4–2A, B, C, D
	3. Bad fan motor bearings	3. Replace bearings or motor	18–1; 18–1B; 47–1; 47–1E

Air Conditioning			
Condition	Possible Cause	Corrective Action	Reference
Compressor hums but will not run	1. Bad starting capacitor	1. Replace capacitor	13–1; 13–1A
	2. Bad starting relay	2. Replace starting relay	14–1; 14–1A, B, C, D, E
	3. Burned out compressor motor	3. Repair or replace compressor	4–1; 4–2
	4. Stuck compressor	4. Replace compressor	4–3; 4–3A, B; 7–1
	5. Defective contactor contacts	5. Replace contacts	5–1; 5–1B
	6. Three-phase compressor single-phasing	6. Replace fuse or reset circuit breaker	3–1
	7. Low voltage	7. Inform power company	12–1
Compressor cycling on overload	1. Bad starting capacitor	1. Replace starting capacitor	13–1; 13–1A
	2. Bad starting relay	2. Replace starting relay	14–1; 14–1A, B, C, D, E
	3. Bad running capacitor	3. Replace running capacitor	13–1; 13–1B
	4. Weak overload	4. Replace overload	4–2E, F, G
	5. Bad contactor contacts	5. Replace contacts	5–1; 5–1B
	6. Low voltage	6. Inform power company	12–1
	7. Burned compressor motor	7. Repair or replace compressor	4–1; 4–2A, B, C, D
	8. Refrigerant overcharge	8. Remove overcharge	18–1; 18–1F
	9. Shortage of refrigerant	9. Repair leak and recharge	19–1; 19–1A
	10. High suction pressure	10. Reduce load or repair compressor	37–1; 37–1A, B, C
	11. Air or noncondensables in system	11. Remove air or noncondensables	18–1; 18–1G
Compressor off on high-pressure control	1. Refrigerant overcharge	1. Remove overcharge	18–1; 18–1F
	2. Dirty outdoor coil	2. Clean coil	18–1; 18–1B
	3. Slipping outdoor fan belt	3. Replace or adjust fan belt	18–1; 18–1B
	4. Outdoor fan motor not running	4. See previous entry "Outdoor fan motor will not start"	4–1; 4–2A, B, C, D; 47–1A, B, C, D, E, F
	5. Air or noncondensables in system	5. Remove air or noncondensables	18–1; 18–1G

Air Conditioning

Condition	Possible Cause	Corrective Action	Reference
Compressor cycling or off on low-pressure control	1. Shortage of refrigerant	1. Repair leak and recharge	19–1; 19–1A
	2. Dirty or defective expansion valve	2. Clean or replace expansion valve	20–1; 20–1A, B, C, D, E, F
	3. Defective expansion valve power element	3. Replace power element	20–1; 20–1E
	4. Dirty indoor filters	4. Clean or replace filters	19–1; 19–1B
	5. Dirty indoor coil	5. Clean coil	15–1; 19–1C
	6. Indoor blower belt slipping	6. Replace or adjust fan belt	18–1; 18–B1
	7. Indoor blower not running	7. See entry "Indoor blower not running"	4–1, 4–2A, B, C, D; 18–1; 18–1B
	8. Restriction in refrigerant system	8. Locate and remove restrictions	22–1; 22–1A, B, C, D, E
Noisy compressor	1. Loose hold-down bolts	1. Tighten bolts	25–1; 26–1
	2. Low oil level in compressor	2. See next entry "Compressor loses oil"	
	3. Defective compressor valves	3. Replace valves and valve plate	4–3; 4–3C
	4. Wrong expansion valve superheat setting	4. Adjust superheat setting	20–1
	5. Expansion valve stuck	5. Repair or replace expansion valve	20–1; 20–1B
	6. Poor contact of expansion valve thermal bulb	6. Improve contact	20–1; 20–1A
	7. Overcharge of refrigerant (cap tube system)	7. Remove overcharge	18–1; 18–1F
Compressor loses oil	1. Shortage of refrigerant	1. Repair leak and recharge system with refrigerant and oil	19–1; 19–1A
	2. Low suction pressure	2. See entry "Low suction pressure"	19–1; 19–1A, B, C, D, E, F
	3. Expansion valves stuck open	3. Repair or replace expansion valve	20–1; 20–1B
	4. Restriction in refrigeration system	4. Locate restriction and remove	22–1; 22–1A, B, C, D, E
	5. Refrigerant piping improperly sized	5. Resize piping	31–1; 31–1A, B, C; 32–1; 32–1A
No cooling, but compressor runs continuously	1. Shortage of refrigerant	1. Repair leak and recharge	19–1; 19–1A
	2. Defective compressor valves	2. Replace valves and valve plate or compressor	4–3; 4–3C

Air Conditioning

Condition	Possible Cause	Corrective Action	Reference
No cooling, but compressor runs continuously—cont'd.	3. High suction pressure	3. See entry "High suction pressure"	
	4. Air or noncondensables in system	4. Remove air or noncondensables	18–1; 18–1G
	5. Wrong expansion valve superheat setting	5. Adjust superheat setting	20–1
	6. Dirty or defective expansion valve	6. Repair or replace expansion valve	20–1; 20–1E
	7. Indoor coil dirty	7. Clean coil	19–1; 19–1C
	8. Indoor air filter dirty	8. Clean or replace filter	19–1; 19–1B
	9. Indoor blower belt slipping	9. Replace or adjust fan belt	18–1; 18–1B
	10. Restriction in refrigerant circuit	10. Locate and remove	22–1; 22–1A, B, C, D, E
	11. Dirty outdoor coil	11. Clean coil	18–1; 18–1B
Too much cooling; compressor runs continuously	1. Thermostat setting low	1. Reset thermostat	8–1; 8–1A, B
	2. Thermostat in wrong location	2. Relocate thermostat	8–1; 8–1A, B
	3. Faulty wiring	3. Repair wiring	10–1
Liquid refrigerant flooding compressor (cap tube system)	1. Refrigerant overcharge	1. Remove overcharge	18–1; 18–1F
	2. High head pressure	2. See entry "High head pressure"	
	3. Indoor coil dirty	3. Clean coil	19–1; 19–1C
	4. Indoor fan belt slipping	4. Replace or adjust belt	18–1; 18–1B
	5. Indoor air filters dirty	5. Clean or replace filters	19–1; 19–1B
	6. Indoor fan not running	6. See entry "Indoor blower not running"	4–1; 4–2A, B, C, D
Liquid refrigerant flooding compressor (expansion valve system)	1. Expansion valve superheat setting wrong	1. Adjust superheat setting	20–1
	2. Expansion valve stuck open	2. Repair or replace expansion valve	20–1; 20–1B
	3. Expansion valve thermal bulb loose	3. Improve contact	20–1; 20–1A
	4. Refrigerant overcharge	4. Remove overcharge	18–1; 18–1F
	5. Too cold indoor air temperature	5. Raise thermostat setting	8–1; 8–1A, B

Air Conditioning

Condition	Possible Cause	Corrective Action	Reference
High head pressure	1. Refrigerant overcharge	1. Remove overcharge	18–1; 18–1F
	2. High ambient temperature	2. Provide cooler air to condenser	49–1
	3. Air or noncondensables in system	3. Remove air or noncondensables	18–1; 18–1G
	4. Excessive loading	4. Reduce load	35–1; 37–1; 37–1A, B, C; 46–1
	5. Dirty outdoor coil	5. Clean coil	18–1; 18–1B
	6. Outdoor fan motor not running	6. See entry "Outdoor fan motor will not start"	4–1; 47–1A, B, C, D, E, F
	7. Outdoor fan belt slipping	7. Replace or adjust fan belt	18–1; 18–1B
Low head pressure	1. Shortage of refrigerant	1. Repair leak and recharge	19–1; 19–1A
	2. Defective compressor valves	2. Replace valves and valve plate or compressor	4–3; 4–3C
	3. Low suction pressure	3. See entry "Low suction pressure"	
	4. Low air temperature over outdoor coil	4. Provide warmer air	34–1; 34–1A, B
High suction pressure	1. Defective compressor valves	1. Replace valves and valve plate or compressor	4–3; 4–3C
	2. Refrigerant overcharge	2. Remove overcharge	18–1; 18–1F
	3. High head pressure	3. See above entry "High head pressure"	
	4. Return air temperature high	4. Provide cooler air	50–1
	5. Excessive load	5. Reduce load	35–1; 37–1; 37–1A, B, C; 46–1
	6. Expansion valve stuck open	6. Clean or replace valve	20–1; 20–1B
Low suction pressure	1. Refrigerant shortage	1. Repair leak and recharge	19–1; 19–1A
	2. Low return air temperature	2. Set thermostat higher	8–1; 8–1A, B
	3. Wrong expansion valve superheat setting	3. Adjust superheat setting	20–1
	4. Dirty or defective expansion valve	4. Clean or replace expansion valve	20–1; 20–1A, B, C, D, E, F
	5. Defective expansion valve power element	5. Replace power element	20–1E
	6. Indoor blower belt slipping	6. Replace or adjust belt	18–1; 18–1B

Air Conditioning			
Condition	Possible Cause	Corrective Action	Reference
Low suction pressure—cont'd.	7. Indoor blower not running	7. See next entry "Indoor blower not running"	
	8. Restriction in refrigerant circuit	8. Locate restriction and remove	22–1; 22–1A, B, C, D, E
	9. Indoor air filters dirty	9. Clean or replace filter	19–1; 19–1B
	10. Indoor coil dirty	10. Clean coil	19–1; 19–1C
	11. Indoor coil icing	11. See entry "Indoor coil icing"	
	12. Restricted cap tube	12. Replace cap tube	22–1; 22–1D
Indoor blower not running	1. Blown fuse	1. Replace fuse and correct cause	3–1
	2. Indoor fan relay defective	2. Replace relay	51–1
	3. Burned indoor motor	3. Replace fan motor	47–1; 47–1A, B, C, D, E, F
	4. Broken belt	4. Replace belt	18–1; 18–1B
	5. Faulty wiring or loose connections	5. Repair wiring or connections	10–1
Indoor coil icing	1. Shortage of refrigerant	1. Repair leak and recharge	19–1; 19–1A
	2. Low suction pressure	2. See entry "Low suction pressure"	
	3. Low return air temperature	3. Raise thermostat setting	18–1; 18–1A, B
	4. Indoor blower not running	4. See entry "Indoor blower not running"	
	5. Indoor blower belt slipping	5. Replace or adjust belt	18–1; 18–1B
	6. Restriction in refrigerant system	6. Locate and remove restriction	22–1; 22–1A, B, C, D, E
	7. Indoor air filter dirty	7. Clean or replace filters	19–1; 19–1B
	8. Dirty indoor coil	8. Clean coil	19–1; 19–1C
	9. Dirty or defective expansion valve	9. Clean or replace valve	20–1; 20–1E
High operating costs	1. Defective compressor valves	1. Replace valves and valve plate or compressor	4–3; 4–3C
	2. Shortage of refrigerant	2. Repair leak and recharge	19–1; 19–1A
	3. Refrigerant overcharge	3. Remove overcharge	18–1; 18–1F
	4. Dirty outdoor coil	4. Clean coil	18–1; 18–1B
	5. Dirty indoor coil	5. Clean coil	19–1; 19–1C

Air Conditioning

Condition	Possible Cause	Corrective Action	Reference
High operating costs—cont'd.	6. Dirty indoor air filters	6. Clean or replace filters	19–1; 19–1B
	7. High head pressure	7. See entry "High head pressure"	
	8. Thermostat in wrong location	8. Relocate thermostat	8–1; 8–1A, B, C
	9. Air ducts not insulated	9. Insulate ducts	52–1
	10. Unit too small	10. Install proper size unit	53–1
	11. Indoor or outdoor fan belt slipping	11. Replace or adjust belt	18–1; 18–1B
	12. Outdoor thermostat set too high	12. Adjust thermostat	8–1; 8–1E

Ice Machine (Cuber)			
Condition	Possible Cause	Corrective Action	Reference
Unit will not run	1. Switch turned off	1. Turn switch on	2–1
	2. Defective operation control switch	2. Replace switch	87–1
	3. Blown fuse	3. Replace fuse and correct cause	3–1
	4. Storage bin switch off	4. Adjust bin switch	82–1
Unit running but produces little or no ice	1. Low refrigerant charge	1. Repair leak and recharge	19–1; 19–1A
	2. Dirty condenser (aircooled)	2. Clean condenser	18–1; 18–1B
	3. Expansion valve plugged	3. Clean or replace valves	20–1B; 22–1B, C
	4. Hot gas check valves leaking	4. Replace check valves	83–1
	5. Hot gas solenoid leaking	5. Replace hot gas solenoid	39–1; 39–1A
	6. Defective compressor	6. Replace compressor	4–1; 4–2; 4–2A, B, C, D; 4–3C
	7. No water over freeze plates	7. Repair or replace water supply float valve. Repair or replace water pump. Clean water distributor. Clear restricted or pinched water hose. Raise water level in reservoir	47–1; 47–1A, B, C, E, F; 71–1; 71–1A
	8. Unit running in defrost cycle	8. Replace ice thickness switch. Replace freeze relay. Replace defrost control relay. Replace or repair inoperative mercury switch.	84–1; 85–1; 86–1
	9. Too thick ice slab	9. Adjust ice thickness control to about $\frac{1}{2}''$ to $\frac{3}{4}''$ slab	84–1
	10. Low head pressure	10. Relocate unit or repair or replace defective water valve	34–1; 34–1A, B
Condensing unit cycles with bin switch closed	1. High discharge pressure	1. Remove air or noncondensables from system.Clean condenser.	18–1; 18–1B, F, G
	2. Low suction pressure	2. Repair leak and recharge system	19–1; 19–1A
	3. Defective pressure control	3. Replace control	9–1; 9–1A, B

Ice Machine (Cuber)

Condition	Possible Cause	Corrective Action	Reference
Condensing unit cycles with bin switch closed—cont'd.	4. Defective condenser water valve	4. Repair or replace valve	18–1; 18–1C, D, E
	5. Defective condenser fan	5. Replace fan motor or free fan blade	18–1; 18–1B
	6. Compressor cycling on overload	6. See Refrigeration section "Compressor starts and runs but short cycles"	4–3; 4–3B
Ice slab stuck to evaporator plate	1. Lime on evaporator plate	1. Clean plate with proper cleaner	81–1; 81–1A
	2. Warped evaporator plate	2. Replace evaporator plate	80–1; 81–1; 81–1A
	3. Low refrigerant charge	3. Repair leak and recharge	19–1; 19–1A
	4. Inoperative hot gas solenoid	4. Repair or replace valve	39–1; 39–1F, N
	5. Low discharge pressure	5. Relocate unit or condenser water valve	34–1; 34–1A, B
	6. Ice thickness switch imbedded into slab	6. Adjust switch to provide proper thickness of ice slab	84–1
Ice slab; uneven slab hollow in center	1. Low refrigerant charge	1. Repair leak and recharge system	19–1; 19–1A
	2. Inoperative expansion valve (In normal operation, a slab will have a slight depression in the center. Slabs less than $\frac{1}{2}'$ thick are more susceptible to hollow center.)	2. Repair or replace valve	20–1; 20–1A, B, C, D, E; 22–1E
Ice slab uneven; slab has hookup back edge with hang-over front edge	1. Leaking hot gas solenoid valve	1. Repair or replace solenoid valve	39–1; 39–1A
	2. Leaking gas check valve	2. Repair or replace check valve	83–1
Ice slab uneven; insufficient water over freeze plate	1. Water pump not running	1. Replace water pump or repair loose electrical connection	47–1; 47–1A, B, C, E, F
	2. Plugged water line	2. Clear water line	
	3. Low water level in sump	3. Adjust or repair float valve	71–1; 71–1A
	4. Plugged water distributor	4. Clean distributor	89–1; 81–1; 81–1A

Ice Machine (Cuber)			
Condition	Possible Cause	Corrective Action	Reference
Cloudy ice cubes	1. High mineral content in water	1. Install water softener	81–1A
	2. Water not siphoning from reservoir at end of freezing cycle	2. Clear siphoning tube or adjust for proper operation	90–1
	3. Insufficient water supply	3. Provide water supply in excess of 20 psi	
	4. Restricted water distributor	4. Clean distributor	81–1; 81–1A; 89–1
	5. Water float valve set too low	5. Adjust float	71–1; 71–1A
Slabs stack up on cutting grid	1. Low voltage to cutting grids	1. Supply proper voltage	91–1
	2. Grid plug, wires, or pins not making good contact	2. Provide good contact	91–1; 81–1A
	3. Blown grid fuse	3. Correct condition and replace fuse	91–1; 91–1A
	4. Grid fuse loose in receptacle	4. Replace receptacle	91–1A
	5. High mineral content in water	5. Install water softener	81–1A
	6. Lime build-up on grid wires	6. Clean grid wires	81–1A; 91–1
	7. Wrong grid wires	7. Replace with proper wires	91–1
Slow harvest	1. Low ambient air	1. Move to provide 50°F air (10°C)	34–1; 34–1A, B
	2. High mineral content in water	2. Install water softener	81–1A
	3. Water valve passing too much water	3. Adjust valve to 125 psig head pressure	18–1D, E
	4. Water valve leaking	4. Repair or replace valve	18–1D, E

Ice Machine (Flaker)

Condition	Possible Cause	Corrective Action	Reference
Unit will not run	1. Power switch off	1. Turn on switch	2–1
	2. Blown fuse	2. Repair cause and replace fuse	3–1
	3. Ice storage bin thermostat keeping unit off too long	3. Adjust bin thermostat	71–1
	4. Contacts stuck in lockout position in power relay	4. Replace power relay	72–1
	5. Ice storage bin thermostat inoperative	5. Repair or replace thermostat	71–1
	6. Loose electrical connection	6. Repair connection	10–1
Compressor cycling	1. Dirty condenser	1. Clean condenser	18–1; 18–1B
	2. Condenser air circulation restricted	2. Remove restriction	18–1; 18–1B
	3. Defective start relay	3. Replace relay	14–1; 14–1A, B, C, D, E
	4. Defective overload protector or lockout relay	4. Replace overload protector or relay	4–2E, F, G; 73–1
	5. Defective start capacitor	5. Replace start capactor	13–1; 13–1A
	6. Inoperative condenser fan	6. Free fan or replace motor	18–1; 18–1B
	7. Loose electrical connection	7. Repair loose connection	10–1
No ice, but unit running	1. Low refrigerant charge	1. Repair leak and recharge system	19–1; 19–1A
	2. "O" ring at evaporator shell leaking	2. Replace "O" ring seal	74–1
	3. Compressor inoperative	3. See entries "Compressor fails to start (no hum)" and "Compressor will not start (hums and trips overload protector," in the Refrigeration Chart.	4–1; 4–2; 4–2A, B, C, D, E, F, G
	4. No water	4. Provide water as required	
Water leaking from unit	1. Float assembly stuck	1. Free or replace	71–1; 71–1A
	2. Bad float ball	2. Replace float ball	71–1
	3. Storage bin drain plugged	3. Clear drain	76–1

Ice Machine (Flaker)			
Condition	Possible Cause	Corrective Action	Reference
Water leaking from unit—cont'd.	4. Evaporator drain plugged	4. Clear drain	76–1
	5. Inlet water connector leaking	5. Repair connections	
	6. "O" ring at evaporator shell leaking	6. Replace "O" ring	74–1
Soft or wet ice	1. Shortage of refrigerant	1. Repair leak and recharge unit	19–1; 19–1A
	2. High discharge pressure	2. Clean condenser or provide cooler condenser air	18–1; 18–1B
	3. Water in float reservoir too high	3. Adjust water float or repair float valve	75–1; 75–1A
	4. Inefficient compressor	4. See entry "Unit low on capacity" in Refrigeration Chart.	4–1; 4–3C
Excessive or chattering noise	1. Intermittent water supply	1. Provide constant water supply	
	2. Loose hold-down on auger gear motor	2. Tighten or replace	77–1
	3. Scale deposits on inside of evaporator shell	3. Remove evaporator shell and clean in proper solution	81–1
	4. Reservoir water level too low	4. Adjust water level	75–1; 75–1A
	5. End play in auger gear motor	5. Repair gear motor	78–1
	6. Worn auger gear bearings	6. Replace bearings	79–1
	7. Air lock in gravity water line to evaporator shell	7. Eliminate air lock	80–1
Unit operating with ice storage bin full of ice	1. Ice storage bin thermostat set too cold	1. Adjust thermostat	71–1
	2. Defective ice storage bin thermostat	2. Replace thermostat	71–1

Heading (Oil)

Condition	Possible Cause	Corrective Action	Reference
Burner will not start	1. Thermostat off or set too low	1. Turn thermostat on or set to higher temperature	8–1; 8–1A, B
	2. Burner motor overload tripped	2. Push motor overload reset button	18–1; 18–1A, B; 4–1; 4–2A, B, C, D
	3. Primary control off on safety	3. Reset safety switch lever	102–1
	4. Dirty thermostat contacts	4. Clean thermostat contacts	8–1; 8–1A, B
	5. Bad thermostat	5. Replace thermostat	8–1; 8–1A, B
	6. Blown fuse or tripped breaker	6. Replace fuse or reset breaker	3–1
	7. Disconnect switch open	7. Close switch	2–1
	8. Shorted flame detector circuit	8. Replace flame detector	97–1
	9. Shorted flame detector leads	9. Separate and insulate leads	98–1
	10. Flame detector exposed to direct light	10. Protect detector from light	97–1
	11. Faulty friction clutch	11. Replace element or control	99–1
	12. Hot contacts stuck	12. Replace element or control	100–1
	13. Dirty cold contacts	13. Clean contacts	100–1
	14. Flame detector bimetal carboned	14. Clean bimetal	101–1
	15. Loose connection or broken wire on flame detector	15. Repair connection or replace wire	10–1
	16. Low line voltage or power failure	16. Notify power company	12–1
	17. Limit control open	17. Set limit control to 200°F (93°C) then jumper control terminals; if burner starts, replace the control	70–1
	18. Open electric circuit to limit control	18. Repair or replace wiring	10–1
	19. Defective internal primary control circuit	19. Replace control	102–1
	20. Dirty burner relay contacts in primary control	20. Clean contacts	102–1

Heading (Oil)			
Condition	Possible Cause	Corrective Action	Reference
Burner will not start—cont'd.	21. Defective burner motor	21. Replace burner motor	18–1; 18–1A, B; 4–1; 4–2A, B, C, D
	22. Binding burner blower wheel	22. Turn off power and rotate blower by hand; if binding free	103–1
	23. Seized fuel pump	23. Turn power off and rotate blower by hand; if binding, replace fuel pump	104–1
Burner starts and fires but short cycles	1. Thermostat in warm draft	1. Relocate thermostat	8–1; 8–1A, B
	2. Heat anticipator set wrong	2. Correct anticipator setting	8–1; 8–1B
	3. Furnace blower running too slow	3. Speed up blower to obtain an 85°F (29.4°C) to 95°F (35°C) temperature	18–1; 18–1A, B
	4. Limit control set too low	4. Reset limit to 200°F (93°C)	70–1
	5. Dirty air filter	5. Clean or replace filter	19–1; 19–1B, C
	6. Return air restriction	6. Clear restriction	19–1; 19–1B; 59–1; 60–1; 60–1A, B
	7. Low or fluctuating voltage	7. Notify power company	12–1
	8. Loose wiring connection	8. Repair connection	10–1
Burner starts and fires but does not heat enough (short cycles)	1. Vibration at thermostat	1. Correct vibration or relocate thermostat	8–1; 8–1A, B
	2. Thermostat in warm draft	2. Relocate thermostat	8–1; 8–1A, B
	3. Heat anticipator set wrong	3. Correct anticipator setting	8–1; 8–1A, B
	4. Furnace blower running too slow	4. Speed up blower to obtain 85°F (29.4°C) to 95°F (35°C) temperature rise	18–1; 18–1A, B
	5. Dirty air filter	5. Clean or replace filter	19–1; 19–1B, C
	6. Defective blower motor bearings	6. Replace motor	18–1; 18–1A, B; 4–1; 4–2A, B, C, D
	7. Defective blower bearings	7. Replace bearings	18–1; 18–1A, B; 4–1; 4–2A, B, C, D
	8. Dirty furnace blower wheel	8. Clean blower wheel	60–1; 60–1A
	9. Wrong blower motor rotation	9. Change rotation or replace motor	18–1; 18–1A, B; 4–1; 4–2A, B, C, D
	10. Return air restricted	10. Clear restriction	52–1

Heading (Oil)

Condition	Possible Cause	Corrective Action	Reference
Burner starts and fires but does not heat enough (short cycles)—cont'd.	11. Limit control set low	11. Reset limit to 200°F (93°C)	70–1
	12. Low or fluctuating voltage	12. Notify power company	12–1
	13. Loose wiring connection	13. Repair connection	10–1
Burner starts and fires; then locks out on safety	1. Too little primary air; long dirty flame	1. Increase combustion air	105–1
	2. Too much primary air; short lean flame	2. Reduce combustion air	105–1
	3. Unbalanced flame	3. Replace nozzle	108–1
	4. Too little or restricted draft	4. Correct draft or remove restriction	105–1
	5. Excessive draft	5. Adjust barometric damper	105–1
	6. Dirty flame detector bimetal element	6. Clean element	101–1
	7. Faulty flame detector friction clutch	7. Replace flame detector control	99–1
	8. Welded or shorted cold contacts in flame detector	8. Replace flame detector control	100–1
	9. Air leaking into flue pipe around flame detector mount	9. Seal air leaks	102–1
	10. Dirty flame detector cad cell face	10. Clean cad cell face	97–1
	11. Loose or defective flame detector cad cell wires	11. Repair or replace cad cell holder and wires	97–1
	12. Faulty flame detector cad cell; resistance exceeds 1,500 ohms	12. Replace cad cell	97–1
	13. Defective primary control circuit	13. Replace primary control	102–1
Burner starts, but no flame is established	1. Oil tank empty	1. Contact oil distributor	
	2. Oil tank shut-off valve closed	2. Open valve	
	3. Water in oil tank	3. Remove water	106–1
	4. Air leak in oil supply line	4. Repair leak	107–1
	5. Oil filter plugged	5. Install new filter	107–1
	6. Oil pump strainer plugged	6. Clean strainer	104–1
	7. Restricted oil line	7. Repair or replace line	107–1
	8. Excessive combustion air	8. Adjust air supply damper	105–1

Heading (Oil)			
Condition	**Possible Cause**	**Corrective Action**	**Reference**
Burner starts, but no flame is established—cont'd.	9. Excessive vent draft	9. Adjust barometric damper to .030 inches to .035 inches water column	105–1
	10. Off-center spray from nozzle	10. Replace nozzle	108–1
	11. Nozzle strainer plugged	11. Replace nozzle	109–1
	12. Nozzle orifice plugged	12. Replace nozzle	108–1
	13. Faulty oil pump	13. Replace pump	104–1
	14. Low fuel pressure	14. Adjust pressure to desired pressure	104–1
	15. Faulty pump coupling	15. Replace coupling	104–1
	16. Low line voltage to transformer primary	16. Notify power company	12–1
	17. Faulty transformer	17. Replace transformer	110–1
	18. No or weak ignition spark	18. Properly ground transformer case	110–1
	19. Dirty or shorted ignition electrodes	19. Clean electrodes	110–1
	20. Improper position or gap of ignition electrodes	20. Correctly position and reset electrode gap	110–1
	21. Cracked or burned lead insulation	21. Replace electrode leads	110–1
	22. Loose or disconnected electrode leads	22. Repair or replace leads	110–1
	23. Defective electrode lead insulators	23. Replace electrodes	110–1
	24. Oil pump or blower overloading motor	24. Remove overload condition	103–1; 104–1
	25. Faulty oil pump motor	25. Replace motor	18–1; 18–1A, B; 4–1; 4–2A, B, C, D
	26. Low voltage	26. Notify power company	12–1
Burner starts and fires but loses flame and locks out on safety	1. Dirty face and cad cell	1. Clean cad cell face	97–1
	2. Faulty cad cell; resistance exceeds 1500 ohms	2. Replace cad cell	97–1
	3. Loose or defective cad cell wires	3. Repair or replace wires	98–1
	4. Stack control bimetal dirty	4. Clean bimetal element	101–1
	5. Faulty friction clutch in stack control	5. Replace stack control	99–1

Heating (Oil)

Condition	Possible Cause	Corrective Action	Reference
Burner starts and fires but loses flame and locks out on safety—cont'd.	6. Air leaking into vent pipe around stack control mount	6. Seal air leaks	102–1
	7. Defective stack control cold contacts	7. Replace stack control	100–1
	8. Too much combustion air	8. Adjust combustion air damper	105–1
	9. Too little combustion air	9. Adjust combustion air damper	105–1
	10. Unbalanced flame	10. Replace nozzle	108–1
	11. Excessive vent draft	11. Adjust barometric damper	105–1
	12. Too little vent draft	12. Adjust barometric damper	105–1
	13. Vent restricted	13. Clear restriction	105–1
	14. Oil pump loses prime	14. Prime pump at bleed port	104–1
	15. Air leak in oil supply line	15. Repair leaks	107–1
	16. Partially plugged nozzle	16. Replace nozzle	108–1
	17. Partially plugged nozzle strainer	17. Replace nozzle	108–1
	18. Water in oil storage tank	18. Remove water from storage tank	106–1
	19. Oil too heavy	19. Change to Number 1 oil	111–1
	20. Plugged fuel pump strainer	20. Clean strainer or replace pump	104–1
	21. Restricted oil line	21. Clear restriction	107–1
Too much heat; burner runs continuously	1. Defective thermostat	1. Repair or replace thermostat	8–1; 8–1A, B
	2. Shorted thermostat wires	2. Repair or replace wires	10–1
	3. Thermostat in cold location	3. Relocate thermostat	8–1; 8–1A, B
	4. Thermostat not level	4. Level thermostat	8–1; 8–1A, B
	5. Defective primary control	5. Replace control	102–1
Too little heat; burner runs continuously	1. Too much combustion air	1. Reduce combustion air	105–1
	2. Air leaking into heat exchanger	2. Repair leaks	105–1
	3. Excessive vent draft	3. Adjust barometric damper	105–1
	4. Wrong burner head adjustment	4. Correct burner head setting	105–1

Heading (Oil)			
Condition	Possible Cause	Corrective Action	Reference
Too little heat; burner runs continuously—cont'd.	5. Plugged heat exchanger	5. Clean heat exchanger and adjust burner	105–1
	6. Too little combustion air	6. Increase combustion air	105–1
	7. Insufficient vent draft	7. Adjust barometric damper	105–1
	8. Insufficient indoor air	8. Speed blower to obtain 85°F (29.4°C) to 95°F (35°C) temperature rise	18–1; 18–1A, B
	9. Dirty indoor blower	9. Clean blower	60–1
	10. Dirty furnace filter	10. Clean or replace filter	19–1; 19–1B, C
	11. Partially plugged nozzle	11. Replace nozzle	108–1
	12. Nozzle too small	12. Replace with larger nozzle	108–1
	13. Low oil pressure	13. Increase to proper pressure	104–1

8

Useful Engineering Data

Courtesy of Copeland Refrigeration Corp.

The following reference tables and charts cover miscellaneous engineering data and conversion factors frequently required in engineering calculations. Data specifically pertaining to refrigeration has been included where appropriate in previous sections.

TEMPERATURE SCALES

Absolute temperature Rankin = °F. + 459.6°

Absolute temperature Kelvin = °C. + 273.2°

Rankin = 1.8 Kelvin

Centigrade (Celsius) = 5/9 (°F. — 32°)

Fahrenheit = 9/5 °C. + 32°

INTERNATIONAL RATING CONDITIONS

CENTIGRADE - FAHRENHEIT

Evaporating Temperature		Condensing Temperature		Ambient Temperature	
°C.	°F.	°C.	°F.	°C.	°F.
12.5	55	30	86	21	69.8
10	50	32	90	27	80.6
7	45	35	95	32	89.6
5	41	40	104	38	100.4
0	32	45	113	43	109.4
-5	23	50	122		
-10	14	55	131		
-15	5	60	140		
-20	-4				
-25	-13				
-30	-22				
-40	-40				

THERMAL UNITS

Latent heat of ice	= 144 BTU/lb	= 288,000 BTU/ton
1 ton refrigeration	= 12,000 BTU/hr	= 288,000/24 hours
1 British Thermal Unit (BTU)		= .252 kcal
1 kilo - calorie (kcal)		= 3.97 BTU
1 BTU/lb		= 0.555 kcal/kg
1 kcal/kg		= 1.8 BTU/lb
1 BTU/lb/°F.		= 1 kcal/kg/°C.
1 watt		= 3.413 BTU/hr

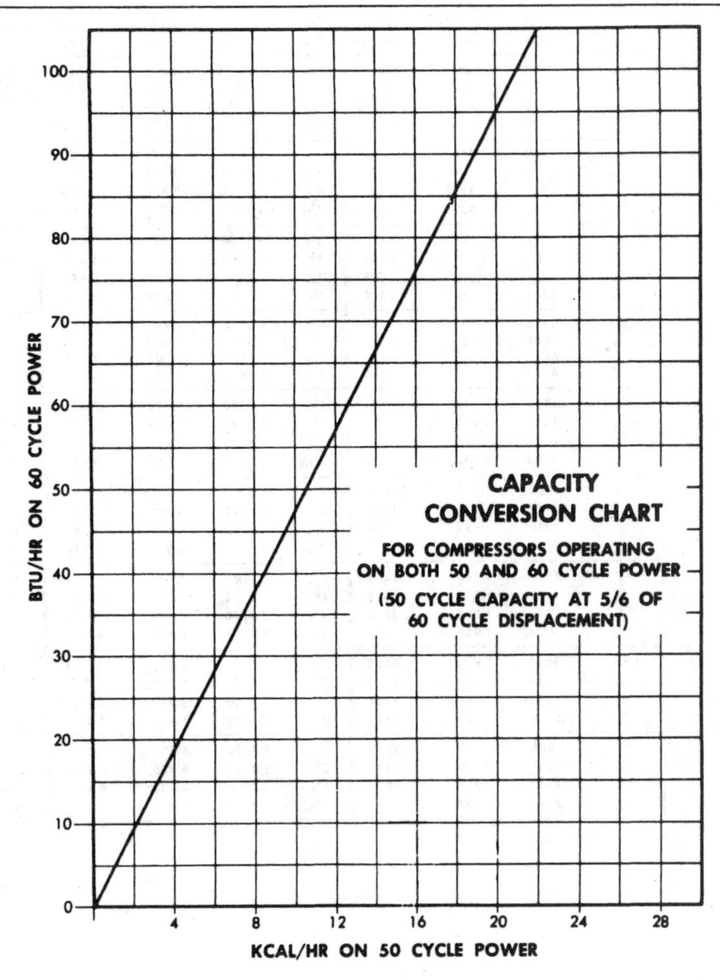

DECIMAL EQUIVALENTS, AREAS AND CIRCUMFERENCES OF CIRCLES

Diameter	Decimal Equivalent	Circumference	Area	Diameter	Decimal Equivalent	Circumference	Area	Diameter	Decimal Equivalent	Circumference	Area
				3/4	.7500	2.356	.4418	3	3.000	9.425	7.069
1/64	.0156	.04909	.00019	49/64	.7656	2.405	.4604	3 1/16	3.0625	9.621	7.366
1/32	.0312	.09817	.00077	25/32	.7812	2.454	.4794	3 1/8	3.1250	9.817	7.670
3/64	.0468	.1473	.00173	51/64	.7969	2.503	.4987	3 3/16	3.1875	10.01	7.980
1/16	.0625	.1963	.00307	13/16	.8125	2.553	.5185	3 1/4	3.2500	10.21	8.296
5/64	.0781	.2454	.00479	53/64	.8281	2.602	.5386	3 5/16	3.3125	10.41	8.618
3/32	.0937	.2945	.00690	27/32	.8437	2.651	.5591	3 3/8	3.3750	10.60	8.946
7/64	.1093	.3436	.00940	55/64	.8594	2.700	.5800	3 7/16	3.4375	10.80	9.281
1/8	.1250	.3927	.01227	7/8	.8750	2.749	.6013	3 1/2	3.5000	11.00	9.621
9/64	.1406	.4418	.01553	57/64	.8906	2.798	.6230	3 9/16	3.5625	11.19	9.968
5/32	.1562	.4909	.01917	29/32	.9062	2.847	.6450	3 5/8	3.6250	11.39	10.32
11/64	.1718	.5400	.02320	59/64	.9219	2.896	.6675	3 11/16	3.6875	11.58	10.68
3/16	.1875	.5890	.02761	15/16	.9375	2.945	.6903	3 3/4	3.7500	11.78	11.04
13/64	.2031	.6381	.03241	61/64	.9531	2.994	.7135	3 13/16	3.8125	11.98	11.42
7/32	.2187	.6872	.03758	31/32	.9687	3.043	.7371	3 7/8	3.8750	12.17	11.79
15/64	.2343	.7363	.04314	63/64	.9844	3.093	.7610	3 15/16	3.9375	12.37	12.18
1/4	.2500	.7854	.04909	1	1.000	3.142	.7854	4	4.000	12.57	12.57
17/64	.2656	.8345	.05542	1 1/16	1.0625	3.338	.8866	4 1/16	4.0625	12.76	12.96
9/32	.2812	.8836	.06213	1 1/8	1.1250	3.534	.9940	4 1/8	4.1250	12.96	13.36
19/64	.2968	.9327	.06920	1 3/16	1.1875	3.731	1.108	4 3/16	4.1875	13.16	13.77
5/16	.3125	.9817	.07670	1 1/4	1.2500	3.927	1.227	4 1/4	4.2500	13.35	14.19
21/64	.3281	1.031	.08456	1 5/16	1.3125	4.123	1.353	4 5/16	4.3125	13.55	14.61
11/32	.3437	1.080	.09281	1 3/8	1.3750	4.320	1.485	4 3/8	4.3750	13.74	15.03
23/64	.3593	1.129	.1014	1 7/16	1.4375	4.516	1.623	4 7/16	4.4375	13.94	15.47
3/8	.3750	1.178	.1104	1 1/2	1.5000	4:712	1.767	4 1/2	4.5000	14.14	15.90
25/64	.3906	1.227	.1198	1 9/16	1.5625	4.909	1.917	4 9/16	4.5625	14.33	16.35
13/32	.4062	1.276	.1296	1 5/8	1.6250	5.105	2.074	4 5/8	4.6250	14.53	16.80
27/64	.4218	1.325	.1398	1 11/16	1.6875	5.301	2.237	4 11/16	4.6875	14.73	17.26
7/16	.4375	1.374	.1503	1 3/4	1.7500	5.498	2.405	4 3/4	4.750	14.92	17.72
29/64	.4531	1.424	.1613	1 13/16	1.8125	5.694	2.580	4 13/16	4.8125	15.12	18.19
15/32	.4687	1.473	.1726	1 7/8	1.8750	5.890	2.761	4 7/8	4.8750	15.32	18.67
31/64	.4844	1.522	.1843	1 15/16	1.9375	6.087	2.948	4 15/16	4.9375	15.51	19.51
1/2	.5000	1.571	.1963	2	2.000	6.238	3.142	5	5.000	15.71	19.63
33/64	.5156	1.620	.2088	2 1/16	2.0625	6.480	3.341	5 1/16	5.0625	15.90	20.13
17/32	.5312	1.669	.2217	2 1/8	2.1250	6.676	3.547	5 1/8	5.1250	16.10	20.63
35/64	.5468	1.718	.2349	2 3/16	2.1875	6.872	3.758	5 3/16	5.1875	16.30	21.14
9/16	.5625	1.767	.2485	2 1/4	2.2500	7.069	3.976	5 1/4	5.2500	16.49	21.65
37/64	.5781	1.816	.2626	2 5/16	2.3125	7.265	4.200	5 5/16	5.3125	16.69	22.17
19/32	.5937	1.865	.2769	2 3/8	2.3750	7.461	4.430	5 3/8	5.3750	16.89	22.69
39/64	.6094	1.914	.2916	2 7/16	2.4375	7.658	4.666	5 7/16	5.4375	17.08	23.22
5/8	.6250	1.963	.3068	2 1/2	2.5000	7.854	4.909	5 1/2	5.5000	17.28	23.76
41/64	.6406	2.013	.3223	2 9/16	2.5625	8.050	5.157	5 9/16	5.5625	17.48	24.30
21/32	.6562	2.062	.3382	2 5/8	2.6250	8.247	5.412	5 5/8	5.6250	17.67	24.85
43/64	.6719	2.111	.3545	2 11/16	2.6875	8.443	5.673	5 11/16	5.6875	17.87	25.41
11/16	.6C75	2.160	.3712	2 3/4	2.7500	8.639	5.940	5 3/4	5.7500	18.06	25.97
45/64	.7031	2.209	.3883	2 13/16	2.8125	8.836	6.213	5 13/16	5.8125	18.26	26.53
23/32	.7187	2.258	.4057	2 7/8	2.8750	9.032	6.492	5 7/8	5.8750	18.46	27.11
47/64	.7344	2.307	.4236	2 15/16	2.9375	9.228	6.777	5 15/16	5.9375	18.65	27.69

CONVERSION TABLE
INCHES INTO MILLIMETERS

Inches	Milli-meters	Inches	Milli-meters	Inches	Milli-meters	Inches	Milli-meters	Inches	Milli-meters	Inches	Milli-meters
1/64	0.3969	55/64	21.8281	2 13/32	61.1189	4 3/32	103.981	5 25/32	146.844	8 15/16	227.013
1/32	0.7937	7/8	22.2250	2 7/16	61.9126	4 1/8	104.775	5 13/16	147.638	9	228.600
3/64	1.1906	57/64	22.6219	2 15/32	62.7064	4 5/32	105.569	5 27/32	148.432	9 1/16	230.188
1/16	1.5875	29/32	23.0187	2 1/2	63.5001	4 3/16	106.363	5 7/8	149.225	9 1/8	231.775
5/64	1.9844	59/64	23.4156	2 17/32	64.2939	4 7/32	107.156	5 29/32	150.019	9 3/16	233.363
3/32	2.3812	15/16	23.8125	2 9/16	65.0876	4 1/4	107.950	5 15/16	150.813	9 1/4	234.950
7/64	2.7781	61/64	24.2094	2 19/32	65.8814	4 9/32	108.744	5 31/32	151.607	9 5/16	236.538
1/8	3.1750	31/32	24.6062	2 5/8	66.6751	4 5/16	109.538	6	152.400	9 3/8	238.125
9/64	3.5719	63/64	25.0031	2 21/32	67.4689	4 11/32	110.331	6 1/16	153.988	9 7/16	239.713
5/32	3.9687	1	25.4001	2 11/16	68.2626	4 3/8	111.125	6 1/8	155.575	9 1/2	241.300
11/64	4.3656	1 1/32	26.1938	2 23/32	69.0564	4 13/32	111.919	6 3/16	157.163	9 9/16	242.888
3/16	4.7625	1 1/16	26.9876	2 3/4	69.8501	4 7/16	112.713	6 1/4	158.750	9 5/8	244.475
13/64	5.1594	1 3/32	27.7813	2 25/32	70.6439	4 15/32	113.506	6 5/16	160.338	9 11/16	246.063
7/32	5.5562	1 1/8	28.5751	2 13/16	71.4376	4 1/2	114.300	6 3/8	161.925	9 3/4	247.650
15/64	5.9531	1 5/32	29.3688	2 27/32	72.2314	4 17/32	115.094	6 7/16	163.513	9 13/16	249.238
1/4	6.3500	1 3/16	30.1626	2 7/8	73.0251	4 9/16	115.888	6 1/2	165.100	9 7/8	250.825
17/64	6.7469	1 7/32	30.9563	2 29/32	73.8189	4 19/32	116.681	6 9/16	166.688	9 15/16	252.413
9/32	7.1437	1 1/4	31.7501	2 15/16	74.6126	4 5/8	117.475	6 5/8	168.275	10	254.001
19/64	7.5406	1 9/32	32.5438	2 31/32	75.4064	4 21/32	118.269	6 11/16	169.863	10 1/16	255.588
5/16	7.9375	1 5/16	33.3376	3	76.2002	4 11/16	119.063	6 3/4	171.450	10 1/8	257.176
21/64	8.3344	1 11/32	34.1313	3 1/32	76.9939	4 23/32	119.856	6 13/16	173.038	10 3/16	258.763
11/32	8.7312	1 3/8	34.9251	3 1/16	77.7877	4 3/4	120.650	6 7/8	174.625	10 1/4	260.351
23/64	9.1281	1 13/32	35.7188	3 3/32	78.5814	4 25/32	121.444	6 15/16	176.213	10 5/16	261.938
3/8	9.5250	1 7/16	36.5126	3 1/8	79.3752	4 13/16	122.238	7	177.800	10 3/8	263.526
25/64	9.9219	1 15/32	37.3063	3 5/32	80.1689	4 27/32	123.031	7 1/16	179.388	10 7/16	265.113
13/32	10.3187	1 1/2	38.1001	3 3/16	80.9627	4 7/8	123.825	7 1/8	180.975	10 1/2	266.701
27/64	10.7156	1 17/32	38.8938	3 7/32	81.7564	4 29/32	124.619	7 3/16	182.563	10 9/16	268.288
7/16	11.1125	1 9/16	39.6876	3 1/4	82.5502	4 15/16	125.413	7 1/4	184.150	10 5/8	269.876
29/64	11.5094	1 19/32	40.4813	3 9/32	83.3439	4 31/32	126.206	7 5/16	185.738	10 11/16	271.463
15/32	11.9062	1 5/8	41.2751	3 5/16	84.1377	5	127.000	7 3/8	187.325	10 3/4	273.051
31/64	12.3031	1 21/32	42.0688	3 11/32	84.9314	5 1/32	127.794	7 7/16	188.913	10 13/16	274.638
1/2	12.7000	1 11/16	42.8626	3 3/8	85.7252	5 1/16	128.588	7 1/2	190.500	10 7/8	276.226
33/64	13.0969	1 23/32	43.6563	3 13/32	86.5189	5 3/32	129.382	7 9/16	192.088	10 15/16	277.813
17/32	13.4937	1 3/4	44.4501	3 7/16	87.3127	5 1/8	130.175	7 5/8	193.675	11	279.401
35/64	13.8906	1 25/32	45.2438	3 15/32	88.1064	5 5/32	130.969	7 11/16	195.263	11 1/16	280.988
9/16	14.2875	1 13/16	46.0376	3 1/2	88.9002	5 3/16	131.763	7 3/4	196.850	11 1/8	282.576
37/64	14.6844	1 27/32	46.8313	3 17/32	89.6939	5 7/32	132.557	7 13/16	198.438	11 3/16	284.163
19/32	15.0812	1 7/8	47.6251	3 9/16	90.4877	5 1/4	133.350	7 7/8	200.025	11 1/4	285.751
39/64	15.4781	1 29/32	48.4188	3 19/32	91.2814	5 9/32	134.144	7 15/16	201.613	11 5/16	287.338
5/8	15.8750	1 15/16	49.2126	3 5/8	92.0752	5 5/16	134.938	8	203.200	11 3/8	288.926
41/64	16.2719	1 31/32	50.0063	3 21/32	92.8689	5 11/32	135.732	8 1/16	204.788	11 7/16	290.513
21/32	16.6687	2	50.8001	3 11/16	93.6627	5 3/8	136.525	8 1/8	206.375	11 1/2	292.101
43/64	17.0656	2 1/32	51.5939	3 23/32	94.4564	5 13/32	137.319	8 3/16	207.963	11 9/16	293.688
11/16	17.4625	2 1/16	52.3876	3 3/4	95.2502	5 7/16	138.113	8 1/4	209.550	11 5/8	295.276
45/64	17.8594	2 3/32	53.1814	3 25/32	96.0439	5 15/32	138.907	8 5/16	211.138	11 11/16	296.863
23/32	18.2562	2 1/8	53.9751	3 13/16	96.8377	5 1/2	139.700	8 3/8	212.725	11 3/4	298.451
47/64	18.6531	2 5/32	54.7688	3 27/32	97.6314	5 17/32	140.494	8 7/16	214.313	11 13/16	300.038
3/4	19.0500	2 3/16	55.5626	3 7/8	98.4252	5 9/16	141.288	8 1/2	215.900	11 7/8	301.626
49/64	19.4469	2 7/32	56.3564	3 29/32	99.2189	5 19/32	142.082	8 9/16	217.488	11 15/16	303.213
25/32	19.8437	2 1/4	57.1501	3 15/16	100.013	5 5/8	142.875	8 5/8	219.075	12	304.801
51/64	20.2406	2 9/32	57.9439	3 31/32	100.806	5 21/32	143.669	8 11/16	220.663		
13/16	20.6375	2 5/16	58.7376	4	101.600	5 11/16	144.463	8 3/4	222.250		
53/64	21.0344	2 11/32	59.5314	4 1/32	102.394	5 23/32	145.257	8 13/16	223.838		
27/32	21.4312	2 3/8	60.3251	4 1/16	103.188	5 3/4	146.050	8 7/8	225.425		

CONVERSION TABLE
DECIMALS OF AN INCH INTO MILLIMETERS

Inches	Milli-meters	Inches	Milli-meters	Inches	Milli-meters	Inches	Milli-meters	Inches	Milli-meters
0.001	0.025	0.140	3.56	0.360	9.14	0.580	14.73	0.800	20.32
0.002	0.051	0.150	3.81	0.370	9.40	0.590	14.99	0.810	20.57
0.003	0.076	0.160	4.06	0.380	9.65	0.600	15.24	0.820	20.83
0.004	0.102	0.170	4.32	0.390	9.91	0.610	15.49	0.830	21.08
0.005	0.127	0.180	4.57	0.400	10.16	0.620	15.75	0.840	21.34
0.006	0.152	0.190	4.83	0.410	10.41	0.630	16.00	0.850	21.59
0.007	0.178	0.200	5.08	0.420	10.67	0.640	16.26	0.860	21.84
0.008	0.203	0.210	5.33	0.430	10.92	0.650	16.51	0.870	22.10
0.009	0.229	0.220	5.59	0.440	11.18	0.660	16.76	0.880	22.35
0.010	0.254	0.230	5.84	0.450	11.43	0.670	17.02	0.890	22.61
0.020	0.508	0.240	6.10	0.460	11.68	0.680	17.27	0.900	22.86
0.030	0.762	0.250	6.35	0.470	11.94	0.690	17.53	0.910	23.11
0.040	1.016	0.260	6.60	0.480	12.19	0.700	17.78	0.920	23.37
0.050	1.270	0.270	6.86	0.490	12.45	0.710	18.03	0.930	23.62
0.060	1.524	0.280	7.11	0.500	12.70	0.720	18.29	0.940	23.88
0.070	1.778	0.290	7.37	0.510	12.95	0.730	18.54	0.950	24.13
0.080	2.032	0.300	7.62	0.520	13.21	0.740	18.80	0.960	24.38
0.090	2.286	0.310	7.87	0.530	13.46	0.750	19.05	0.970	24.64
0.100	2.540	0.320	8.13	0.540	13.72	0.760	19.30	0.980	24.89
0.110	2.794	0.330	8.38	0.550	13.97	0.770	19.56	0.990	25.15
0.120	3.048	0.340	8.64	0.560	14.22	0.780	19.81	1.000	25.40
0.130	3.302	0.350	8.89	0.570	14.48	0.790	20.07		

CONVERSION TABLE
MILLIMETERS INTO INCHES

Milli-meters	Inches	Milli-meters	Inches	Milli-meters	Inches	Milli-meters	Inches	Milli-meters	Inches	Milli-meters	Inches
1	0.0394	33	1.2992	65	2.5590	97	3.8189	129	5.0787	161	6.3386
2	0.0787	34	1.3386	66	2.5984	98	3.8583	130	5.1181	162	6.3779
3	0.1181	35	1.3779	67	2.6378	99	3.8976	131	5.1575	163	6.4173
4	0.1575	36	1.4173	68	2.6772	100	3.9370	132	5.1968	164	6.4567
5	0.1968	37	1.4567	69	2.7165	101	3.9764	133	5.2362	165	6.4960
6	0.2362	38	1.4961	70	2.7559	102	4.0157	134	5.2756	166	6.5354
7	0.2756	39	1.5354	71	2.7953	103	4.0551	135	5.3149	167	6.5748
8	0.3150	40	1.5748	72	2.8346	104	4.0945	136	5.3543	168	6.6142
9	0.3543	41	1.6142	73	2.8740	105	4.1338	137	5.3937	169	6.6535
10	0.3937	42	1.6535	74	2.9134	106	4.1732	138	5.4331	170	6.6929
11	0.4331	43	1.6929	75	2.9527	107	4.2126	139	5.4724	171	6.7323
12	0.4724	44	1.7323	76	2.9921	108	4.2520	140	5.5118	172	6.7716
13	0.5118	45	1.7716	77	3.0315	109	4.2913	141	5.5512	173	6.8110
14	0.5512	46	1.8110	78	3.0709	110	4.3307	142	5.5905	174	6.8504
15	0.5905	47	1.8504	79	3.1102	111	4.3701	143	5.6299	175	6.8897
16	0.6299	48	1.8898	80	3.1496	112	4.4094	144	5.6693	176	6.9291
17	0.6693	49	1.9291	81	3.1890	113	4.4488	145	5.7086	177	6.9685
18	0.7087	50	1.9685	82	3.2283	114	4.4882	146	5.7480	178	7.0079
19	0.7480	51	2.0079	83	3.2677	115	4.5275	147	5.7874	179	7.0472
20	0.7874	52	2.0472	84	3.3071	116	4.5669	148	5.8268	180	7.0866
21	0.8268	53	2.0866	85	3.3464	117	4.6063	149	5.8661	181	7.1260
22	0.8661	54	2.1260	86	3.3858	118	4.6457	150	5.9055	182	7.1653
23	0.9055	55	2.1653	87	3.4252	119	4.6850	151	5.9449	183	7.2047
24	0.9449	56	2.2047	88	3.4646	120	4.7244	152	5.9842	184	7.2441
25	0.9842	57	2.2441	89	3.5039	121	4.7638	153	6.0236	185	7.2834
26	1.0236	58	2.2835	90	3.5433	122	4.8031	154	6.0630	186	7.3228
27	1.0630	59	2.3228	91	3.5827	123	4.8425	155	6.1023	187	7.3622
28	1.1024	60	2.3622	92	3.6220	124	4.8819	156	6.1417	188	7.4016
29	1.1417	61	2.4016	93	3.6614	125	4.9212	157	6.1811	189	7.4409
30	1.1811	62	2.4409	94	3.7008	126	4.9606	158	6.2205	190	7.4803
31	1.2205	63	2.4803	95	3.7401	127	5.0000	159	6.2598	191	7.5197
32	1.2598	64	2.5197	96	3.7795	128	5.0394	160	6.2992	192	7.5590

CONVERSION TABLE (Cont'd)
MILLIMETERS INTO INCHES

Milli-meters	Inches	Milli-meters	Inches	Milli-meters	Inches	Milli-meters	Inches	Milli-meters	Inches	Milli-meters	Inches
193	7.5984	261	10.2756	329	12.9527	397	15.6299	465	18.3070	533	20.9842
194	7.6378	262	10.3149	330	12.9921	398	15.6693	466	18.3464	534	21.0236
195	7.6771	263	10.3543	331	13.0315	399	15.7086	467	18.3858	535	21.0629
196	7.7165	264	10.3937	332	13.0708	400	15.7480	468	18.4252	536	21.1023
197	7.7559	265	10.4330	333	13.1102	401	15.7874	469	18.4645	537	21.1417
198	7.7953	266	10.4724	334	13.1496	402	15.8267	470	18.5039	538	21.1811
199	7.8346	267	10.5118	335	13.1889	403	15.8661	471	18.5433	539	21.2204
200	7.8740	268	10.5512	336	13.2283	404	15.9055	472	18.5826	540	21.2598
201	7.9134	269	10.5905	337	13.2677	405	15.9448	473	18.6220	541	21.2992
202	7.9527	270	10.6299	338	13.3071	406	15.9842	474	18.6614	542	21.3385
203	7.9921	271	10.6693	339	13.3464	407	16.0236	475	18.7007	543	21.3779
204	8.0315	272	10.7086	340	13.3858	408	16.0630	476	18.7401	544	21.4173
205	8.0708	273	10.7480	341	13.4252	409	16.1023	477	18.7795	545	21.4566
206	8.1102	274	10.7874	342	13.4645	410	16.1417	478	18.8189	546	21.4960
207	8.1496	275	10.8267	343	13.5039	411	16.1811	479	18.8582	547	21.5354
208	8.1890	276	10.8661	344	13.5433	412	16.2204	480	18.8976	548	21.5748
209	8.2283	277	10.9055	345	13.5826	413	16.2598	481	18.9370	549	21.6141
210	8.2677	278	10.9449	346	13.6220	414	16.2992	482	18.9763	550	21.6535
211	8.3071	279	10.9842	347	13.6614	415	16.3385	483	19.0157	551	21.6929
212	8.3464	280	11.0236	348	13.7008	416	16.3779	484	19.0551	552	21.7322
213	8.3858	281	11.0630	349	13.7401	417	16.4173	485	19.0944	553	21.7716
214	8.4252	282	11.1023	350	13.7795	418	16.4567	486	19.1338	554	21.8110
215	8.4645	283	11.1417	351	13.8189	419	16.4960	487	19.1732	555	21.8503
216	8.5039	284	11.1811	352	13.8582	420	16.5354	488	19.2126	556	21.8897
217	8.5433	285	11.2204	353	13.8976	421	16.5748	489	19.2519	557	21.9291
218	8.5827	286	11.2598	354	13.9370	422	16.6141	490	19.2913	558	21.9685
219	8.6220	287	11.2992	355	13.9763	423	16.6535	491	19.3307	559	22.0078
220	8.6614	288	11.3386	356	14.0157	424	16.6929	492	19.3700	560	22.0472
221	8.7008	289	11.3779	357	14.0551	425	16.7322	493	19.4094	561	22.0866
222	8.7401	290	11.4173	358	14.0945	426	16.7716	494	19.4488	562	22.1259
223	8.7795	291	11.4567	359	14.1338	427	16.8110	495	19.4881	563	22.1653
224	8.8189	292	11.4960	360	14.1732	428	16.8504	496	19.5275	564	22.2047
225	8.8582	293	11.5354	361	14.2126	429	16.8897	497	19.5669	565	22.2440
226	8.8976	294	11.5748	362	14.2519	430	16.9291	498	19.6063	566	22.2834
227	8.9370	295	11.6141	363	14.2913	431	16.9685	499	19.6456	567	22.3228
228	8.9764	296	11.6335	364	14.3307	432	17.0078	500	19.6850	568	22.3622
229	9.0157	297	11.6929	365	14.3700	433	17.0472	501	19.7244	569	22.4015
230	9.0551	298	11.7323	366	14.4094	434	17.0866	502	19.7637	570	22.4409
231	9.0945	299	11.7716	367	14.4488	435	17.1259	503	19.8031	571	22.4803
232	9.1338	300	11.8110	368	14.4882	436	17.1653	504	19.8425	572	22.5196
233	9.1732	301	11.8504	369	14.5275	437	17.2047	505	19.8818	573	22.5590
234	9.2126	302	11.8897	370	14.5669	438	17.2441	506	19.9212	574	22.5984
235	9.2519	303	11.9291	371	14.6063	439	17.2834	507	19.9606	575	22.6377
236	9.2913	304	11.9685	372	14.6456	440	17.3228	508	20.0000	576	22.6771
237	9.3397	305	12.0078	373	14.6850	441	17.3622	509	20.0393	577	22.7165
238	9.3701	306	12.0472	374	14.7244	442	17.4015	510	20.0787	578	22.7559
239	9.4094	307	12.0866	375	14.7637	443	17.4409	511	20.1181	579	22.7952
240	9.4488	308	12.1260	376	14.8031	444	17.4803	512	20.1574	580	22.8346
241	9.4882	309	12.1653	377	14.8425	445	17.5196	513	20.1968	581	22.8740
242	9.5275	310	12.2047	378	14.8819	446	17.5590	514	20.2362	582	22.9133
243	9.5669	311	12.2441	379	14.9212	447	17.5984	515	20.2755	583	22.9527
244	9.6063	312	12.2834	380	14.9606	448	17.6378	516	20.3149	584	22.9921
245	9.6456	313	12.3228	381	15.0000	449	17.6771	517	20.3543	585	23.0314
246	9.6850	314	12.3622	382	15.0393	450	17.7165	518	20.3937	586	23.0708
247	9.7244	315	12.4015	383	15.0787	451	17.7559	519	20.4330	587	23.1102
248	9.7638	316	12.4409	384	15.1181	452	17.7952	520	20.4724	588	23.1496
249	9.8031	317	12.4803	385	15.1574	453	17.8346	521	20.5118	589	23.1889
250	9.8425	318	12.5197	386	15.1968	454	17.8740	522	20.5511	590	23.2283
251	9.8819	319	12.5590	387	15.2362	455	17.9133	523	20.5905	591	23.2677
252	9.9212	320	12.5984	388	15.2756	456	17.9527	524	20.6299	592	23.3070
253	9.9606	321	12.6378	389	15.3149	457	17.9921	525	20.6692	593	23.3464
254	10.0000	322	12.6771	390	15.3543	458	18.0315	526	20.7086	594	23.3858
255	10.0393	323	12.7165	391	15.3937	459	18.0708	527	20.7480	595	23.4251
256	10.0787	324	12.7559	392	15.4330	460	18.1102	528	20.7874	596	23.4645
257	10.1181	325	12.7952	393	15.4724	461	18.1496	529	20.8267	597	23.5039
258	10.1575	326	12.8346	394	15.5118	462	18.1889	530	20.8661	598	23.5433
259	10.1968	327	12.8740	395	15.5511	463	18.2283	531	20.9055	599	23.5826
260	10.2362	328	12.9134	396	15.5905	464	18.2677	532	20.9448	600	23.6220

CONVERSION TABLE
HUNDREDTHS OF A MILLIMETER INTO INCHES

Millimeters	Inches	Millimeters	Inches	Millimeters	Inches	Millimeters	Inches	Millimeters	Inches	Millimeters	Inches
0.01	0.0004	0.18	0.0071	0.35	0.0138	0.52	0.0205	0.69	0.0272	0.86	0.0339
0.02	0.0008	0.19	0.0075	0.36	0.0142	0.53	0.0209	0.70	0.0276	0.87	0.0343
0.03	0.0012	0.20	0.0079	0.37	0.0146	0.54	0.0213	0.71	0.0280	0.88	0.0346
0.04	0.0016	0.21	0.0083	0.38	0.0150	0.55	0.0217	0.72	0.0283	0.89	0.0350
0.05	0.0020	0.22	0.0087	0.39	0.0154	0.56	0.0220	0.73	0.0287	0.90	0.0354
0.06	0.0024	0.23	0.0091	0.40	0.0157	0.57	0.0224	0.74	0.0291	0.91	0.0358
0.07	0.0028	0.24	0.0094	0.41	0.0161	0.58	0.0228	0.75	0.0295	0.92	0.0362
0.08	0.0031	0.25	0.0098	0.42	0.0165	0.59	0.0232	0.76	0.0299	0.93	0.0366
0.09	0.0035	0.26	0.0102	0.43	0.0169	0.60	0.0236	0.77	0.0303	0.94	0.0370
0.10	0.0039	0.27	0.0106	0.44	0.0173	0.61	0.0240	0.78	0.0307	0.95	0.0374
0.11	0.0043	0.28	0.0110	0.45	0.0177	0.62	0.0244	0.79	0.0311	0.96	0.0378
0.12	0.0047	0.29	0.0114	0.46	0.0181	0.63	0.0248	0.80	0.0315	0.97	0.0382
0.13	0.0051	0.30	0.0118	0.47	0.0185	0.64	0.0252	0.81	0.0319	0.98	0.0386
0.14	0.0055	0.31	0.0122	0.48	0.0189	0.65	0.0256	0.82	0.0323	0.99	0.0390
0.15	0.0059	0.32	0.0126	0.49	0.0193	0.66	0.0260	0.83	0.0327	1.00	0.0394
0.16	0.0063	0.33	0.0130	0.50	0.0197	0.67	0.0264	0.84	0.0331		
0.17	0.0067	0.34	0.0134	0.51	0.0201	0.68	0.0268	0.85	0.0335		

METRIC PREFIXES

micro	= 10^{-6}	(example)	micron
milli	= 10^{-3}		millimeter
centi	= 10^{-2}		centimeter
deci	= 10^{-1}		decimeter
deka	= 10		decaliter
hecto	= 10^{2}		hectoliter
kilo	= 10^{3}		kilometer
mega	= 10^{6}		megaton

LENGTH

1 inch (in.)	=	2.54 centimeters	=	25400 microns
1 foot (ft)	=	12 inches	=	.3048 meter
1 yard (yd)	=	3 feet	=	.9144 meter
1 mile	=	5280 feet	=	1.609 kilometers
1 nautical mile	=	6080 feet	=	1.853 kilometers
1 millimeter (mm)	=	1000 microns	=	.0394 inch
1 centimeter (cm)	=	10 millimeters	=	.3937 inch
1 decimeter	=	10 centimeters	=	3.937 inches
1 meter (m)	=	100 centimeters	=	3.281 feet
1 kilometer	=	1000 meters	=	.6214 mile

AREA

1 sq in.	=		=	6.45 sq cm
1 sq ft	=	144 sq in.	=	.0929 sq meter
1 sq. yard	=	9 sq ft	=	.836 sq meter
1 acre	=	43560 sq ft	=	.4047 hectare
1 sq mile	=	640 acres	=	259 hectares
1 sq cm	=	100 sq mm	=	.155 sq in.
1 sq. meter	=	10,000 sq cm	=	10.764 sq ft
1 hectare	=	10,000 sq meters	=	2.471 acres
1 sq km	=	100 hectares	=	.3861 sq mile

WEIGHT, AVOIRDUPOIS

1 ounce (oz)	=	473.5 grains	=	28.35 grams
1 pound (lb)	=	16 ounces	=	453.59 grams
1 pound	=	7000 grains	=	.454 kilograms
1 short ton	=	2000 pounds	=	907 kilograms
1 cu ft water @ 4° C.	=	62.42 lb	=	28.31 kilograms
1 gallon water @ 4° C.	=	8.34 lb	=	3.78 kilograms
1 gram (g)	=	1000 milligrams	=	15.43 grains
1 kilogram (kg)	=	1000 grams	=	2.205 pounds
1 metric ton	=	1000 kilograms	=	2204.6 pounds
1 cu cm water @ 4° C.	=	1 gram	=	.035 ounces
1 liter water @ 4° C.	=	1 kilogram	=	2.205 pounds

VOLUME, DRY

1 cu in.	=	= 16.39 cu cm
1 cu ft	= 1728 cu in.	= .0238 cu meter
1 cu yard	= 27 cu ft	= .7646 cu meter
1 quart, U.S.	= .0389 cu ft	= 1101 cu cm
1 gallon, U.S.	= 4 quarts	= 4.405 cu decimeters
1 peck	= 2 gallons	= 8.810 cu decimeters
1 bushel, U.S.	= 4 pecks	= 35.239 cu decimeters
1 bushel, U.S.	= 1.244 cu ft	
1 bushel, Imperial	= 1.032 U.S. bushels	
1 cord	= 128 cu ft	
1 cu cm	=	= .061 cu in.
1 cu. decimeter	= 1000 cu cm	= 61.02 cu in.
1 cu meter	= 1000 cu decimeters	= 1.308 cu yd
1 cu meter		= 35.314 cu ft

VOLUME, LIQUID

1 pint	= 16 fluid ounces	= .473 liters
1 quart	= 2 pints	= .946 liter
1 gallon, U.S.	= 4 quarts	= 3.785 liters
1 gallon, U.S.	= 231 cu in.	= 3.785 cu decimeters
1 cu ft	= 7.48 gallon, U.S.	= 28.32 cu decimeters
1 gallon, Imperial	= 1.201 gallon, U.S.	=
1 liter	= 1000 cu cm	= 1.057 quarts
1 liter	= 1 cu decimeter	= .0353 cu ft
1 decaliter	= 10 liters	= 2.64 gallon, U.S.
1 cu meter	= 1000 liters	= 264.18 gallon, U.S.

DENSITY

1 lb/cu in.	= 1728 lb/cu ft	= 27.68 gram/cu cm
1 lb/cu ft	= 27 lb/cu yd	= 16.018 kg/cu m
1 gram/cu cm	= 1000 kg/cu m	= 62.43 lb/cu ft

PRESSURE

1 lb/sq in.	= 144 lb/sq ft	= .0703 kg/sq cm
1 lb/sq in.	= 2.036 in. Hg	= 2.307 ft water @ 4° C.
1 kg/sq cm	= 735.51 mm Hg	= 14.22 lb/sq in.
1 kg/sq cm	= 10 m water @ 4° C.	= .968 standard atmospheres
1 kg/sq cm	= 1 at	
1 in. Hg	= .491 lb/sq in.	= 1.133 ft water at 4° C.
1 in. water	= 5.20 lb/sq ft	= .0361 lb/sq in.
1 ft water @ 4° C.	= 62.43 lb/sq ft	= .0305 kg/sq cm
1 ft water @ 4° C.	= .433 lb/sq in.	= .883 in. Hg
1 standard atmosphere (Atm)	= 14.7 lb/sq in.	= 29.92 in. Hg
1 standard atmosphere	= 33.9 ft water @ 4° C.	= 760 mm Hg
1 standard atmosphere	= 1.033 kg/sq cm	= 1.0133 bars
1 cm Hg	= .0136 kg/sq cm	= .1934 lb/sq in.
1 Bar	= 750 mm Hg	= 14.5 lb/sq in.
1 ata	= 1 kg/sq cm absolute	

VELOCITY

1 ft/sec	=	.682 miles/hr	=	.3048 m/sec
1 mile/hr	=	1.467 ft/sec	=	.447 m/sec
1 mile/hr	=	.868 knots	=	1.609 km/hr
1 m/sec	=	3.6 km/hr	=	3.28 ft/sec
1 km/hr	=	.2778 m/sec	=	.621 miles/hr
1 knot	=	1.152 miles/hr	=	1 nautical mile/hr

HEAT, ENERGY, WORK

1 ft lb	=	.001285 BTU	=	0.13826 kg-meter
1 joule	=	1 watt - second	=	.000948 BTU
1 BTU	=	778.1 ft lb	=	.252 kcal
1 KCAL	=	3.968 BTU	=	1000 cal
1 hp-hr	=	.746 kw-hr	=	2544.7 BTU
1 kw-hr	=	1.341 hp-hr	=	3413 BTU
1 boiler horsepower	=	33479 BTU/hr	=	Evaporation of 34.5 water/hr at 212° F.

SOLID AND LIQUID EXPENDABLE REFRIGERANTS

Evaporating Temperature of dry ice (solid CO_2) at 1 atmosphere	=	-109° F.
Heat of sublimation of dry ice at -109° F.	=	246.3 BTU/lb
Specific heat of CO_2 gas	=	.2 BTU/lb/°F.
Refrigerating effect of solid CO_2 to gas at 32° F. (246.3 + .2 [109 + 32])	=	274.5 BTU/lb
Evaporating Temperature of liquid carbon dioxide (CO_2) at 1 atmosphere	=	-70° F.
Heat of vaporization of liquid CO_2 at -70° F.	=	149.7 BTU/lb
Specific heat of CO_2 gas	=	.2 BTU/lb/°F.
Refrigerating effect of liquid CO_2 to gas at 32° F. (149.7 + .2 [70 + 32])	=	170.1 BTU/lb
Evaporating Temperature of liquid nitrogen (N_2) at 1 atmosphere	=	-320° F.
Heat of vaporization of liquid N_2 at -320° F.	=	85.67 BTU/lb
Specific heat of N_2 gas	=	.248 BTU/lb/°F.
Refrigerating effect of liquid nitrogen to gas at 32° F. (85.67 + .248 [320 + 32])	=	172.97 BTU/lb

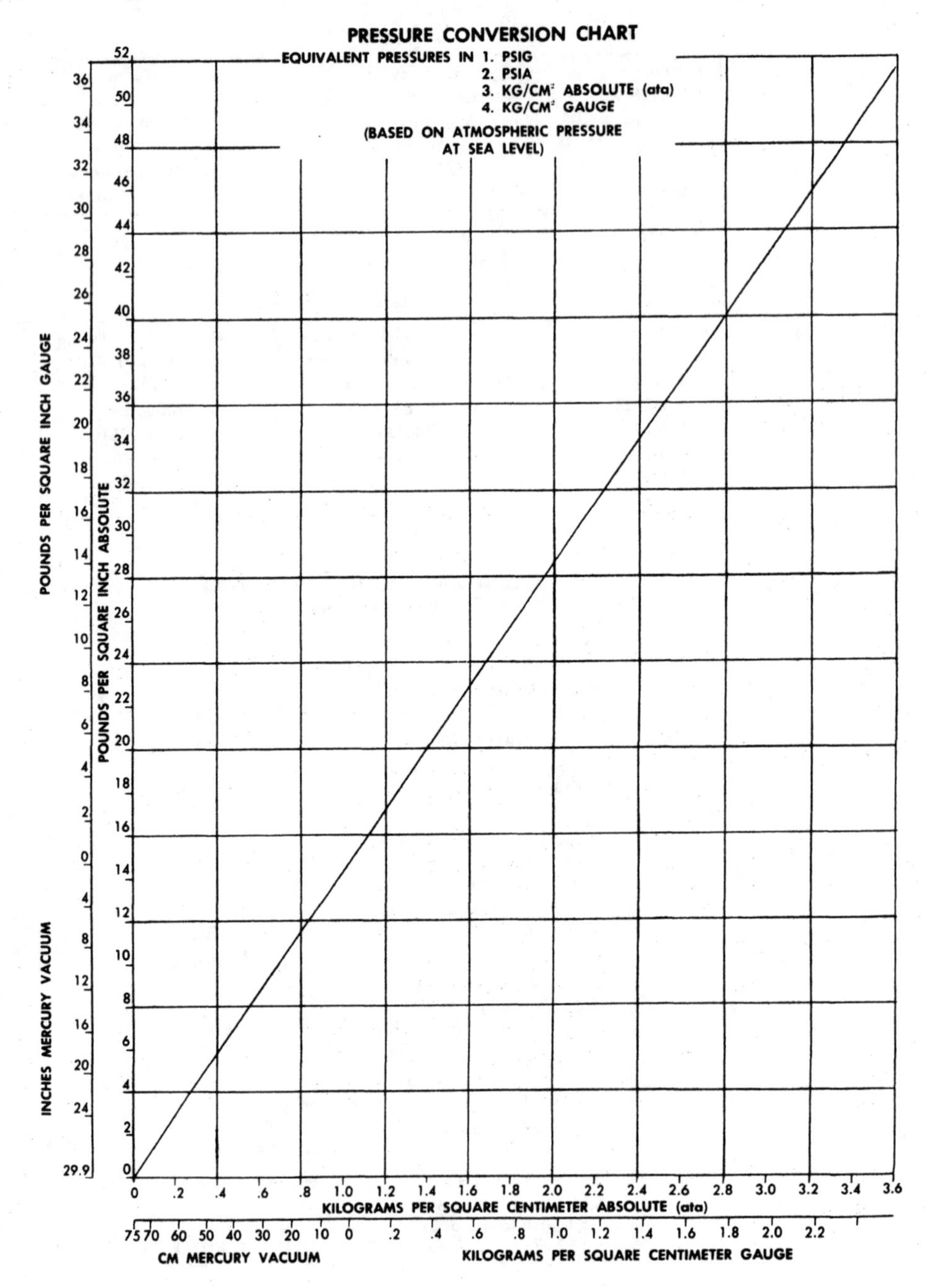
PRESSURE CONVERSION CHART
EQUIVALENT PRESSURES IN 1. PSIG
2. PSIA
3. KG/CM² ABSOLUTE (ata)
4. KG/CM² GAUGE
(BASED ON ATMOSPHERIC PRESSURE AT SEA LEVEL)
POUNDS PER SQUARE INCH GAUGE
INCHES MERCURY VACUUM
POUNDS PER SQUARE INCH ABSOLUTE
KILOGRAMS PER SQUARE CENTIMETER ABSOLUTE (ata)
CM MERCURY VACUUM
KILOGRAMS PER SQUARE CENTIMETER GAUGE

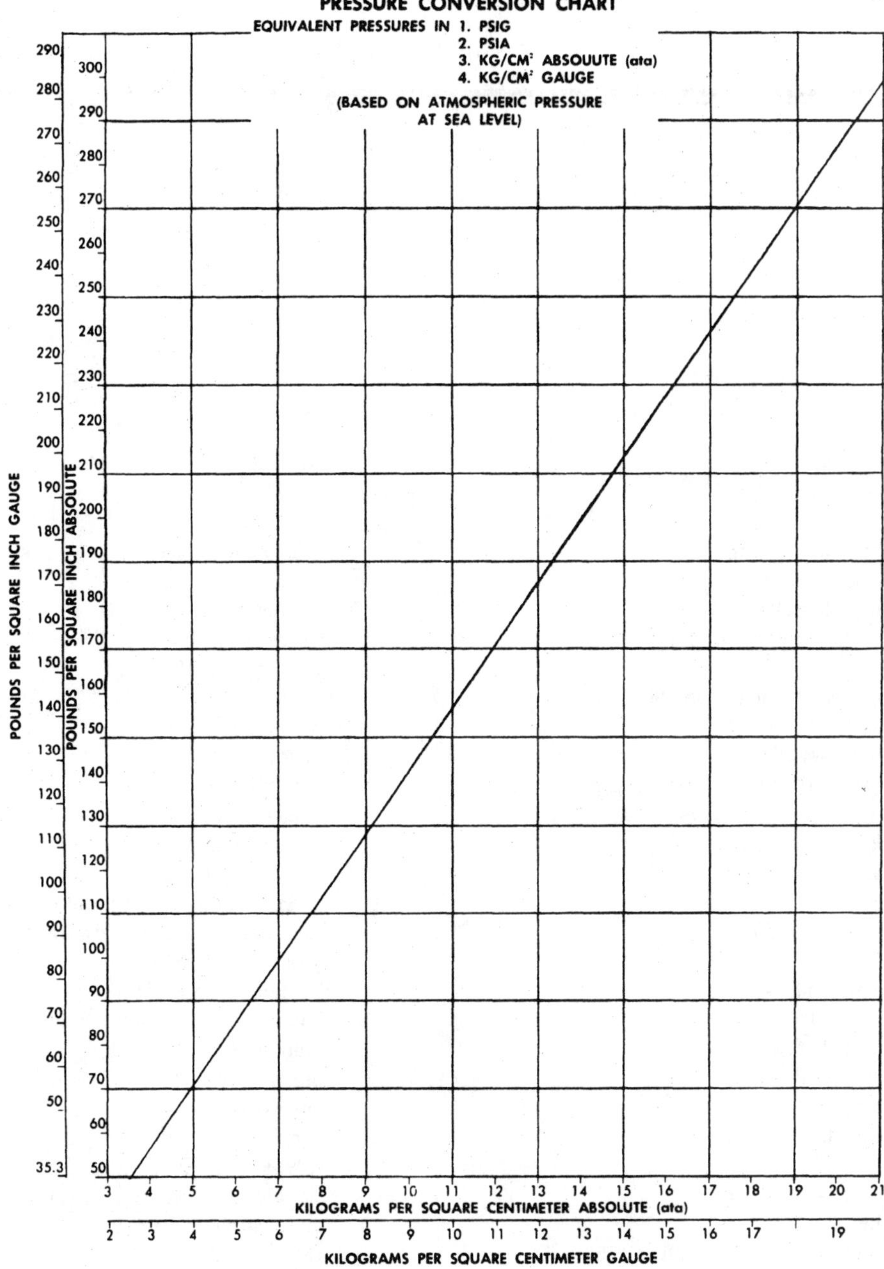
PRESSURE CONVERSION CHART
EQUIVALENT PRESSURES IN 1. PSIG
2. PSIA
3. KG/CM² ABSOUUTE (ata)
4. KG/CM² GAUGE
(BASED ON ATMOSPHERIC PRESSURE
AT SEA LEVEL)
POUNDS PER SQUARE INCH GAUGE
POUNDS PER SQUARE INCH ABSOLUTE
KILOGRAMS PER SQUARE CENTIMETER ABSOLUTE (ata)
KILOGRAMS PER SQUARE CENTIMETER GAUGE

Index